Ch. 16
Pn. 3
P. 461 Determine that there is
a relationship
& how strong the
correl
The R test
what % of
y is strong

SMALL-SAMPLE INFERENCE
($n < 30$)

Parameter (Population Value)	Statistic (Sample Value)	Hypothesis (Example)	Standard Error	Test Statistic	Confidence Interval
μ (Mean)	$\bar{X}$	$H_0:\ \mu \le 100$ $H_A:\ \mu > 100$	$S_{\bar{X}} \approx \dfrac{S}{\sqrt{n}}$	$t = \dfrac{\bar{X} - \mu}{\frac{S}{\sqrt{n}}}$	$\bar{X} \pm t\,\dfrac{S}{\sqrt{n}}$
$\mu_1 - \mu_2$ (Two means)	$\bar{X}_1 - \bar{X}_2$	$H_0:\ \mu_1 - \mu_2 = 38$ $H_0:\ \mu_1 - \mu_2 \ne 38$	$S_{\bar{X}_1 - \bar{X}_2} \approx S_{pooled}\sqrt{\dfrac{1}{n_1} + \dfrac{1}{n_2}}$ where $S_{pooled} = \sqrt{\dfrac{S_{\bar{X}_1}^2(n_1 - 1) + S_{\bar{X}_2}^2(n_2 - 1)}{n_1 + n_2 - 2}}$	$t = \dfrac{(\bar{X}_1 - \bar{X}_2) - (\mu_1 - \mu_2)}{S_{pooled}\sqrt{\dfrac{1}{n_1} + \dfrac{1}{n_2}}}$	$(\bar{X}_1 - \bar{X}_2) \pm t\,S_{pooled}\sqrt{\dfrac{1}{n_1} + \dfrac{1}{n_2}}$
μ_d (Paired difference)	$\bar{d}$	$H_0:\ \mu_d = 0$ $H_A:\ \mu_d \ne 0$	$\dfrac{S_d}{\sqrt{n}}$	$t = \dfrac{\bar{d} - \mu_d}{\frac{S_d}{\sqrt{n}}}$	$\bar{d} \pm t\,\dfrac{S_d}{\sqrt{n}}$

STATISTICAL INFERENCE (OTHER TESTS)

Parameter (Population Value)	Statistic (Sample Value)	Hypothesis (Example)	Test Statistic
$\sigma_{\bar{X}}^2$ (Variance)	S^2	$H_0:\ \sigma^2 \le 30$ $H_A:\ \sigma^2 > 30$	$\chi^2 = \dfrac{(n-1)\,S_1^2}{\sigma^2}$
$\sigma_1^2 - \sigma_2^2$ (Two variances)	$S_1^2 - S_2^2$	$H_0:\ \sigma_1^2 = \sigma_2^2$ $H_A:\ \sigma_1^2 \ne \sigma_2^2$	$F = \dfrac{S_1^2}{S_2^2}$
β_1 (Regression slope)	b_1	$H_0:\ \beta_1 = 0$ $H_A:\ \beta_1 \ne 0$	$t = \dfrac{b_1 - \beta_1}{S_{b_1}}$
ρ (Correlation)	r	$H_0:\ \rho = 0$ $H_A:\ \rho \ne 0$	$t = \dfrac{r}{\sqrt{(1 - r^2)/(n - 2)}}$

BUSINESS STATISTICS

A Decision-Making Approach

DAVID F. GROEBNER
PATRICK W. SHANNON
Boise State University

CHARLES E. MERRILL PUBLISHING COMPANY
A Bell & Howell Company
Columbus Toronto London Sydney

Published by Charles E. Merrill Publishing Co.
A Bell & Howell Company
Columbus, Ohio 43216

This book was set in Optima and Benguiat.
Production Coordination and Text Design: Ann Mirels
Cover Design Coordination: Will Chenoweth
Cover photograph by Photofile.

Library of Congress Catalog Card Number: 80-81967
International Standard Book Number: 0–675–08083-5

2   3   4   5   6   7   8   9   10—86   85   84   83   82   81

Printed in the United States of America

To Jane and Kathy

*Your patience and understanding
made this project possible*

preface

Business statistics is a subject that exposes students to concepts and techniques that are used extensively in business decision-making activities, both in the private and public sectors. However, as our years of university teaching have shown us, despite its practical value, business statistics is viewed with disfavor by many students. Although we don't claim to know with certainty the reason for this, we feel the problem lies in the fact that the tools and underlying theory are all too often presented without insight into their value to the business decision maker. It is the decision-making applications that will enable students to understand *why* their statistics course is important, and therefore help demystify the subject matter for them. Overcoming student fear can only serve to make the educational process easier and more rewarding for student and instructor alike.

In addition, in most universities and colleges, business statistics is a service course, and generally a core course. As such, the major test is not whether the students can understand and apply the material while in the statistics class, but whether they will be able to apply it in the future. To help ensure that a business statistics course passes this test, our experience indicates that presenting business applications with an emphasis on understanding *why* a particular statistical procedure is employed is better than a rigorous presentation of the theory behind the procedure.

Business Statistics: A Decision-Making Approach represents our effort to provide a meaningful statistics book for students and professionals who are interested in learning not only how the statistical techniques are used, but why decision makers need to use them. To this end, each chapter begins with a section entitled "Why Decision Makers Need to Know." Throughout this book, we have attempted to present decision situations followed by easy-to-read-and-understand presentations of the statistical tools that would be helpful in making the decisions. We have included business applications from all the functional areas of business.

We have designed this text to satisfy the requirements of a two-semester introductory business statistics sequence. The content is similar to that covered by many other statistics books. However, we present it in a nonmathematical manner, but not simplified to the point where it would be useful only for elementary applications.

While this text is complete enough for a two-semester coverage, the instructor may easily tailor it for a one-semester course by not covering some topics in later chapters, such as nonparametric statistics, multiple regression analysis, and decision making under uncertainty.

The classical statistical material covered extends from descriptive statistics and hypothesis testing through multiple regression, correlation, and time series analysis. This coverage is necessary because of course offerings in the functional business disciplines:

1. A finance major must understand regression and correlation analysis to understand investment theory and capital budgeting.

2. A marketing major needs to understand data-gathering and projection techniques to forecast future sales and assess the value of a marketing campaign.

3. An economics major must understand regression analysis to be able to understand econometrics and economic forecasting.

This text also includes an introductory treatment of Bayesian statistics. We feel managers should have a firm grasp of the concepts and techniques involved in decision making under uncertainty, since uncertainty most accurately describes the manager's true decision-making environment.

We also firmly believe students should learn to utilize the computer in statistical applications. Therefore, we have included computer-related applications and exercises.

In addition to business cases, this book contains several other teaching aids. Chapters begin with listings of Chapter and Student Objectives and, in most cases, end with a Chapter Glossary and Chapter Formulas. The majority of chapters also include a Solved Problems section, where business applications of statistical tools are worked out in full. Following this is the regular Problems section, which contains general review questions along with special exercises that require additional effort.

An additional feature that sets this text apart is the inclusion of business cases in most chapters. We have always been concerned that students might leave a statistics course thinking that actual business decision situations will always come to them in concise, directly stated questions. Because this is far from the truth, our case situations are purposefully vague in their requirements. These cases and others the instructor may assign will help prepare students for what it's "really" like in a business environment.

Many individuals provided comments and constructive criticisms of our writing style and our technical approach. First, Jean Gibbons of The University of Alabama deserves our deepest gratitude for reviewing the entire manuscript line by line on three separate occasions. Other colleagues whose careful reviewing assistance made this a better text are Wayne W. Daniel, Georgia State University; Thomas R. Hawk, Community College of Philadelphia; Kenneth A. Dunning, University of Akron; Marvin J. Karson, The University of Alabama; Rod J. Lievano, University of New Mexico; and Arthur Reitsch, Eastern Washington University.

Our special thanks and appreciation go to V. Lyman Gallup, who was always available for consultation on ideas and approaches and who, along with Michael T. Lyon, helped class-test our manuscript. V. Lyman Gallup also provided us with a most comprehensive review of the entire text.

The final manuscript typing was skillfully performed by Ann Lee, Norma McKinnon, and Kathy Shannon. We thank them all very much for somehow transferring our handwriting into typed words.

Finally, we are grateful to Linda Johnstone for her thoughtful copy-editing and attention to detail.

We have learned a great deal while writing this text, both about ourselves and about our attitudes toward education. We know we are better off for having written this book; we hope that after students have read it they will understand the importance of statistics as a tool in the life of the decision maker.

contents

6 DISCRETE PROBABILITY DISTRIBUTIONS   121

12 INTRODUCTION TO HYPOTHESIS TESTING   279

13 ADDITIONAL TOPICS IN HYPOTHESIS TESTING   311

1

The Role of Statistics in Decision Making

The regional vice-president of a large retail chain was recently approached by a development company representing a western city. This development company wanted the vice-president to build a new department store in the city's downtown area. The decision whether to build or not to build is typical of those decisions facing businesses today. To make this type of decision, decision makers need to analyze carefully all possible alternatives in light of any information they have.

The premise of this book is that decision makers make "better" decisions when they utilize all available information in an effective and meaningful way. The primary role of statistics is to provide decision makers with methods for obtaining and analyzing information to help make these "better" decisions.

In this text we introduce many fundamental statistical techniques that have application in the functional areas of business, including accounting, finance, marketing, management, real estate, and economics. Just as decision makers can

1

benefit from understanding these functional business areas, they also need to understand the statistical methods available to help make decisions in these areas.

chapter objectives

The objectives of Chapter 1 are to introduce the subject of statistics and to emphasize the role of statistics in the decision-making process.

student objectives

After reading the material in this chapter, you should:

1. Have a basic idea of the role of statistics in business decision making.
2. Have reviewed the mathematical skills required to deal with the statistical methods introduced in this text.

1-1
WHAT DOES STATISTICS INVOLVE?

Many of you no doubt have already formed an opinion of what statistics is. All too often individuals think of statistics as only numerical measures of some item of interest. For example, the unemployment rate is a statistic published monthly by the U.S. Department of Labor. The Dow Jones Industrial Average is a statistic interesting to many stock market investors. Another important statistic is the inflation rate. Although the numbers in these examples are a type of statistics, the philosophy of statistics introduced in this text is something quite different.

Statistics, as described in *Webster's New Collegiate Dictionary,* is "the science of the collection and classification of facts on the basis of relative number or occurrence as a ground for induction; systematic compilation of instances for the inference of general truths." This definition, although very precise, can be reduced to more understandable terms. These terms are: (1) **data description,** (2) **probability,** and (3) **inference.**

1. **Data description,** or **descriptive statistics,** is one important area of statistics. Descriptive statistics consists of techniques and measures that help decision makers describe data. Graphs, charts, tables, and summary numerical measures are tools that decision makers use to help turn data into meaningful information. For example, Figure 1–1 illustrates two approaches for presenting education-related data in a meaningful way. In Chapters 3 and 4 we introduce some basic methods and procedures of descriptive statistics.

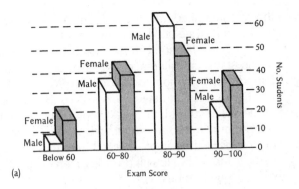

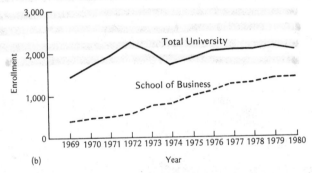

FIGURE 1–1 Descriptive statistics.

2. **Probability** is important to statistics because it measures a decision maker's level of uncertainty about some outcome occurring. For example, suppose an investment manager for a bank in California is considering investing $10 million of her bank's money in a government bond. The manager recognizes that this investment may earn an acceptable rate of return, or because of increasing inflation, may result in an actual loss to the bank. Although the investment manager is uncertain as to which will occur, she has analyzed the available information and feels there is a 75 percent chance of the investment being profitable and a 25 percent chance of a loss. These percentages, or **probabilities**, reflect her attitude about the chances of the two outcomes occurring and thus measure her uncertainty about which outcome will result. The fact that she has assigned a 0.75 probability to the profit outcome indicates that she feels a profit is three times as likely as a loss.

3. **Inference,** or **inferential statistics,** is an area in which conclusions about a large body of data are reached by observing only part of those data. For example, an accountant in charge of auditing a client's financial records may select a subset, or **sample,** of accounts from the ledger of all accounts. Then, based on the accuracy rate of

the sampled accounts, the accountant can make inferences about the accuracy rate of all accounts.

In another instance, quality control testing relies heavily on statistical inference to accept or reject production output. Suppose a semiconductor production process is designed to produce less than 1 percent defective items. To determine if the production process is in control, the quality control manager may select a sample of items from the production process. If he finds "too many" defectives in the sample, he will infer that the defective rate of all items is too high. He will probably then decide that the production process needs to be adjusted.

Statistics, **as defined in this book, is a collection of techniques to describe data and provide a means for making inferences about a total group based on observations from part of the total.** The primary purpose of statistics is to provide information for the decision process. If you keep this in mind, you should understand why we say **statistics is a partner in decision making.**

1-2
STATISTICS AS AN AID TO DECISION MAKING

Many people have the unfortunate, although understandable, tendency to pigeon-hole statistics according to certain predetermined applications. For instance, many decision makers recognize that statistics has important applications in market research, quality control, political poll taking, acceptance sampling, and other areas. However statistics should not be thought of in terms of a set of specific applications but rather as a general aid to the managerial decision-making process. To demonstrate what we mean, let's first consider the concept of managerial problem solving and then discuss how statistics can be employed as part of this process.

Many advances in today's technological tools and equipment can be attributed, in varying degrees, to the use of the *scientific method* in attacking problems. This same methodology can, and should, be used in approaching many managerial problems. The scientific method to a large extent minimizes the "gut feel" approach to solving problems and instead concentrates on using information to help make decisions.

The basic problem-solving process consists of the following five steps:

1. **Defining the Problem.** Although seemingly obvious, this step is often overlooked, perhaps because it is often the hardest to take. A problem can be defined as the difference between something as it exists and what is viewed as a satisfactory situation. To determine this difference requires both that the present situation be defined in

terms that are clear and indisputable, and that a realistic and attainable satisfactory situation be agreed on by those who are affected by the problem.

2. **Searching for Causes.** A difference does exist between a problem and its causes, and often this distinction is hard to make. All individuals should be able to agree on the problem definition, but not all will agree on the problem's causes. Most people define causes in terms of their particular areas of interest. However decision makers must remain flexible in their thinking and look for as many potential causes as possible within time and cost constraints. At some point the decision makers must call off the search for causes. A problem can be "overstudied," that is, considered to the point where searching becomes dysfunctional. Deciding when to stop searching for causes is a difficult decision, and one that some simply cannot make.

3. **Listing Alternative Solutions.** Once the probable causes are determined, solutions must be sought. When searching for solutions, evaluation should not be allowed. As many solutions as possible should be sought since the most unusual often turns out to be the best, or it triggers a new line of thought that leads to an unusual new solution. The methods used to search for solutions differ depending on the organization. Some organizations use brainstorming sessions, in which people locked in a room "toss around" ideas until those in charge are satisfied that all effective approaches have been determined. Other organizations use a much more structured approach to arrive at possible solutions. Here again, the search must be conducted within time and cost constraints. At some point further searching becomes dysfunctional.

4. **Evaluating the Solutions.** Once the possible solutions have been determined, they must be evaluated. However they should all be evaluated using the same criteria. Often individuals with pet solutions attack other possibilities without recognizing potential weaknesses in their own solutions. For instance, one solution may require too much time to implement and be attacked by a person whose personal favorite may cost too much to implement. One good suggestion is to evaluate all solutions on cost, time to implement, acceptability to those involved, and how much the solution will contribute to solving the problem.

5. **Choosing a Course of Action and Measurement Criteria.** Once the solutions have been evaluated, one or a combination should be chosen and an action plan developed. The steps to implement the solution should be determined, but then specific expected improvements should also be identified. Often solutions are implemented with no regard for measuring the solution's effectiveness. Is the gap

found in the problem definition step being closed? If not, future action will be needed.

1-3
USING STATISTICS IN THE PROBLEM-SOLVING PROCESS

The process just described is certainly not new with this text. As mentioned previously, it is the scientific method applied to the managerial process, and has been suggested in one form or another by many management writers. However statistics and statistical techniques fit directly into the process at many points:

1. Statistics is a direct aid in defining problems. Often problems are so vaguely defined that they are not open to any clear line of attack. Using statistics involves working with quantifiable data. Some management writers have gone as far as to suggest that if a problem cannot be defined in measurable terms, it should not be considered.*

2. When evaluating problem causes, a difficulty often arises in distinguishing between major and minor causes. Statistics can be directly used to determine which organization areas are most affected by some problem and which areas are not affected. Statistics is also directly applied in designing the information systems needed to measure any unknown values. Statistics can also determine how much information is needed to reduce the uncertainty in the problem to a workable level and how this information should best be gathered.

3. Since the raw data available to attack many problems are often overwhelming in magnitude, particularly since the advent of computerized information systems, some format is needed to present data in a condensed and useful form. Statistics is directly involved in the process of organizing and presenting data in useful and understandable forms. Often clear data presentation will spell the difference between an idea or proposal being accepted or rejected.

4. Statistics is an invaluable aid in the process of analyzing possible solutions and measuring the outcome of any solution implementation. For instance, is the problem area different after implementing the proposed solution? Is the difference significant?

We will use applications to both introduce and reinforce the specific statistical techniques covered in the text. However, please remember that statistics is not limited to these applications, but instead is a general and flexible managerial aid.

---

*For a general discussion of measuring problem variables in a managerial environment, see George S. Odiorne, *Management Decisions by Objectives*, pp. 20–35.

1-4
IS STATISTICS A MATH COURSE?

Probably the greatest misconception about statistics is that it is "just another math course." Certainly, if statistics is approached from a theoretical viewpoint, as is the case in many mathematical statistics courses, the required mathematics can be considerable. However, the concepts introduced in this text can be discussed with a minimum of math. In fact, one of our objectives has been to present the statistical material at a precalculus level. We emphasize how statistics is applied in decision-making settings and stress the intuitive logic of the techniques. This is not to say, however, that some basic math skills are not required. This section contains a short review of the concepts you will encounter in this text.

Basic Algebra

RULE

Numbers can be substituted for letters.

EXAMPLE 1

Given

$$X = 6Y + 10$$

Let

$$Y = 5$$

Solve for X.

$$X = 6(5) + 10 = 30 + 10$$
$$= 40$$

EXAMPLE 2

Given

$$A = 20X + 7Y + 3Z$$

Let

$$X = 2 \quad Y = 6 \quad Z = 8$$

Solve for A.

$$A = 20(2) + 7(6) + 3(8) = 40 + 42 + 24$$
$$= 106$$

EXAMPLE 3

Given

$$A = 3\frac{X}{Y} + Z$$

Let

$$X = 7 \quad Y = 4 \quad Z = 8$$

Solve for A.

$$A = 3\left(\frac{7}{4}\right) + 8 = 5.25 + 8$$
$$= 13.25$$

RULE

To add or subtract similar terms, add or subtract their coefficients.

EXAMPLE 4

$$6XY + 3XY = 9XY$$

EXAMPLE 5

$$18.5X - 13X = 5.5X$$

RULE

When numbers are contained within parentheses, the enclosed quantity is considered as one term.

EXAMPLE 6

$$10(8 - 3) = 10(5)$$
$$= 50$$

EXAMPLE 7

$$X\left(14 - \frac{33}{11}\right) = X(14 - 3) = X(11)$$
$$= 11X$$

RULE

To change a percentage to a decimal, move the decimal point two places to the left and drop the percent sign.

EXAMPLE 8

Change 10 percent to a decimal.

$$10\% = 0.10$$

EXAMPLE 9

Change 93.6 percent to a decimal.

$$93.6\% = 0.936$$

RULE

To convert a fraction to a decimal, just carry out the division.

EXAMPLE 10

Convert 3/4 to a decimal.

$$\frac{3}{4} = 0.75$$

EXAMPLE 11

Convert 4/5 to a decimal.

$$\frac{4}{5} = 0.80$$

RULE

The square of a number can be found by multiplying the number by itself.

EXAMPLE 12

Given

$$X = Y^2$$
$$= (Y)(Y)$$

Let

$$Y = 8$$

Solve for X.

$$X = Y^2 = (8)(8)$$
$$= 64$$

EXAMPLE 13

Given

$$X = (Y - Z)^2$$
$$= (Y - Z)(Y - Z)$$

Let

$$Y = 12 \quad Z = 15$$

Solve for X.

$$X = (12 - 15)^2 = (12 - 15)(12 - 15) = (-3)(-3)$$
$$= 9$$

One Equation, One Unknown

RULE

You can do anything you want to an equation as long as you do the same thing to both sides.

EXAMPLE 14

$$3X = 9$$
$$\frac{3X}{3} = \frac{9}{3}$$
$$X = 3$$

EXAMPLE 15

Given

$$e = \frac{ZX}{\sqrt{n}}$$

Let

$$Z = 2 \quad X = 4 \quad e = 0.5$$

Solve for n.

$$0.5 = \frac{2(4)}{\sqrt{n}}$$
$$\sqrt{n}(0.5) = 2(4)$$
$$\sqrt{n} = \frac{2(4)}{0.5}$$
$$n = \frac{(2)^2(4)^2}{(0.5)^2}$$
$$= 256$$

EXAMPLE 16

Given

$$e = \frac{ZX}{\sqrt{n}}$$

Let

$$e = 0.5 \qquad n = 400 \qquad X = 5$$

Solve for Z.

$$0.5 = \frac{Z(5)}{\sqrt{400}}$$

$$0.5\sqrt{400} = Z(5)$$

$$\frac{0.5\sqrt{400}}{5} = Z$$

$$Z = 2$$

Inequalities

$>$ is read "is greater than."
$<$ is read "is less than."
$\geq$ is read "is greater than or equal to."
$\leq$ is read "is less than or equal to."

EXAMPLE 17

a. $1.80 < 2.50$ is read "1.80 is less than 2.50."
b. $3.80 \geq X$ is read "3.80 is greater than or equal to X."
c. $Z_1 \leq Z_2$ is read "Z_1 is less than or equal to Z_2."
d. $-2.8 < -1.5$ is read "negative 2.8 is less than negative 1.5."

Summation Notation

Statistics relies heavily on *summation notation* to simplify equations involving sums of numbers. Consider the following sequence of numbers in examples 18–20:

$$1, 2, 3, 4, 5, 6, 7, \ldots$$

EXAMPLE 18

The notation summing the first five numbers is

$$\sum_{X=1}^{5} X = 1 + 2 + 3 + 4 + 5$$

$$= 15$$

Note that we read this as "the sum of X values where X goes from 1 through 5 is equal to 15." The capital form of the Greek letter sigma (Σ) stands for the summation of these numbers.

EXAMPLE 19

$$\sum_{X=1}^{4} X^2 = 1 + 4 + 9 + 16$$

$$= 30$$

This is read "the sum of the squared X values where X ranges from 1 through 4."

EXAMPLE 20

$$\sum_{X=1}^{3} (X - 3) = (1 - 3) + (2 - 3) + (3 - 3)$$

$$= -3$$

Subscripts and Summation Notation

Suppose we have

i	X_i
1	5
2	7
3	3
4	1
5	12
6	5

where: $i = 1, 2, 3, 4, 5, 6$

$X_i = i$th value of X (i.e., $X_1 = 5$, $X_2 = 7$, etc.)

EXAMPLE 21

$$\sum_{i=1}^{3} X_i = X_1 + X_2 + X_3 = 5 + 7 + 3$$

$$= 15$$

EXAMPLE 22

$$\sum_{i=1}^{5} X_i^2 = X_1^2 + X_2^2 + X_3^2 + X_4^2 + X_5^2 = 5^2 + 7^2 + 3^2 + 1^2 + 12^2$$

$$= 25 + 49 + 9 + 1 + 144$$
$$= 228$$

EXAMPLE 23

$$\sum_{i=1}^{3} X_i - 4 = X_1 + X_2 + X_3 - 4 = 5 + 7 + 3 - 4$$

$$= 11$$

RULE ──

If c is a constant (an element that does not contain a subscript), the summation of the constant is

$$\sum_{i=1}^{n} c = nc$$

──

EXAMPLE 24

$$\sum_{i=1}^{5} c = c + c + c + c + c$$

$$= 5c$$

Note that the c value has no subscript.

RULE ──

If c is a constant, then

$$\sum_{i=1}^{n} cX_i = c\sum_{i=1}^{n} X_i$$

──

EXAMPLE 25

$$\sum_{i=1}^{5} cX_i = cX_1 + cX_2 + cX_3 + cX_4 + cX_5 = c(X_1 + X_2 + X_3 + X_4 + X_5)$$

$$= c\sum_{i=1}^{5} X_i$$

EXAMPLE 26

Suppose

i	X_i
1	11
2	14
3	3
4	9
5	15

Then

$$\sum_{i=1}^{3} 5X_i = 5\sum_{i=1}^{3} X_i = 5(X_1 + X_2 + X_3) = 5(11 + 14 + 3)$$

$$= 140$$

RULE

> The summation of two or more different elements can be found by summing each element and adding the summed quantities. That is,
>
> $$\sum_{i=1}^{n}(X_i + Y_i) = \sum_{i=1}^{n}X_i + \sum_{i=1}^{n}Y_i$$

EXAMPLE 27

Given

i	X_i	Y_i
1	3	14
2	8	6
3	5	9
4	2	7

Then

$$\sum_{i=1}^{4}(X_i + Y_i) = (3 + 8 + 5 + 2) + (14 + 6 + 9 + 7)$$

$$= \quad\quad 18 \quad\quad + \quad\quad 36$$

$$= 54$$

Factorials

Factorials are convenient means of representing a product of decreasing integers in the form $n!$, read "n factorial."

EXAMPLE 28

a. $3! = (3)(2)(1)$
   $= 6$
b. $4! = (4)(3)(2)(1)$
   $= 24$

RULE

> $0! = 1$

EXAMPLE 29

$$\frac{3!}{(3-3)!} = \frac{(3)(2)(1)}{0!} = \frac{(3)(2)(1)}{1}$$

$$= 6$$

The preceding examples have been a review of the basic math skills that are required in the text. If you are comfortable with these operations, you should have no problems doing the math portion of this statistics course.

1-5
CONCLUSIONS

Statistics is a discipline that includes methods and techniques for data description and inference. The primary objective of the statistical tools presented in the remaining chapters is to provide decision makers with a means of transferring data into meaningful information that can be used to make "better" decisions. However, it must be emphasized that statistical techniques do not make decisions; individuals make decisions. Statistics is a collection of well-defined tools that individuals can use to help make these decisions.

Finally, statistics is not a math course. Only a basic understanding of mathematical operations is required to apply statistical methods in a decision-making context.

===problems

Problems marked with an asterisk (*) require extra effort.

1. Look up the course description of your college statistics course. Relate this description to the basic description of statistics presented in this chapter.

2. List some examples of television and radio commentators referring to or giving descriptive statistics. When are they discussing statistical inference?

3. In connection with other classes, discuss instances when you have been exposed to descriptive statistics and when you have seen statistical inference used.

4. Discuss cases when a decision maker, perhaps a politician, has neglected to clearly define a problem before looking for possible solutions.

5. Discuss some cases when a decision maker would find it necessary to project trends into the future. Suppose you are:
   a. An aid to the governor
   b. A corporation president
   c. A defense planner
   d. The parent of several small children

The following problems are for students who feel they need an additional review in the basic math concepts introduced in this chapter.

✓6. Let

$$X = 6$$
$$Y = -2$$
$$Z = 1$$
$$W = 3X + (4 - 2XY + Z^2) - 12YZ$$

Solve for W.

✓**7.** Let

$$A = 2.6$$
$$B = -1.7$$
$$C = -3.8$$
$$X = 2 - 3(2A - BC) + 6AC - C^2$$

Solve for X.

8. Let

$$S = 17.8$$
$$T = 3.0$$
$$U = 11.6$$
$$V = -7.0$$
$$G = 6SV - 17T^2 + TV$$

Solve for G.

9. Change each of the following percentages to a decimal:
   a. 17.87 percent
   b. 316 percent
   c. 0.05 percent
   d. 20 percent
   e. 1.6783 percent

10. Convert each of the following fractions to a decimal and a percentage:
   a. $\dfrac{17}{18}$
   b. $\dfrac{386}{6}$
   c. $\dfrac{16}{48}$
   d. $\dfrac{1}{112}$
   e. $\dfrac{5}{6}$
   f. $\dfrac{116}{32}$

✓**11.** Solve each of the following equations:
   a. $7X - 6 = 6 - 3X$
   b. $5Y - 3(Y - 6) = 3(4 - 2Y) + 6 - 4$
   c. Solve the following for X:

$$4 - 3(Y - 6) = 2(X + Y) - 3(X - 4)$$

d. Solve the following for Y:

$$4\left(\frac{1}{Y} - 6X\right) + 2X^4 - X^2 = 0$$

e. Solve the following for Z:

$$A(Z - X^2) + 4YX = -6 + 4(A - Z)$$

✓ **12.** Given

$$X_1 = 3$$
$$X_2 = 5$$
$$X_3 = 7$$

Perform the following operations:

a. $\displaystyle\sum_{i=1}^{3} X_i$

b. $\displaystyle\sum_{i=1}^{3} X_i^2$

c. Given

$$\bar{X} = \frac{\displaystyle\sum_{i=1}^{3} X_i}{3}$$

Compute

$$\sum_{i=1}^{3}(X_i - \bar{X})$$

d. $\displaystyle\sum_{i=1}^{3}(X_i - \bar{X})^2$ where $\bar{X}$ is defined as in part c.

e. $\displaystyle\left(\sum_{i=1}^{3} X_i\right)^2$

✓ **13.** Suppose you are given the following tabulated values:

i	X_i	Y_i
1	4	7
2	5	8
3	9	10
4	2	3

Compute each of the following:

a. $\displaystyle\sum_{i=1}^{4} X_i Y_i$

b. $\displaystyle\sum_{i=1}^{4} X_i \sum_{i=1}^{4} Y_i$

c. Given

$$\bar{X} = \frac{\sum\limits_{i=1}^{4} X_i}{4}$$

$$\bar{Y} = \frac{\sum\limits_{i=1}^{4} Y_i}{4}$$

Find

$$\sum_{i=1}^{4} (X_i - \bar{X})(Y_i - \bar{Y})$$

d. Find r if

$$r = \frac{4\sum\limits_{i=1}^{4} X_i Y_i - \sum\limits_{i=1}^{4} X_i \sum\limits_{i=1}^{4} Y_i}{\sqrt{\left[4\sum\limits_{i=1}^{4} X_i^2 - \left(\sum\limits_{i=1}^{4} X_i\right)^2\right]\left[4\sum\limits_{i=1}^{4} Y_i^2 - \left(\sum\limits_{i=1}^{4} Y_i\right)^2\right]}}$$

$\sqrt{}$**14. a.** Solve for X.

$$X = \frac{8!}{4!4!}$$

b. Solve for Z.

$$Z = \frac{8!}{4!}$$

c. Suppose

$$n = 5$$
$$r = 2$$

Find

$$\frac{n!}{r!\,(n-r)!}$$

d. Given

$$Y = \frac{5!}{3!2!(3-3)!}$$

Solve for Y.

*15. Verify

$$\sum_{i=1}^{n} (X_i - \bar{X}) = 0$$

where: $\quad \overline{X} = \dfrac{\sum\limits_{i=1}^{n} X_i}{n}$

*16. Verify

$$\sum_{i=1}^{n}(X_i - \overline{X})^2 = \sum_{i=1}^{n} X_i^2 - \dfrac{\left(\sum\limits_{i=1}^{n} X_i\right)^2}{n}$$

*17. Given

$$\overline{X} = \dfrac{\sum\limits_{i=1}^{n} X_i}{n}$$

$$\overline{Y} = \dfrac{\sum\limits_{i=1}^{n} Y_i}{n}$$

Verify

$$\sum_{i=1}^{n}(X_i - \overline{X})(Y_i - \overline{Y}) = \sum_{i=1}^{n} X_i Y_i - \dfrac{\left(\sum\limits_{i=1}^{n} X_i\right)\left(\sum\limits_{i=1}^{n} Y_i\right)}{n}$$

references

HUFF, D. and I. GEIS. *How to Lie with Statistics*. New York: Norton Press, 1954.

HYLAND, E. *Business Mathematics*. Reston, Va.: Reston, 1978.

ODIORNE, GEORGE S. *Management Decisions by Objectives*. Englewood Cliffs, N.J.: Prentice-Hall, 1969.

TANUR, JUDITH M. *Statistics: A Guide to Business and Economics*. San Francisco: Holden-Day, 1976.

2

Data Collection

Statistics, like most other branches of organized knowledge, has its own terminology. As a student and potential decision maker, you must learn the language of statistics before you can converse in any detail about the subject. For example, the term *data* was used several times in Chapter 1. **Data are simply values, facts, or observations and may come in many different forms depending on the decision situation.** In this chapter we discuss some methods of collecting data and illustrate that data collection is the source of many problems for decision makers. Without good data, the results of a statistical analysis may be misleading.

chapter objectives

Two objectives of this chapter are to introduce some terminology associated with statistics and to discuss the problems and methodology associated with data collection. In addition, we shall discuss some common sources of error found when statistical data are used.

student objectives

After studying the material in this chapter, you should be able to:

1. Discuss the four levels of data measurement.
2. Identify different methods used to gather statistical data.
3. Identify the strengths and weaknesses of the different data-gathering techniques.
4. Discuss some common errors found when analyzing data.

2-1
DATA LEVELS

The statistical techniques introduced in this text deal with some form of data. The data, in turn, come from observations that can be measured. However the level of measurement may vary greatly from application to application. In some cases the observed value can be expressed in purely numerical terms, for instance, dollars, pounds, inches, percentages, and so forth. In other cases the observation may be quantifiable only to the extent of saying it has more of some characteristic than does a second observation. There are thus observations that can only be ranked in a greater-than, an equal-to, or a less-than form. As an example, we may be firmly convinced that the quality of one product is greater than that of another product but be unable to state that the quality is, say, twice as great. In still other cases we may be only able to count the number of observations having a particular characteristic. For instance, we may count the number of males and females in a statistics class.

To use statistical tools, we must, in one manner or another, be able to attach some quantifiable measure to the observations we are making. Thus the measurement technique used to gather data becomes important in determining how the data can be analyzed. We shall discuss four levels of measurements and provide some examples of each. These four levels are *nominal, ordinal, interval,* and *ratio.*

Nominal

Nominal measurement is the weakest data measurement technique. **In nominal measurements, numbers or other symbols are used to describe an item or characteristic.** For example, when developing a computerized payroll system, designers will generally assign a number to each employee. Most of the data processing is then performed by employee number, not name. Employee numbers make up a nominal measurement scale; they are used to represent the employees. Most colleges and universities require each student to have a unique student number—another example of nominal measurement.

Ordinal

Ordinal or *rank* measurement is one notch above nominal data in the measurement hierarchy. At this level, the data elements can be rank-ordered based on some relationship between them. For example, a typical market-research technique is to offer potential customers the chance to use two unidentified brands of a product. The customers are then asked to indicate which brand they prefer. The brand eventually offered to the general public depends on how often it was the most preferred test brand. The fact that an ordering of items took place makes this an ordinal measure.

A corporate management hierarchy offers another example of ordinal measurement. The president, vice-president, division manager, plant manager, and so on form an ordinal scale: the president is "greater than" the vice-president, who is "greater than" the division manager, who is "greater than" the plant manager, and on down the line.

Ordinal measurement allows decision makers to equate two or more observations or to rank-order the observations. In contrast, nominal data can only be compared for equality. You cannot order nominal measurements. Thus a primary difference between ordinal data and nominal data is that ordinal data contain both an equality ("=") and a greater-than (">") relationship, whereas the nominal scale contains only an equality ("=") relationship.

Interval

If you have data with ordinal properties—">" and "="—and can also measure the distance between two data items, you have an *interval* measurement. Interval measurements are preferred over ordinal measurements because, with them, decision makers can precisely determine the difference between two observations.

Frozen-food packagers have daily contact with a common interval measurement—temperature. Both the Fahrenheit and Celsius temperature scales have equal distances between points on the scales, which is an interval scale property, along with the ordinal properties of ">" and "=." For example, 32°F > 30°F, and 8°C > 4°C. The difference between 32°F and 30°F is the same as the difference between 80°F and 78°F: two degrees in either case.

Our ordinal example about the corporate hierarchy—president > vice-president > division manager > plant manager—could not be interval unless we could measure the difference between the various titles using some equal scale.

Ratio

Data that have all the characteristics of interval data but that also have a unique or true zero point (where zero means "none") are called *ratio* data. Ratio measurement is the highest level of measurement.

Packagers of frozen foods encounter ratio measures when they pack their products by weight. Weight, whether measured in pounds or grams, is a ratio measurement because it has a unique zero point. Many other types of data encountered in business environments involve ratio measurements, for example, distance, money, and time.

Figure 2–1 illustrates the four levels of data measurement and summarizes the properties of each.

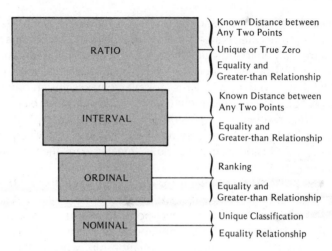

FIGURE 2–1 Levels of data measurement.

2-2
POPULATIONS AND SAMPLES

Two of the most important terms in statistics are *population* and *sample*. A *population* **is a collection of all the items or observations of interest to the decision makers.** Decision makers can specify the population in any manner they choose. **A *sample* is a subset of the population.** An example should clarify what is meant by a population and a sample.

One of the problems a certified public accounting firm faces when auditing a business is to determine the number of accounts to examine. Not too many years ago, good accounting practice dictated that the auditors verify the balance of every account and trace through each financial transaction. Though this is still done in some audits, the size and complexity of most businesses has forced accountants to select only some accounts and some transactions to audit.

The accountants' first problem is to determine just what they wish to examine. Suppose one part of the financial audit is to verify the accounts receivable balances. Recall that a population includes all items of interest to the data gatherer. The accountants define the population of interest as all receivable accounts on record. Next they select from this population a representative group of accounts and determine their level of accuracy or inaccuracy. This representative group is the sample. Remember, a sample is a collection of data gathered from a population.

The accountants use the audit results from the sample to make *infer-ences,* or conclusions, about the population. (You may want to review the discussion of statistical inference in Chapter 1.) How these inferences are drawn will be discussed at great length in later chapters.

Populations can be classified into one of two types: *finite* and *infinite.* **A population is *finite* if it contains a countable number of elements or items.** For the accounts receivable, there is no doubt that the population could be counted. Thus this population is a finite population.

A population is *infinite* if it contains an uncountable number of items. For instance, in an audit of a business, the accountants may want to verify the physical inventory. Suppose the business being audited is a brick and concrete block company. Its inventory consists of materials and finished products. With enough time, the accountants might count every brick and concrete block in inventory; however, in practice, a complete count would not be made. Unless the company stopped both production and sales until the audit is finished, the actual inventory would be continually changing while the count is taking place. The chances of an accurate tally under even the best conditions are small. Therefore the accountants face an infinite population.

In many cases, even if we could measure the characteristic of interest in the entire population, we would decide not to. For instance, if our company manufactures school desks, we would likely not want to weigh all grade school children in our market area to determine how strong to make our desks. Rather, we would be content with taking a sample of students and base any decision on information obtained from that sample.

The problems of sampling from populations will consume much of our time in this text and much of your time after you have mastered this course. Although in general the larger the sample the better the information, sample sizes are constrained by real-world considerations of time and money. As the sample size is increased, so are the cost and time of gathering the sample. Also, in many situations an item must be destroyed in the process of sampling. For example, to determine the life of a light bulb, the decision maker must use the bulb until it burns out. The more expensive the item being tested, the more costly this destructive sampling will be. The trade-off is always to get the most data in the least time for the money you have available.

2-3
DATA SOURCES

Statistical tools cannot be used unless data are available. Data are available to an organization either from *internal* or *external* sources.

Internal Data Sources

Almost all organizations generate a substantial number of data in the course of their ordinary operations. Financial reports and the controls necessary to generate these reports cause data to be kept on manufacturing costs, direct labor costs, inventory costs, return by product line, and an almost unending line of financial variables. Operating reports are generated for quality control purposes, inventory control records are kept, product control documents and personnel files are maintained, and on and on. Government regulatory bodies such as the IRS and OSHA have played an important role in determining the quantity and variety of data that businesses are required to collect and maintain. Unfortunately, many internal data are generated for reasons other than those germane to managers. Decision makers must be certain the data they use are appropriate for the decision being made. If the data are inappropriate, new data should be generated.

In spite of possible problems, internally generated data are usually the most important source of decision-making information. This is particularly true if the organization has a computerized data-processing system. If a computerized system is used, information in extremely detailed form can be generated. In fact, organizations often generate too many data instead of not enough. Remember that data and information are two different things. **Data are transformed into useful information only when they are presented in an organized and meaningful format with the detail needed to assist managers in making decisions.** Within an organization, the management level will generally determine how detailed the presented information should be.

External Data Sources

Sometimes information required by organizational decision makers is simply not available from internal sources. For instance, if a business firm is considering expanding into a new market, no amount of internal data can tell if this is a growing market, if sufficient suppliers of raw materials are available, what the present and future states of the local economy are, or the future market depth for the product in the new area. To supply information in these areas, decision makers have to rely on data from external sources. The sources of external data are many and varied, and to be effective, decision makers should have at least a general idea of some easily available external data sources.

The single largest source of external data is the federal government. An excellent data source is the Department of Commerce's monthly publication, *Survey of Current Business*. This periodical contains data on prices, production, inventories, wages, sales, company incomes, and employment in industries, as well as general composite data on the state of the national economy. The Bureau of Labor Statistics in the *Monthly Labor Review* publishes data concerning labor-force levels, industry employment, hours worked, wages, labor turnover, lost hours, and other labor-related activities. The Bureau of the Census annually publishes the *Statistical Abstract of the United States,* which contains a wide variety of information on living conditions and trends in the United States. One of the principal responsibilities of the U. S. Department of Commerce field officers is to assist businesses in need of data. The data-gathering and -reporting activities of the federal government are immense. Because of the recent passage of the Freedom of Information Act, gathering data from federal sources should become easier.

Most statistical data published by the federal government can be found in one of the many federal government publications depositories. These are often located on university campuses. We suggest you spend some time in your library and become acquainted with the various sources of external data.

Government agencies are not the only groups gathering and reporting data. Many industrial trade organizations report data gathered from organization members. This is done, for instance, in the oil, construction, and insurance industries. The Federal Reserve System banks issue numerous reports on the banking industries' condition and also on the national monetary situation. They also report on and make projections of the economic conditions in the areas they serve.

In addition to the sources already mentioned, much information, both statistical and background, can be gathered from sources such as the *Wall Street Journal, Business Week, Fortune,* and local and regional newspapers. A vast number of journals from both trade organizations and scholarly institutions often contain data on specific problems that may confront decision makers.

Decision makers using external sources of data should make sure the assumptions and methods used to collect the data are consistent with their purposes.

Alternative Methods of Obtaining Data

Much information gathered from internal and external sources is originally generated in the normal day-to-day operations of either a firm or the economy in general. This type of data is extremely useful for resolving normal operational problems in an organization. However many important decisions made in any organization involve actions that will occur in the future or that have future consequences. And, specifically, much of the information needed by decision makers can be found only in the actions or thoughts of individuals. The problems of gathering information on human actions are important for almost all managers and decision makers. This kind of information can be gathered in many ways. In the following paragraphs we will outline some basic techniques.

Observation. Conceptually, the simplest method to gather data on human behavior is to watch people. If you are trying to decide whether a new method of displaying your product at the supermarket will be more pleasing to customers, change a few displays and watch the customers' reactions. If you would like some information on whether your new movie will be a success, have a preview screening and listen to the comments of patrons as they exit.

The discipline of time and motion study is actually a detailed method for watching people and determining their responses to environmental conditions. Much of what has been learned by psychologists has come from watching individuals' reactions when confronted with a new situation or a series of stimuli.

The conceptually simple method of watching human reactions or behavior patterns is not, however, without problems. Again, the major constraints are time and money—a person or an organization never has enough of either. When time and money are considered, personal observation is a very expensive method of gathering data. For this method to be effective, trained observers must be used, and this naturally increases the cost. Personal observation is also time consuming, and to be certain that enough observations are taken, a long observation period is generally necessary. Not only is beauty in the eye of the beholder, but so is any personal perception. You have no guarantee that two observers will see a situation the same, much less report it the same.

Personal Interviews. Primarily for reasons of cost and time, personal interviews are used much more frequently than observation as a means of data collection. Individuals are asked a series of questions with the hope that they will supply useful information to the decision makers. Almost everyone has at one time or another been asked to fill out a questionnaire giving information on such items as product preference, activities, work habits, transportation choices, and so on. If these questions were asked by an interviewer, this was a ***personal interview***.

The questions asked in personal interviews fall into two categories: ***closed-ended*** questions and ***open-ended*** questions. **If the question is phrased in such a way that only certain answers are possible, it is a** *closed-ended* **question.** Examples of this would be:

1. Are you considering buying a new car this year?

   Yes_____   No_____

2. How would you rate the food at the student union?

   a. Excellent   b. Good   c. Average   d. Poor   e. Flushable

If the question is phrased so that the respondents are required to formulate their own answers, it is *open-ended***.** For instance:

1. What qualitites do you look for in a car?
2. Why did you decide to major in business administration?

In general, open-ended questions will gather more accurate information than closed-ended questions, but statistically analyzing open-ended answers is often a nightmare. When open-ended questions are analyzed, the answers are usually grouped into subjective categories. There is always room for disagreement, both about the categories and about which responses belong in which category. **Data collected from closed-ended questions are much easier to analyze statistically.**

Although the time and money spent on personal interviews are generally less than for personal observations, both may still be high. This method of data gathering has additional problems:

1. Interviewers may unconsciously predetermine the answer to a question by indicating the socially acceptable answer. Consequently the interviewer can influence the response given. This problem is compounded the greater the number of interviewers used. Each interviewer may indicate different acceptable answers.

2. Many people are uncomfortable answering certain questions in person. For instance, most individuals are reluctant to give information about their incomes. However, these people may give the information in an anonymous questionnaire.

3. An increasingly common problem with using personal interviews is that many door-to-door salespeople use questionnaires as a ploy to gain entry to a house. Individuals who have answered questions involving their children's education only to find themselves listening to a sales pitch for encyclopedias may have little sympathy for the person asking legitimate questions for the local school board.

Telephone Interviews.   A common method of gathering data is the telephone interview. The main advantage of a telephone interview is that it is a rapid method of gathering data. It has the added benefit of relatively low cost, certainly when compared with personal observations and personal interviews.

In spite of the advantages of low cost and short time, the telephone interview has several disadvantages. The first is that people are not as likely to answer questions over the phone as they are in person. A second disadvantage is that people are able to lie easily over the phone, and unfortunately often do. During a personal interview, the interviewer can do some screening to determine if the answers given appear to be in the correct ballpark. This screening can be done for questions involving relative socioeconomic status, age, sex, and so forth. This screening is generally not possible over the telephone. The telephone interview generally limits the sample to people listed in the phone book. This eliminates that segment of the population who do not have phones or who are not listed in the directory.

Thus, when telephone questionnaires are used, the time and cost advantages must be weighed against the disadvantages of potentially poorer information. Since the hardest phase of telephone interviewing is holding a person's attention, only short-answer questions should be asked.

Mail Questionnaires. Certainly the cheapest method of gathering data is a mail questionnaire. Many questionnaires can be printed and sent for the cost of one personal interview. The mail questionnaire also eliminates the possibility of interviewer bias. However mail questionnaires have definite disadvantages. Probably the largest is a characteristically poor response. Most response rates to mail questionnaires run less than 10 percent. Also, the individuals who do answer are often those who have a particular interest in the subject, and consequently may not accurately represent the overall population. In addition, individuals will often interpret the same question differently, and so similar answers may not mean the same thing.

A poor questionnaire response can be improved upon by following up the initial mailing with letters, phone calls, or even personal visits. However these procedures add greatly to the cost of data gathering.

A basic point to remember whenever gathering data from people is that the needed information is in their minds, but the problems of confidentiality are great. If the questions are asked incorrectly or concern sensitive information, many people will refuse to answer, or will give wrong answers.

2-4
CONCLUSIONS

Decision makers are in a continual bind. They have to make decisions and, as an aid, they gather information. However all methods used to gather data have inherent problems. In this chapter we have presented some methods used to gather data and some problems that can occur when data are gathered. Because there are problems does not mean that decision makers should throw up their hands and use the old dart method. Rather, they should view statistical techniques as tools to be used in making decisions. These tools give decision makers added information to use in arriving at decisions, but the degree to which the information is useful depends on the degree to which the problems associated with data gathering have been avoided.

chapter glossary

data A set of values, facts, or observations.

finite population A population with a countable number of elements or items.

infinite population A population with an uncountable number of items.

interval data Data with the " = " and " > " comparative relationships and known distance between any two items or numbers.

nominal data The weakest form of data. Numbers or other symbols are used to describe an item or characteristic. The " = " relationship is the highest level of comparison.

ordinal data Data elements that can be placed in a rank ordering based on some specific relationship between them. The relationships for comparison are " = " and " > ."

population All items of interest to the data gatherer.

ratio data Data with all the characteristics of interval data but with a unique zero point. Considered to be the highest data level.

sample A collection or subset of data gathered from the population.

problems

1. You have been hired by a convenience-food manufacturer to determine whether its new product will be successful in your area.
   a. In what population are you interested?
   b. How will you determine your sample?
   c. What technique will you use to select your sample?
   d. How will you gather information from your sample?
   e. Comment on any problems you see in drawing conclusions based on information from your sample.

2. Many congressional representatives send questionnaires to individuals they represent. Do you see any problems with gathering data in this manner?

3. The newspapers have recently printed several articles comparing present test scores of high school students with scores from past years and with scores from other parts of the country. What statistical difficulties do you see in these types of comparisons?

4. What problems do you see for both utility companies and regulating agencies in deciding on future demand and supply of energy? In particular, how would you decide what the population in a given service area will be ten years from now? What type of data would you use, and how might you go about gathering it?

5. Ford Motor Company has been advertising a series of comparisons between its cars and competitors' cars. Into which class of quantifiable data do each of the following tests fall?
   a. Measuring sound levels inside the cars
   b. Having drivers compare the handling of the cars
   Do you see any possible data-gathering problems in these tests?

6. Assume you want to open a factory in your home town to manufacture fiberglass pipe for both water and electrical cables. Locate information on:
   a. Likely demand for this product
   b. Sources of raw materials
   c. Competitors in the area
   d. Transportation available
   e. Work force available
   f. Likely hourly wage rate for your workers

7. Select a business in your community. Develop a mail questionnaire to determine customer attitudes about the business's products and services. Also design a telephone survey to accomplish the same objective. How should the two data-collection devices differ? In what ways are they the same?

8. Many colleges and universities use some form of faculty evaluation by students. If yours does, obtain a copy of the form and write a critique based on your understanding of this chapter. If your school does not have one, try to develop a good form.

9. In this chapter we presented four levels of data: nominal, ordinal, interval, and ratio. List several examples of each. If you have trouble, try the library.

===**references**

KISH, LESLIE. *Survey Sampling*. New York: Wiley, 1965.

LOETHER, H. J. and D. MCTAVISH. *Descriptive Statistics for Sociologists*. Boston: Allyn and Bacon, 1974.

NETER, JOHN, WILLIAM WASSERMAN, and G. A. WHITMORE. *Applied Statistics*. Boston: Allyn and Bacon, 1978.

SIEGEL, S. *Nonparametric Statistics for the Behavioral Sciences*. New York: McGraw-Hill, 1956.

3

Organizing and Presenting Data

The element most common to the decision makers' environment is uncertainty. This is true, at least in part, because other people make up much of the managerial environment. Whenever you interact with other individuals, you can never be certain of their actions. Other parts of the managerial environment are also characterized by uncertainty. Such unpredictable events as machine breakdowns and weather changes affect output or productivity, making them anything but consistent.

The problem most decision makers must resolve is how to deal with the uncertainty that is inherent in almost all aspects of their jobs. In making decisions with uncertain results, decision makers should use all available data to analyze the possible alternatives. In many cases the data available are in *raw* form. Unfortunately, raw data are often not meaningful and cannot be considered to be information. Take the personnel manager who is concerned that employee absenteeism is becoming excessive. The data he has available consist of absentee rates for each of the last 500 days. These raw data provide little, if any, information about the pattern or growth of absenteeism in the company.

Decision makers need a means of converting the raw data into useful information. **A first step in transforming data into information is to organize and**

present the data in a meaningful way. Therefore decision makers must be aware of the basic techniques for effectively organizing and presenting data.

chapter objectives

The objective of this chapter is to introduce some methods for handling the uncertainty that managers continually face. Practical managers are often not interested in attempting to eliminate the uncertainty or variation they face since they recognize that variation will always exist. Rather, managers are often mainly interested in determining if the uncertainty they face is what would normally be expected, or if somehow the extent of the uncertainty or variation has changed. Production managers are interested in whether their production processes are continuing in a normal fashion or whether they have been altered. A change may indicate the need for corrective action. Personnel managers are interested in knowing if overall employee skill levels have changed over time. A change may indicate that the managers need to institute a training procedure or that a new training program is paying off.

In this chapter we shall introduce some tools decision makers need for examining the uncertainty they face. The first tools managers need are those necessary to "picture" the situation around them. To this end, we shall discuss several frequently used methods of presenting data.

student objectives

Upon studying the material in this chapter, you should be able to:

1. Determine how to separate large volumes of data into more workable forms.
2. Describe several different methods to graph and present data in order to transform raw data into usable information.

3-1

METHODS OF REPRESENTING DATA

As mentioned earlier, uncertainty, though at times upsetting, should not be particularly surprising. What is upsetting is when the uncertainty or variation in a manager's world changes and the manager is unaware a change has occurred.

The process of gathering the information necessary to cope with variation is often complicated not by a lack of potential information, but because the body of raw data is too large to analyze. Consider the problems facing merchandising managers interested in sales data from a series of stores, or even from one store. They are interested in which product lines are selling best, what the inventory levels are, which store areas have the highest sales levels, and numerous related questions. Imagine the managers being given a box of sales-slip copies each day and being asked to answer the questions just presented. The data cannot be effectively used in that form and therefore must be converted to a more useful form. In this section we introduce several useful techniques of data presentation.

Frequency Distribution

Perhaps the easiest method of organizing data is to construct a *frequency distribution.* To do this, we first establish *classes*, where a class is simply a category of interest. In a study of weather the classes might be "clear days" and "cloudy days." In another example the categories of loan balances might be "$0 and under $500," "$500 and under $1,000," and "$1,000 and under $2,000."

The second step in developing a frequency distribution is to count the number of observations that fall into each class. Table 3–1 presents the raw data and a frequency distribution of salaries for 134 employees in a hypothetical company. We see that 39 persons have salaries in the class "$10,000 and under $12,500" and that only 6 individuals earn $17,500 or more. Note how confusing the salary data are in their raw form.

TABLE 3–1 Employee salaries

RAW DATA							
$ 2,525	$ 4,852	$ 4,900	$ 2,800	$ 3,300	$ 4,000	$ 7,704	$ 9,120
3,346	4,750	3,950	4,300	3,100	2,950	8,000	8,900
3,850	4,425	5,105	5,250	7,204	7,000	9,305	7,905
6,250	6,000	5,900	5,403	6,250	7,005	7,700	8,200
5,950	5,700	6,300	5,500	7,100	6,300	7,950	8,300
6,450	7,100	7,480	6,475	6,000	7,620	8,750	9,250
9,200	9,650	9,000	8,800	9,300	9,000	9,600	10,100
9,400	8,300	8,900	8,605	9,200	9,800	10,003	11,900
11,000	12,210	12,175	11,290	12,000	12,210	12,400	11,000
12,000	11,000	10,800	10,500	10,250	11,400	12,000	12,450
11,000	11,000	12,400	11,000	12,000	11,000	12,100	12,175
10,700	10,900	11,175	12,150	11,275	10,283	11,240	13,700
11,400	12,200	10,983	12,800	13,250	14,100	14,200	14,750
13,350	12,950	13,000	13,000	13,000	13,475	14,900	15,300
14,200	14,750	14,000	14,000	13,750	13,800	15,250	19,200
15,400	17,000	16,900	17,200	16,000	16,500	17,800	
19,500	20,000	21,005	19,000	10,700	11,750	8,120	

FREQUENCY DISTRIBUTION	
Salary Class	Frequency of Employees
$ 2,500 and under $ 5,000	14
$ 5,000 and under $ 7,500	21
$ 7,500 and under $10,000	27
$10,000 and under $12,500	39
$12,500 and under $15,000	19
$15,000 and under $17,500	8
$17,500 and over	6
Total	134

Although developing a frequency distribution is fairly straightforward, two factors must be considered: how many classes should be used and the limits for each class. These are considered in the following paragraphs.

Decision makers want to organize data in the first place so that they can picture what the data look like. However they are immediately on the horns of a dilemma. The data must be grouped into separate classes. If the number of classes is too small, much detail available in the data is lost, and the amount of potential information is reduced. On the other hand, having too many classes often makes the representation as confusing as the raw data.

Consider a company that wants some information about the age of its employees. (Many companies use this information for future hiring practices and for pension planning.) The employees would likely range in age from 20 or less to 65 or more. Separating the data into single-year groups would give at least 45 categories—too many to digest and analyze. The company could, for instance, have no one in the 43-year category, but several people in the 42-year category. On the other hand, if the data were broken into two classes—those over 40 and those 40 and under—much potential information would be lost. **In practice, a frequency distribution usually has from 5 to 20 classes.** Having more than 20 classes will often reflect peculiarities in a particular set of data and may hide general trends. Having fewer than 5 classes requires grouping many data that should be separated. Both the number of observations and their intended use must be considered in determining the number of classes.

Determining class size and class limits is essentially arbitrary, and two individuals will often come up with different ways of arranging data into classes. Nevertheless, certain rules and guidelines are available. **The only firm rule in grouping data is that the classes be both *mutually exclusive* and *all-inclusive*. By *mutually exclusive* we mean that the classes must be arranged so that every piece of data can be placed in only one class.** In particular, this means selecting class limits so that an item cannot fall into two classes. ***All-inclusive* classes are classes that together contain all the data.**

Every class in a frequency distribution has *expressed* class limits and *real* class limits. **Expressed class limits are those shown in the frequency distribution.** For example, a frequency distribution of employee ages might have the following expressed class limits:

$$\text{Expressed lower limits} \left\{ \begin{array}{l} 21\text{--}30 \\ 31\text{--}40 \\ 41\text{--}50 \\ 51\text{--}60 \\ 61\text{--}70 \end{array} \right\} \text{Expressed upper limits}$$

The real class limits extend one-half unit above and one-half unit below the expressed limits, which in our example would be as follows:

21 & under 31

$$\text{Real lower limits} \left\{ \begin{array}{l} 20.5\text{--}30.5 \\ 30.5\text{--}40.5 \\ 40.5\text{--}50.5 \\ 50.5\text{--}60.5 \\ 60.5\text{--}70.5 \end{array} \right\} \text{Real upper limits}$$

The *class width* is the difference between the upper and lower real class limits. In this age distribution example, the width of the first class is

Width = real upper limit − real lower limit = 30.5 − 20.5
= 10 yr

To find the center of a class, called the *class midpoint,* add half the class width to the real lower class limit. For example,

Expressed Class Limits	Real Class Limits	Midpoint
21–30	20.5–30.5	25.5
31–40	30.5–40.5	35.5
41–50	40.5–50.5	45.5
51–60	50.5–60.5	55.5
61–70	60.5–70.5	65.5

Thus, for the first class, the value 25.5 is midway between the expressed class limits and is therefore the midpoint of this class.

We mentioned that the classes of a frequency distribution need to be developed such that they are mutually exclusive and all-inclusive. In addition, it is desirable to select class limits so that the actual observations are evenly distributed throughout the class interval. This means that if an age class has limits of "21–30," the ages within this class should be evenly spread between 21 and 30 inclusive. If most of the people are younger than 25, the limits "21–30" are not very representative of the data in this class.

An easy way to determine an approximate class size is to take the smallest and largest values in the raw data, subtract the two, and divide by the desired number of classes.

$$\text{Class size} = \frac{\text{largest value} - \text{smallest value}}{\text{number of classes}}$$

This will give a starting figure for class widths. You should also apply some common sense in developing the frequency distribution. For example, if the division gives class sizes of 4.817 years, by all means round this off to 5 years. Also start your class limits at some easy-to-use value. If you have data on automobile weights, don't have 100-pound intervals starting at 1,805.6 pounds; start with 1,800 pounds.

If possible, have class intervals of equal size. Frequency distributions with equal class intervals are not only easier to understand, but will make later statistical analysis much simpler. However, constant intervals are not possible in many managerial applications because the data are *skewed.* Data are skewed when most observations are located relatively close together, but a few points are located in one direction far from the majority. Data that are not skewed are *symmetrical.* Examples of skewed and symmetrical distributions are shown in Figure 3–1.

Many data of interest to managers are skewed. Incomes in the United States form a skewed frequency distribution. The vast majority of incomes are less than $40,000, but a few individuals make much more than that. Class intervals wide enough to contain all observations using even 20 classes would be hundreds of thousands of dollars wide. Thus almost everyone would fall in the first interval, totally

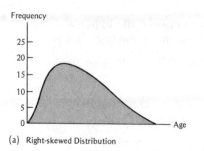

(a) Right-skewed Distribution

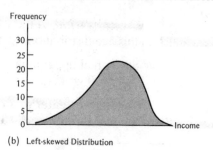

(b) Left-skewed Distribution

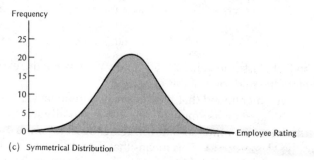

(c) Symmetrical Distribution

FIGURE 3–1 Skewed and symmetrical frequency distributions.

obscuring information in the data. However, if the intervals were $2,000, $5,000, or even $10,000 wide, many more than 20 intervals would be needed to contain all data points. The way out of this dilemma is to use **open-ended** intervals, or intervals of unequal sizes. **Open-ended intervals occur when the first data class contains no lower limit or the last data class contains no upper limit.** Table 3–1 contains an open-ended interval, that is, "$17,500 and over."

Data may be skewed in either direction; thus open-ended intervals or unequal intervals are common. Once again, they are necessary when the majority of data are contained within a limited range of values and a relatively few observations exist with extreme values.

Constructing a Frequency Distribution

The manager of a local department store is interested in what her store's sales values look like. Table 3–2 presents the raw data from a sample of 64 sales. What do the data in this form indicate to you?

The manager decides to construct a frequency distribution of the sample data. She arbitrarily decides to divide the data into ten classes. To determine the approximate class widths, she divides the difference between the two extreme observations, $74.95 and $0.97, by 10 and finds

$$\text{Class size} = \frac{\$74.95 - \$0.97}{10}$$

$$= \$7.40$$

TABLE 3–2 Raw sales data, department store example

$ 6.49	$ 8.90	$22.95	$16.30
7.19	11.97	74.95	13.39
18.63	4.44	24.99	11.99
1.29	13.88	69.99	4.44
34.98	8.12	8.99	61.98
12.95	64.88	35.95	6.99
21.25	24.99	1.26	11.99
68.99	9.97	3.97	9.97
9.98	3.99	4.35	7.49
0.97	14.99	5.99	12.49
2.19	7.75	11.99	21.50
41.69	67.29	5.25	1.69
4.65	4.50	9.85	49.99
3.19	34.95	6.50	13.89
10.45	7.49	29.97	19.97
23.85	5.69	3.57	2.77

She rounds off to $7.50. She then picks a starting point of $0.00 and forms the following expressed classes:

$$\begin{array}{r} \$\ 0.00–\$\ 7.49 \\ 7.50–\ 14.99 \\ 15.00–\ 22.49 \\ 22.50–\ 29.99 \\ 30.00–\ 37.49 \\ 37.50–\ 44.99 \\ 45.00–\ 52.49 \\ 52.50–\ 59.99 \\ 60.00–\ 67.49 \\ 67.50–\ 74.99 \end{array}$$

Note that the class limits are mutually exclusive and that no overlap exists between classes. Notice also that the classes are all-inclusive since the smallest and largest elements are included. Since money is the data measure, two decimals are sufficient to eliminate overlap, and the interval "$0.00–$7.49" is, in fact, $7.50 wide.

With the class limits determined, the manager counts the number of observations that fall into each class. The resulting frequency distribution is shown in Table 3–3. **The midpoint is the center point of each class in a frequency distribution** and has been recorded for future use. The midpoint of the second class, for example, is found by first defining the real class limits as

$$7.495 \quad \text{and} \quad 14.995$$

Then the true interval width is 7.5. Adding half this width (3.75) to the lower real limit yields

$$7.495 + 3.75 = 11.245 = \text{midpoint}$$

Referring again to Table 3–3, the frequency column shows the number of sales found in each class. **The *relative frequency* is the ratio of observations in a**

TABLE 3–3 Frequency distribution, department store sales

Sales Class	Midpoint	Tally	Frequency	Relative Frequency
$ 0.00–$ 7.49	$ 3.745	||||||||||||||||||||||||	24	0.375
7.50– 14.99	11.245	||||||||||||||||||	19	0.297
15.00– 22.49	18.745	|||||	5	0.078
22.50– 29.99	26.245	|||||	5	0.078
30.00– 37.49	33.745	|||	3	0.047
37.50– 44.99	41.245	|	1	0.016
45.00– 52.49	48.745	|	1	0.016
52.50– 59.99	56.245		0	0
60.00– 67.49	63.745	||||	4 3	0.062
67.50– 74.99	71.245	||	2 3	0.031
Total			64	1.000

class to the total observations in all classes. Thus 29.7 percent of the observations fall in the second class.

3-2
PRESENTING DATA

The usefulness of a body of data depends on whether the individuals working with it are able to readily understand the information it contains. Data must be arranged in the most understandable manner possible. In addition, the information should be arranged in an eye-catching way. Unfortunately frequency distributions are often neither totally meaningful nor eye-catching. Because of this, **graphical representations** are often used to display statistical information. There are almost as many ways to present data as there are persons doing the presenting. In the following sections we discuss some of the most common forms of graphical data representation.

Pie Charts

An effective manner of presenting statistical data is with a **pie chart.** This method is often used, particularly by newspapers and magazines when national, state, or local government budgets are being determined. Pie charts are also used to show how a total has been used or divided. Perhaps the major advantage of a pie diagram is that it is extremely easy to understand. The entire circle, or "pie," represents the total amount available, and the pieces are proportional to the amount of the total they represent. Figure 3–2 illustrates how a property-tax dollar was distributed in Boise, Idaho, in 1975.

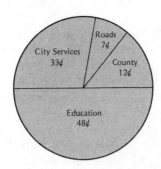

FIGURE 3–2 Pie chart.

Histograms

A *histogram* is a graphical picture of a frequency (or relative frequency) distribution. The number of observations in each class is represented by a rectangle whose base is equal to the class width and whose height is proportional to the frequency (or relative frequency) of cases belonging to that class. The vertical axis represents the class frequency and begins at the zero point. The horizontal axis represents the measure of interest, and the scale can begin with any conveniently low value.

The purpose of a histogram is to provide a visual picture of a frequency distribution. Relative differences in the areas of the rectangles correspond to relative differences in the number of observations between different classes. *if & only if there are equal class sizes*

When constructing a histogram, the measure for the item being considered is plotted on the horizontal axis, and the appropriate frequency measure on the vertical axis. Figure 3–3 shows a histogram representing the frequency distribution of sales data shown in Table 3–2.

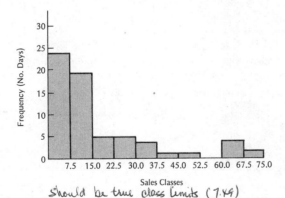

Should be true class limits (7.49)

FIGURE 3–3 Histogram showing the distribution of daily department store sales.

Using Histograms in Decision Making. The following example demonstrates how histograms can assist a manager in a decision-making process.

Although you may not think of hospitals as business enterprises, in today's environment they are very much businesses and must be operated like one to remain financially solvent. Some hospitals have decreasing patient admissions, whereas others are faced with overcrowded conditions. Those hospitals with decreasing admissions find themselves searching for new patients while at the same time trying to reduce costs without sacrificing quality health care. Overcrowded hospitals search for ways to satisfy their growing demand while at the same time maintaining high-quality care.

      The Magic Valley Memorial Hospital is currently experiencing steady growth. By analyzing the financial records and patient admissions summaries, the administrator has concluded that the hospital is approaching full utilization. However he also knows his ability to effectively allocate some resources (for instance patient rooms) is restricted. For example, if the Pediatrics ward is full, the administrator *cannot* transfer children to the unused beds in the obstetrics department. To overcome the crowding problem, the administrator has decided to attempt to schedule admissions more effectively. Because of past resistance to advance scheduling by both patients and doctors, the administrator has decided to begin his efforts with those departments where the problem is greatest. To help identify these departments, the administrator has developed the frequency histograms shown in Figure 3–4 from

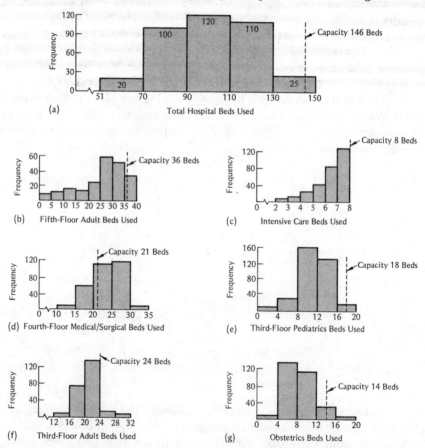

FIGURE 3–4   Histograms of patient census data, Magic Valley Memorial Hospital.

patient data collected during the past year. Note that for each histogram, the horizontal axis shows the number of beds used, and the vertical axis shows the number of days each number of beds was used. The capacity (total beds available) is also indicated on each histogram.

Figure 3–4(a) is the frequency histogram for total hospital bed use (all departments combined). Based only upon this distribution, it appears that the hospital is being utilized well below its capacity. However an examination of the bed-use distributions for each department points out the administrator's problem with overcrowding [see Figures 3–4(b)–(g)].

The graphical view of the departmental patient data clearly indicates which departments are most overcrowded. Using this information, the administrator has decided to begin his patient-scheduling campaign in the fifth-floor Adult Department and the fourth-floor Medical/Surgical Department.

Histograms with Unequal Class Intervals. Sometimes data can be most clearly and meaningfully presented by using *unequal-size class intervals.* When this is the case, a slight modification must be made in the procedure for developing a histogram. For example, Table 3–4 shows a frequency distribution for department store sales with unequal-size class intervals. Note that the interval "15.00–29.99" is twice the width of the interval "7.50–14.99," and that the interval "45.00–74.99" is four times as wide. In developing a histogram for the relative frequency distribution, we must account for the extra width by reducing the height of the histogram proportionately. That is, for an interval having a width twice the normal size, the height of the associated frequency rectangle must be cut in half. Figure 3–5 illustrates this concept for the department store sales data.

TABLE 3–4 Frequency distribution with unequal class intervals, department store data

Sales Class	Frequency	Relative Frequency
$ 0.00–$ 7.49	24	0.375
7.50– 14.99	19	0.297
15.00– 24.99*	10	0.156
30.00– 37.49	3	0.047
37.50– 44.99	1	0.016
45.00– 74.99*	7	0.109
Total	64	1.000

*Intervals having widths greater than $7.50.

Bar Charts

Bar charts are variations of histograms and are often used to illustrate data or to emphasize classes or categories of interest. For example, Figure 3–6 shows a barchart representation of data collected in a recent survey of college students. In this example, each bar is the same length, representing 100 percent of those taking the course. From this we can visualize the relative rate of students passing and failing

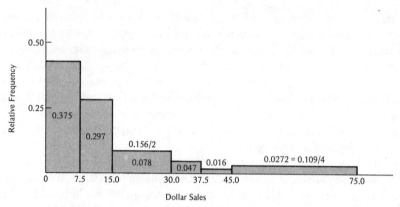

FIGURE 3–5 Relative frequency histogram with unequal class intervals, department store sales data.

specific courses. Figure 3–7 uses a bar chart to illustrate the growth in number of homes heated by electricity in a western state. The bar chart is an effective means of presenting data to demonstrate a particular point. It is easy to see that electricity use has been rising rapidly in this western state since 1974.

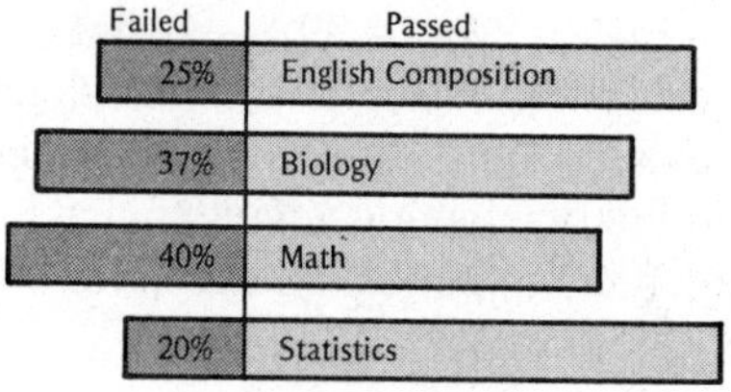

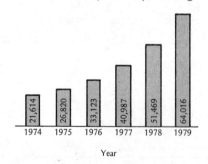

FIGURE 3–6 Bar chart of pass/fail percentages by course.

FIGURE 3–7 Bar chart of annual electricity use, number of homes with electrical heat.

Cumulative Frequency Distributions

Sometimes a person is not as interested in the data's frequency distribution as in the number, or percentage, of data that lie below or above particular values. The personnel director of a steel plant may be interested in the total number of workers with less than ten years seniority. In another instance a decision maker may be interested in the percentage of the total population that earn above a certain annual income.

In looking at sales data from a particular store, a store manager may be interested in the percentage of sales coming from items of less than a predetermined dollar value. The graph that illustrates data in a form needed by these decision makers is called a ***cumulative frequency distribution*** or a ***cumulative relative frequency distribution.*** **Cumulative frequency distributions are used if we are interested in the number of observations above or below a point. Cumulative relative frequency distributions are used if we are interested in percentages.** Both distributions can be in either a less-than or greater-than form.

To construct "less than" distributions, we can start with the frequencies shown in Table 3–5, which is a condensed version of Table 3–3. The right column in Table 3–5 shows the relative frequency for each class interval. We use the information in Table 3–5 to construct the "less than" cumulative frequency distribution and the cumulative relative frequency distribution shown in Table 3–6. The "less than" cumulative frequency for the first class ("less than $7.50") is the same as the frequency for that class. However the cumulative frequency for the second class ("less than $15.00") is found by summing the frequency for the "$0.00–$7.49" class and the "$7.50–$14.99" class. Likewise, the cumulative frequency for the "less than

TABLE 3–5   Frequency distribution, department store sales data

Sales Class	Frequency	Relative Frequency
$ 0.00–$ 7.49	24	0.375
7.50– 14.99	19	0.297
15.00– 22.49	5	0.078
22.50– 29.99	5	0.078
30.00– 37.49	3	0.047
37.50– 44.99	1	0.016
45.00– 52.49	1	0.016
52.50– 59.99	0	0
60.00– 67.49	4	0.063
67.50– 74.99	2	0.031
Total	64	1.000

TABLE 3–6   Cumulative distribution, department store sales data

Sales Class	Cumulative Frequency	Cumulative Relative Frequency
Less than $ 7.50	24	0.375
Less than $15.00	43	0.672
Less than $22.50	48	0.750
Less than $30.00	53	0.828
Less than $37.50	56	0.875
Less than $45.00	57	0.891
Less than $52.50	58	0.907
Less than $60.00	58	0.907
Less than $67.50	62	0.970
Less than $75.00	64	1.000

$22.50" class is found by summing the frequencies in the first three classes. The process is continued until all classes have been included.

The cumulative relative frequency distribution is formed in the same manner, except the relative frequencies are summed.

3-3
CONCLUSIONS

Uncertainty or variation is always present in decision-making situations. Managers must be able to make operational decisions in situations where such uncertainty exists. In dealing with uncertainty, a first step is to arrange available data in a consistent and understandable manner, and often the second step is to represent the data in graphical form. In this chapter we have introduced some common methods used both to organize and to graph large collections of data, and have given some examples of their use.

A quick look at some of the leading business journals such as *Fortune, Forbes,* and *Business Week* should convince you that the descriptive techniques discussed here are used extensively. Remember, the primary purpose of the descriptive techniques is to help decision makers transform data into meaningful information.

chapter glossary

all-inclusive classes   The frequency classes that include all observations in a set of data.

bar chart   A variation of a histogram used to visually display data or to emphasize the categories into which the data have been divided.

class midpoint   The center point between the upper and lower real class limits of a frequency class.

class width   The numerical distance between the lower real class limit and upper real class limit of a frequency class.

cumulative frequency distribution   A distribution that represents the frequency of observations equal to or less than a particular class limit or value.

frequency distribution   A way in which data are arranged, showing the number of cases in each category or class.

frequency histogram   A graphical representation of a frequency or relative frequency distribution. The frequency of cases in each class is represented by a rectangle with a base equal to the class width and a height proportional to the frequency of cases belonging to the class.

mutually exclusive classes   Frequency classes that have boundaries that do not overlap.

relative frequency   The percentage of observations falling within a particular class.

skewed distribution   A distribution that has more than half of the observations falling above or below the midpoint of the center class.

1. Describe briefly each of the following:
   a. Class limits
   b. Histogram
   c. Relative frequency distribution
   d. Cumulative frequency distribution

2. Describe the difference between a frequency distribution and a relative frequency distribution.

3. Assume you are trying to construct a frequency distribution for the weights of people in this class. Describe the steps you would take.

4. Think of examples of skewed distributions. How would you select class limits for these distributions?

5. Go to the library, and from the *Statistical Abstract of the United States,* construct a frequency distribution for unemployment rate by state.
   a. Justify your choice of class limits and number of classes.
   b. Present your data as a histogram.
   c. Write a paragraph analyzing this graph.

6. Suppose that the loan officer at Money First National Bank wants to obtain information about the loans she has made over the past five years. She has decided to develop a distribution showing the loan frequency by size of loan. A quick look at the data indicates that the smallest loan she made was for $1,000, and the largest loan was $25,000. She has decided to have ten classes in her distribution. Define the ten classes in terms of lower and upper limits.

7. Determine the midpoints for each class developed in problem 6.

8. Suppose the loan officer in problem 6 finds that the frequencies are as follows:

Class	Frequency
1	40
2	80
3	120
4	250
5	500
6	700
7	500
8	400
9	200
10	10
Total	2,800 loans

From this information, and using the class intervals you developed in problem 6, produce the appropriate histogram.

9. Develop the cumulative frequency distribution for the information given in problem 8.

10. Develop the relative frequency distribution for the information given in problems 8 and 9.

11. Develop a cumulative relative frequency distribution for the loan data in problem 8.

12. The following data are a sample of 60 accounts receivable balances selected from accounts at the Wallingford Department Store:

$ 39.93	$ 72.04	$ 69.04	$ 87.00	$ 55.55	$ 33.33
107.56	146.93	107.33	80.00	7.50	29.59
98.05	27.50	141.88	68.00	15.00	11.05
24.88	105.19	70.00	96.07	150.00	9.47
25.00	11.41	37.73	44.09	80.05	99.99
19.95	53.72	125.00	75.55	97.94	47.09
72.50	16.18	33.97	56.25	12.11	19.58
20.00	126.12	16.47	110.00	8.00	49.00
30.72	14.50	11.01	76.47	19.33	62.50
90.05	19.33	49.99	52.52	27.05	66.05

Decide how many classes would be appropriate for these data and justify why you selected that number.

13. Using the number of classes you selected in problem 12, develop a frequency distribution for the accounts receivable.

14. Construct a bar chart from the frequency distribution in problem 13.

15. Write a one-paragraph statement describing the accounts receivable balances as reflected by the sample in problem 12. (Remember, in business, report writing is an important way of conveying information.)

16. Comment on using the following classes in connection with the data given in problem 12:
   a. $ 9.47–$19.46
      19.47– 29.46
         etc.
   b. $ 5–$15
      15– 25
      25– 30
         etc.
   c. $ 5 to under $35
      35 to under 45
      45 to under 55
         etc.
   d. $16 to under $30
      31 to under 45
      46 to under 60
         etc.

17. You work for the State Industrial Development Council. You are presently working on a financial services brochure to send to out-of-state companies. You are given the following data on banks in your state and are to present it in an eye-catching manner with a two-paragraph summary of what the data would mean to a company considering moving:

Deposit Size (× $1 million)	No. Banks	Total Deposits (× $1 million)
Less than 5	2	7.2
5 to less than 10	7	52.1
10 to less than 25	6	111.5
25 to less than 50	3	95.4
50 to less than 100	2	166.6
100 to less than 500	2	529.8
Over 500	2	1,663.0

Your boss has just said you need to include relative frequencies in your presentation.

18. Your company is about to introduce a complete line of steel-belted radial tires. Because you are late entering the market, you were supposed to develop a new, hard-hitting advertising approach. You decided to put test tires on 100 taxis along the Alaskan pipeline. The following is a listing of how far the 100 taxis ran before one of the four tires did not meet minimum federal standards (rounded to the nearest thousand miles):

38	24	12	36	41	40	45	41	40	47
26	15	48	44	29	43	28	29	37	10
37	45	29	31	23	49	41	47	41	42
61	40	40	45	37	55	47	42	28	38
38	48	18	16	39	50	14	52	33	32
51	10	49	21	44	31	43	34	49	48
28	39	28	36	56	54	39	31	35	36
32	20	54	25	39	44	25	42	50	41
9	34	32	34	42	40	43	32	30	45
20	29	14	19	38	46	46	39	40	47

Although you don't want to be dishonest in presenting the results of your test, at the same time your job depends on showing the results in as favorable a light as possible. Organize and present the data using the methods discussed in this chapter and decide which method would be most favorable for selling the tires.

Your boss apparently delights in putting people on the spot. Make sure you can defend your choice, including how you determined the interval size and interval limits.

TABLE P–1   Five-year financial and operating summaries

	1976	1975	1974	1973	1972
Revenues					
Net sales	$5,218.0	$4,977.2	$5,381.0	$4,137.6	$3,113.6
Interest, dividends, and other income	56.7	51.1	67.7	41.3	26.6
Total	$5,301.7	$5,028.3	$5,448.7	$4,178.9	$3,140.2
Costs and expenses					
Operating charges					
Employment costs	$2,313.6	$2,139.2	$2,072.0	$1,759.3	$1,418.6
Materials and services	2,373.0	2,340.1	2,112.7	1,765.1	1,208.1
Depreciation	275.6	234.2	210.9	196.1	180.8
Taxes other than employment and income taxes	70.8	68.1	63.1	58.8	60.5
Total	$5,033.0	$4,681.9	$4,788.7	$3,779.3	$2,898.3
Interest and other debt charges	77.7	63.4	11.0	43.0	10.3
Taxes on income	26.0	41.0	271.0	150.0	67.0
Total	$5,136.7	$4,786.3	$5,106.7	$3,972.3	$3,005.6
Net income					
Amount	$ 168.0	$ 242.0	$ 342.0	$ 206.6	$ 134.6
Percentage of revenues	3.2%	4.8%	6.3%	4.9%	4.3%
Percentage of stockholders' equity at beginning of year	6.4%	9.7%	15.3%	9.7%	6.5%
Per share	$ 3.85	$ 5.51	$ 7.85	$ 4.72	$ 3.02
Dividends					
Amount	$ 87.1	$ 120.1	$ 100.2	$ 72.4	$ 53.1
Per share	$ 2.00	$ 2.75	$ 2.30	$ 1.65	$ 1.20
Capital expenditures					
Property, plant, and equipment	$ 396.2	$ 671.3	$ 521.2	$ 248.8	$ 172.7
Investments in associate enterprises	$ 10.4	$ 13.5	$ 17.3	$ 21.3	$ 19.9
Long-term debt at year end	$1,023.4	$ 856.9	$ 648.3	$ 663.0	$ 612.1
Stockholders' equity at year end	$2,692.6	$2,612.0	$2,490.1	$2,242.3	$2,136.7
Common stock at year end					
Shares outstanding in hands of public*	43,666	43,665	43,666	43,467	44,469
Stockholders'	165,000	179,000	188,000	193,000	202,000

Note: Dollars in millions, except per-share data.
*In thousands.

19. Data are often presented in a graphical format to show important trends. The data in Table P–1 were taken from the 1977 annual report of a large U.S. corporation. Choose a graphical method to show trends in the following factors:
  a. Net sales
  b. Employment costs
  c. Materials and services
  d. Earnings per share
  e. Dividends per share
  f. Investment in property, plant, and equipment

═══cases

3A   CITY ELECTRIC COMPANY

Tom Reynolds, the manager of City Electric Company's research and development group, has just reported the development of a new electrical motor that will save users a considerable amount in electrical energy costs. Tom Reynolds has had his department estimate the annual cost savings that could be expected with use of this new motor. The potential savings are shown in Table C–1 for two different motor sizes. Tom is understandably proud of this development and thinks the potential savings will be a major selling factor.

Howard King, the marketing manager, has just received the savings estimates from Reynolds. Although Howard agrees the cost savings are considerable,

TABLE C–1   Annual savings

15-hp MOTOR			
Power Cost ($/kWh)	40-hr. Week (1 shift)	80-hr Week (2 shifts)	168-hr Week (continuous)
0.03	$36.33/yr	$ 72.66/yr	$158.70/yr
0.04	48.44/yr	96.88/yr	211.60/yr
0.05	60.55/yr	121.10/yr	264.50/yr
0.06	72.66/yr	145.32/yr	317.40/yr
0.07	84.77/yr	169.54/yr	370.30/yr
40-hp MOTOR			
Power Cost ($/kWh)	40-hr Week (1 shift)	80-hr Week (2 shifts)	168-hr Week (continuous)
0.03	$109.38/yr	$218.77/yr	$ 477.80/yr
0.04	145.84/yr	291.69/yr	637.00/yr
0.05	182.31/yr	364.61/yr	796.30/yr
0.06	217.77/yr	437.54/yr	955.60/yr
0.07	255.23/yr	510.46/yr	1,114.80/yr

he feels a more effective presentation of the potential cost savings must be made. He has asked his administrative assistant to use the cost savings developed by the research and development department and prepare a more effective way to demonstrate the cost savings.

references

HUNTSBERGER, DAVID V., PATRICK BILLINGSLEY, and D. JAMES CROFT. *Statistical Inference for Management and Economics*. Boston: Allyn and Bacon, 1975.

LOETHER, H. J. and D. G. MCTAVISH. *Descriptive Statistics for Sociologists*. Boston: Allyn and Bacon, 1974.

MENDENHALL, WILLIAM and JAMES REINMUTH. *Statistics for Management and Economics*. 3rd ed. North Scituate, Mass.: Duxbury, 1978.

PARSONS, ROBERT. *Statistical Analysis: A Decision Making Approach*. New York: Harper and Row, 1978.

Measures of
Location and Spread

Measures for ungrouped data
Mean, Median, Mode

The methods for graphically presenting data discussed in Chapter 3 provide a starting point for analyzing data. However these methods do not reveal all the information contained in a set of data. Managers who want to know as much as possible about their companies' sales divisions most likely will not be satisfied with frequency distributions showing the distribution of daily sales. Even frequency histograms or other graphical techniques probably will not provide all the required information. The managers will likely want to make comparisons between sales divisions to discern whether major differences exist. Although frequency distributions and histograms provide some basis for making this type of comparison, they may be misleading. In fact, histograms from two quite different sets of data may appear very similar due to the number and width of the class intervals selected.

To overcome some limitations in the methods discussed in Chapter 3, decision makers need to become acquainted with some additional tools of descriptive statistics. Specifically, managers need to know some statistical measures that can be determined from any set of data. There are two broad categories within which these measures fall: *measures of location* and *measures of spread.*

chapter objectives

Statistics is often used to decide whether a true difference exists between two sets of data. Many statistical techniques used to test for differences between data groups attempt to determine whether the data from the two groups have the same distribution. That is, do the groups have the same central location and the same spread?

In this chapter we shall discuss some techniques for measuring the central location of a data distribution. Although many ways of measuring data location exist, we shall concentrate on the three most common: the arithmetic mean, the median, and the mode.

We shall also discuss techniques for measuring the spread, or dispersion, in a set of data. The measures that will receive the greatest emphasis are the range, the variance, and the standard deviation.

student objectives

After studying the material in this chapter, you should be able to:

1. Calculate the mean and median for both grouped and ungrouped data.
2. Determine the variance and standard deviation for grouped and ungrouped data.

4-1
MEASURES OF LOCATION

As is implied by the name, **the central location is the middle, or center, of a set of data.** In this section we shall consider several common measures of central location. We shall discuss the important statistical properties and introduce some business applications for each of the measures. The measures of location that will be covered are the *mode,* the *median,* and the *arithmetic mean.*

The Mode

The *mode* is the observation that occurs most frequently in a data set. Suppose a small businessperson has just opened a men's clothing store and is interested in determining how many of each size men's shirts to stock. Men's shirts are generally ordered by neck size in inches, for example 14½, 15, 15½, 16, 16½, and so forth. To help in deciding the inventory levels for each size shirt, the manager has selected a sample of 50 potential customers and has recorded their shirt sizes as shown in Table 4–1.

Initially, the store manager might use this sample to estimate which size shirt will be demanded most often. Recalling that the mode is the value occurring most often, the mode is 16½ inches since this size was observed 13 times in the sample—more than any other size.

TABLE 4–1 Men's shirt sizes
(neck measurements in inches)

16	16	16	14½	14½
16½	16	16½	16½	15
16½	15½	15	16½	15½
15	15½	17	16	15½
15½	15	16½	15	16
15	14½	14½	15	16
15	17	15½	15½	16½
14½	15½	15½	16½	16½
17	15½	16	16½	16½
16½	16	16½	17	14½

A set of data may have two or more values that tie for the most frequently occurring. When this happens, the distribution is **multimodal.** Also, a data set may have no mode if no one value occurs more frequently than another.

The following example illustrates how the mode can assist the decision-making process. Fox Corporation executives saw that the existing corporate jets such as Lear and Gulfstream were expensive ($1 million or more apiece). Some of this cost was due to the fact these jets were usually built to carry eight or more passengers. A study of passenger loads for a large number of trips for different corporate jets revealed that although sometimes the jets were full, in a vast majority of trips there were empty seats. In fact, for the data studied, the number of passengers observed most often was three, including the pilot. Thus the value 3, the mode for the observed data, figured strongly into Fox Corporation's plans to build a "small," lower-priced, four-passenger jet.

The Median

The *median* is the middle observation in data that have been arranged in ascending or descending numerical sequence. A numerical sequence of data is called an ***array.*** For example, suppose the following data represent the mileage ratings for nine new cars:

$$18.0 \text{ m.p.g.}$$
$$19.5$$
$$22.0$$
$$24.5$$
$$29.0 \leftarrow \text{Median}$$
$$32.0$$
$$33.0$$
$$33.0$$
$$33.0$$

When the array has an *odd* number of observations, the median is $(N + 1)/2$ observations from either end. Thus the median mileage rating for the nine new cars is the $(9 + 1)/2 = $ fifth observation from the top or bottom. Therefore the median is 29.0 m.p.g.

If the array contains an *even* number of observations, the median is any point between the two middle values. Generally the median is found by taking the average of the middle two values. For example, a company that supplies heating oil to residential customers has collected the following data on the quantity of oil purchased by eight households in December:

42 gal
51
53
53
59
61
75
100

Since there are eight observations, the median is the average of the fourth and fifth observations.

$$\text{Median} = \frac{53 + 59}{2} = 56 \text{ gal}$$

The median is most useful as a measure of central tendency, or central location, in situations where the data contain some extreme observations. For instance, incomes in a market area are often characterized by a few very high incomes, and the majority under $25,000. Although the few high income earners are worth noting, a person contemplating opening a drive-in restaurant would like to have a measure of central income that represents the majority of people in the area. The median might be this measure since it is relatively unaffected by extreme values.

To illustrate that the median is not sensitive to extreme cases, suppose the highest mileage for the nine new cars discussed earlier was 85.0 m.p.g. rather than 35.0 m.p.g. The median is still 29.0 m.p.g. The extreme case has no effect.

Finding the Median for Grouped Data. If the data have been grouped into a frequency distribution, calculating the median requires *interpolation.* For example, Table 4–2 gives the distribution of 4,000 parts stocked by a large automobile

TABLE 4–2   Distribution of prices for auto parts

Class	Frequency f_i	Cumulative Frequency
$ 0 and under $ 5	100	100
$ 5 and under $10	400	500
$10 and under $15	600	1,100
$15 and under $20	1,300	2,400
$20 and under $25	900	3,300
$25 and under $30	500	3,800
$30 and under $35	100	3,900
$35 and under $40	100	4,000
Total	4,000	

dealership that have retail prices under $40. **The *median class* contains the middle value in the array.** Thus, if the frequency distribution has N items, the median class contains the N/2 item (that is, the 2,000th item). Starting from either the top or bottom, item 2,000 is in the class "$15 and under $20." Therefore that class contains the median. But does the median lie near $15, or is it closer to $20? This question is answered by interpolating the required distance from the lower limit of the median class as shown in the median formula

$$\text{Median} = L_{median} + \left(\frac{\frac{N}{2} - \Sigma f_{prec}}{f_{median}}\right)(w) \qquad \textbf{(4–1)}$$

where: L_{median} = lower limit of the median class
   N = total number of observations
   Σf_{prec} = sum of observations in classes preceding the median class
   w = width of the median class
   f_{median} = observations in the median class

Therefore

$$\text{Median} = \$15 + \left(\frac{\frac{4,000}{2} - 1,100}{1,300}\right)(\$5)$$

$$= \$15 + \left(\frac{900}{1,300}\right)(\$5)$$

$$= \$18.46$$

We should discuss the rationale behind this procedure. Notice in Table 4–2 that there are 1,100 observations in the classes with prices less than $15. Because we want to find item 2,000, we need to go 900 items into the next interval, "$15 and under $20." Since 1,300 items are contained in this interval, and the items are assumed evenly spread through the interval, we need to move 900/1,300 of the way through this class. Because the class width is $5.00, we will go 900/1,300 of this $5.00 width. This is $3.46 through the $5.00 interval. Since the interval starts at $15.00, the median price is

$$\$15.00 + \$3.46 = \$18.46$$

Remember, the median is not sensitive to extreme values in a data set and thus is particularly useful as a measure of location for skewed distributions.

The Mean

By far the most common statistical measure of location is the *mean*. **The mean is often called the *arithmetic average* in nonstatistical applications and is found by summing all the observations and dividing the sum by the number of observations.**

The notation used to represent the process differs depending on whether the data represent an entire population or a sample from a population. If the data comprise an entire population, the mean is found by

$$\mu_x = \frac{\sum\limits_{i=1}^{N} X_i}{N} \tag{4-2}$$

where: μ_x = population mean (μ is pronounced *mu*)
             N = population size
             Σ = summation symbol
             X_i = individual observations

Recall from the review of basic mathematical concepts in Chapter 1 that $\sum\limits_{i=1}^{N}$ requires us to sum all the elements with an *i* subscript from $i = 1$ to $i = N$ inclusive. If the data are from a sample, the sample mean is found by

$$\overline{X} = \frac{\sum\limits_{i=1}^{n} X_i}{n} \tag{4-3}$$

where: $\overline{X}$ = sample mean (pronounced X *bar*)
             n = sample size
             X_i = individual observations

Now, suppose the population of interest is the total dollar sales for the last ten days at the Hillcrest Golf Pro Shop. Table 4–3 lists these ten values. The mean of the sales for these ten days is

$$
\begin{aligned}
\mu_x &= \frac{\sum\limits_{i=1}^{N} X_i}{N} \\
&= \frac{\$88 + \$102 + \$90 + \$120 + \$50 + \$60 + \$130 + \$100 + \$95 + \$115}{10} \\
&= \frac{\$950}{10} \\
&= \$95
\end{aligned}
$$

We must use equation (4–2) since we are dealing with the entire *population* of values.
    Suppose we select a *sample* of five days' sales from this population, say, $90, $100, $60, $95, and $50. The sample mean is

$$\overline{X} = \frac{\sum\limits_{i=1}^{n} X_i}{n} = \frac{\$90 + \$100 + \$60 + \$95 + \$50}{5} = \frac{\$395}{5}$$

$$= \$79$$

TABLE 4–3 Daily golf sales, Hillcrest Golf Pro Shop

$ 88	$ 60
102	130
90	100
120	95
50	115

Finding the Mean for Grouped Data. If the data have been grouped into classes and a frequency distribution has been formed, we apply the following formula to find the mean:

$$\mu_x = \frac{\sum_{i=1}^{c} f_i M_i}{N} \qquad (4\text{–}4)^*$$

where: μ_x = population mean

c = number of classes

f_i = frequency in the ith class

M_i = midpoint of the ith class

N = population size

Using the grouped distribution of automobile part prices in Table 4–2, the mean price can be found as follows:

Class	Midpoint M_i	Frequency f_i	$f_i M_i$
$ 0 and under $ 5	$ 2.50	100	$ 250
$ 5 and under $10	7.50	400	3,000
$10 and under $15	12.50	600	7,500
$15 and under $20	17.50	1,300	22,750
$20 and under $25	22.50	900	20,250
$25 and under $30	27.50	500	13,750
$30 and under $35	32.50	100	3,250
$35 and under $40	37.50	100	3,750

$$N = \sum_{i=1}^{c} f_i = 4{,}000 \qquad \sum_{i=1}^{c} f_i M_i = \$74{,}500$$

$$\mu_x = \frac{\sum_{i=1}^{8} f_i M_i}{\sum_{i=1}^{8} f_i = N} = \frac{\$74{,}500}{4{,}000}$$

$$= \$18.63$$

*Equation (4–4) can also be used to define a weighted mean, or weighted average, where

f_i = weights assigned to each category or class

N = sum of all the weights

Recall that the median for these data was $18.46. **For any set of data, the mean and median need not be the same. Also, when the data are in grouped form, the calculations for the median and mean provide only approximations.** When the data are grouped, we either assume that all observations fall at the class midpoint or that the observations are evenly spread throughout each class interval and can be represented by the midpoint. The closer these assumptions are to being true, the better the approximations will be.

The mean is sensitive to extreme observations and can be a misleading measure of central location when the data are skewed. For example, suppose that we reexamine the Hillcrest Golf sales shown in Table 4–3. Instead of the largest daily sales being $130, suppose the largest value was $930. The mean would be

$$\mu_x = \frac{\$1,750}{10}$$

$$= \$175$$

For these data, the mean is not a good measure of central tendency since it is higher than all but one observation.

The Mean, Median, and Mode for Skewed Data

The mean, median and mode usually are not the same for a given set of data. Yet each of the three measures is an attempt to describe the central position of the data being considered. The only time the mean, median, and mode will all have the same

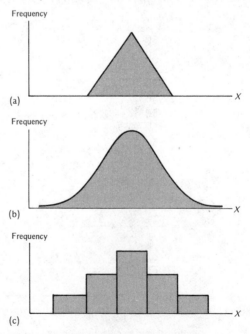

FIGURE 4–1   Unimodal and symmetrical distributions.

value is when the data distribution is both unimodal *and* symmetrical. Several uni-modal and symmetrical distributions are shown in Figure 4–1.

The three measures of location do not have the same value if the data distribution is skewed. As mentioned in Chapter 3, data can be skewed to the left or right depending on the direction of the tail in the distribution. Skewed distributions affect the three measures of location differently. The mode is the value that occurs most frequently. In the distribution shown in Figure 4–2, the mode is at the highest point of the distribution. The median, which is the center observation, lies to the skewed side of the mode. The mean—the measure most affected by extreme values in the distribution's tail—lies beyond the median. **Of the three common measures of location, the mean is most affected by a skewed distribution. For this reason, the mean, although the most commonly used statistical measure, is not always the best measure of location.**

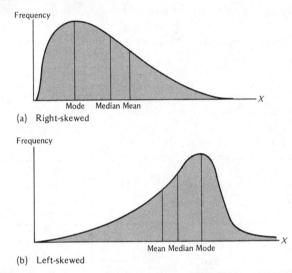

(a) Right-skewed

(b) Left-skewed

FIGURE 4–2 Skewed distributions.

4-2
MEASURES OF SPREAD

As shown in the previous section, the mean, median, and mode describe the central location of a distribution. However, the location is only one data characteristic of interest to decision makers. Measures of *spread* are also important.

For example, the Fine-Soup Company is considering a new machine for filling 12-ounce cans with soup. The machine is guaranteed to put an average of 12 ounces of soup in each can. If the machine always put in 12 ounces, the Fine-Soup Company would have no worries. However, as in any production process, there will be variation, here in the amount of soup going into each can. Even though the average may be 12 ounces, sometimes the fill may be slightly more than 12 ounces and other times less than 12 ounces. Because of the legal ramifications of putting too little soup in cans, and of the lost profit if too much soup is used, the Fine-Soup Company is concerned with the variation in the filling process.

Statistical measures of variation are also called measures of *spread*, or measures of *dispersion*. In this section we shall discuss the measures of dispersion most commonly used in statistical analysis. These measures are the ***range, variance,*** and ***standard deviation.***

The Range

The simplest measure of spread in a set of data is the ***range.* The *range* is the difference between the largest and smallest observations in the data.** For example, in Table 4–4 we have listed the number of new car sales by a Ford dealer during the 20 days the dealership has been open. The range for these data is found as follows:

$$\text{Range} = \text{maximum} - \text{minimum}$$

Therefore

$$\text{Range} = 12 - 1$$
$$= 11 \text{ cars}$$

TABLE 4–4   New car sales for 20 days

8	6	7	2	6
1	2	7	4	4
1	3	5	1	12
5	4	3	4	2

Although the range is the easiest measure of dispersion to calculate, it also conveys the least information. Since the range is determined by only the two extreme values, it provides no indication about the spread of the other values. The range is therefore very sensitive to extreme values in the data. For example, if the car dealer had on the one day sold 20 cars instead of 12 cars, the range would have been 19 rather than 11, even though no other values had changed.

The range is also sensitive to the number of observations in the data set. In general, if more observations are included, there is a greater chance of having extreme values in the data set.

The Variance and the Standard Deviation

To a large extent, the measure of location used in an application determines the measure of spread to be used. As we mentioned earlier, the mean is the most frequently used measure of location for a set of data. When the mean is used as the measure of location, the ***variance*** and ***standard deviation*** are the most appropriate measures of spread.

The methods for calculating both the variance and the standard deviation use all data points in measuring the spread of the data around the mean. However these measures are calculated differently depending on whether we are finding the measures of spread for an entire population, or determining their values for a sample from the population.

If we are dealing with a finite population,

$$\text{Variance} = \sigma_X^2 = \frac{\sum_{i=1}^{N}(X_i - \mu_X)^2}{N} \tag{4–5}$$

$$\text{Standard deviation} = \sigma_X = \sqrt{\frac{\sum_{i=1}^{N}(X_i - \mu_X)^2}{N}} \tag{4–6}$$

where: σ_X^2 = population variance (pronounced *sigma sub X squared*)

σ_X = population standard deviation

X_i = individual population values

μ_X = population mean

N = population size

If we are dealing with sample data,

$$\text{Variance} = S_X^2 = \frac{\sum_{i=1}^{n}(X_i - \overline{X})^2}{n - 1} \tag{4–7}\dagger$$

$$\text{Standard deviation} = S_X = \sqrt{\frac{\sum_{i=1}^{n}(X_i - \overline{X})^2}{n - 1}} \tag{4–8}$$

where: S_X^2 = variance from the sample

S_X = standard deviation from the sample

X_i = individual sample values

$\overline{X}$ = sample mean

n = sample size

Up until now we have been using formal summation notation utilizing the subscripts for which the summation applies. For the remainder of the text we will drop these subscripts whenever all data are to be summed. Thus

---

†Note that the divisor in the formula for S_X^2 is $(n - 1)$. We will discuss in Chapter 9 why $(n - 1)$ is used rather than simply the sample size, n. For now, remember that if you are dealing with a population, the variance is found by dividing by N, the population size. If you are using a sample to estimate the population variance, divide by $(n - 1)$.

$$\sigma_X^2 = \frac{\Sigma(X - \mu_X)^2}{N}$$

and

$$S_X^2 = \frac{\Sigma(X - \overline{X})^2}{n - 1}$$

These equations are equivalent to equations (4–5) and (4–6), respectively. You should find this relaxed notation less confusing.

The variance and standard deviation are closely related. Once you have determined one, the other is determined quite easily: **the *standard deviation* is simply the positive square root of the variance.**

The equations for the variance and the standard deviation are not as complicated as they initially may appear. Both start out with finding the distance between the mean and individual observation values, as follows:

For populations:

$$X - \mu_X$$

For samples:

$$X - \overline{X}$$

However, one property of the mean is that the total distance between it and the values above the mean just equals the total distance between it and the values below the mean. That is:

For populations:

$$\Sigma(X - \mu_X) = 0$$

For samples:

$$\Sigma(X - \overline{X}) = 0$$

Since a value of zero provides no information in determining the spread in a set of data, we eliminate this problem by *squaring* the individual distance measures. We then sum the squared differences and divide by either N or $(n - 1)$ depending upon whether we are dealing with a population or a sample.

Returning to the car dealer example in Table 4–4, σ_X^2 is found as follows. First we calculate the mean, μ_X.

$$\mu_X = \frac{\Sigma X}{N} = \frac{87}{20}$$

$$= 4.35$$

Then we can set up the following calculation table:

X	$(X - \mu_x)$	$(X - \mu_x)^2$
8	$(8 - 4.35) = 3.65$	13.32
1	$(1 - 4.35) = -3.35$	11.22
1	$(1 - 4.35) = -3.35$	11.22
5	$(5 - 4.35) = 0.65$	0.42
6	$(6 - 4.35) = 1.65$	2.72
2	$(2 - 4.35) = -2.35$	5.52
3	$(3 - 4.35) = -1.35$	1.82
4	$(4 - 4.35) = -0.35$	0.12
7	$(7 - 4.35) = 2.65$	7.02
7	$(7 - 4.35) = 2.65$	7.02
5	$(5 - 4.35) = 0.65$	0.42
3	$(3 - 4.35) = -1.35$	1.82
2	$(2 - 4.35) = -2.35$	5.52
4	$(4 - 4.35) = -0.35$	0.12
1	$(1 - 4.35) = -3.35$	11.22
4	$(4 - 4.35) = -0.35$	0.12
6	$(6 - 4.35) = 1.65$	2.72
4	$(4 - 4.35) = -0.35$	0.12
12	$(12 - 4.35) = 7.65$	58.52
2	$(2 - 4.35) = -2.35$	5.52
		$\Sigma = 146.50$

Then

$$\sigma_X^2 = \frac{146.50}{20}$$

$$= 7.33 \text{ cars squared}$$

$$\sigma_X = \sqrt{7.33}$$

$$= 2.71 \text{ cars}$$

Although the variance of 7.33 *cars squared* causes no problems statistically, it is difficult to realistically interpret. In fact, we are not quite sure what a *car squared* is. The standard deviation of 2.71 cars is easier to understand. Thus an important reason for using the standard deviation as a measure of spread is that it is expressed in the same units as the mean.

Shortcut Formulas for Calculating the Variance and the Standard Deviation.   In cases with many observations, using the previous equations for calculating the standard deviation and variance becomes tedious. Fortunately, shortcut equations exist for calculating both the standard deviation and variance.

For populations:

$$\sigma_X^2 = \frac{\Sigma X^2 - \dfrac{(\Sigma X)^2}{N}}{N} \qquad (4\text{--}9)$$

$$\sigma_X = \sqrt{\frac{\Sigma X^2 - \dfrac{\Sigma (X)^2}{N}}{N}} \qquad (4\text{--}10)$$

For samples:

$$S_X^2 = \frac{\Sigma X^2 - \dfrac{(\Sigma X)^2}{n}}{n - 1} \qquad (4\text{--}11)$$

$$S_X = \sqrt{\frac{\Sigma X^2 - \dfrac{(\Sigma X)^2}{n}}{n - 1}} \qquad (4\text{--}12)$$

Using formulas (4–11) and (4–12) on the automobile dealer data, we get

$$\Sigma X^2 = 525$$
$$\Sigma X = 87$$

Then

$$\sigma_X^2 = \frac{525 - \dfrac{(87)^2}{20}}{20} = \frac{525 - 378.45}{20}$$

$$= 7.33 \text{ cars squared}$$

$$\sigma_X = \sqrt{7.33}$$

$$= 2.71 \text{ cars}$$

These are, of course, the same results found earlier for the standard deviation and variance.

What the Standard Deviation Means. The standard deviation is a measure of the average spread around the mean. The standard deviation can be most easily understood if you think of it as a *distance measure*. Whereas 1 foot always equals 12 inches and 1 meter always equals 100 centimeters, one standard deviation may equal *any* value depending on the data being analyzed. Figure 4–3 illustrates the mean and standard deviation for our car dealer example. Notice that the distance can be measured in either direction from the mean. A distance of one standard deviation above the mean reaches the value 7.06, and a distance of one standard deviation below the mean reaches the value of 1.64. Two standard deviations above and below the mean are at the points 9.77 and −1.07, respectively.

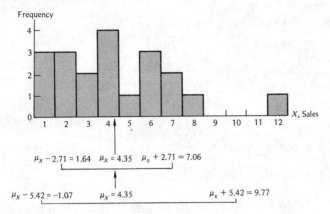

FIGURE 4–3 Standard deviation as a measure of spread, car dealer sales data.

Suppose the standard deviation had been 6 instead of 2.71. Then the distance measures would be

$$\mu_x + 1\sigma_x = 4.35 + 6 = 10.35$$
$$\mu_x - 1\sigma_x = 4.35 - 6 = -1.65$$
$$\mu_x + 2\sigma_x = 4.35 + 12 = 16.35$$
$$\mu_x - 2\sigma_x = 4.35 - 12 = -7.65$$

Thus, regardless of the value σ_x, we think of it as a measure of distance along the horizontal axis on either side of the mean.

The Variance and the Standard Deviation for Grouped Data.   As we have illustrated, data are often best presented by grouping into classes. The variance and standard deviation can be calculated for grouped data in much the same manner as the mean is calculated for grouped data. The formulas for calculating the variance and standard deviation for grouped data are as follows:

For populations:

$$\sigma_X^2 = \frac{\Sigma f(M - \mu_x)^2}{N} \tag{4–13}$$

$$\sigma_X = \sqrt{\frac{\Sigma f(M - \mu_x)^2}{N}} \tag{4–14}$$

where: f = number of values in each class

M = midpoint of each class

μ_x = mean of the population = $\dfrac{\Sigma f M}{N}$

N = population size = Σf

For samples:‡

$$S_X^2 = \frac{\Sigma f(M - \bar{X})^2}{n-1}$$

(4–15)

$$S_X = \sqrt{\frac{\Sigma f(M - \bar{X})^2}{n-1}}$$

(4–16)

where: $\bar{X}$ = sample mean = $\dfrac{\Sigma fM}{n}$

n = sample size = Σf

Table 4–5 illustrates the frequency distribution from a sample of 100 days of production at a lumber mill. The standard deviation for this sample is found as follows. First

$$\bar{X} = \frac{\Sigma fM}{n} = \frac{5,850}{100}$$

$$= 58.50$$

Then

M	f	$(M - \bar{X})$	$(M - \bar{X})^2$	$f(M - \bar{X})^2$
20	5	(20 − 58.50) = − 38.50	1,482.25	7,411.25
30	8	(30 − 58.50) = − 28.50	812.25	6,498.00
40	15	(40 − 58.50) = − 18.50	342.25	5,133.75
50	20	(50 − 58.50) = − 8.50	72.25	1,445.00
60	15	(60 − 58.50) = 1.50	2.25	33.75
70	10	(70 − 58.50) = 11.50	132.25	1,322.50
80	17	(80 − 58.50) = 21.50	462.25	7,858.25
90	5	(90 − 58.50) = 31.50	992.25	4,961.25
100	5	(100 − 58.50) = 41.50	1,722.25	8,611.25
	$n = 100$			$\Sigma = 43,275.00$

which makes the standard deviation

$$S_X = \sqrt{\frac{43,275}{100 - 1}}$$

$$= 20.90$$

Now, based on these calculations, the production manager has estimates for both the central location and dispersion in the daily lumber production. Without the information contained in the mean and standard deviation, the production manager would be hard pressed to reasonably assess and monitor the mill's daily

---

‡The equations presented here can be put in shortcut form for ease of computation. You are encouraged to look at the formulas at the end of the chapter for the shortcut methods of determining the variance and standard deviation for grouped data.

TABLE 4–5 Lumber production sample (board feet × 1,000)

Class	M	f	f M
15 and under 25	20	5	100
25 and under 35	30	8	240
35 and under 45	40	15	600
45 and under 55	50	20	1,000
55 and under 65	60	15	900
65 and under 75	70	10	700
75 and under 85	80	17	1,360
85 and under 95	90	5	450
95 and under 105	100	5	500
		$n = 100$	$\Sigma = 5{,}850$

lumber production. For example, the large standard deviation in this example indicates high variability in production output. High variability in production is generally undesirable and is an indication that better production control is needed.

4-3
COEFFICIENT OF VARIATION

The standard deviation measures the variation in a set of data. For distributions having the same mean, the distribution with the largest standard deviation has the greatest relative spread. This is illustrated in Figure 4–4, where the distribution in (b) has

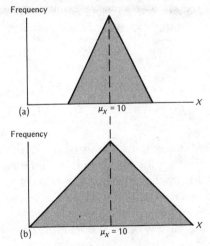

Note: The distribution in part (a) and the distribution in part (b) both have the same mean, but that in (b) has a larger standard deviation than that in (a) and is thus more spread out around the mean.

FIGURE 4–4 Distribution spread.

greater dispersion than that in (a). For decision makers, the standard deviation indicates how spread out, or uncertain, a distribution is. Given two distributions with the same mean, decision makers should feel more certain about the decisions made from the distribution with the smaller standard deviation. However, when considering distributions with different means, decision makers cannot compare the uncertainty in distributions only by comparing standard deviations. **The *coefficient of variation* is often used to indicate the relative uncertainty of distributions with different means.**

The coefficients of variation for populations and samples are as follows:

For populations:

$$CV = \frac{\sigma_x}{\mu_x}(100) \tag{4-17}$$

where: CV = coefficient of variation

For samples:

$$CV = \frac{S_x}{\overline{X}}(100) \tag{4-18}$$

The coefficients of variation for different distributions are compared, and the distribution with the largest CV value possesses the greatest relative spread.

The coefficient of variation is an excellent example of both the advantages and disadvantages of many statistical aids to decision making. When the mean and standard deviation of a distribution are combined to form the coefficient of variation, the effect of different mean values tends to be eliminated. However, many decisions must be based on both central location *and* spread of distributions. By combining the measure of location and measure of spread into one value, decision makers lose some information. The point we want to make is that **any statistical tool should be used as an aid to good managerial judgment, not in place of managerial judgment.** Cases 4B and 4C at the end of the chapter reinforce this point.

4-4
CONCLUSIONS

Measures of central location and measures of variation, or spread, are statistical tools for transforming data into meaningful information. When used together, measures of location and dispersion allow managerial decision makers to compare distributions and determine whether they are statistically the same or whether a difference exists between them. Although we have discussed several measures of both location and spread for a distribution, the most widely used measures are the mean and standard deviation.

Throughout the chapter we have discussed new statistical concepts in general managerial terms. In the chapter glossary we present a list of more formal statistical definitions for these concepts. To test your understanding, reconcile the managerial discussion with these definitions.

The important equations introduced in this chapter are presented in a special section following the glossary.

chapter glossary

coefficient of variation   A measure that can be used to compare relative dispersion between two or more data sets. It is formed by dividing the standard deviation by the mean of a data set.

mean   The mean of a set of n measurements is equal to the sum of the measurements divided by n. The mean is the most commonly used measure of location.

median   A measure of location. The middle item in a set of measurements that have been arranged according to magnitude.

mode   A measure of location. The value in a series of measurements that appears more frequently than any other. The mode may not be unique.

range   The difference between the largest and the smallest values in a set of data.

standard deviation   The positive square root of the variance.

variance   A measure of data dispersion. The average of the squared differences between the individual measurements and the mean.

chapter formulas

Median

$$\text{Median} = L_{\text{median}} + \left(\frac{\frac{N}{2} - \Sigma f_{\text{prec}}}{f_{\text{median}}} \right)(w)$$

Mean

Population

Ungrouped

$$\mu_x = \frac{\Sigma X}{N}$$

Grouped

$$\mu_x = \frac{\Sigma fM}{N}$$

Sample

Ungrouped

$$\overline{X} = \frac{\Sigma X}{n}$$

Grouped

$$\overline{X} = \frac{\Sigma fM}{n}$$

Variance

Population

Ungrouped

$$\sigma_X^2 = \frac{\Sigma(X - \mu_x)^2}{N}$$

Grouped

$$\sigma_X^2 = \frac{\Sigma f(M - \mu_x)^2}{N}$$

Sample

Ungrouped

$$S_X^2 = \frac{\Sigma(X - \bar{X})^2}{n - 1}$$

Grouped

$$S_X^2 = \frac{\Sigma f(M - \bar{X})^2}{n - 1}$$

Standard deviation

Population

Ungrouped

$$\sigma_X = \sqrt{\frac{\Sigma(X - \mu_x)^2}{N}}$$

Grouped

$$\sigma_X = \sqrt{\frac{\Sigma f(M - \mu_x)^2}{N}}$$

Sample

Ungrouped

$$S_X = \sqrt{\frac{\Sigma(X - \bar{X})^2}{n - 1}}$$

Grouped

$$S_X = \sqrt{\frac{\Sigma f(M - \bar{X})^2}{n - 1}}$$

Shortcut formulas for variance of ungrouped data

Population

$$\sigma_X^2 = \frac{\Sigma X^2 - \dfrac{(\Sigma X)^2}{N}}{N}$$

Sample

$$S_X^2 = \frac{\Sigma X^2 - \dfrac{(\Sigma X)^2}{n}}{n-1}$$

Shortcut formulas for standard deviation of ungrouped data

Population

$$\sigma_X = \sqrt{\frac{\Sigma X^2 - \dfrac{(\Sigma X)^2}{N}}{N}}$$

Sample

$$S_X = \sqrt{\frac{\Sigma X^2 - \dfrac{(\Sigma X)^2}{n}}{n-1}}$$

Shortcut formulas for variance of grouped data

Population

$$\sigma_X^2 = \frac{\Sigma fM^2 - \dfrac{(\Sigma fM)^2}{N}}{N}$$

Sample

$$S_X^2 = \frac{\Sigma fM^2 - \dfrac{(\Sigma fM)^2}{n}}{n-1}$$

Shortcut formulas for standard deviation of grouped data

Population

$$\sigma_X = \sqrt{\frac{\Sigma fM^2 - \dfrac{(\Sigma fM)^2}{N}}{N}}$$

Sample

$$S_x = \sqrt{\frac{\Sigma f M^2 - \frac{(\Sigma f M)^2}{n}}{n - 1}}$$

Coefficient of variation

Population

$$\frac{\sigma_x}{\mu_x}(100)$$

Sample

$$\frac{S_x}{\overline{X}}(100)$$

solved problems

1. A retail store manager has selected a sample of sales slips during the past month and now needs your assistance in answering some questions about the sample. The 64 sales values have been numerically ordered as follows:

$0.97	$5.69	$10.48	$22.95
1.23	5.99	11.97	23.85
1.29	6.49	11.99	24.99
1.69	6.50	11.99	24.99
2.19	6.99	11.99	29.97
2.77	7.19	12.49	34.95
3.19	7.44	12.95	34.98
3.57	7.49	13.39	35.95
3.97	7.75	13.88	41.69
3.99	8.12	13.89	49.99
4.35	8.90	14.99	61.98
4.44	8.99	16.30	64.88
4.44	9.85	18.63	67.29
4.50	9.97	19.97	68.99
4.65	9.97	21.25	69.99
5.25	9.98	21.50	74.96

a. What is the mean sales level for the 64 sales values?
b. What is the median sales level for these data?
c. What is the mode?
d. Note that the median and mode are fairly close but that the mean is much larger. Why is this?

e. What is the standard deviation from the sample of sales values?

f. Suppose the manager has grouped the data into ten classes as follows:

Sales	Midpoint M	Frequency f
$ 0.00–$ 7.49	$ 3.75	24
7.50– 14.99	11.25	19
15.00– 22.49	18.75	5
22.50– 29.99	26.25	5
30.00– 37.49	33.75	3
37.50– 44.99	41.25	1
45.00– 52.49	48.75	1
52.50– 59.99	56.25	0
60.00– 67.49	63.75	4
67.50– 74.99	71.25	2
		$\Sigma f = n = 64$

Find the mean and median for these grouped data.

g. What is the sample variance calculated from the grouped data in part f?

Solutions:

a. The formula for finding the mean is

$$\overline{X} = \frac{\Sigma X}{n}$$

If we sum the 64 values, we get

$$\Sigma X = \$1,149.86$$

Dividing by $n = 64$,

$$\overline{X} = \frac{\$1,149.86}{64}$$

$$= \$17.97$$

Thus the mean sale for the sample is $17.97.

b. The median is the center item when the data have been rank-ordered. When the number of observations is an even number, the median is the average of the two middle observations. Thus the median for these data is midway between the thirty-second and thirty-third sales values.

$$\text{Median} = \frac{\$9.98 + \$10.48}{2}$$

$$= \$10.23$$

c. The mode is the value in the data that appears most often. In this data set, the value $11.99 occurred three times, which is more often than any other value. Thus the mode is $11.99.

d. Whenever the mean, median, and mode differ in value, the data distribution is skewed. The further the mean is from the median and the mode, the greater the skewness. The mean is sensitive to extreme observations, and in this example, although the majority of the observations are less than $15.00, the few high sales values over $60.00 have pulled the mean up to $17.97.

e. The sample standard deviation is found using the following formula:

$$S_x = \sqrt{\frac{\Sigma(X - \bar{X})^2}{n - 1}}$$

Recalling that $\bar{X} = 17.97$, we can find the standard deviation as follows:

X	$(X - \bar{X})$	$(X - \bar{X})^2$
0.97	$(0.97 - 17.97) = -17.00$	289.00
1.23	$(1.23 - 17.97) = -16.74$	280.23
1.29	$(1.29 - 17.97) = -16.68$	278.22
1.69	$(1.69 - 17.97) = -16.28$	265.04
.	. .	.
.	. .	.
.	. .	.
74.95	$(74.95 - 17.97) = 56.98$	3,246.72
		$\Sigma = 23,384.57$

$$S_x = \sqrt{\frac{23,384.57}{64 - 1}} = \sqrt{371.18}$$
$$= \$19.27$$

Thus the sample standard deviation is $19.27.

f. The mean for grouped data is

$$\bar{X} = \frac{\Sigma fM}{n}$$

Thus we must multiply the frequency in each class by the class midpoint.

fM
90.00
213.75
93.75
131.25
101.25
41.25
48.75
0
255.00
142.50
$\Sigma = 1,117.50$

Then

$$\bar{X} = \frac{1,117.50}{64}$$

$$= \$17.46$$

The median for the grouped data is found with the following formula:

$$\text{Median} = L_{\text{median}} + \left(\frac{\frac{n}{2} - f_{\text{prec}}}{f_{\text{median}}}\right)(w)$$

where L_{median} is the lower limit of the median class and is equal to 7.50. In this problem,

$$\text{Median} = 7.50 + \left(\frac{\frac{64}{2} - 24}{19}\right)(7.50) = 7.50 + \left(\frac{32 - 24}{19}\right)(7.50)$$

$$= \$10.66$$

g. We use the following formula to find the variance of a set of data that have been grouped into a frequency distribution:

$$S_{\bar{X}}^2 = \frac{\Sigma f(M - \bar{X})^2}{n - 1}$$

Then

Class Midpoint M	Class Frequency f	$(M - \bar{X})$	$(M - \bar{X})^2$	$f(M - \bar{X})^2$
3.75	24	$(3.75 - 17.46)$	187.96	4,511.14
11.25	19	$(11.25 - 17.46)$	38.56	732.72
18.75	5	$(18.75 - 17.46)$	1.66	8.32
26.25	5	$(26.25 - 17.46)$	77.26	386.32
33.75	3	$(33.75 - 17.46)$	256.36	796.09
41.25	1	$(41.25 - 17.46)$	565.96	565.96
48.75	1	$(48.75 - 17.46)$	979.06	979.06
56.26	0	$(56.25 - 17.46)$	1,504.66	0
63.75	4	$(63.75 - 17.46)$	2,142.76	8,571.06
71.25	2	$(71.25 - 17.46)$	2,893.36	5,786.73
	$\Sigma = 64$			$\Sigma = 22,337.40$

$$S_{\bar{X}}^2 = \frac{22,337.40}{63}$$

$$= \$354.56 \text{ Squared}$$

Thus the variance is $354.56 squared. The standard deviation is

$$S_x = \sqrt{S_x^2} = \sqrt{354.56}$$
$$= \$18.83$$

problems

Problems marked with an asterisk (*) require additional thought.

1. Discuss how to locate the median of a group of observations.

2. Discuss the relationship of the three measures of location in a skewed distribution. In particular, draw graphs showing the relationship of the three measures of location in a right-hand skewed distribution and a left-hand skewed
distribution.

3. Go to the library, and from the *Statistical Abstract of the United States,* gather
data on unemployment rates by state. Using these data, find the mean, median,
and mode. Find these measures of location using both the individual and
grouped formulas for the mean and median. Is this distribution skewed? If so,
in what direction?

4. Using the data from problem 3, find the standard deviation and range. Find the
standard deviation for both grouped and ungrouped data.

5. Discuss in your own words the advantages of using the standard deviation
instead of the range as a measure of spread for a set of data.

6. From *Time, U.S. News and World Report,* or the *Wall Street Journal,* find six
examples of information presented that use a measure of location or a measure
of spread.

7. Discuss some problems you see in using the mean and standard deviation as
the only measures of output for a production process. What factors would you
want to consider when using these measures?

8. Joan Horton is a building contractor whose company builds many homes every
year. In planning for each job, Joan needs some idea about the direct labor
hours required to build a home. She has collected sample information on the
labor hours for ten jobs during the past year.

245 h	402 h
291	351
353	242
118	595
152	175

a. Calculate the mean for this sample and explain what it means.
b. Calculate the median for this sample.
c. If Joan had to select the mean or the median as the measure of location for
direct labor hours, what factors about each should she consider before making the decision?

   d. Calculate the sample variance and standard deviation. What do these values measure?
   e. Calculate the range of this distribution.
   f. Compare the measures of spread given by the standard deviation and range. What are the advantages of each?

√**9.** I. M. Sawyer owns another home-construction company and is in competition with Joan Horton's company. Sawyer's company has averaged 600 hours per house with a standard deviation of 50 hours. Sawyer claims that his company is more consistent in the direct labor hours applied to houses than is Horton's company. Horton disagrees. Who is right and why? Use values calculated in problem 8.

10. The president of the Holiday Motel chain is concerned about occupancy rates during weekdays. A sample of 1,000 motels was selected, and from these a sample of five weekday occupancy numbers was recorded for each motel. Suppose the frequency distribution is as follows:

Rooms Occupied	Frequency
0– 30	250
31– 60	1,200
61– 90	1,900
91–120	1,100
121–150	550
Total	5,000

   a. What is the average number of full rooms per night per motel for the Holiday Motel chain?
   b. What is the median value?
   c. Determine the variance and standard deviation for this distribution and indicate what each measures.

*__11.__ Suppose the cost of staying a night in a Holiday Motel is $20 regardless of the location. What is the average of nightly receipts per motel based upon the information in problem 10? What is the standard deviation of nightly receipts per motel?

12. Suppose you are considering purchasing an apartment house. There would certainly be many things to consider in the buy/don't buy analysis. One of these would be the maintenance costs you might incur. The following frequency distribution reflects the historical weekly costs (rounded to the nearest dollar) based upon the current owner's records for the past 30 weeks:

Expense Class	Frequency
$ 0–$100	12
101– 200	5
201– 300	6
301– 400	4
401– 500	3
Total	30 weeks

a. Calculate the mean weekly expense.

b. Calculate the variance and standard deviation.

c. Calculate and interpret the coefficient of variation for this distribution.

13. Referring to problem 12, suppose you have made a deal with the current owner to keep the maintenance records yourself for one week. You find that $600 was spent. Assuming this was a typical week, what would you conclude about the current owner's records? Why?

14. In the past few years many companies using trees as raw material have become interested in growing trees. One company has collected the following information on tree growth per year for samples from three species of trees:

Growth Class (feet)	Pine Frequency	Redwood Frequency	Cedar Frequency
0 and under 1	10	15	5
1 and under 2	5	10	10
2 and under 3	9	2	4
3 and under 4	4	2	4
4 and under 5	2	1	7
Total	30 trees	30 trees	30 trees

Write a report to management analyzing this information. Be sure to discuss the location and spread in the growth for each type of tree. Also, compare the relative dispersion in growth between the three types of trees.

15. Explain why the mean calculated for a set of ungrouped data might differ from the mean if the same data were grouped into a frequency distribution. Would this also be the case for the standard deviation? Why?

$\sqrt{}$***16.** Suppose a trucking company has a straight-line route between New York and Los Angeles. On the route are five warehouses spaced as follows:

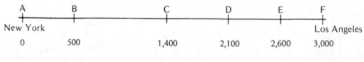

Miles from New York

Determine the best possible location for the trucking company headquarters if the objective is to locate the headquarters so that the total distance between the warehouses and the headquarters is minimized. (*Hint:* What measure of location is required?)

$\sqrt{}$ **17.** A survey of local airline passengers shows that the mean height of male passengers is 69.5 inches, with standard deviation 2.5 inches. The mean weight is 177 pounds, and the standard deviation is 12 pounds. Which of the two distributions has the greater variability?

4A THE ASSOCIATION OF INDEPENDENT HOMEOWNERS

The Association of Independent Homeowners has been spurred into action by the success of tax-limitation efforts in many states. However, although members are not particularly happy with the property-tax levels in their states, they are more concerned with the possibility of assessment errors. They do not think the county assessors are doing anything illegal, but they have all heard stories of how some properties are assessed at a much lower rate than other properties when in fact they are of equal value.

The property-tax level in each county depends on the assessed valuation of all property in the county. One person's or business's property tax will depend on the levy set by the county and on the property value. Since the value of property changes from year to year, the association is particularly concerned that the assessed value accurately reflect the changing market value as specified by state law.

Ruth Powers, an active member of the association, has heard of a recent study performed by the State Tax Commission. The commission gathers information on the sale price of recently marketed property and compares the true market price with the assessed value. Using the gathered data, the commission computes a mean assessment-to-market-value ratio for each county in the state. Since property is broken into four categories for assessment purposes ("residential," "rural investment," "business and industrial," and "other rural"), the commission also computes an adjusted mean value based on the number of parcels in each of these categories in each county. The State Tax Commission determined these values, plus the standard deviation of the individual ratios, for each county in the state. Ruth gathers the values for the seven counties nearest hers and for her own. They are as follows:

County	Ratio Mean Value	Ratio Weighted Mean	Ratio Standard Deviation
Box	18.38%	18.36%	1.96%
Elm	12.93	13.39	3.48
Canyon	15.93	15.90	6.91
Valley	11.92	13.45	7.97
Rice	15.42	16.90	6.24
Washington	20.44	19.69	11.74
Grant	13.82	17.07	2.25
Sandstone*	13.72	14.50	6.50

*Ruth's county.

Ruth wants to discuss the implications of these figures at the next meeting of the association's directors.

√ 4B DELPHI INVESTMENT COMPANY, PART 1

The Delphi Investment Company offers financial advice for individual and corporate clients. Its clients are of course concerned with the return on investment. Jacqueline

Morton, managing partner at Delphi Investment, is considering three separate investment alternatives. Summary statistics for these three are as follows:

	Alternative 1	Alternative 2	Alternative 3
μ_x	12% return	14% return	16% return
σ_x	6%	5%	6%

Jacqueline Morton needs some assistance in presenting these three investments to a client. She is particularly concerned with analyzing the relative risks of the investments.

4C   DELPHI INVESTMENT COMPANY, PART 2

At 10:10 A.M Jacqueline Morton received word about two investment opportunities. The first investment will return an average of 8 percent with a standard deviation of 0.5 percent. The second investment will return an average of 20 percent with a standard deviation of 5 percent. Jacqueline has quickly computed the coefficient of variation of each investment and has concluded that the second investment has the greater relative dispersion in the possible returns on investment. At 10:35 A.M Jacqueline calls a client and recommends the first investment alternative based on the relative risk between the two investments.

references

HAMBURG, MORRIS. *Statistical Analysis for Decision Making.* 2nd ed. New York: Harcourt Brace Jovanovich, 1977.

MCALLISTER, HARRY E. *Elements of Business and Economic Statistics: Learning by Objectives.* New York: Wiley, 1975.

PFAFFENBERGER, ROGER C. and JAMES H. PATTERSON. *Statistical Methods for Business and Economics.* Homewood, Ill.: Irwin, 1977.

NEWTON, BYRON L. *Statistics for Business.* Chicago: Science Research Associates, 1973.

5

Probability Concepts

& Ch. 2

Decision making means selecting between two or more alternatives. To make "good" decisions, managers must establish general criteria for deciding among these alternatives. Certainly the criteria must somehow be related to the objective of the decision-making situation. This objective may involve a profit level, a sales level, or even creating an orderly situation from near chaos.

Once the objective has been established, managers often assess the chances that each alternative will reach this objective. The managers may make this assessment while saying something like, "The chances are good that if we build a parking garage, we will make a higher return on our investment than if we build an office building. On the other hand, the chances are fair that building a shopping mall will produce a higher return than a parking garage."

When managers refer to the chances of something occurring, they are using *probability* in the decision-making process. The decision makers are establishing in their minds the probability that some result will occur if a particular action is taken. If the business world were a place of certainty, decision makers would have little need to understand probability, but as emphasized in the previous chapters, the world is not certain. In practice, probability assists managers in dealing with uncertainty.

In Chapter 2 we pointed out that managers must often operate with sample information collected from the population of interest. They are uncertain about the population but know a great deal about the sample. Probability theory allows managers to make inferences about the population based upon knowledge of the sample and to have confidence in these inferences. As we shall see in later chapters, because statistical inference is based upon probability, decision makers must have a solid grasp of basic probability theory.

chapter objectives

In this chapter we shall present the fundamentals of probability needed to understand statistical inference. We shall begin by discussing the differences between subjective probability, classical probability, and the relative frequency of occurrence approach to probability.

We shall also cover some rules and concepts associated with probability theory. Such topics as sample space, the addition rule, the multiplication rule, independent events, conditional probability, and Bayes' rule will be included in this discussion.

student objectives

We have all too often found that decision makers, when dealing with probability, rely on intuition rather than on well-established probability principles. This intuitive approach often proves faulty. After studying the material in this chapter, you should be able to:

1. Discuss the reasoning behind, and the differences between, three approaches to assessing probability.
2. Discuss and use the common rules of probability.
3. Determine situations where Bayes' rule could be used and know how to use it.
4. Discuss and use three methods of counting large numbers of possible outcomes.

In studying the material in this chapter, concentrate initially on how probability helps decision makers rather than on the formulas necessary to use probability. The formulas are useless unless they are applied in the correct situation, and concentrating on applications often will remove the "mystery" sometimes associated with probability.

5-1
WHAT IS PROBABILITY?

Before we can apply probability to the decision-making process, we must first discuss just what we mean by **probability**. The mathematical study of probability originated over three hundred years ago. A Frenchman named Gombauld, who today would probably be a dealer in Las Vegas, began asking questions about games of chance. He was mostly interested in the probability of observing various outcomes (probably 7s and 11s) when a pair of dice was repeatedly rolled. A French mathematician named Pascal (you may remember studying Pascal's triangle in a mathematics class) was able to answer Gombauld's questions. Of course Pascal began asking more and more complicated questions of himself and his colleagues, thus beginning the *formal* study of probability.

Several explanations of what probability is have come out of this mathematical study. Although probabilities will help managers in a decision-making process, managers often have trouble determining how to assign probabilities. Later we shall discuss three methods managers can use to assign probabilities to possible outcomes. These three methods are **relative frequency of occurrence, subjective probability,** and **classical probability.** However, first we introduce some basic probability terminology.

As we discussed in Chapter 2, data come in many forms and are gathered in many ways. In a business environment, when a sample is selected or a decision is made, there are generally many possible outcomes. **A sample space is the collection of all possible outcomes that can result from the selection or decision. In probability language, the process that produces the outcomes is an experiment.** In business situations, the experiment can range from an investment decision to a personnel decision to a choice of warehouse location. **The individual outcomes from an experiment are called elementary events.** Thus the sample space for an experiment consists of all the elementary events that the experiment can produce. **A collection of elementary events is called an event.** An example will help clarify these terms.

Suppose the personnel manager for the Treelight Manufacturing Company decides to remove the time clock at his manufacturing plant. Now, instead of the workers punching in and out and getting paid for the exact time worked, they will be paid for eight hours. The personnel manager is interested in the effect this change will have on employee arrival times.

Suppose the personnel manager defines the elementary events for one employee as follows:

$$e_1 = \text{employee arrives early}$$
$$e_2 = \text{employee arrives on time}$$
$$e_3 = \text{employee arrives late}$$

The sample space for the experiment for a single employee is

$$SS = (e_1, e_2, e_3)$$

where: SS = notation for the sample space

If the experiment is expanded to include two employees, the sample space is

$$SS = (e_1, e_2, e_3, e_4, e_5, e_6, e_7, e_8, e_9)$$

where the events are defined as follows:

Elementary Event	Employee 1	Employee 2
e_1	early	early
e_2	early	on time
e_3	on time	early
e_4	early	late
e_5	late	early
e_6	late	on time
e_7	on time	late
e_8	late	late
e_9	on time	on time

Here, each elementary event consists of the combined outcomes of employee 1 and employee 2.

The manager might be interested in the event "At least one employee arrives late." This event (E) is

$$E = (e_4, e_5, e_6, e_7, e_8)$$

In another example, the Wright National Bank classifies all outstanding installment loans into one of four categories. The experiment in this case is the loan classification. The elementary events for each loan are

$$e_1 = \text{very solid}$$
$$e_2 = \text{solid}$$
$$e_3 = \text{doubtful}$$
$$e_4 = \text{uncollectable}$$

Then the sample space for a loan is

$$SS = (e_1, e_2, e_3, e_4)$$

The manager in charge of installment loans might be interested in the event "The loan is doubtful *or* uncollectable." If too many loans are in one or both of these categories, the bank examiners will issue an unfavorable report. This event is

$$E = (e_3, e_4)$$

Keeping in mind the definitions for *experiment, sample space, elementary events,* and *events,* we introduce two additional concepts. The first is the concept of **mutually exclusive** events. **Two events are mutually exclusive if the occurrence of one event precludes the occurrence of the other.** For example, consider again the Treelight Manufacturing Company example. The possible elementary events for two employees are

$$e_1 = \text{early, early}$$
$$e_2 = \text{early, on time}$$
$$e_3 = \text{on time, early}$$

$$e_4 = \text{early, late}$$
$$e_5 = \text{late, early}$$
$$e_6 = \text{late, on time}$$
$$e_7 = \text{on time, late}$$
$$e_8 = \text{late, late}$$
$$e_9 = \text{on time, on time}$$

Suppose we define one event as consisting of the elementary events in which at least one employee is late.

$$E_1 = (e_4, e_5, e_6, e_7, e_8)$$

Further, suppose we define two more events as follows:

$$E_2 = \text{neither employee is late}$$
$$= (e_1, e_2, e_3, e_9)$$
$$E_3 = \text{both employees arrive at the same time}$$
$$= (e_1, e_8, e_9)$$

Events E_1 and E_2 are mutually exclusive since if E_1 occurs, E_2 cannot occur, and conversely. That is, if at least one employee is late, then it is not also possible for neither employee to be late. This can be verified by observing that no elementary events in E_1 appear in E_2. This provides another way of defining mutually exclusive events: **two events are mutually exclusive if they have no common elementary events.**

The second additional probability concept is that of *independent* versus *dependent* events. **Two events are independent if the occurrence of one event in no way influences the occurrence of the other event.** For instance, again consider the Treelight Manufacturing example. The experiment is to check the arrival times of two employees at the factory. Each day that the two employees are observed would be one trial of the experiment.

Suppose that two employees are checked on Tuesday and that the elementary event

$$e_8 = \text{late, late}$$

occurred. If on Wednesday (a new trial) a different pair of employees are observed and the elementary event

$$e_1 = \text{early, early}$$

occurred, the two trials would be *independent* if e_1 occurring on Wednesday was in no way influenced by elementary event e_8 on Tuesday. This might well be the case if the employees did not know each other and took separate means of transportation to work. On the other hand, if the outcome on Wednesday was influenced by what happened by the result of the trial on Tuesday, the trials would be considered *dependent*.

Another example of dependence and independence might be an assembly-line operation. Each item produced could be an experimental trial. On each trial the outcome is either a *good* or a *defective* item. Thus the sample space is

$$SS = (\text{good, defective})$$

As long as the machine is properly adjusted, it may produce some good outcomes and some defective outcomes with no apparent pattern, or dependency, between trials. That is, one item being good has no influence on the outcome of subsequent trials. However, if the machine goes out of adjustment, problems begin. A defective item may cause still further adjustment problems and increase the chances that subsequent items will be defective. In this case the trials are dependent because the outcome on one trial is in some way influenced by the outcome of a previous trial.

5-2
METHODS OF ASSIGNING PROBABILITY

Part of the confusion surrounding probability may be due to the fact that probability means different things to different people. In this section we present three ways to assign probability to events. These three methods are **relative frequency of occurrence, subjective probability assessment,** and **classical probability assessment.**

Relative Frequency of Occurrence

The **relative frequency of occurrence** approach is based on actual observation. For instance, if we were interested in the probability of eight or fewer customers arriving at our tobacco shop before 10:00 A.M., we would pick a trial number of days (*n*) and count how often eight or fewer customers actually did arrive before 10:00 A.M. Our probability assessment would be the ratio of days when eight or fewer customers arrived to the total number of days observed. More formally, the probability "assessment" for event E_1 using the relative frequency of occurrence approach is given by

$$RF(E_1) = \frac{\text{number of times } E_1 \text{ occurs}}{n}$$

where: $RF(E_1)$ = relative frequency of E_1 occurring

  n = number of trials

We can use $RF(E_1)$ to represent $P(E_1)$, the probability of E_1 occurring. However some dangers exist when using a relative frequency to represent a probability. If *n* is small, as it often must be in business situations because of dollar and time constraints, the estimated probability may be quite different from the true probability.

Although the relative frequency approach has some limitations, it is a valuable tool for decision makers because it can be used to quantitatively represent experience. For example, if 90 percent of the ventures a firm has undertaken have proven profitable, a starting point for estimating the probability of success of a similar new venture might be 0.90.

Subjective Probability Assessment

Unfortunately, even though managers may have past experience, there will always be new factors in a decision that make that experience only an approximate guide to the future. In other cases managers may have no past experience and therefore not be able to use a relative frequency of occurrence as even a starting point in assessing the desired probability. When past experience is not available, decision makers must make a *subjective probability assessment*. **A subjective probability is a measure of a personal conviction that an outcome will occur. Subjective probability rests in a person's mind and not with the physical event.**

Classical Probability Assessment

The final method of probability measurement is **classical probability,** or **a priori probability.** Although classical probability is not as directly applicable to business decision making as are the subjective and relative frequency methods, it is an interesting and useful tool for discussing the rules of probability. You are probably already familiar with classical probability. It had its beginning with games of chance and is still most often discussed in those terms.

If all possible outcomes are *equally likely*, the classical probability measurement is defined as

$$P(E_1) = \frac{\text{number of ways } E_1 \text{ can occur}}{\text{total number of ways anything can occur}} \qquad \textbf{(5–1)}$$

For example, if a "fair" coin is tossed one time, the probability of heads is

$$P(\text{heads}) = \frac{\text{number of ways heads can result}}{\text{number of possible outcomes from one flip}} = \frac{1 \text{ (heads)}}{2 \text{ (heads or tails)}}$$
$$= 0.50$$

Drawing cards from a "fair" deck of cards is another example of a classical probability experiment. The probability of drawing an ace from the 52 well-shuffled cards is

$$P(\text{ace}) = \frac{\text{number of aces}}{\text{total number of cards}} = \frac{4}{52}$$
$$= \frac{1}{13}$$

The probability of drawing an ace of hearts is

$$P(\text{ace of hearts}) = \frac{\text{number of aces of hearts}}{\text{total number of cards}}$$
$$= \frac{1}{52}$$

As you can see, the classical approach to probability measurement is fairly straightforward. However it is difficult to apply to most business situations. The major problem is determining values for the numerator and denominator in formula (5–1). For example, if we were interested in the probability a particular W & A

Restaurant would have gross sales next year over $50,000, we really couldn't use the classical approach

$$P(\text{sales over } \$50,000) = \frac{\text{number of ways sales can exceed } \$50,000}{\text{total sales levels possible}}$$

We could not possibly define either the total ways to have sales over $50,000 or the total possible sales levels. There are too many factors not within our control. Unlike with flipping a coin, the probability of sales over $50,000 depends on the persons involved and the operating environment.

Another problem with the classical approach to probability assessment is that for equation (5–1) to apply, all possible events must be *equally likely* to occur. Although decision makers may be faced with such a situation, rarely will this be the case. In the W & A Restaurant example, the possible ways of obtaining a sales level over $50,000, even if they could all be identified, certainly do not have equal chances of occurring.

As we have indicated, each of the three means by which probabilities are assigned to events has special advantages and applications. However, regardless of how decision makers arrive at a probability assessment, the rules by which these probabilities are used to assist in decision making are the same.

5-3
PROBABILITY RULES

The probability attached to an event represents the likelihood of that event occurring on a specified trial of an experiment. This probability also measures the perceived uncertainty about whether the event will occur. If we are not uncertain at all, we will assign the event a probability of zero or 1.0: $P(E_1) = 0.0$ indicates that the event, E_1, will not occur, and $P(E_1) = 1.0$ means that E_1 will definitely occur. These values represent the outside limits on a probability assessment. If, in fact, we are uncertain about the result of an experiment, we measure this uncertainty by assigning a probability between zero and 1.0. Probability rule 1 shows the range for probability assessments.

PROBABILITY RULE 1

$$0.0 \leq P(E_i) \leq 1.0 \quad \text{for all } i$$

and

$$0.0 \leq P(e_i) \leq 1.0 \quad \text{for all } i$$

We have stated that all possible elementary events associated with an experiment form the sample space. Probability rule 2 indicates that the probabilities of all elementary events must add to 1.0.

PROBABILITY RULE 2

$$\sum_{i=1}^{K} P(e_i) = 1.0$$

where: K = number of elementary events in the sample space

e_i = ith elementary event

For example, suppose we use classical probability theory to assign probabilities to each possible outcome of a roll of one die. We can use the classical approach since the possible outcomes are equally likely. Let the sample space be

$$SS = (1, 2, 3, 4, 5, 6)$$

Then the probability the die will turn up "1" is

$$P(1) = \frac{\text{number of ways "1" can occur}}{\text{number of ways any value can occur}}$$

Since there is only one way to roll a "1," and six possible outcomes, the probability is

$$P(1) = \frac{1}{6}$$

The same is true for each elementary event in the sample space.

e_i	$P(e_i)$
1	1/6
2	1/6
3	1/6
4	1/6
5	1/6
6	1/6
	= 1.0

Note that the probabilities of the individual elementary events sum to 1.0.

Closely connected with probability rules 1 and 2 is the **complement** of an event. **The *complement* of an event E is the collection of all possible outcomes that are not contained in event E.** The complement of event E is represented by $\overline{E}$. Thus a corollary to probability rules 1 and 2 is

$$P(\overline{E}_i) = 1 - P(E_i)$$

That is, the probability of the complement of event E is 1.0 minus the probability of event E.

Addition Rules

When making a decision involving probabilities, you will often need to combine elementary event probabilities to find the probability associated with the event of interest. **Combining probabilities requires addition.** As we will show, there are three rules that govern the addition of probabilities.

Channel 16 Television in Newberg, Wisconsin, has recently performed a viewer survey as part of a new marketing effort. Probabilities are often useful in analyzing the results of questionnaires or surveys. One question of particular interest is which station the respondent watches during the early local news time (6:00–6:30 P.M.). Table 5–1 shows the results of the survey for this question.

We define the sample space for the experiment for each respondent as

$$SS = (e_1, e_2, e_3, e_4)$$

where: e_1 = Channel 16

e_2 = Channel 8

e_3 = Channel 12

e_4 = don't watch

Using the relative frequency of occurrence approach, we assign the following probabilities:

$$P(e_1) = 0.42$$
$$P(e_2) = 0.30$$
$$P(e_3) = 0.24$$
$$P(e_4) = \underline{0.04}$$
$$\Sigma = 1.00$$

Suppose we define an event from these elementary events as follows:

E_1 = respondent watches the early local news on
Channel 8 or Channel 12

The elementary events that make up E_1 are

$$E_1 = (e_2, e_3)$$

We can find the probability $P(E_1)$ by using probability rule 3.

PROBABILITY RULE 3

Addition Rule for Elementary Events

The probability of an event E_i is equal to the sum of the probabilities of the elementary events forming E_i. That is, if

$$E_i = (e_1, e_2, e_3)$$

then

$$P(E_i) = P(e_1) + P(e_2) + P(e_3)$$

TABLE 5–1 Television viewer results

Channel Watched	Frequency	Relative Frequency
Channel 16	2,100	0.42
Channel 8	1,500	0.30
Channel 12	1,200	0.24
Don't watch	200	0.04
Total	5,000	1.00

The probability that a respondent watches Channel 8 *or* Channel 12 is

$$P(E_1) = P(e_2) + P(e_3) = 0.30 + 0.24$$
$$= 0.54$$

Suppose the television survey also contained questions about education level. Table 5–2 shows the breakdown of the sample by education level and by news station watched. This table illustrates two important concepts in data analysis: **joint frequencies** and **marginal frequencies. Joint frequencies are represented by the values inside the table, and the *marginal frequencies* are the row and column totals.** For example, 1,500 people watch Channel 16 *and* have a B.A. degree. Thus 1,500 is a joint frequency. In this table the joint frequencies are elementary events. However 2,400 people in the survey have a B.A. degree. This column total is a marginal frequency. Table 5–3 shows the relative frequencies for the data in Table 5–2.

TABLE 5–2 Channel watched by education level

Channel Watched	Education Level			
	E_5 *High School*	E_6 *B.A.*	E_7 *Graduate School*	
E_1 Channel 16	e_1 0	e_2 1,500	e_3 600	2,100
E_2 Channel 8	e_4 900	e_5 600	e_6 0	1,500
E_3 Channel 12	e_7 1,000	e_8 200	e_9 0	1,200
E_4 Don't watch	e_{10} 100	e_{11} 100	e_{12} 0	200
	2,000	2,400	600	5,000

Now we let event E_1 be that the viewer watches Channel 16.

$$E_1 = (e_1, e_2, e_3)$$

Further, let event E_6 be a B.A. educational level.

$$E_6 = (e_2, e_5, e_8, e_{11})$$

TABLE 5–3   Relative frequencies of television channel watched by education level

Channel Watched	Education Level			
	E_5 High School	E_6 B.A.	E_7 Graduate School	
E_1 Channel 16	e_1 $0/5{,}000 = 0.00$	e_2 $1{,}500/5{,}000 = 0.30$	e_3 $600/5{,}000 = 0.12$	$2{,}100/5{,}000 = 0.42$
E_2 Channel 8	e_4 $900/5{,}000 = 0.18$	e_5 $600/5{,}000 = 0.12$	e_6 $0/5{,}000 = 0.00$	$1{,}500/5{,}000 = 0.30$
E_3 Channel 12	e_7 $1{,}000/5{,}000 = 0.20$	e_8 $200/5{,}000 = 0.04$	e_9 $0/5{,}000 = 0.00$	$1{,}200/5{,}000 = 0.24$
E_4 Don't watch	e_{10} $100/5{,}000 = 0.02$	e_{11} $100/5{,}000 = 0.02$	e_{12} $0/5{,}000 = 0.00$	$200/5{,}000 = 0.04$
	$2{,}000/5{,}000 = 0.40$	$2{,}400/5{,}000 = 0.48$	$600/5{,}000 = 0.12$	$5{,}000/5{,}000 = 1.00$

Using rule 3, we find the probabilities for E_1 and E_6 as follows (we are using the relative frequency approach for assigning probabilities):

$$P(E_1) = P(e_1) + P(e_2) + P(e_3)$$
$$= 0.42$$
$$P(E_6) = P(e_2) + P(e_5) + P(e_8) + P(e_{11})$$
$$= 0.48$$

Now suppose we wish to find the probability of a respondent watching news on Channel 16 *or* having a B.A. degree. That is,

$$P(E_1 \text{ or } E_6) = ?$$

To find this probability, we must use probability rule 4.

PROBABILITY RULE 4

Addition Rule for Any Two Events E_1 and E_2

$$P(E_1 \text{ or } E_2) = P(E_1) + P(E_2) - P(E_1 \text{ and } E_2)$$

The key word in knowing when to use rule 4 is *or*. **The word *or* indicates addition.**
Table 5–4 shows the relative frequencies with the events of interest shaded. The overlap corresponds to the *joint occurrence* of Channel 16 *and* a B.A. degree. The relative frequency of this overlap is $P(E_1 \text{ and } E_6)$ and must be subtracted out to avoid double counting when calculating $P(E_1 \text{ or } E_6)$. Thus

$$P(E_1 \text{ or } E_6) = 0.42 + 0.48 - 0.30$$
$$= 0.60$$

Therefore the probability is 0.60 that a respondent will either watch Channel 16 news or have a B.A. degree.

TABLE 5–4 Joint occurrence of Channel 16 and B.A. degree

Channel Watched	Education Level			
	E_5 *High School*	E_6 *B.A.*	E_7 *Graduate School*	
E_1 Channel 16	e_1 0/5,000 = 0.00	e_2 1,500/5,000 = 0.30	e_3 600/5,000 = 0.12	2,100/5,000 = 0.42
E_2 Channel 8	e_4 900/5,000 = 0.18	e_5 600/5,000 = 0.12	e_6 0/5,000 = 0.00	1,500/5,000 = 0.30
E_3 Channel 12	e_7 1,000/5,000 = 0.20	e_8 200/5,000 = 0.04	e_9 0/5,000 = 0.00	1,200/5,000 = 0.24
E_4 Don't watch	e_{10} 100/5,000 = 0.02	e_{11} 100/5,000 = 0.02	e_{12} 0/5,000 = 0.00	200/5,000 = 0.04
	2,000/5,000 = 0.40	2,400/5,000 = 0.48	600/5,000 = 0.12	5,000/5,000 = 1.00

What is the probability a respondent will favor Channel 16 *or* have a high school education? We can use rule 4 again.

$$P(E_1 \text{ or } E_5) = P(E_1) + P(E_5) - P(E_1 \text{ and } E_5)$$

Table 5–5 shows the relative frequencies for these events. We have

$$P(E_1 \text{ or } E_5) = 0.42 + 0.40 - 0.00$$
$$= 0.82$$

As can be seen, no one both favors Channel 16 news and has only a high school education. The joint relative frequency is zero.* **When the joint probability of two events is zero, the events are mutually exclusive.** This is consistent with our earlier definition of mutually exclusive events. When events are mutually exclusive, a special form of rule 4 applies.

PROBABILITY RULE 5

Addition Rule for Mutually Exclusive Events E_1 and E_2

$$P(E_1 \text{ or } E_2) = P(E_1) + P(E_2)$$

In this section we have presented three rules for adding probabilities. You should become very familiar with these rules and understand how they are used. To test your understanding, use the information in Table 5–5 and let

$$E_8 = (E_1, E_2)$$
$$E_9 = (E_5, E_6)$$

---

*This example illustrates a potential weakness in using relative frequencies to represent probabilities. In this case no person both favored Channel 16 and had only a high school education. Thus, using relative frequencies, we conclude the probability of the joint event is zero. A larger sample may well have included one or more persons in this joint category, in which case the true probability is not zero. However, for the purposes of this example, we will assume these events are mutually exclusive.

TABLE 5–5 Joint occurrence of Channel 16 and high school education

Channel Watched	Education Level			
	E_5 High School	E_6 B.A.	E_7 Graduate School	
E_1 Channel 16	0/5,000 = 0.00	1,500/5,000 = 0.30	600/5,000 = 0.12	2,100/5,000 = 0.42
E_2 Channel 8	900/5,000 = 0.18	600/5,000 = 0.12	0/5,000 = 0.00	1,500/5,000 = 0.30
E_3 Channel 12	1,000/5,000 = 0.20	200/5,000 = 0.04	0/5,000 = 0.00	1,200/5,000 = 0.24
E_4 Don't watch	100/5,000 = 0.02	100/5,000 = 0.02	0/5,000 = 0.00	200/5,000 = 0.04
	2,000/5,000 = 0.40	2,400/5,000 = 0.48	600/5,000 = 0.12	5,000/5,000 = 1.00

Find

$$P(E_8 \text{ or } E_9)$$

Your answer should be 1.0. Do you know why?

Conditional Probability

In dealing with probabilities, you will often need to determine the chances of two or more events occurring either at the same time or in succession. For example, a quality control manager for a manufacturing company may be interested in the probability of selecting two successive defectives from an assembly line. If the probability is low, this manager would be surprised at such a result and might readjust the production process.

In other instances the decision maker may know that an event has occurred and may want to know the chances of a second event occurring. For instance, suppose an oil company geologist feels oil will be found at a certain drilling site. The oil company exploration vice-president might well be interested in the probability of finding oil, given the favorable report.

These situations require tools different from those presented in the section on addition rules. Specifically, you need to become acquainted with rules for **conditional probability** and **multiplication** of probabilities.

West-Air, Inc., a regional airline, has performed a study of its customers' traveling habits. Among the information collected are the data shown in Table 5–6. Lee Hansel, the operations manager, is aware the average traveler has changed over the years. Given the recent increase in discount fares, Lee is particularly interested in maintaining good relations with business travelers. However, since a business traveler is no longer necessarily dressed in a suit, or even always a man, Lee has trouble telling a business traveler from a nonbusiness traveler. Yet he wants to know the present composition of people traveling on his company's airline.

TABLE 5–6 West-Air, Inc. data

Trips per Year	E_4 Female	E_5 Male	
E_1 1 or 2	e_1 $f = 450$ $RF = 450/2,500 = 0.18$	e_2 $f = 500$ $RF = 500/2,500 = 0.20$	$950/2,500 = 0.38$
E_2 3–10	e_3 $f = 300$ $RF = 300/2,500 = 0.12$	e_4 $f = 800$ $RF = 800/2,500 = 0.32$	$1,100/2,500 = 0.44$
E_3 Over 10	e_5 $f = 100$ $RF = 100/2,500 = 0.04$	e_6 $f = 350$ $RF = 350/2,500 = 0.14$	$450/2,500 = 0.18$
	$850/2,500 = 0.34$	$1,650/2,500 = 0.66$	$2,500$

Suppose Lee knows a traveler is female and wants to know the chances this woman will travel between three and ten times a year. We let

$E_2 = (e_3, e_4)$ = event: person travels 3–10 times per year

$E_4 = (e_1, e_3, e_5)$ = event: traveler is female

Then Lee needs to know the probability of E_2 given E_4. Table 5–7 shows the frequencies and relative frequencies of interest shaded. One way to find the desired probability is as follows:

1. We know E_4 has occurred. There are 850 females in the survey.
2. Of the 850 females, 300 travel between three and ten times per year.
3. Then

$$P(E_2|E_4) = \frac{300}{850}$$

$$= 0.3529$$

The notation $P(E_2|E_4)$ is read, "probability of E_2 given E_4."

TABLE 5–7 Joint occurrence of 3–10 trips and female

Trips per Year	E_4 Female	E_5 Male	
E_1 1 or 2	e_1 $f = 450$ $RF = 450/2,500 = 0.18$	e_2 $f = 500$ $RF = 500/2,500 = 0.20$	$950/2,500 = 0.38$
E_2 3–10	e_3 $f = 300$ $RF = 300/2,500 = 0.12$	e_4 $f = 800$ $RF = 800/2,500 = 0.32$	$1,100/2,500 = 0.44$
E_3 Over 10	e_5 $f = 100$ $RF = 100/2,500 = 0.04$	e_6 $f = 350$ $RF = 350/2,500 = 0.14$	$450/2,500 = 0.18$
	$850/2,500 = 0.34$	$1,650/2,500 = 0.66$	$2,500$

Although this approach produces the desired probability, probability rule 6 offers a general rule for conditional probability.

PROBABILITY RULE 6

Conditional Probability for Any Two Events E_1 and E_2 $[P(E_2) \neq 0]$

$$P(E_1|E_2) = \frac{P(E_1 \text{ and } E_2)}{P(E_2)}$$

Rule 6 uses a **joint probability,** $P(E_1 \text{ and } E_2)$, and a **marginal probability,** $P(E_2)$, to calculate the conditional probability, $P(E_1|E_2)$.

Applying rule 6 in our previous problem,

$$P(E_2|E_4) = \frac{0.12}{0.34}$$

$$= 0.3529$$

where: $P(E_2 \text{ and } E_4) = 0.12$

$P(E_4) = 0.34$

In an earlier section we said that two events are independent if one event occurring has no bearing on whether the second event occurs. Therefore, when two events are independent, the rule for conditional probability takes a special form, as indicated in probability rule 7.

PROBABILITY RULE 7

Conditional Probability for Independent Events E_1 and E_2

$$P(E_1|E_2) = P(E_1)$$

and

$$P(E_2|E_1) = P(E_2)$$

As rule 7 shows, the conditional probability of one event occurring, given a second has already occurred, is simply the probability of the first occurring.

Table 5–8 shows some more data from the West-Air passenger survey which we can use to demonstrate rule 7.

Suppose Lee Hansel wants to know the probability a passenger will pay cash for the airline ticket given the passenger is a male. To find this probability, let

$E_1 = (e_1, e_2) = $ event: pay with cash

$E_4 = (e_4, e_4) = $ event: passenger is male

TABLE 5–8 Payment method by sex of traveler

Payment Method	Sex		
	E_3 Female	E_4 Male	
E_1 Cash	e_1 f = 272 RF = 272/2,500 = 0.1088	e_2 f = 528 RF = 528/2,500 = 0.2112	$\frac{800}{2,500} = 0.3200$
E_2 Credit card	e_3 f = 578 RF = 578/2,500 = 0.2312	e_4 f = 1,122 RF = 1,122/2,500 = 0.4488	$\frac{1,700}{2,500} = 0.6800$
	850/2,500 = 0.3400	1,650/2,500 = 0.6600	$\frac{2,500}{2,500} = 1.0000$

Then, using rule 7, $P(E_1|E_4)$ is as follows:

$$P(E_1|E_4) = \frac{P(E_1 \text{ and } E_4)}{P(E_4)} = \frac{0.2112}{0.6600}$$

$$= 0.32$$

But, from Table 5–8, $P(E_1) = 0.32$. Thus

$$P(E_1|E_4) = P(E_1)$$

Therefore these two events are independent.

You should become comfortable with the rules for conditional probability since they are used heavily in statistical decision making.

Multiplication Rules

We needed the joint probability of two events in the preceding discussion. We were able to find $P(E_1 \text{ and } E_4)$ simply by examining the frequencies in Table 5–7. However, often we need to find $P(E_1 \text{ and } E_2)$ when we do not know the joint relative frequencies. To illustrate how to find a joint probability, consider an example involving classical probability.

Suppose a game of chance involves selecting 2 cards from a 52-card deck. If both cards are aces, the participant receives a specified payoff. What is the probability of selecting two aces? To answer this question, we must recognize that two events are required to form the desired outcome. Therefore, let

$$A_1 = \text{event: ace on the first draw}$$

$$A_2 = \text{event: ace on the second draw}$$

The question really being asked is, What are the chances of observing both A_1 and A_2? The key word here is *and*, as contrasted with the addition rule, where the key word is *or*. The *and* signifies that we are interested in the joint probability of two events, as noted by

$$P(A_1 \text{ and } A_2)$$

To find this probability, we employ rule 8, the multiplication rule.

PROBABILITY RULE 8

Multiplication Rule for Two Events E_1 and E_2

$$P(E_1 \text{ and } E_2) = P(E_1)P(E_2|E_1)$$

and

$$P(E_2 \text{ and } E_1) = P(E_2)P(E_1|E_2)$$

Note that rule 8 is an algebraic rearrangement of rule 7.

We use the classical approach to probability assessment to find the value of $P(A_1 \text{ and } A_2)$ as follows:

$$P(A_1) = \frac{\text{number of aces}}{\text{number of possible cards}}$$

$$= \frac{4}{52}$$

Then, since we are not replacing the first card, we find $P(A_2|A_1)$ by

$$P(A_2|A_1) = \frac{\text{number of remaining aces}}{\text{number of remaining cards}}$$

$$= \frac{3}{51}$$

Thus, by rule 8,

$$P(A_1 \text{ and } A_2) = P(A_1)P(A_2|A_1) = \left(\frac{4}{52}\right)\left(\frac{3}{51}\right)$$

$$= 0.0045$$

Therefore there are slightly more than 4 chances in 1,000 of selecting two successive aces from a 52-card deck *without replacement*.

Note that rule 8 requires that conditional probability be used since the result on the second draw depends on the card selected on the first draw. The chance of obtaining an ace on the second draw was lowered from 4/52 to 3/51 given that the first card was an ace. However, if the two events of interest are *independent*, the conditional aspect is not important, and the multiplication rule takes the form shown in probability rule 9.

PROBABILITY RULE 9

Multiplication Rule for Independent Events E_1 and E_2

$$P(E_1 \text{ and } E_2) = P(E_1)P(E_2)$$

Thus the joint probability of two independent events is simply the product of the marginal probabilities of the two events.

The probability rules presented in this section are vital to managers who will use statistical decision-making techniques. Remember the key words *or* and *and*. Know with what rules they are associated, and you should have little trouble with basic probability theory.

5-4
BAYES' RULE

Conditional probabilities provide a lot of good information for decision makers. For instance, say that medical researchers are interested in determining the probability of a person getting cancer given that the person was exposed to hazardous chemicals. That is,

$$P(\text{cancer}|\text{hazardous chemicals})$$

Educators are interested in the probability a student learns math skills given that the student completed certain background work.

$$P(\text{learning skills}|\text{background work})$$

Managers are interested in the probability a worker will develop superior skills given that the worker receives a certain job advancement sequence.

$$P(\text{work skills}|\text{advancement sequence})$$

Such conditional probabilities are calculated using probability rule 6.

$$P(E_1|E_2) = \frac{P(E_1 \text{ and } E_2)}{P(E_2)}$$

However, in many practical applications, decision makers may know for certain an event has occurred but not know what the chances were of that event occurring before the fact, and thus will not be able to use probability rule 6 directly. For example, suppose an oil exploration company wants to determine the probability that its geologist's report will be favorable given there is oil at the site; that is,

$$P(\text{favorable report}|\text{oil}) = ?$$

If the company applied probability rules 6 and 8, this probability would be

$$P(\text{favorable report}|\text{oil}) = \frac{P(\text{favorable report and oil})}{P(\text{oil})}$$

$$= \frac{P(\text{favorable report})P(\text{oil}|\text{favorable report})}{P(\text{oil})}$$

In this case the exploration company might have information to supply the probabilities in the numerator but would not know the probability of oil needed in the denominator.

For these types of problems, an extension of conditional probability called **Bayes' rule** can be used. Bayes' rule might best be developed through the following example.

Winner Bakery makes and distributes frozen bread loaves. The company has one production plant and one central warehouse where all products are stored until they can be moved to sales distribution points. The production plant has two lines, A and B. Line A produces 60 percent of all loaves, and line B produces the remaining 40 percent. Winner has had quality control problems. The production manager, after extensive testing, has determined that 5 percent of the loaves produced on line A are defective, and that 10 percent of the loaves produced on line B are defective.

The two production lines use different combinations of equipment and workers, and the accounting department keeps cost records on each line. Winner Bakery offers a money-back guarantee on any defective loaves, and the plant manager wants to allocate these costs to the two production lines. However, she cannot identify which line produced any specific defective loaf and therefore decides to prorate the cost of honoring guarantees between the two lines. She wants to determine the following:

If a defective loaf is returned, what is the probability it came from line A? What is the probability the loaf came from line B?

She can then allocate the cost of defective loaves based on these probabilities.

This problem can be solved using conditional probabilities in the following manner:

1. Using conditional probability statements, the manager wants to determine
   a. P(loaf came from line A|loaf is defective)
   b. P(loaf came from line B|loaf is defective)

2. Using the conditional probability rule (rule 6),

$$P(\text{line A}|\text{defective loaf}) = \frac{P(\text{line A and defective loaf})}{P(\text{defective loaf})}$$

Although we don't know any of these probabilities directly, we know

$$P(\text{line A}) = 0.6$$
$$P(\text{defective loaf}|\text{line A}) = 0.05$$

And so

$$P(\text{line A}|\text{defective loaf}) = \frac{(0.6)(0.05)}{P(\text{defective loaf})}$$

3. The only trouble now is that we don't know the overall probability of a loaf being defective. To find P(defective loaf), we ask, How can a defective loaf be produced? The answer is that a loaf can be

defective *and* produced by line A *or* defective *and* produced by line B. The next step is to find the probability of each.

$$P(\text{line A and defective loaf}) = P(\text{line A})\,P(\text{defective loaf}|\text{line A})$$
$$= (0.6)(0.05)$$
$$= 0.03$$
$$P(\text{line B and defective loaf}) = P(\text{line B})\,P(\text{defective loaf}|\text{line B})$$
$$= (0.4)(0.1)$$
$$= 0.04$$

Then, because we can get a defective loaf from either line A *or* line B, we use the addition rule (rule 5) to get

$$P(\text{defective loaf}) = P(\text{defective loaf and line A})$$
$$+ P(\text{defective loaf and line B})$$
$$= 0.03 + 0.04$$
$$= 0.07$$

4. Thus the probability that a defective loaf was actually produced on line A is

$$P(\text{line A}|\text{defective loaf}) = \frac{P(\text{line A and defective loaf})}{P(\text{defective loaf})}$$
$$= \frac{(0.6)(0.05)}{0.07}$$
$$= 0.428$$

and, since a defective loaf must either come from line A or line B,

$$P(\text{line B}|\text{defective loaf}) = 1 - P(\text{line A}|\text{defective loaf}) = 1 - 0.428$$
$$= 0.572$$

This conditional probability application has used Bayes' rule, which we now formally define.

BAYES' RULE

Let A_i $(i = 1, 2, 3, \ldots, n)$ be a complete set of mutually exclusive events.
Let B be another event which is preceded by an A_i event.
And $P(A_i)$ and $P(B|A_i)$ must be known.
Then

$$P(A_1|B) = \frac{P(A_1)P(B|A_1)}{\sum\limits_{i=1}^{n} P(A_i)P(B|A_i)}$$

Using Bayes' rule directly in our bread example, let

$$A_1 = \text{event: loaf came from line A}$$
$$A_2 = \text{event: loaf came from line B}$$
$$B = \text{event: loaf is defective}$$

The manager knows the following probabilities:

$$P(A_1) = 0.6$$
$$P(B|A_1) = 0.05$$
$$P(A_2) = 0.4$$
$$P(B|A_2) = 0.1$$

Therefore

$$P(A_1|B) = \frac{P(A_1)P(B|A_1)}{P(A_1)P(B|A_1) + P(A_2)P(B|A_2)} = \frac{(0.6)(0.05)}{(0.6)(0.05) + (0.4)(0.1)} = \frac{0.03}{0.03 + 0.04}$$
$$= 0.428$$

Remember, we began with the plant manager who was trying to allocate costs of defective loaves between two production lines. Her problem has been solved. She should allocate 42.8 percent of all costs to line A and 57.2 percent to line B.

Bayes' rule is also applied in situations where new information changes the probability of the original events. These changes in probability are called **revisions,** and as we will show in Chapter 20, Bayes' rule is very important in allowing managers to make the correct revisions.

5-5
COUNTING TECHNIQUES

In our earlier discussion of classical probability we defined the probability that event A will occur, for equally likely outcomes, as

$$P(A) = \frac{\text{number of ways } A \text{ can occur}}{\text{number of ways any event can occur}}$$

As we saw, finding $P(A)$ is straightforward when values in the numerator and denominator are known. However in many business situations they will not be known. For example, a local organization has been in trouble recently over hiring male applicants rather than female applicants. The last three employees hired were males even though the list of six qualified applicants included three females. (All applicants were available beginning with the first hire.) The company's personnel director claims that in each case, all applicants were given an equal chance because the person hired was drawn from a hat. One female applicant claims that if the selections were made in this fashion, the process was "rigged." She says that the probability of fairly selecting three males is so low it could only have happened with assistance from the company. How do we find this probability?

First, let A be the event of hiring three males and no females. (We must stipulate that we started with three qualified males and three qualified females.) Then

$$P(A) = \frac{\text{number of ways } A \text{ can occur}}{\text{number of ways three people can be hired}}$$

Our problem is to determine values for the numerator and denominator. However the "number of ways" for each may not be immediately apparent. We might approach this problem in several ways. The most elementary approach, assuming the firm has six applicants, three males and three females, is to list the *sample space* (see Table 5–9). Notice that the order of hiring is not important. We are interested only in the three jobs as a group. We find by listing the entire sample space that there are 20 possible ways of hiring three people from six applicants. Of these, only one satisfies the requirements of event A: all three hires are men, and that is (M_1, M_2, M_3). Thus

$$P(A) = \frac{1}{20}$$
$$= 0.05$$

TABLE 5–9 Sample space

Elementary Event	Outcome		
1	M_1	M_2	M_3
2	M_1	M_2	F_1
3	M_1	M_2	F_2
4	M_1	M_2	F_3
5	M_1	M_3	F_1
6	M_1	M_3	F_2
7	M_1	M_3	F_3
8	M_2	M_3	F_1
9	M_2	M_3	F_2
10	M_2	M_3	F_3
11	M_1	F_1	F_2
12	M_1	F_1	F_3
13	M_1	F_2	F_3
14	M_2	F_1	F_2
15	M_2	F_1	F_3
16	M_2	F_2	F_3
17	M_3	F_1	F_2
18	M_3	F_1	F_3
19	M_3	F_2	F_3
20	F_1	F_2	F_3

where: M_1 = male number 1

M_2 = male number 2

M_3 = male number 3

F_1 = female number 1

F_2 = female number 2

F_3 = female number 3

The probability of hiring three males and no females using the hat scheme is 0.05. Since you are studying to become a decision maker, decide whether the personnel manager was telling the truth or whether the selection process was "rigged" if he had six qualified applicants.

Permutations and Combinations

Listing the entire sample space is fine for a limited situation. However, suppose that instead of 6 applicants, there were 12, 5 male and 7 female. Now, what is the probability of all three hires being males? Although listing the sample space is possible, it would be time consuming and there would be a good chance that one or more sample points would accidentally be omitted. To avoid listing the entire sample space, you may use one of two **counting techniques.** When the order in which the simple events occur is not important, the **combinations** method should be used. When the order is important, the **permutations** method should be employed. Table 5–10 lists the formula for each counting method.

TABLE 5–10   Counting techniques

Combinations (order is not important)

$$_nC_r = \frac{n!}{r!(n - r)!}$$

where: n = number of items in a group
          r = number of items to be selected from n

Permutations (order is important)

$$_nP_r = \frac{n!}{(n - r)!}$$

where: n = number of items in a group
          r = number of items to be selected from n

In the example involving six applicants, we should use the *combinations* method because order *is not* important. Recall that

$$P(A) = \frac{\text{number of ways three males can be hired}}{\text{number of ways any three people can be hired}}$$

Using combinations, we get

$$P(A) = \frac{_3C_3}{_6C_3} = \frac{\dfrac{3!}{3!(3 - 3)!}}{\dfrac{6!}{3!(6 - 3)!}} = \frac{\dfrac{6}{6}}{\dfrac{720}{36}} = \frac{1}{20}$$

$$= 0.05$$

Therefore we have obtained the same result that previously required listing the entire sample space using a much easier method.

For the more complex case with five male and seven female applicants, we can also find the probability without too much difficulty.

$$P(A) = \frac{_5C_3}{_{12}C_3} = \frac{\dfrac{5!}{3!(5-3)!}}{\dfrac{12!}{3!(12-3)!}} = \frac{10}{220} = \frac{1}{22}$$

$$= 0.045$$

Thus, if all applicants received an equal chance, the probability is 0.045 that all three jobs would go to males.

Consider the situation where five contractors have all bid on three different jobs with a state building agency. If no contractor can receive more than one job, how many different ways can the state building director award the three contracts? The order in which the contractors are selected is important because the three contracts are for different jobs, and we are considering five different contractors. Thus the *permutations* method should be used to determine the alternatives available. From Table 5–10,

$$_5P_3 = \frac{5!}{(5-3)!} = \frac{5!}{2!} = \frac{(5)(4)(3)(2)(1)}{(2)(1)}$$

$$= 60$$

Thus 60 alternatives are available to the building director.

Note that if the three contracts were identical, we would employ combinations to determine the number of alternatives.

$$_5C_3 = \frac{5!}{3!(5-3)!}$$

$$= 10 \text{ alternatives}$$

We will always find fewer combinations than permutations for any $r > 0$. Verify for yourself that this is true.

Distinct Permutations

A final method of counting is **distinct permutations.** This method is used when order is important but the simple events within categories are basically indistinguishable. A good example is found on sailing ships which have long used flags to pass messages to shore and to other ships. A ship might have flags of four colors—say, red, yellow, green, and white—with 3 flags of each color. Suppose a message is sent by arranging the 12 flags in a particular order. For example, the order "red, red, red, white, white, white, green, green, green, yellow, yellow, yellow" may indicate that the ship is carrying an injured person. Since we don't distinguish between flags of the same color, just between different colors, how many different messages are possible with the 12 flags?

The distinct permutations calculation is as follows:

$$\text{Distinct permutations} = \frac{n!}{X_1! X_2! X_3! \ldots X_K!}$$

where: $X_1!$, $X_2!$, $X_3!$, etc. = number of items belonging to a particular category

K = number of categories

n = total number of items

In our flag example,

$$\text{Number of messages} = \frac{12!}{3!3!3!3!}$$
$$= 369,600$$

Just a few! We shall have a real need for distinct permutations when we reach Chapter 6 and will illustrate its application more extensively there.

5-6
CONCLUSIONS

Probability provides decision makers a quantitative measure of the chance an environmental outcome will occur. It allows decision makers to quantify uncertainty. The objectives of this chapter have been to discuss the various types of probability and to provide the basic rules that govern probability operations.

We have discussed many probability concepts from a managerial perspective. In the chapter glossary we list a set of strict definitions for important probability concepts introduced in this chapter. To test your understanding of these concepts, reconcile the managerial definitions with the more formal definitions.

The list of chapter formulas, including Bayes' rule for conditional probability, follows the glossary.

chapter glossary

classical probability   A method of determining probability based on the ratio of the number of ways the event of interest can occur to the number of ways any event can occur.

combinations   The method of counting possible selections from a set of elementary events when order is not important.

$$_nC_r = \frac{n!}{r!(n-r)!}$$

distinct permutations   The method of counting possible selections from a set of elementary events when order is important but the simple events within particular categories are indistinguishable.

$$\text{Distinct permutations} = \frac{n!}{X_1! X_2! X_3! \ldots X_K!}$$

elementary events   The single outcomes resulting from an experiment.

permutations   The method of counting possible arrangements from a set of elementary events when order is important.

$$_nP_r = \frac{n!}{(n - r)!}$$

relative frequency of occurrence   The method that defines probability as the number of times an event occurs divided by the total number of times an experiment is performed.

sample space   The set of all possible elementary events, or outcomes, that can result from a single trial or experiment.

subjective probability   The method that relates probability to a decision maker's state of mind regarding the likelihood a particular event will occur.

chapter formulas

Probability rule 1

$$0.0 \le P(E_i) \le 1.0 \quad \text{for all } i$$

and

$$0.0 \le P(e_i) \le 1.0 \quad \text{for all } i$$

Probability rule 2

$$\sum_{i=1}^{K} P(e_i) = 1.0$$

Probability rule 3
Addition rule for elementary events

The probability of an event E_i is equal to the sum of the probabilities of the elementary events forming E_i. That is, if

$$E_i = (e_1, e_2, e_3)$$

then

$$P(E_i) = P(e_1) + P(e_2) + P(e_3)$$

Probability rule 4
Addition rule for any two events E_1 and E_2

$$P(E_1 \text{ or } E_2) = P(E_1) + P(E_2) - P(E_1 \text{ and } E_2)$$

Probability rule 5
Addition rule for mutually exclusive events E_1 and E_2

$$P(E_1 \text{ or } E_2) = P(E_1) + P(E_3)$$

Probability rule 6
Conditional probability for any two events E_1 and E_2 [$P(E_2) \neq 0$]

$$P(E_1|E_2) = \frac{P(E_1 \text{ and } E_2)}{P(E_2)}$$

Probability rule 7
Conditional probability for independent events E_1 and E_2

$$P(E_1|E_2) = P(E_1)$$

and

$$P(E_2|E_1) = P(E_2)$$

Probability rule 8
Multiplication rule for two events E_1 and E_2

$$P(E_1 \text{ and } E_2) = P(E_1)P(E_2|E_1)$$

and

$$P(E_2 \text{ and } E_1) = P(E_2)P(E_1|E_2)$$

Probability rule 9
Multiplication rule for independent events E_1 and E_2

$$P(E_1 \text{ and } E_2) = P(E_1)P(E_2)$$

Bayes' rule

Let A_i ($i = 1, 2, 3, \ldots, n$) be a complete set of mutually exclusive events.
Let B be another event which is preceded by an A_i event.
And $P(A_i)$ and $P(B|A_i)$ must be known.
Then

$$P(A_1|B) = \frac{P(A_1)P(B|A_1)}{\sum\limits_{i=1}^{n} P(A_i)P(B|A_i)}$$

solved problems

1. There are four defective transistors in a package of ten. If two transistors are selected randomly one after another, what is the probability of each of the following?
   a. One defective and one good transistor will be selected
   b. Two defectives will be selected
   c. At least one defective will be selected
   d. Three good transistors will be selected

Solutions:

a. We will solve this problem two ways. First we list the sample space as follows:

$$G, G$$
$$G, D$$
$$D, G$$
$$D, D$$

where: G = good transistor
            D = defective transistor

Then we attach probabilities to each simple event.

$$P(G_1) = P(\text{good on first draw})$$
$$= \frac{6}{10}$$
$$P(D_1) = P(\text{defective on first draw})$$
$$= \frac{4}{10}$$

Notice that the probabilities on the second draw depend on what took place on the first draw. Thus

$$P(G_2|G_1) = \frac{5}{9}$$
$$P(G_2|D_1) = \frac{6}{9}$$
$$P(D_2|G_1) = \frac{4}{9}$$
$$P(D_2|D_1) = \frac{3}{9}$$

Then we find the probability of one defective transistor and one good transistor in a sample of two by using both the addition rule and the conditional probability rule.

$$P(G_1 \text{ and } D_2) + P(D_1 \text{ and } G_2) = P(G_1)P(D_2|G_1) + P(D_1)P(G_2|D_1)$$
$$= \left(\frac{6}{10}\right)\left(\frac{4}{9}\right) + \left(\frac{4}{10}\right)\left(\frac{6}{9}\right)$$
$$= \frac{24}{90} + \frac{24}{90} = \frac{48}{90}$$
$$= \frac{8}{15}$$

A second method utilizes the combinations counting method. We are looking for the following sample event:

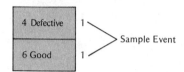

$$P(1 \text{ defective and 1 good}) = \frac{\text{number of ways to get 1 defective and 1 good}}{\text{number of ways to draw 2 transistors}}$$

$$= \frac{_4C_1 \text{ and } _6C_1}{_{10}C_2} = \frac{\left(\dfrac{4!}{1!3!}\right)\left(\dfrac{6!}{1!5!}\right)}{\dfrac{10!}{2!8!}}$$

$$= \frac{(4)(6)}{45} = \frac{24}{45}$$

$$= \frac{8}{15}$$

b. We want the following situation:

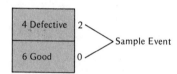

Again using combinations, we obtain

$$P(2 \text{ defective}) = \frac{\text{number of ways to draw 2 defective and 0 good}}{\text{number of ways to draw 2 transistors}}$$

$$= \frac{_4C_2 \text{ and } _6C_0}{_{10}C_2} = \frac{\left(\dfrac{4!}{2!2!}\right)\left(\dfrac{6!}{0!6!}\right)}{45} = \frac{6}{45}$$

$$= \frac{2}{15}$$

Remember, $0! = 1$.

c. We are looking for the following sample event:

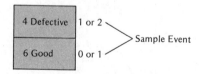

We know that

$$P(0 \text{ defective}) + P(1 \text{ defective}) + P(2 \text{ defective}) = 1$$

or

$$P(0 \text{ defective}) + P(1 \text{ } or \text{ } more \text{ defective}) = 1$$

So

$$P(1 \text{ or more defective}) = 1 - P(0 \text{ defective})$$

To find $P(1$ or more defective), we find $P(0$ defective). Using combinations,

$$P(0 \text{ defective}) = \frac{(_4C_0)(_6C_2)}{_{10}C_2} = \frac{(1)(15)}{45}$$

$$= \frac{1}{3}$$

Therefore

$$P(1 \text{ or more defective}) = 1 - \frac{1}{3}$$

$$= \frac{2}{3}$$

d. $P(3 \text{ good}) = 0$ since only two transistors are being selected.

2. A small town has two ambulances. Records indicate that the first ambulance is in service 60 percent of the time and the second one is in service 40 percent of the time. What is the probability that when an ambulance is needed, one will not be available?

Solution:

The sample space is as follows:

$$A, A = \text{both available}$$
$$A, B = \text{one available}$$
$$B, A = \text{one available}$$
$$B, B = \text{both busy}$$

If we assume that the ambulances being busy are *independent* events (no large accidents), the probability of both ambulances being busy is

$$P(B \text{ and } B) = (0.60)(0.40)$$
$$= 0.24$$

3. For the information given in problem 2, what is the probability that at least one ambulance will be available?

Solution:

Since

$$P(0 \text{ available}) + P(1 \text{ or more available}) = 1$$

then

$$P(1 \text{ or more available}) = 1 - P(0 \text{ available}) = 1 - 0.24$$
$$= 0.76$$

4. A cement company has two suppliers of the raw materials used in making cement. Vendor 1 supplies 30 percent of the raw materials, and vendor 2 supplies 70 percent. Tests have shown that 40 percent of vendor 1's materials are poor quality whereas 5 percent of vendor 2's materials are poor quality. The cement company's manager has just found some poor-quality materials in inventory. Which company most probably supplied these materials?

Solution:

We begin by listing the information given in probability form.

$$P(1) = 0.30$$
$$P(2) = 0.70$$
$$P(\text{poor}|1) = 0.40$$
$$P(\text{poor}|2) = 0.05$$

Then, using Bayes' rule,

$$P(1|\text{poor}) = \frac{P(1)P(\text{poor}|1)}{P(1)P(\text{poor}|1) + P(2)P(\text{poor}|2)}$$

$$= \frac{(0.30)(0.40)}{(0.30)(0.40) + (0.70)(0.05)} = \frac{0.12}{0.155}$$

$$= 0.77$$

$$P(2|\text{poor}) = \frac{P(2)P(\text{poor}|2)}{0.155} = \frac{0.035}{0.155}$$

$$= 0.23$$

Thus vendor 1 most likely supplied the poor materials found in inventory.

problems

Problems marked with an asterisk (*) require extra effort.

1. In terms of the definitions of probability given in this chapter, discuss each of the following statements:
   a. The probability of rain tomorrow is 0.30.
   b. The National League team is listed as a 6 to 5 favorite in the World Series.
   c. The probability is high a car engine will need major repair before 100,000 miles.
   d. A new product is given a 50/50 chance of success.

2. Define
   a. Mutually exclusive events
   b. Independent events
   List five business examples of each.

3. Suppose $P(A) = 0.50$, $P(B) = 0.40$, and $P(A \text{ and } B) = 0.20$.
   a. Are A and B mutually exclusive? Why or why not?
   b. Are A and B independent events? Why or why not?

4. Alex and Alice are discussing their plans for a family. They decide they would like two children: one boy and one girl. Assume that male and female children are equally likely and successive births are independent.
   a. What is the probability that Alex and Alice will have one child of each sex?
   b. They decide to have a third child if their plans don't work out. What is the probability they will have a third child?
   c. After having two sons, Alex and Alice feel certain their next child will be a daughter. Comment.
   d. What is the probability Alex and Alice could have four girls in a row?

5. A board of directors consists of ten members, six of whom are loyal to the current company president and four of whom want to fire the president. If the chairman of the board, who is a loyal supporter of the president, decides to randomly select four other board members to serve with him on a committee to decide the president's fate, what is the probability all five members will vote to keep the president if no one changes sides?

6. Referring to problem 5, what is the probability that a majority of the five members will vote to keep the president if no one changes sides? A majority vote will decide the issue.

7. Referring to problem 5, what is the probability that the vote will be 4 to 1 in favor of firing the current president if no one on the board changes sides?

8. The probability an Avon salesperson, following up a special sale campaign advertisement, will make a sale is 0.30. Assuming independence, what is the probability she will make two sales if she calls on two customers?

9. Referring to problem 8, what is the probability of making one sale?

10. Referring to problem 8, what is the probability of making at least one sale?

11. Dave Newbury has just left an interview with a prospective employer. The employer has told Dave that she will put up with one mistake during his first year but will fire him if he makes two mistakes. Based upon Dave's analysis of the job, he will have to make five critical decisions during the year, and Dave feels his chances are 0.80 of making any one of them correctly. Dave does not want to run more than a 25 percent chance of getting fired. Assuming the five decisions are independent, should Dave take the job if it is offered?

12. Referring to problem 11, suppose the employer thinks that Dave has a 70 percent chance of making a correct decision. She does not want to hire Dave if there is over a 40 percent chance he will be fired. Should she offer Dave the job?

*13. Referring to problems 11 and 12, what is the probability that Dave will be offered the job and that he will accept? Assume independence.

14. In the sales business, repeat calls to finally make a sale are common. Suppose a particular salesperson has a 0.70 probability of selling on the first call and that the probability of selling drops by 0.10 on each successive call. If the salesperson is willing to make up to four calls on any client, what is the probability of a sale?

15. Of a batch of 20 television picture tubes, 5 are known to be defective. What is the probability that a sample of 5 without replacement will result in each of the following?
   a. Exactly 1 defective
   b. No defectives
   c. Two or fewer defectives

*16. Recreational developers are considering opening a skiing area near a western town. They are trying to decide whether to open an area catering to family skiers or to some other group. To help make their decision, they gather the following information. If

A_1 = family will ski

A_2 = family will not ski

B_1 = family has children but none in the 8–16 age group

B_2 = family has children in the 8–16 age group

B_3 = family has no children

then, for this location,

$$P(A_1) = 0.40$$
$$P(B_2) = 0.35$$
$$P(B_3) = 0.25$$
$$P(A_1|B_2) = 0.70$$
$$P(A_1|B_1) = 0.30$$

   a. Use the probabilities given to construct a joint probability distribution table.
   b. What is the probability a family will ski *and* have children not in the 8–16 age group? How do you write this probability?
   c. What is the probability a family with children in the 8–16 age group will not ski?
   d. Are the two categories—"skiing" and "family composition"—independent?

17. A company is considering changing its starting hour from 8:00 A.M. to 7:30 A.M. A census of the company's 1,200 office and production workers shows

370 of its 750 production workers against the change, and a total of 715 workers in favor of the change. To further assess worker opinion, the region manager decides to randomly talk with workers.

a. What is the probability a randomly selected worker will be in favor of the change?

b. What is the probability a randomly selected worker will be against the change *and* an office worker?

c. Is the relationship between job type and opinion independent? Why?

18. A nationwide retailer has just had a change of top management. The new president wants to analyze the relationship between age of store and size of city for the distribution outlets. The following data are gathered:

		Age of Store			
Population of City		Less Than 2 Years X_1	2 Years to Less Than 5 Years X_2	5 Years to Less Than 15 Years X_3	Over 15 Years X_4
Under 20,000	Y_1	6	5	3	6
20,000 to under 100,000	Y_2	11	13	15	22
100,000 to under 250,000	Y_3	38	17	12	42
Over 250,000	Y_4	30	45	35	65

a. Using these data, construct a relative frequency distribution.

b. What is the probability of randomly selecting the following?

 i. A store less than 2 years old

 ii. A store over 15 years old in a city over 250,000

 iii. A store in the 2–5-year category in a city less than 100,000

c. Explain each of the following probability statements:

 i. $P(Y_1|X_4)$

 ii. $P(Y_3 \text{ or } Y_4)$

 iii. $P(X_3|Y_2)$

 iv. $P(Y_2 \text{ and } X_4)$

 v. $P(X_2|X_4)$

d. Find the probabilities for the statements given in part c.

e. The new president wants you to answer the following questions:

 i. "Has our policy of adding new stores favored any size city?"

 ii. "Are our older stores proportionately concentrated in a particular size city or are they spread representatively?"

 iii. "Are city size and store age independent?"

19. An investment advisor has a portfolio of 80 stocks: 50 blue-chip and 30 growth stocks. Of the 50 blue-chip stocks, 30 have increased in price during the past month, and 20 of the 30 growth stocks have increased in price. If a stock is selected at random from the portfolio, what is the probability it will be a blue-chip stock that has not increased in price?

20. Referring to problem 19, what is the probability of selecting a stock that *has* increased in price?

21. Referring to problem 19, if the stock selected has not increased in price, what is the probability it is a growth stock?

*22. Bill Jones and Herman Smith are long-time business associates. They know that regular exercise improves their productivity and have made a practice of either playing tennis or golf every Saturday for the past ten years. Jones enjoys tennis, and Smith prefers golf. Each Saturday they flip a coin to decide what sport to play. Jones beats Smith at tennis 80 percent of the time, whereas he beats Smith at golf only 30 percent of the time. Suppose Jones walks into the Monday morning staff meeting and announces he beat Smith on Saturday. What sport do you think they played, and why?

23. Referring to problem 22, suppose open tennis courts are hard to find on Saturday, so instead of flipping a coin, Smith and Jones always first look for a tennis court. If they find one open, they play tennis; if not, they play golf. Further, suppose the chance of finding an open court is 30 percent. Given this, what sport do you think they played on Saturday given that Jones won?

24. A marketing research team is considering using a mailing list for an advertising campaign. They know that 40 percent of the people on the list have only a Master Charge card and that 10 percent have only an American Express card. Another 20 percent hold both Master Charge and American Express. Finally, 30 percent of those on the list have neither credit card.
    Suppose a person on the list is known to have a Master Charge card. What is the probability that person also has an American Express card?

25. The M-K Construction Company is considering bidding on a contract. The president feels the probability of winning the contract is 0.50 if Lite Construction does not submit a bid and 0.20 if it does. There is a 0.75 probability that Lite Construction will enter a bid. What is the probability that M-K will win the bid?

26. Referring to problem 25, suppose M-K has just won the bid. What is the probability Lite submitted a bid?

27. As a manager at a crucial point in a baseball game, you feel your present pitcher has a 70 percent chance of getting the next batter out. You could replace him with a relief pitcher who has a 90 percent chance of getting the batter out if he is at his best, but only a 40 percent chance if he is not. Your pitching coach in the bullpen states the relief pitcher has a 70 percent of being at his best. Do you change pitchers? Why or why not?

28. Referring to problem 27, suppose the bullpen coach has said there is a 50 percent chance the relief pitcher will be at his best. Would the manager's decision change? Why or why not?

29. Your neighbor is an avid fisher. He is planning a Canadian fishing trip where he will be flown into a northern lake. He shows you the brochures from Tree Top Airlines promoting the trip he is planning. Tree Top claims its planes are

very safe, and as proof states that only 5 percent of its flights experience any difficulty. On the flights with difficulty, half the time the difficulty is caused by mechanical circumstances, and half the time by electrical problems. If the plane has mechanical difficulty, there is a 40 percent chance the flight will be cut short. The chance increases to 50 percent for electrical problems.

   a. What is the probability your neighbor will be on a flight that is unexpectedly cut short?

   b. Assume your neighbor goes and the flight is cut short. What is the probability the cut was due to electrical problems?

30. Explain why there should be more permutations than combinations when the two methods are used to determine the number of ways events can occur.

31. If a finance subcommittee for an electrical utility's board of directors consists of ten members, in how many ways can a chairperson, vice-chairperson, and secretary be selected?

32. Baseball manager Sparky Martin's team is in the midst of a terrible slump. Sparky is willing to try almost anything to change their luck.

   a. If he decides to put the names of all 25 team members in a hat and randomly draw a group of 9 players, how many different teams of 9 can he choose?

   b. How many different batting orders can he make from the 25 team members?

   c. If Sparky simply decides to change the batting order of his present starting lineup, how many different batting orders does he have to choose from?

33. A management consulting firm needs a team of four to analyze the operations of a new client. The team should contain an accountant, a production specialist, a finance specialist, and a management specialist. On its staff the consulting firm has available six accountants, five production experts, three finance specialists, and eight management specialists. How many different teams can be formed from the individuals available?

34. A major firm has recently changed the maintenance procedures in the fleet of cars at one of its regional offices. This office maintains 27 company cars. At the end of an 18-month period, the firm decides to tear down 6 cars to look for changes in wear. How many different groups of 6 cars can be selected for test purposes?

35. States rely on some combination of letters and numbers when making car license plates. If a state uses six numbers, how many separate license plates can it issue? If the state relies on two letters and four numbers? Three letters and three numbers? Answer the last two questions assuming that the letters must come first and the numbers second.

36. The Germaine L. Jones Manufacturing Company produces automobile windshields at three plants in the Detroit area. The windshields are sent to a centralized warehouse where they are so mixed up that the identity of the particular plant is lost.

      Plant A makes 50 percent of the windshields, of which 10 percent contain some defect. Plant B accounts for 30 percent of the windshields, and of these 12 percent are defective. Finally, plant C makes 20 percent of the windshields, and of these only 5 percent are defective.

      Last year the Germaine L. Jones Company incurred a replacement cost of $400,000 on windshields produced. The cost accountants wish to divide this cost between the three plants based on the ratio of defectives produced by each plant. Provide this breakdown in costs for the accountants.

37. The Upland Paving Company has three locations where it mixes its paving material. Forty percent of all material is mixed at plant A, 50 percent at plant B, and 10 percent at plant C. Although Upland would like to believe it has a good quality control system at each plant, sometimes bad batches do get mixed. For instance, in the past, 15 percent of plant A's batches, 10 percent of plant B's batches, and 25 percent of plant C's batches have been bad.

      Suppose a bad batch has just been discovered at a construction project. Which plant most likely provided the bad batch?

references

BLYTH, C. R. "Subjective vs. Objective Methods in Statistics." *American Statistician* 26 (June 1972): 20–22.

KING, W. R. *Probability for Management Decisions.* New York: Wiley, 1968.

MOSTELLER, FREDERICK, ROBERT E. ROURKE, and GEORGE B. THOMAS, JR. *Probability with Statistical Applications.* 2nd ed. Reading, Mass.: Addison-Wesley, 1970.

RAIFFA, HOWARD. *Decision Analysis.* Reading, Mass.: Addison-Wesley, 1968.

6

Discrete Probability Distributions

why decision makers need to know

Thus far we have seen that a frequency distribution transforms ungrouped data into a more meaningful form. Therefore frequency distributions help decision makers deal with the uncertainty in their decision environments. We also learned, in Chapter 5, that probability is a fundamental part of statistics. Because all managers operate in an uncertain environment, they must be able to make the connection between descriptive statistics and probability. This connection is made by moving from frequency distributions to **probability distributions.**

Constructing and analyzing a frequency distribution for every decision-making situation would be time consuming. Just deciding on the correct data-gathering procedures, the appropriate class intervals, and the right methods of presenting the data is not a trivial problem. Fortunately many physical events that appear to be unrelated have the same underlying characteristics and can be described by the same probability distribution. If decision makers are dealing with an application described by a predetermined **theoretical** probability distribution, they can utilize a great deal of developmental statistical work already known and save considerable personal effort in analyzing their situation. Therefore decision makers need to become comfortable with probability distributions if they are to apply them effectively in the decision-making process.

chapter objectives

In this chapter we shall introduce general discrete probability distributions. We then shall consider two common discrete probability distributions: the binomial distribution and the Poisson distribution. Both distributions describe situations with discrete values for the variable of interest. Many physical and economic events are described by one of these two distributions.

We also shall discuss how discrete probability distributions are developed and indicate the type of events these distributions describe. In addition, we will cover several descriptive measures that help define these distributions.

student objectives

After studying the material in this chapter, you should be able to:

1. Identify the type of processes that are represented by discrete distributions in general and by the binomial and the Poisson distributions in particular.
2. Find the probabilities associated with particular outcomes in discrete distributions.
3. Determine the mean and standard deviation for general discrete distributions and for the binomial and Poisson distributions.

6-1
DISCRETE RANDOM VARIABLES

As we discussed in Chapter 5, when a random experiment or trial is performed, some outcome, or event, must occur. When the trial or experiment has a quantitative characteristic, we can associate a number with each outcome. For example, suppose the quality control manager at the American Plywood Plant examines three pieces of plywood. Letting "G" stand for a good piece of plywood and "D" stand for a defective piece, the sample space is

G, G, G

G, G, D

G, D, G

D, G, G

G, D, D

D, G, D

D, D, G

D, D, D

We can let X be the number of *good* pieces of plywood in the sample of three pieces. Then X can only be 0, 1, 2, or 3, depending on how many defectives are found. Although the quality control manager knows these are the possible values of X before she samples, she would be uncertain about which would occur on any given trial. Not only that, but the value of X may vary from trial to trial. Under these conditions we say that X is a ***random variable*. Formally, a *random variable* is a variable whose numerical value is determined by the outcome of a random experiment or trial.**

In another example, if an accountant randomly examines 15 accounts, the number of inaccurate account balances can be represented by the variable X. Then X is a random variable with the following values:

$$0$$
$$1$$
$$2$$
$$\cdot$$
$$\cdot$$
$$\cdot$$
$$15$$

Two classes of random variables exist: *discrete* random variables and *continuous* random variables. **A *discrete* random variable is a random variable that can assume only distinct values.** The two previous examples illustrate discrete random variables. The pieces of defective plywood could assume only the values 0, 1, 2, or 3, and the number of incorrect account balances had to be one of the values 0, 1, 2, . . ., 15.

On the other hand, *continuous* random variables are random variables that may assume any value on a continuum. For example, time is often thought to be continuous. The time it takes a trainee to perform a job task may be any value between two points, say one minute and ten minutes.

In this chapter we discuss discrete random variables and introduce the concept of discrete probability distributions. Chapter 7 covers continuous random variables.

6-2
DISCRETE PROBABILITY DISTRIBUTIONS

The discrete probability distribution is actually an extension of the relative frequency distribution first introduced in Chapter 3. For example, the Colman Ford dealership in Salem, Oregon, each week offers specials on four specific cars as part of a sales campaign. For a period of 30 weeks the sales manager has recorded how many of the four cars were sold each week (see Table 6-1). Note that X is a discrete random variable whose value equals the number of cars sold. The possible values of X are 0, 1, 2, 3, and 4.

TABLE 6–1   Car Sales, Colman Ford Dealership

Cars Sold X	Frequency	Relative Frequency
0	12	12/30 = 0.40
1	8	8/30 = 0.27
2	4	4/30 = 0.13
3	4	4/30 = 0.13
4	2	2/30 = 0.07
	$\Sigma = 30$	$\Sigma = 1.00$

As you can see in Table 6–1, the relative frequencies for each value of X have been computed. For instance, during this 30-week period the dealership sold none of its "special" cars during 12, or 40 percent, of the weeks. During 8, or 27 percent, of the weeks, it sold one "special" car. If you recall from Chapter 5, one way to assess probability is to use the relative frequency of occurrence. That is, the probability of an outcome (or value of a random variable) occurring can be assessed by the relative frequency of that outcome.

The probability distribution for a discrete random variable shows each value of the random variable and its associated probability. Thus the Colman Ford probability distribution is

X	P(X)
0	0.40
1	0.27
2	0.13
3	0.13
4	0.07
	$\Sigma = 1.00$

Since X represents all possible sales values, the probability distribution must add to 1.0. Figure 6–1 shows this probability distribution in graphical form.

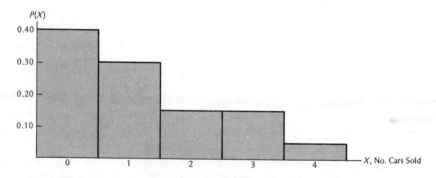

FIGURE 6–1   Probability distribution, Colman Ford dealership.

6-3

MEAN AND STANDARD DEVIATION OF A DISCRETE PROBABILITY DISTRIBUTION

A probability distribution, like a frequency distribution, can only be partially described by a graph. Often decision makers will need to calculate the distribution's _mean_ and _standard deviation_. These values measure the _central location_ and _spread_, respectively, of the probability distribution.

Mean of a Discrete Probability Distribution

The mean of a discrete probability distribution is also called the _expected value_ of the discrete random variable. The expected value is actually a **weighted average** of the random variable values where the weights are the probabilities assigned to the values. In equation form, the expected value is

$$E(X) = \Sigma XP(X) \tag{6–1}$$

where: $E(X)$ = expected value of X

X = values of the random variable

$P(X)$ = probability of each value of X

The mean of the random variable X for the Colman Ford example is found as follows:

X	$P(X)$	$XP(X)$
0	0.40	0.00
1	0.27	0.27
2	0.13	0.26
3	0.13	0.39
4	0.07	0.28
	$\Sigma = 1.00$	$\Sigma = 1.20$

$$\mu_x = E(X) = 1.2 \text{ cars}$$

Therefore, in the "long run," the average number of cars sold per week is 1.2. Again, the expected value is just a weighted average of random variable values.

Standard Deviation of a Discrete Probability Distribution

The standard deviation measures the spread, or dispersion, in a set of data. The standard deviation also measures the spread in the values of a random variable. To calculate the standard deviation for a discrete probability distribution, we use the following equation:

$$\sigma_x = \sqrt{\Sigma [X - E(X)]^2 P(X)} \tag{6–2}$$

where: X = values of the random variable

$E(X)$ = expected value of X

$P(X)$ = probability of each value of X

As you can see, **the standard deviation is a weighted average of squared differences between each value of the random variable and the expected value.** The weights are the respective probabilities. For the Colman Ford example the standard deviation is computed as follows:

X	$P(X)$	$X - E(X)$	$[X - E(X)]^2$	$[X - E(X)]^2 P(X)$
0	0.40	$0 - 1.2 = -1.2$	1.44	$(1.44)(0.40) = 0.576$
1	0.27	$1 - 1.2 = -0.2$	0.04	$(0.04)(0.27) = 0.010$
2	0.13	$2 - 1.2 = 0.8$	0.64	$(0.64)(0.13) = 0.083$
3	0.13	$3 - 1.2 = 1.8$	3.24	$(3.24)(0.13) = 0.421$
4	0.07	$4 - 1.2 = 2.8$	7.84	$(7.84)(0.07) = \underline{0.549}$
				$\Sigma = 1.639$

$$\sigma_x = \sqrt{1.639}$$
$$= 1.28$$

Thus the standard deviation of the random variable, cars sold, is 1.28 cars.

6-4
CHARACTERISTICS OF THE BINOMIAL DISTRIBUTION

Managers could face an uncountable number of discrete probability distributions such as those illustrated in the previous sections. Fortunately there are several *theoretical* discrete distributions that have extensive application in business decision making. **By *theoretical* we mean that the probability distribution is well defined and that the probabilities associated with values of the random variable can be computed from a well-established equation.** In this section we introduce the first of two such distributions presented in this chapter: the *binomial probability distribution.*

The simplest probability distribution we will consider is one that describes processes with only *two* possible outcomes. The physical events described by this type of process are widespread. For instance, a quality control system in a manufacturing plant labels each tested item either defective or acceptable. A firm bidding for a contract will either get the contract, or it won't. A marketing research firm may receive responses to a questionnaire in the form of "Yes, I will buy" or "No, I will not buy." The personnel manager in an organization is faced with a two-state process each time he offers a job: the applicant will either accept the offer or not accept.

Suppose the management of a firm that makes radio transistors feels its production process is operating correctly if 10 percent of the transistors produced are defective and 90 percent are acceptable. In a random sample of ten transistors, how often would we expect to find no defectives? Exactly one defective? Two defectives? The quality control manager may have a real reason for asking this type of question. He depends upon the sample to provide the information necessary to decide whether to let the production process continue as is or to take corrective action.

A process in which each trial or observation can assume only one of two states is called a *binomial* or *Bernoulli process.* In a true Bernoulli process, several conditions are necessary.

1. The process has only two possible outcomes: successes and failures.

2. There are n identical trials or experiments.

3. The trials or experiments are independent of each other. In a production process, this means that if one item is found defective, this fact does not influence the chances of another being found defective.

4. The process must be stationary in generating successes and failures. That is, the probability associated with a success, p, remains constant from trial to trial.

5. If p represents the probability of a success, then $(1 - p) = q$ is the probability of a failure.

6-5
DEVELOPING A BINOMIAL PROBABILITY DISTRIBUTION

We introduced you to the Winner Bakery Company in Chapter 5. Winner produces and distributes frozen bread dough to supermarkets in the western United States. Frozen bread loaves are sensitive to both the production process and the storage process. If a batch of bread dough is improperly mixed it can be ruined, and if a loaf is allowed to thaw after it has been produced it will be ruined.

Winner has completed an extensive study of its production and distribution process. The information shows that if the company is operating at *standard quality,* 10 percent (0.10) of its loaves will be defective by the time they reach the consumer's home. Assuming that the production, inventory, and distribution systems are such that the Bernoulli process applies, the following conditions are true:

1. There are only two possible outcomes when a loaf is produced: the loaf is good or the loaf is defective.*

2. Each of the loaves is made by the same production process.

3. The outcome of a loaf (defective or good) is independent of whether the preceding loaf was good or defective.

---

*Students are often confused about the definition of **success** and **failure.** A success occurs when we observe the outcome of interest. If we are looking for defective loaves, finding one is a success.

4. The probability of a defective loaf, $p = 0.10$, remains constant from loaf to loaf.

5. The probability of a good loaf, $q = (1 - p) = 0.90$, also remains constant from loaf to loaf.

Suppose the quality assurance people at Winner Bakery have devised a plan for sampling four bread loaves each week to help determine whether the company is maintaining its quality standard. Because this is a physical process, we would expect the number of defective loaves to vary from sample to sample. Table 6–2 shows the results of the first 20 weeks of sampling. Notice that the number of defectives is limited to discrete values: 0, 1, 2, 3, or 4.

TABLE 6–2 Frequency distribution of defective loaves, Winner Bakery

No. Defectives ($n = 4$)	Frequency
0	4
1	6
2	4
3	4
4	2
	$\Sigma = 20$ weeks

If we let the number of defective loaves be the random variable of interest, we can determine the probability that the random variable will have any of the discrete values. One way of finding these probabilities is to list the sample space as shown in Table 6–3. We can find the probability of zero defectives, for instance, by employing the *multiplication rule for independent events*.

$$P(0 \text{ defective}) = P(G, G, G, G)$$

where: G = loaf is good (not defective)

   G, G, G, G = four good loaves

Since

$$P(G) = 0.90$$

and we have assumed the loaves are independent, then

$$P(G, G, G, G) = P(G)P(G)P(G)P(G) = (0.9)(0.9)(0.9)(0.9) = 0.9^4$$
$$= 0.6561$$

Note that when we have more than two independent events, the joint probability is determined by multiplying each individual probability.

We can find the probability of exactly one defective loaf in a sample of $n = 4$ using both the multiplication rule for independent events and the addition rule for mutually exclusive events.

$$P(1 \text{ defective}) = P(G, G, G, D) + P(G, G, D, G) + P(G, D, G, G) + P(D, G, G, G)$$

TABLE 6–3   Sample space of defective loaves, Winner Bakery

Results	No. Defectives ($n = 4$)	No. Ways
G,G,G,G	0	1
G,G,G,D G,G,D,G G,D,G,G D,G,G,G	1	4
G,G,D,D G,D,G,D D,G,G,D G,D,D,G D,G,D,G D,D,G,G	2	6
D,D,D,G D,D,G,D D,G,D,D G,D,D,D	3	4
D,D,D,D	4	1

where: G = good loaf
            D = defective loaf

where

$$P(G, G, G, D) = P(G)P(G)P(G)P(D) = (0.9)(0.9)(0.9)(0.1)$$
$$= (0.9^3)(0.1)$$

Likewise,

$$P(G, G, D, G) = (0.9^3)(0.1)$$
$$P(G, D, G, G) = (0.9^3)(0.1)$$
$$P(D, G, G, G) = (0.9^3)(0.1)$$

Then

$$P(1 \text{ defective}) = (0.9^3)(0.1) + (0.9^3)(0.1) + (0.9^3)(0.1) + (0.9^3)(0.1)$$
$$= (4)(0.9^3)(0.1)$$
$$= 0.2916$$

Using the same method, we can find the probabilities of two, three, and four defectives.

$$P(2 \text{ defective}) = (6)(0.9^2)(0.1^2)$$
$$= 0.0486$$
$$P(3 \text{ defective}) = (4)(0.9)(0.1^3)$$
$$= 0.0036$$
$$P(4 \text{ defective}) = (0.1^4)$$
$$= 0.0001$$

TABLE 6–4 Binomial probability distribution, Winner Bakery
($n = 4$, $p = 0.1$)

No. Defectives X	Probability $P(X)$
0	$0.9^4 = 0.6561$
1	$(4)(0.9^3)(0.1) = 0.2916$
2	$(6)(0.9^2)(0.1^2) = 0.0486$
3	$(4)(0.9)(0.1^3) = 0.0036$
4	$0.1^4 = \underline{0.0001}$
	$\Sigma = 1.0000$

Table 6–4 gives the binomial probability distribution for the number of defective loaves in a random sample of four if the probability of any loaf being defective is 0.1. If you look closely at the distribution in Table 6–4, you will see that each probability can be determined by: (1) finding the probability of one way a particular event can occur and (2) multiplying this probability by the number of different ways that event can occur. For example,

$$P(2 \text{ defective}) = (6)(0.9^2)(0.1^2)$$
$$= 0.0486$$

From the sample space in Table 6–2, one way of getting two defectives in four loaves is

$$G, G, D, D$$

and the probability of this happening is

$$P(G, G, D, D) = (0.9)(0.9)(0.1)(0.1)$$
$$= (0.9^2)(0.1^2)$$

Now, how many ways can we get two defectives in a sample of four? Again, from the sample space in Table 6–3 we find there are six ways, each having exactly the same probability. Thus

$$P(2 \text{ defective}) = (6)(0.9^2)(0.1^2)$$

Therefore the key to developing the probability distribution for a binomial process is first to determine the probability of any one way the event of interest can occur and then to multiply this probability by the number of ways that event can occur.

For small sample sizes, the binomial distribution can be generated by listing the sample space and proceeding as we have just described. However, when the sample size is large, this method becomes tedious because the number of different events becomes very large. Fortunately we can develop any binomial probability distribution without actually listing the sample space. To do this, we use the **binomial formula**

$$P(X_1) = \frac{n!}{X_1!X_2!} p^{X_1}q^{X_2} \tag{6–3}$$

where: n = sample size

X_1 = number of successes (where a success is what we are looking for)

X_2 = number of failures

p = probability of a success

$q = 1 - p$ = probability of a failure

You should recognize the quantity $n!/X_1!X_2!$ as the expression for the number of *distinct permutations* discussed in Chapter 5. This expression determines the number of ways that X_1 successes can occur in a sample of size n. The expression $p^{X_1}q^{X_2}$ represents the probability of one way that X_1 successes can occur.

By applying the binomial formula to the Winner Bakery example with a sample of size $n = 8$ rather than $n = 4$, the binomial distribution shown in Table 6–5 is found.

TABLE 6–5   Binomial probability distribution, Winner Bakery $(n = 8, p = 0.1)$

No. Defectives X_1	Binomial Formula $\dfrac{n!}{X_1!X_2!}p^{X_1}q^{X_2}$		Probability $P(X_1)$
0	$\dfrac{8!}{0!8!}(0.1^0)(0.9^8)$	=	0.4305
1	$\dfrac{8!}{1!7!}(0.1^1)(0.9^7)$	=	0.3826
2	$\dfrac{8!}{2!6!}(0.1^2)(0.9^6)$	=	0.1488
3	$\dfrac{8!}{3!5!}(0.1^3)(0.9^5)$	=	0.0331
4	$\dfrac{8!}{4!4!}(0.1^4)(0.9^4)$	=	0.0046
5	$\dfrac{8!}{5!3!}(0.1^5)(0.9^3)$	=	0.0004
6	$\dfrac{8!}{6!2!}(0.1^6)(0.9^2)$	=	0.0000*
7	$\dfrac{8!}{7!1!}(0.1^7)(0.9^1)$	=	0.0000*
8	$\dfrac{8!}{8!0!}(0.1^8)(0.9^0)$	=	0.0000*
		$\Sigma =$	1.0000

*The probability is very small. Rounded to four decimal places, $P(X_1) = 0.0000$.

6-6

USING THE BINOMIAL DISTRIBUTION TABLE

Using equation (6–3) to develop the binomial distribution is not difficult, but it can be time consuming. To make binomial probabilities easier to find, you can use the binomial probability table in Appendix A. This table is constructed to give individual probabilities for different sample sizes and p values. Within the table for each specified sample size you will find columns of probabilities. Each column is headed by a probability value, p—the probability associated with a success. The column headings correspond to p values ranging from 0.01 to 0.50. At the bottom of each column are p values corresponding to probabilities of successes ranging from 0.50 to 0.99. Down both sides of the table are integer values which correspond to the number of successes. The X_1 values on the *left side* are used with p values between 0.01 and 0.50. The X_1 values on the *right side* are used for p values greater than 0.50. Note that the values on the extreme right also correspond to the number of failures, X_2, for p values between 0.01 and 0.50.

Instead of using equation (6–3), you can find the appropriate binomial probability by turning to the part of the table with the correct sample size. Then look down the column headed by the appropriate p value until you locate the probability corresponding to the desired X_1 value. For example, if $n = 5$ and $p = 0.5$, the probability of exactly three successes is seen to be 0.3125. This is the same value we find using the binomial formula

$$P(X_1 = 3) = \frac{5!}{3!2!} (0.5^3)(0.5^2)$$

$$= 0.3125$$

Using the binomial table, we can find the probabilities of selecting zero, one, two, and three defectives in a sample of ten from a production process producing 10 percent defectives. We go to the table for $n = 10$, $p = 0.10$ and find

$$P(X_1 = 0) = 0.3487$$
$$P(X_1 = 1) = 0.3874$$
$$P(X_1 = 2) = 0.1937$$
$$P(X_1 = 3) = 0.0574$$

As another example, assume you are working for an automobile manufacturer. Based upon engineering reports, 2 percent of the cars your company produces will receive a "below-standard" rating from the Environmental Protection Agency (EPA) on the pollution control devices. Thus even a correctly operating production process will have variation, and not all cars will be produced exactly as designed. If $n = 20$ cars are selected at random from the inventory in Detroit, what is the probability of each of the following?

1. Finding no below-standard cars
2. Finding 2 or 3 below-standard cars
3. Finding more than 3 below-standard cars

The answers to these questions can be found directly from the binomial table in Appendix A if we assume a binomial distribution applies.

Using the binomial table with $n = 20$ and $p = 0.02$, we find

$$P(X_1 = 0) = 0.6676$$

And

$$P(X_1 = 2 \text{ or } 3) = P(X_1 = 2) + P(X_1 = 3) = 0.0528 + 0.0065$$
$$= 0.0593$$

There are two ways to find the probability of *more than* 3 below-standard cars in a sample of 20. The first way is to add the probabilities of

$$P(X_1 = 4) + P(X_1 = 5) + P(X_1 = 6) + \ldots + P(X_1 = 20) = 0.0006$$

Or we could find the probability of selecting 3 or fewer below-standard cars and subtract this probability from 1. That is,

$$P(X_1 = 3 \text{ or fewer}) = P(X_1 = 3) + P(X_1 = 2) + P(X_1 = 1) + P(X_1 = 0)$$
$$= 0.0065 + 0.0528 + 0.2725 + 0.6676$$
$$= 0.9994$$

and

$$P(X_1 = \text{more than } 3) = 1 - 0.9994$$
$$= 0.0006$$

Suppose the EPA has been given a mandate by Congress to determine if automobiles manufactured in the United States meet pollution standards. The EPA wishes to allow no more than 2 percent of the cars produced by any manufacturer to receive a substandard rating. A western state with strict pollution control standards has decided to base its enforcement policy on the EPA 2 percent standard. Since the state enforcement agency cannot test every car sold in the state, it randomly samples 20 cars of each make and model. If it finds more than 1 car in the sample with a substandard pollution rating, the manufacturer receives a stiff fine and is ordered to recall all cars of that make and model sold in the state. The state's rule says that if more than 1 substandard car is found, the conclusion is that the automobile company is exceeding the 2 percent limit.

Of course the automobile manufacturers are concerned about the chances of being unjustly accused. That is, a company may in fact be producing 2 percent or fewer cars with substandard pollution control devices but the state could find more than 1 such car in its sample of 20. The binomial probability table can be used to find the probability of this happening. The company wants

$$P(X_1 > 1) = 1 - P(X_1 \le 1)$$

Go to the binomial table with $n = 20$ and $p = 0.02$ and find

$$P(X_1 > 1) = 1 - [P(0) + P(1)] = 1 - (0.6676 + 0.2725) = 1 - 0.9401$$
$$= 0.0599$$

This means that, under the proposed sampling plan, there is just under a 6 percent chance the auto makers will be unjustly accused of making too many substandard cars (by pollution standards).

Because of the high potential costs of being unjustly accused, the man-ufacturers would likely challenge the sampling plan. They would probably argue that more cars be sampled and that the cutoff point for recalling be altered to reduce the probability of being unjustly accused.

6-7
MEAN AND STANDARD DEVIATION OF THE BINOMIAL DISTRIBUTION

Recall that the mean of a discrete probability distribution is referred to as its *expected value*. The expected value of a discrete random variable X is found using equation (6–1).

$$E(X) = \Sigma XP(X)$$

where: $E(X)$ = expected value of X

X = value of the random variable

$P(X)$ = probability of each value of X

Suppose we wish to find the expected number (value) of defective frozen bread loaves in a sample of $n = 4$ with a 0.10 probability of a single loaf being defective. This probability distribution for Winner Bakery was given in Table 6–4. We find the expected value as follows:

No. Defectives X	P(X)	XP(X)
0	0.6561	0
1	0.2916	0.2916
2	0.0486	0.0972
3	0.0036	0.0108
4	0.0001	0.0004
		$\Sigma = 0.4000$

Thus

$$\mu_x = E(X) = \Sigma XP(X)$$

$$= 0.4000$$

Therefore, if the probability of a single loaf being defective is 0.10, the average number of defectives found in repeated samples of size four is 0.4000. Of course for any single sample we could not find 0.4000 defective since defectives must occur in discrete values, in this case 0, 1, 2, 3, or 4.

If the sample size were increased to $n = 8$, we would use Table 6–5 and calculate the mean for this probability distribution as follows:

No. Defectives X	P(X)	XP(X)
0	0.4305	0
1	0.3826	0.3826
2	0.1488	0.2976
3	0.0331	0.0993
4	0.0046	0.0184
5	0.0004	0.0021
6	0.0000	0.0000
7	0.0000	0.0000
8	0.0000	0.0000
		$\Sigma = 0.8000$

$$\mu_x = E(X) = \Sigma XP(X)$$
$$= 0.8000$$

Therefore, with repeated samples of eight loaves we would expect an average of 0.8000 defective.

Mean of the Binomial Distribution

The mean, or expected value, of any discrete random variable can be found by

$$E(X) = \Sigma XP(X)$$

However, if we are working with a binomial distribution, the mean can be found much more easily. If the discrete distribution is binomial, we use

$$\mu_x = np \qquad \text{(6–4)}$$

where: n = sample size
p = probability of a success

Using the Winner Bakery example with a sample of four selected from a population with 10 percent defective loaves, the distribution mean is

$$\mu_x = np = (4)(0.10)$$
$$= 0.40$$

Notice that this is the same value we found earlier using the expected value equation. When $n = 8$ and $p = 0.10$, the mean of the binomial distribution is

$$\mu_x = np = (8)(0.10)$$
$$= 0.80$$

Again, the mean agrees with the value found previously.
For the binomial distribution,

$$\mu_x = E(X) = \Sigma XP(X) = np$$

Standard Deviation of the Binomial Distribution

Recall that to calculate the standard deviation for a discrete probability distribution, we use equation (6–2).

$$\sigma_x = \sqrt{\Sigma(X - \mu_x)^2 P(X)}$$

where: X = value of the random variable

σ_x = standard deviation of X

μ_x = mean = $E(X)$

$P(X)$ = probability of X

Continuing with the Winner Bakery example for a sample of $n = 4$ and $p = 0.10$, we find the standard deviation for the distribution of defective loaves as follows:

$$\mu_x = E(X) = \Sigma XP(X) = np = (4)(0.10)$$
$$= 0.4000$$

X	$P(X)$	$X - \mu_x$	$(X - \mu_x)^2$	$(X - \mu_x)^2 P(X)$
0	0.6561	−0.4000	0.16	0.1049
1	0.2916	0.6000	0.36	0.1049
2	0.0486	1.6000	2.56	0.1245
3	0.0036	2.6000	6.76	0.0244
4	0.0001	3.6000	12.96	0.0013
				$\Sigma = 0.3600$

$$\sigma_x^2 = 0.3600$$
$$\sigma_x = \sqrt{0.3600}$$
$$= 0.60$$

Thus the mean and standard deviation for this distribution are 0.4000 and 0.60, respectively.

Using the distribution for Winner Bakery where $n = 8$ and $p = 0.10$, we find a different standard deviation.

$$\mu_x = E(X) = \Sigma XP(X) = np = (8)(0.10)$$
$$= 0.8000$$

X	$P(X)$	$X - \mu_x$	$(X - \mu_x)^2$	$(X - \mu_x)^2 P(X)$
0	0.4305	−0.8	0.64	0.275
1	0.3826	0.2	0.04	0.015
2	0.1488	1.2	1.44	0.214
3	0.0331	2.2	4.84	0.161
4	0.0046	3.2	10.24	0.048
5	0.0004	4.2	17.64	0.007
6	0.0000	5.2	27.04	0.000
7	0.0000	6.2	38.44	0.000
8	0.0000	7.2	51.84	0.000
				$\Sigma = 0.720$

$$\sigma_x^2 = 0.720$$
$$\sigma_x = \sqrt{0.720}$$
$$= 0.848$$

Thus the mean and standard deviation for this distribution are 0.8000 and 0.848, respectively.

If a discrete probability distribution meets the binomial distribution conditions, the standard deviation is defined as

$$\sigma_x = \sqrt{npq} \qquad\qquad (6\text{–}5)$$

where: n = sample size

p = probability of a success

q = probability of a failure

For the Winner Bakery example with $n = 4$ and $p = 0.10$, the standard deviation is

$$\sigma_x = \sqrt{npq} = \sqrt{(4)(0.10)(0.90)} = \sqrt{0.3600}$$
$$= 0.60$$

The standard deviation when $n = 8$ and $p = 0.10$ is

$$\sigma_x = \sqrt{(8)(0.10)(0.90)} = \sqrt{0.7200}$$
$$= 0.848$$

We see that these values agree with those found using

$$\sigma_x = \sqrt{\Sigma(X - \mu_x)^2 P(X)}$$

6-8
SOME COMMENTS ABOUT THE BINOMIAL DISTRIBUTION

The binomial distribution has many applications. We have already presented an elementary quality control example. In later chapters on decision making under uncertainty, we will use this distribution for applications in the other functional areas of business.

At this point, several comments about the binomial distribution are worth making. If p, the probability of a success, is 0.5, the binomial distribution is *symmetrical* regardless of the sample size. This is illustrated in Figure 6–2, which shows frequency histograms for samples of $n = 5$, $n = 10$, and $n = 15$. Notice that all three distributions are centered at the expected value, μ_x.

When the value of p differs from 0.5 in either direction, the binomial distribution is *skewed*. This was the case for the Winner Bakery examples with $n = 4$ or $n = 8$ and $p = 0.10$. The skewness will be most pronounced when n is small and p approaches zero or 1.0. However, the binomial distribution approaches symmetry as n increases. The frequency histograms shown in Figure 6–3 bear this out. This fact will be used in Chapter 7, where we will show that the ***normal distri-***

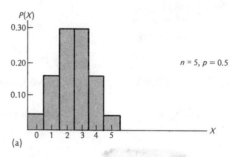

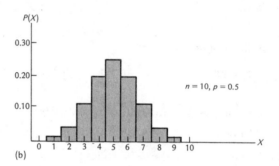

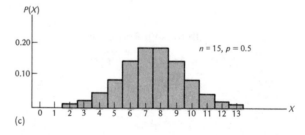

FIGURE 6–2 The binomial distribution with varying sample sizes.

bution (a symmetrical continuous distribution) can be used to approximate probabilities from a binomial distribution.

6-9

THE POISSON PROBABILITY DISTRIBUTION

To use the binomial distribution, we must be able to count the number of successes and the number of failures. Whereas in many applications you may be able to count the number of successes, you often cannot count the number of failures. For example, suppose a company builds freeways in Vermont. The company could count the number of chuckholes that develop per mile (here a chuckhole is a success since it's what we are looking for), but how could it count the number of non-chuckholes? Or

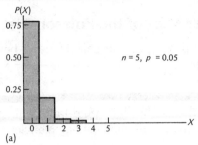

(a)

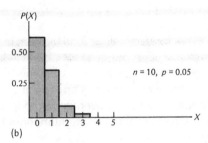

(b)

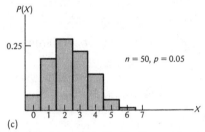

(c)

FIGURE 6–3   The binomial distribution with varying sample sizes.
($p = 0.05$)

what about the emergency medical service in Los Angeles? They could easily count the number of emergencies their units respond to in one week, but how could they determine how many calls they did not receive? Obviously, in these cases the number of possible outcomes (successes + failures) is difficult, if not impossible, to determine. If the total number of possible outcomes cannot be determined, the binomial distribution cannot be applied as a decision-making aid.

Fortunately, the *Poisson distribution* can be applied in these situations without knowing the total possible outcomes. To apply the Poisson distribution, we need only know the *average* number of successes for a given segment. For instance, we could use the Poisson distribution for our freeway construction company if we were able to determine the average number of chuckholes per mile. Likewise, the emergency medical service in Los Angeles could apply the Poisson distribution if they could find the average number of responses per week. Of course, before the Poisson distribution can be applied, certain conditions must be satisfied.

Characteristics of the Poisson Distribution

A physical situation must possess certain characteristics before it can be described
by the Poisson distribution:

1. The physical events must be considered *rare events*. Considering
   all the chuckholes that could form, only a few actually do form.
   Considering all the medical emergencies that might result, only a
   few do occur.
2. The physical events must be *random* and *independent* of each
   other. That is, an event occurring must not be predictable, nor can
   it influence the chances of another event occurring.

**The Poisson distribution is described by a single parameter, λ (lambda),
which is the average occurrence per segment. The value of λ depends on the situation
being described.** For instance, λ could be the average number of machine breakdowns
per month or the average number of customers arriving at a checkout stand in a ten-
minute period. Lambda could also be the average number of emergency responses
for the emergency medical service or the average number of chuckholes in a section
of freeway.

Once λ has been determined, we can calculate the average occurrence
for any multiple segment, t. This is λt. Lambda and t must be in compatible units. If
we have $\lambda = 20$ arrivals per hour, we cannot multiply this by a time period measured
in minutes. That is, if we have

$$\lambda = 20/h \quad \text{and} \quad t = 30 \text{ min}$$

we must set

$$t = \text{one-half h or } \frac{1}{2}$$

Then

$$\lambda t = 10$$

The average number of occurrences is not necessarily the number we
will see if we observe the process one time. We might expect an average of 20 people
to arrive at a checkout stand in any given hour, but we don't expect to find that exact
number arriving every hour. The actual arrivals will form a distribution with an
expected value, or mean, equal to λt. So, for the Poisson distribution,

$$\mu_x = \lambda t$$

The Poisson distribution is a discrete distribution. This means that if
we are dealing with airplane arrivals at the San Francisco International Airport, in
any given hour only 0, 1, 2, . . . airplanes can arrive; 1.5 airplanes cannot land.

The Poisson Distribution Formula

Once λt has been specified, the probability for any discrete value in the Poisson
distribution can be found using the following formula:

$$P(X_1) = \frac{(\lambda t)^{X_1} e^{-\lambda t}}{X_1!} \qquad \text{(6–6)}$$

where: X_1 = number of successes in segment t

λt = expected number of successes in segment t

e = base of the natural number system (2.71828)

Consider the case where the predetermined average number of airplane arrivals at the San Francisco International Airport is 60 per hour. Thus $\lambda = 60$. We can find the probability of exactly 7 planes arriving in a ten-minute period as follows:

$$t = \text{one 10-min period} = \frac{1}{6}\text{h}$$

$$\lambda = 60/\text{h}$$

$$\lambda t = (60)\left(\frac{1}{6}\right)$$

$$= 10$$

Then

$$P(X_1 = 7) = \frac{(10)^7 e^{-10}}{7!}$$

$$= 0.0901$$

The probability of 5, 6, or 7 airplanes arriving in the ten-minute period is calculated as follows:

$$P(X_1 = 5) = \frac{(10)^5 e^{-10}}{5!}$$

$$= 0.0378$$

$$P(X_1 = 6) = \frac{(10)^6 e^{-10}}{6!}$$

$$= 0.0631$$

$$P(X_1 = 7) = 0.0901$$

Then

$$P(5 \leq X_1 \leq 7) = P(5) + P(6) + P(7) = 0.0378 + 0.0631 + 0.0901$$

$$= 0.1910$$

Applying the Poisson Probability Distribution

Consider the problem facing a Boise Cascade Corporation lumber mill manager. Logs arrive by truck and are scaled (measured to determine the number of board feet) before they are dumped into the log pond. Figure 6–4 illustrates the basic flow. The mill manager must determine how many scale stations to have open during various times of the day. If he has too many stations open, the scalers will have excessive

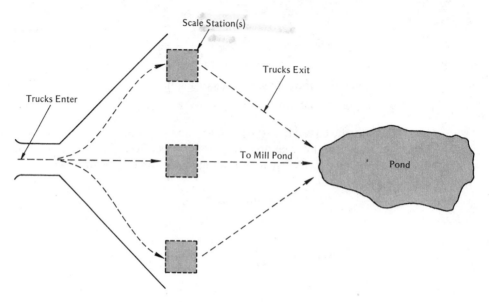

FIGURE 6–4   Truck flow, Boise Cascade mill.

idle time and the cost of scaling will be unnecessarily high. On the other hand, if too few scale stations are open, some log trucks will have to wait.

The manager has studied the truck arrival patterns and has determined that during the first open hour (7:00–8:00 A.M.), the trucks randomly arrive at six per hour. He knows that if eight or fewer arrive during an hour, two scale stations can keep up with the work. If more than eight trucks arrive, three scale stations are required. The manager recognizes that the distribution of log truck arrivals during this hour can be represented by a Poisson distribution with $\lambda = 6.0$ per hour. The mill manager can use the Poisson distribution to determine the probability that three scale stations will be needed.

As was the case for the binomial distribution, Poisson probability tables have been developed for different values of λt and X_1. A Poisson probability table is located in Appendix B. To use this table, turn to the page with the appropriate λt value and find the probability corresponding to the desired number of successes, X_1. For example, the Boise Cascade mill manager would select $\lambda t = 6.0$ and find the probability that more than 8 trucks will arrive during the 7:00–8:00 A.M. time slot. She could add the probabilities for $X_1 = 9, 10, 11, \ldots$, as follows:

$$P(9 \text{ or more trucks}) = P(9) + P(10) + P(11) + \ldots + P(17)$$
$$= 0.0688 + 0.0413 + 0.0225 + \ldots + 0.0001$$
$$= 0.1526$$

Notice that the probability of more than 17 trucks is so small that it has been rounded to zero.

The probability of needing three scale stations is 0.1526, and the probability that only two stations will be needed is 0.8474 (1 − 0.1526). The manager must now make the decision. She must balance the cost of an additional scale station against the potential dissatisfaction of both log truck drivers and the companies they represent.

Suppose this Boise Cascade manager studies the log truck arrivals in the hour between 11:00 A.M. and 12:00 noon and finds the average number of trucks arriving is 3.5. Assuming truck arrivals can be represented by the Poisson distribution, what is the probability of 8 or fewer trucks arriving? We can determine this probability using the column under $\lambda t = 3.5$ in the Poisson probability table. We sum the individual probabilities from $X_1 = 0$ to $X_1 = 8$.

$$\begin{aligned} P(8 \text{ or fewer trucks}) &= P(0) + P(1) + P(2) + \ldots + P(8) \\ &= 0.0302 + 0.1057 + 0.1850 + \ldots + 0.0169 \\ &= 0.9902 \end{aligned}$$

This indicates there is a 99 percent chance of needing only two scale stations from 11:00 A.M. to 12:00 noon.

6-10
VARIANCE AND STANDARD DEVIATION OF THE POISSON DISTRIBUTION

The mean of the Poisson distribution is

$$\mu_x = \lambda t$$

Thus, for the Boise Cascade lumber mill example, the mean number of arrivals was 6.0 trucks between 7:00 A.M. and 8:00 A.M. Since the Poisson distribution is a discrete distribution, we can calculate the mean, or expected value, using

$$\mu_x = E[X] = \Sigma X P(X)$$

and the variance using

$$\sigma_x^2 = \Sigma (X - \mu_x)^2 P(X)$$

The appropriate calculations are shown in Table 6–6.

Note that both the mean and the variance equal 6.0. This is no accident. **The variance for the Poisson distribution always equals the mean.**

$$\sigma_x^2 = \lambda t \qquad\qquad (6\text{–}7)$$

The standard deviation of the Poisson distribution is the square root of the mean.

$$\sigma_x = \sqrt{\lambda t} \qquad\qquad (6\text{–}8)$$

Thus, for those processes that can be assumed to follow a Poisson distribution, variance can be controlled directly by controlling the mean. If, as noted in Chapter 4, the variance can be considered a measure of uncertainty, then for those applications where a Poisson distribution applies, the uncertainty can be controlled by controlling the mean. Of course the mean must be within the decision maker's control. Oftentimes this control can be exercised by somehow scheduling occurrences.

TABLE 6–6 Poisson distribution, log truck arrivals
($\lambda t = 6.0$)

X (1)	$P(X)$ (2)	$XP(X)$ (3)	$X - \mu_x$ (4)	$(X - \mu_x)^2$ (5)	$(X - \mu_x)^2 P(X)$ (6)
0	0.0025	0	− 6	36	0.0900
1	0.0149	0.0149	− 5	25	0.3725
2	0.0446	0.0892	− 4	16	0.7136
3	0.0892	0.2676	− 3	9	0.8028
4	0.1339	0.5356	− 2	4	0.5356
5	0.1606	0.8030	− 1	1	0.1606
6	0.1606	0.9636	0	0	0
7	0.1337	0.9359	1	1	0.1337
8	0.1033	0.8264	2	4	0.4132
9	0.0688	0.6192	3	9	0.6192
10	0.0413	0.4130	4	16	0.6608
11	0.0225	0.2475	5	25	0.5625
12	0.0113	0.1356	6	36	0.4068
13	0.0052	0.0676	7	49	0.2548
14	0.0022	0.0308	8	64	0.1408
15	0.0009	0.0135	9	81	0.0729
16	0.0003	0.0045	10	100	0.0300
17	0.0001	0.0017	11	121	0.0121
		$\Sigma = 5.9696 \approx 6.0^*$			$\Sigma = 5.9900 \approx 6.0^*$

$$\mu_x = \Sigma XP(X) = 6.0 = \lambda t$$
$$\sigma_x^2 = \Sigma(X - \mu_x)^2 P(X) = 6.0 = \lambda t$$

*Difference due to rounding.

6-11
CONCLUSIONS

In this chapter we have introduced discrete random variables and showed how a probability distribution is developed for a discrete random variable. Additionally, we have showed how to compute the mean and standard deviation for any discrete distribution.

As we indicated in this chapter, there is virtually no end to the possible discrete probability distributions decision makers might use. However the binomial and Poisson distributions represent two of the most commonly used theoretical distributions. In spite of some seemingly strong restrictions on the situations these distributions represent, they can be used in a surprising number of managerial applications.

In this chapter we have discussed some concepts connected with discrete distributions from a managerial perspective. In the chapter glossary we define these concepts in more formal terms. Test your understanding of the material in this chapter by comparing our managerial discussion with the more formal definitions.

The important statistical equations presented in the chapter are summarized following the glossary.

chapter glossary

Bernoulli process   A sequence of random experiments such that the outcome of each trial has one of two complementary outcomes (success or failure), each trial is independent of the preceding trials, and the probability of a success remains constant from trial to trial.

binomial distribution   A probability distribution that gives the probability of X_1 successes in n trials of a Bernoulli process. The distribution is specified by

$$P(X_1) = \frac{n!}{X_1!X_2!} p^{X_1}q^{X_2}$$

discrete variable   A variable that can assume only integer or specific fractional values.

expected value   A measure of location for a probability distribution. The expected value of an experiment is the weighted average of the values the outcomes of the experiments may assume. The weighting factors are the probabilities associated with each outcome.

$$E[X] = \Sigma XP(X)$$

Poisson probability distribution   A probability distribution that gives the probability of X_1 occurrences from a Poisson process when the average number of occurrences is λt. The distribution is specified by

$$P(X_1) = \frac{(\lambda t)^{X_1}e^{-\lambda t}}{X_1!}$$

Poisson process   A process describing the random occurrences of independent rare events.

random variable   A function or rule that assigns numerical values to the possible outcomes of a trial or experiment.

sample space   A set representing the universe of all possible outcomes from a statistical experiment.

chapter formulas

Binomial formula

$$P(X_1) = \frac{n!}{X_1!X_2!} p^{X_1}q^{X_2}$$

Expected value of a discrete probability distribution

$$E[X] = \Sigma XP(X)$$

Standard deviation of a discrete probability distribution

$$\sigma_X = \sqrt{\Sigma(X - \mu_X)^2 P(X)}$$

Mean of a binomial distribution

$$\mu_X = np$$

Standard deviation of a binomial distribution

$$\sigma_X = \sqrt{npq}$$

Poisson formula

$$P(X_1) = \frac{(\lambda t)^{X_1} e^{-\lambda t}}{X_1!}$$

Mean of a Poisson distribution

$$\mu_X = \lambda t$$

Variance of a Poisson distribution

$$\sigma_X^2 = \lambda t$$

solved problems

1. The Ace Electronics Corporation produces electronic calculators and markets them on a national basis. These calculators are sent to the division warehouses in lots of 1,000. You as the warehouse manager will not accept a lot of calculators if you think more than 5 percent are defective. You sample $n = 20$ calculators and test to see how many are defective and how many are good. Using the binomial distribution, find the following if $n = 20$ and $p = 0.05$:
   a. The probability of exactly one defective.
   b. The probability of finding more than one defective.
   c. The probability of finding from one to three defectives.
   d. The variance and standard deviation.
   e. Suppose you have just found three defectives in the sample. Would your recommendation be to keep the lot of 1,000 calculators or return them?
   f. How many defectives would you expect to find in a sample of $n = 20$ if there is actually 0.05 defective in a lot?

Solutions:

a. Using the binomial equation,

$$P(1 \text{ defective}) = \frac{20!}{1!19!}(0.05)^1(0.95)^{19}$$
$$= 0.3774$$

From the binomial table in Appendix A, with $n = 20$ and $p = 0.05$,

$$P(1) = 0.3774$$

b. $P(2 \text{ or more defective}) = P(2) + P(3) + P(4) + \ldots + P(20)$

or, by using the complement,

$$P(2 \text{ or more defective}) = 1 - [P(1) + P(0)]$$

From the binomial table,

$$P(2 \text{ or more defective}) = 1 - (0.3774 + 0.3585)$$
$$= 0.2641$$

c. $$P(1 \leq X_1 \leq 3) = P(1) + P(2) + P(3)$$

From the binomial table,

$$P(1 \leq X_1 \leq 3) = 0.3774 + 0.1887 + 0.0596$$
$$= 0.6257$$

d. $$\sigma_X^2 = npq = (20)(0.05)(0.95)$$
$$= 0.95$$
$$\sigma_X = \sqrt{0.95}$$
$$= 0.9747$$

e. The recommendation should depend upon the probability of observing three or more defective calculators if, in fact, only 5 percent are defective. If this probability is very small, we would conclude that the lot must actually contain more than 5 percent defectives and should be returned. If the probability is not small, the recommendation should be to keep the calculators. Thus, with $n = 20$ and $p = 0.05$,

$$P(3 \text{ or more defective}) = 1 - P(2 \text{ or fewer defective})$$
$$= 1 - P(2) + P(1) + P(0)$$
$$= 1 - (0.1887 + 0.3774 + 0.3585) = 1 - 0.9246$$
$$= 0.0754$$

Thus 0.0754 is the probability of finding three or more defectives in a sample of $n = 20$ when the population is supposed to contain, at most, 5 percent defective calculators. We will let you make the final decision. (This type of question and the complete method for analysis will be discussed in Chapters 12 and 13, which cover inference and hypothesis testing.)

f. The number of expected defectives is

$$E[X] = \Sigma X P(X)$$

For a binomial distribution,

$$\mu_X = E[X] = np$$

Therefore

$$\mu_X = (20)(0.05)$$
$$= 1.0$$

2. In a study for a local hospital you have found that the average number of arrivals at the emergency room between 6 P.M. and 9 P.M. on Friday night is five patients. Using a Poisson distribution, answer the following:

   a. What is the probability distribution describing emergency room arrivals?
   b. Is this distribution skewed? If so, in what direction? Are all Poisson distributions skewed?
   c. What is the probability that only one or two patients will arrive during the three-hour period on any Friday night?
   d. What is the probability exactly five patients will arrive?
   e. What is the probability more than eight patients will arrive?
   f. What are the mean and standard deviation of this probability distribution?
   g. Would you expect the same distribution to apply between 6 A.M. and 9 A.M. on Friday morning? Why or why not?

Solutions:

   a. The probability distribution relates the discrete arrivals to the probability of each arrival occurring. The following distribution is taken from the table of Poisson probabilities in Appendix B.

X_1	$P(X_1)$	X_1	$P(X_1)$
0	0.0067	8	0.0653
1	0.0337	9	0.0363
2	0.0842	10	0.0181
3	0.1404	11	0.0082
4	0.1755	12	0.0034
5	0.1755	13	0.0013
6	0.1462	14	0.0005
7	0.1044	15	0.0002

   b. Yes, the distribution is skewed to the right. Looking at the probability columns in Appendix B, we see that all Poisson distributions are skewed. However, as λt becomes very large, the Poisson distribution approaches a symmetrical distribution.
   c. From the Poisson table,

$$P(1) + P(2) = 0.0337 + 0.0842$$
$$= 0.1179$$

   d. Again, from the Poisson distribution table,

$$P(5) = 0.1755$$

   e.
$$P(\text{more than 8 arrive}) = P(9) + P(10) + P(11) + P(12)$$
$$+ P(13) + P(14) + P(15)$$
$$= 0.0363 + 0.0181 + 0.0082 + 0.0034$$
$$+ 0.0013 + 0.0005 + 0.0002$$
$$= 0.068$$

   f. The mean is given as five per three hours. The standard deviation, by definition, is the square root of the mean; that is,

$$\sigma_x = \sqrt{\mu_x} = \sqrt{5}$$
$$= 2.236$$

g. Since the underlying conditions that give rise to the Friday night distribution would not remain constant, we would expect a different distribution to apply during the morning hours.

3. Assume you are responsible for operating a large fleet of taxicabs. The average number of cabs under repair during any day is four. The Poisson distribution applies.

   a. What is the probability that during any day fewer than three cabs will be under repair?

   b. What is the probability that eight or more cabs will be under repair?

   c. How many spare cabs must you have if you want the probability of not being able to assign a driver a spare cab to be a maximum of 5 percent? Assume it takes an entire day to repair a cab.

Solutions:

a. Again, from the Poisson table, in the column under a λt of 4.00,

$$P(\text{fewer than 3}) = P(0) + P(1) + P(2) = 0.0183 + 0.0733 + 0.1465$$
$$= 0.2381$$

b. $P(8 \text{ or more}) = P(8) + P(9) + P(10) + P(11) + P(12) + P(13) + P(14)$

$$= 0.0298 + 0.0132 + 0.0053 + 0.0019 + 0.0006$$
$$+ 0.0002 + 0.0001$$
$$= 0.0511$$

c. To answer this part, start at the bottom of the $\lambda t = 4.00$ probability column and work toward the top in the following manner. If you have 13 spare cabs, you will be short only when 14 or more break down. The probability of this happening is 0.0001. If you have 12 spare cabs, you will be short if 13 or more cabs break down. The probability of this happening is

$$P(13) + P(14) = 0.0001 + 0.0002$$
$$= 0.0003$$

If you have 7 spare cabs, the probability of being short is the probability of 8 or more cabs being under repair. This is

$$P(8 \text{ or more}) = 0.0511 \qquad (\text{See part b})$$

Since having 7 cabs gives a probability of being short of *more* than 5 percent, you will have 8 spare cabs with a probability of being caught short equal to

$$P(9 \text{ or more}) = 0.0213$$

====problems

Problems marked with an asterisk (*) require extra effort.

1. Identify three business-related examples other than those discussed in this chapter where the binomial distribution applies. Indicate the conditions that must

be satisfied for the binomial distribution to be used and show how each example meets these conditions.

2. What is meant by a discrete probability distribution?

3. Identify three business-related examples other than those discussed in this chapter where the Poisson distribution applies.

4. Discuss the basic differences between a binomial process and a Poisson process.

5. In what ways are the Poisson and binomial distributions alike?

6. The Ames Collection Agency collects unpaid bills for stores in Kansas City. The manager feels there is a 0.4 probability of collecting any bill. If the collection process follows a binomial distribution, how much should Ames expect to make if it tries to collect 20 accounts and receives $100 per collected account?

7. Referring to problem 6, what is the probability that Ames will collect fewer than four accounts?

8. Referring to problem 6, what is the probability that Ames will collect exactly four accounts?

9. The Diamond Cutter Company cuts diamonds for jewelry. Suppose the probability of ruining a diamond is 0.10 and Diamond will cut four diamonds this week. Develop the probability distribution for the number of ruined diamonds during the week assuming the binomial distribution will apply.

10. Referring to problem 9, find the expected value of the probability distribution.

11. Referring to problem 10, find the distribution variance and standard deviation.

12. The bookkeeper for the Gatewater Company makes an average of four errors per journal page. Assuming errors are Poisson-distributed, what is the probability that a given page will have exactly two errors?

13. Referring to problem 12, what is the probability that there will be fewer than two errors on any two pages?

14. Suppose the number of industrial accidents for workers of the McLaughlin Company is Poisson-distributed with $\lambda = 3$ accidents per week. Develop the probability distribution for the random variable, number of accidents per week.

15. Referring to problem 14, find the expected value of the probability distribution.

16. Referring to problem 15, find the variance and standard deviation.

17. Each week Champion Bakery receives a truckload of yeast sticks. The company that sells the yeast claims the probability is 0.05 that any stick will be defective. Suppose Champion samples 100 sticks and finds 15 defective. What should Champion conclude about the 0.05 probability? Why?

18. Referring to problem 17, how would your response change if Champion found only five defectives? Why?

*19. Suppose under normal conditions a pressure-valve assembly process produces 5 percent defectives which can be represented by a binomial process. The quality control department has discovered 12 defectives in a sample of 100. Based upon the probability of these results, what would you conclude about the pressure-valve assembly process?

20. Ten workers at an engineering company are working late on a contract proposal. They decide to phone ten orders for food to Sam's Sandwiches. From past history, they realize Sam's will make a mistake on 25 percent of all phone orders. Unfortunately Sam's is the only local food operation that will deliver.
   a. What is the probability Sam's will get eight or more orders correct?
   b. What is the probability Sam's will get more than three orders wrong?
   c. What is the probability Sam's will get two or three orders wrong?
   d. One workeer offers an open wager of 20 to 1 against Sam's getting all ten orders right. Would you accept this bet? Why?

*21. A venture-capital group has decided to invest in eight projects based on your recommendation. You assess the probability of success for each project to be 15 percent. You are also presently considering buying a house. You realize that if one or fewer projects are successful, you will be fired and won't be able to afford a house. However, if four or more are successful, you will receive a bonus and could afford a much larger house than you are considering. List your options, assign probabilities, and determine your course of action.

*22. The Never-Fade Mill produces standard bolts of denim cloth. If a bolt has fewer than three defects, it is sold to a company making clothes for small boutiques. Intermediate bolts are sold to a company making clothes for a large retail chain. Bolts with more than eight defects are sold to a discount operation. If defects are described by a Poisson distribution with an average of six defects per bolt, out of 1,000 bolts, how many go to each customer?

*23. In problem 22, what are the expected earnings per bolt if Never-Fade earns the following profits?

$$\text{Boutiques} = \$1.50/\text{bolt}$$
$$\text{Retail chain} = \$0.90/\text{bolt}$$
$$\text{Discount} = \$0.55/\text{bolt}$$

*24. Hash Star, the production manager, is considering changing the manufacturing process in problems 22 and 23 so that the average number of defects per bolt is reduced to four. However the process will cost more and reduce the profit for each bolt by $0.30 for each customer. Should Hash make the change?

=cases

6A   MELLOW YELLOW CAB COMPANY

The Mellow Yellow Cab Company has just received an exclusive contract to supply taxi service for a large western city. This contract amounts to a monopoly for Mellow

Yellow, and was not a popular move on the part of the city council. The council, however, had received many complaints about service when the city allowed many taxi companies and feels it will now have greater control over both equipment and drivers.

When Mellow Yellow currently has 176 cars. However the company is required, as part of the contract, to have 325 taxis on the streets during peak rush hours. Max Winter, Mellow Yellow's managing director, feels 325 cars are too many to have on the street; however he knows the city will monitor Mellow Yellow's performance closely. In particular, he is sure the council will check driver records to see that enough taxis are on the streets during the peak demand hours.

When purchasing additional cars, Max also has to consider replacement, or backup, cars. Although he disputes the frequent claim that his drivers drive without regard for life or limb, he feels they sometimes drive without regard for the company's cars. For instance, a long time driver, Crash Billings, has just driven into the garage with his left front door in the back seat. Frequent minor accidents and mechanical failure seem to be a fact of life in the taxi business.

Max Winter has been managing director of Mellow Yellow for seven years. His policy has been to buy good-quality two- or three-year-old used cars to use as taxis. Mellow Yellow maintains its own repair shop. However the mechanics do only minor body and mechanical repairs. If a taxi requires more than a day to repair, the car will be replaced, not repaired.

Up to now, Max has not kept comprehensive daily records on the number of cars out of commission. He also does not know exactly how many cars he has had available on a daily basis. However he feels that over the last two years, Mellow Yellow has had an average of 150–175 cars available, and an average of 8 cars per day in the repair shop.

Max knows he will have to buy additional cars now that his company has the exclusive contract. However he doesn't really know how many to buy, or the ramifications of buying different numbers.

6B   GREAT PLAINS OIL COMPANY

Margaret Clemonts, operations vice-president of Great Plains Oil, is putting together a proposal for the board of directors. Great Plains has several refineries in the mid-central United States. It has for years relied on crude oil from its fields in Texas and the Gulf area. However these fields are very mature and have declining outputs. Margaret has recently been buying foreign crude delivered to eastern ports. In addition, she has been considering trying to buy some extra Alaska crude from the West Coast. Her main problem up until now has been the extra cost of transporting Alaska oil to Great Plains refineries.

Margaret has recently learned about an unused natural gas pipeline that would drastically reduce the cost of bringing oil from the West Coast to Great Plains refineries. Great Plains can buy the pipeline for a reasonable price, and in fact could use the excess distribution capability to add to its cash flow by selling excess crude. The main drawback to the proposal is that the pipeline was built for gas and, therefore, may not be strong enough for the additional forces generated by transmitting crude oil. The pipe material is strong enough; however the welds may not be.

The vast majority of the pipe is buried approximately 10 feet underground. The pipeline is made of 50-foot sections, and since it is approximately 1,000 miles long, there are some 110,000 separate welds. If the welds were solidly made, they will be strong enough to handle the crude oil. If the welds were not carefully made, they will need to be redone.

An industrial x-ray company has recently developed a small, portable machine that will travel through a pipe and x-ray each weld. However this machine is very expensive to rent and requires the pipe to be cut periodically to extend the machine's power cables. Margaret does not want to use the machine unless she is sure she will recommend buying the pipeline. Once the line is bought, any faulty welds will have to be located and redone.

As part of the negotiation procedure, the pipeline owners have agreed to let Great Plains dig up sections of the pipe and x-ray a sample of welds. Margaret's industrial engineers have estimated that if 1 percent or fewer of the welds need to be redone, the pipeline will be a good investment. If 2 percent or more need to be redone, the pipeline should not be bought. Unfortunately, taking the sample x-rays will not be easy, and Margaret estimates that each x-ray will cost $500.

Margaret recognizes the problems of sampling, but feels a representative random sample can be taken. She would be willing to spend up to $175,000 to determine whether the pipeline should be bought, but would not want to spend this entire amount if the pipeline should obviously not be bought. She is presently devising her sampling plan.

6C   DEARBORN CORPORATION, PART 1

The Dearborn Corporation of Palo Alto, California, makes stereo speakers and sells these speakers wholesale to several electronics companies. These companies install the Dearborn speakers in special cabinets and market them under a variety of brand names.

Dearborn's price quotations are based upon the assumption that at least 95 percent of all speakers it manufactures are acceptable. The potential exists, according to Dearborn, that up to 5 percent of its speakers contain some defect.

The Dearborn production control manager has been getting a lot of pressure lately because of customer complaints. Some customers claim the defective rate has climbed above 5 percent. In fact, one customer recently found 4 defective speakers in a shipment of 50 speakers. In his letter of complaint, the customer implied that something must be wrong at the Dearborn plant if the defective rate has increased to 8 percent. He demanded an adjustment on his purchase price for this shipment (Dearborn would gladly do this since it is part of the sales contract) and that the stated price on future speaker purchases reflect the higher defective rate.

The production control manager wonders how he should respond to this customer.

6D   DEARBORN CORPORATION, PART 2

A few days after receiving the first customer's letter of complaint (see case 6C), the Dearborn production control manager received a phone call from a second customer.

This customer indicated that her quality control people had discovered 6 defective speakers in a shipment of 50 speakers. This customer indicated that Dearborn's defective rate must be higher than the stated 5 percent and that the production manager should do something about it right away.

      The production manager sat down to prepare a letter to this customer but was having trouble assessing the issue.

references

DUNCAN, ACHESON J. *Quality Control and Industrial Statistics.* 3rd ed. Homewood, Ill.: Irwin, 1965.

HARNETT, DONALD L. *Introduction to Statistical Methods.* 2nd ed. Reading, Mass.: Addison-Wesley, 1975.

HAYES, GLENN E. and HARRY G. ROMIG. *Modern Quality Control.* Encinco, Calif.: Bruce, 1977.

HOGG, R. V. and A. T. CRAIG. *Introduction to Mathematical Statistics.* 2nd ed. New York: Macmillan, 1965.

7

Continuous Probability Distributions

In Chapter 6 we introduced discrete random variables and discussed discrete probability distributions as they apply to the decision-making process. We also discussed two useful theoretical discrete probability distributions: the binomial and the Poisson. For discrete distributions, the variable of interest can take on only specified values. For example, the number of defective tires produced by the day shift at General Tire Company can have only integer values (0, 1, 2, . . .).

In many business applications the variable of interest is not restricted to integer values. For example, checkout times through a supermarket checkout stand can take on any value between zero and some large number. And the load weight carried by a freight truck can take on values between zero and some large number. Variables that are measured in units of time, weight, volume, or distance are often assumed to be **continuous** variables. Because of measuring limitations, many persons argue that there is no such thing as a truly continuous variable. These individuals consider all variables discrete even though they can take on decimal values. In this text **we define a *continuous random variable* as a variable that can assume a large number of values between any two points.**

155

Because many business applications involve continuous or quasi-continuous variables, decision makers need to become acquainted with continuous probability distributions and learn how to use them in decision making.

chapter objectives

In this chapter we shall discuss the characteristics of continuous probability distributions. We will emphasize the normal distribution and illustrate how to apply this distribution in a decision-making environment.

student objectives

After studying the material in this chapter, you should be able to:

1. Discuss the important properties of the normal probability distribution.
2. Recognize when the normal distribution might apply in a decision-making process.
3. Calculate probabilities using the normal distribution table.

7-1
CONTINUOUS RANDOM VARIABLES

We have defined a discrete random variable as a variable that can have only a specific finite set of values. In many cases these discrete values are limited to integers. As we showed in Chapter 6, many decision situations can be analyzed using discrete random variables and their associated probability distributions.

In many instances decision makers will be faced with random variables that can take on a seemingly unlimited number of values. Such a random variable is a **continuous** random variable. Figure 7–1 illustrates a continuous random variable and contrasts it with a discrete random variable. Here, the actual waist measurement is a continuous variable because it can assume any value along the scale. However,

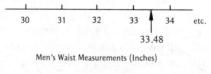

Men's Waist Measurements (Inches)

(a) Continuous Random Variable

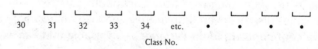

Class No.

(b) Discrete Random Variable

FIGURE 7–1 Classification of random variables.

when the scale is divided into intervals, a discrete value can be assigned to each waist size depending on where the size falls. In business decision making, many variables are defined as continuous, including:

- Time measurements
- Interest rates
- Financial ratios
- Income levels
- Weight measurements
- Distance measures
- Volume measures

In Chapter 6 we discussed the problems facing the Boise Cascade Lumber mill manager when deciding how many log-scaling stations to open during a certain hour. The number of trucks arriving in that hour is a *discrete* random variable, but the time between arrivals (the interarrival time) is a *continuous* random variable. The time it takes to scale a truckload of logs is also a continuous random variable, and the amount of board feet of lumber on a truck still another.

In general, the value of a continuous random variable is found by *measuring*, whereas the value of a discrete random variable is determined by counting.

7-2
CONTINUOUS PROBABILITY DISTRIBUTIONS

In contrast to discrete case situations where the appropriate probability distribution can be represented by the areas of rectangles (histogram), the probability distribution of a continuous random variable is represented by a ***probability density function.*** Figure 7–2 illustates a probability density function and shows its relationship to discrete probability distributions. Note that as the class width in the discrete examples becomes narrower, the top of the histogram approaches a smooth curve. This smooth curve represents the probability density function for a continuous variable.

Remember, discrete probability distributions have two characteristics. First, the rectangle representing each discrete value has an area corresponding to the probability of that value occurring. Second, the areas (probabilities) of all the rectangles must sum to 1.0. These two characteristics generally also apply to probability density functions. First of all, **the total area (probability) under the density function curve must equal 1.0.** In addition, **the probability that the variable will have a value between any two points on the continuous scale equals the area under the curve between these two points.** However, the probability that the variable will have any specific value cannot be determined because that probability would correspond to the area directly above a point. Because the area above a point is a line, and because a line has no width, it has no area.

Since the probability of a single point on a continuous scale is zero, when dealing with continuous random variables we never consider the probability

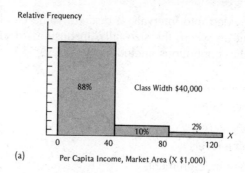

(a)

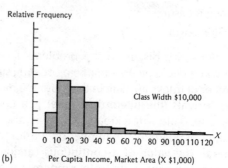

(b)

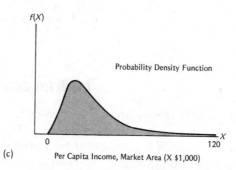

(c)

FIGURE 7–2 Comparison of a probability density function and discrete probability distributions.

of a single value occurring. Rather, we consider the probability of a range of values occurring by finding the area under the density function for this range.

    There are many continuous density functions which may face the decision maker. However the normal distribution is by far the most important. In the next section we introduce this continuous probability distribution.

7-3

CHARACTERISTICS OF THE NORMAL DISTRIBUTION

The most important continuous distribution in statistical decision making is the **normal distribution**. The normal distribution has the following properties:

1. It is *unimodal*. That is, the normal distribution peaks at a single value.

2. It is *symmetrical*, with the mean, median, and mode equal. Symmetry assures us that 50 percent of the area under the curve lies below the center and 50 percent lies above the center.

3. It approaches the horizontal axis on either side of the mean toward plus and minus infinity ($\pm\infty$). In more formal terms, the normal distribution is *asymptotic* to the X-axis.

Figure 7–3 illustrates a typical normal distribution and highlights these three characteristics.

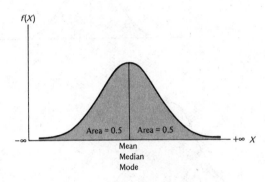

Characteristics:

1. Unimodal
2. Symmetric
3. Asymptotic to the X-axis in both directions

FIGURE 7–3 Characteristics of a normal distribution.

Although all normal distributions have the shape shown in Figure 7–3, their central locations and spreads can vary greatly depending upon the situation being considered. The horizontal-axis scale is determined by the process being represented. It may be pounds, inches, dollars, or any other physical attribute with a continuous or quasi-continuous measurement. Figure 7–4 shows several normal distributions with different centers and different spreads. Note that the area under each normal "curve" equals 1.0.

The normal distribution is defined by two parameters: μ_X, the population mean, and σ_X, the population standard deviation. Given these two values, the normal distribution is described by the following probability density function:

$$f(X) = \frac{1}{\sigma_X \sqrt{2\pi}} e^{-(X-\mu_X)^2/2\sigma_X^2} \tag{7–1}$$

where: X = any value of the continuous random variable

σ_X = population standard deviation

e = base of the natural log ≈ 2.7182

μ_X = population mean

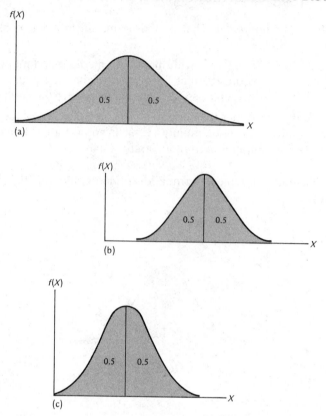

FIGURE 7–4   Normal distributions with different locations and spreads.

Equation (7–1) will determine the height of the normal distribution curve for each possible value of the random variable, X. If we were to substitute values for μ_x and σ_x along with many values for X, the plot of $f(X)$ would be a curve similar to those shown in Figures 7–3 and 7–4.

Determining $f(X)$ requires only basic algebra. However, rarely in business applications will decision makers need to plot the normal distribution.

7-4

FINDING PROBABILITIES FROM A STANDARD NORMAL DISTRIBUTION

As we indicated earlier, the probability of any particular value of a discrete random variable can be represented by an area in a relative frequency histogram. For example, Figure 7–5 shows a histogram for a binomial distribution with $n = 6$ and $p = 0.5$. The probabilities (areas) sum to 1.0, and the area above each value on the horizontal axis represents the probability of that value occurring. For example, the probability of exactly two successes in $n = 6$ trials is 0.2344. We also stated in the preceding section that the probability of observing values between any two

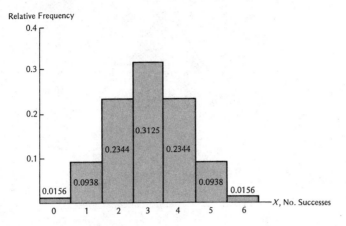

X	P(X)
0	0.0156
1	0.0938
2	0.2344
3	0.3125
4	0.2344
5	0.0938
6	0.0156
	$\Sigma = 1.0001^*$

*Difference due to rounding.

FIGURE 7-5   Binomial distribution.
($n = 6, p = 0.5$)

points in a normal distribution is equal to the area under the normal curve between those two points. Figure 7–6 shows several examples.

Integral calculus is used to find the area under a normal distribution curve. However we can avoid using calculus by transforming all normal distributions to fit the **standard normal distribution. The principle behind the standard normal distribution is that all normal distributions can be converted to a common normal distribution with a common mean and standard deviation.** This conversion is done by **rescaling** the normal distribution axis from its true units (time, weight, dollars, and so forth) to a standard measure referred to as a **Z value.** Thus any value of the normally distributed continuous random variable can be represented by a unique Z value. The Z value is determined by the following formula:

$$Z = \frac{X - \mu_X}{\sigma_X} \tag{7-2}$$

where: Z = scaled value

X = any point on the horizontal axis

μ_X = mean of the normal distribution

σ_X = standard deviation of the normal distribution

Using an overly simplified example, suppose The Boeing Company, which designs and manufactures commercial jet airplanes, has determined that its

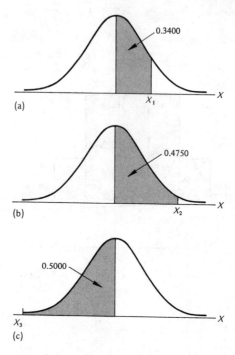

FIGURE 7–6   Areas and probabilities for a normal distribution.

newest model can take off in an average of 4,000 feet with a standard deviation of 500 feet. Further, the distribution of takeoff distances can be represented by a normal distribution. Figure 7–7 shows this normal distribution for the Boeing jet. Suppose we select a point $X = 4,000$. (Note, 4,000 is also μ_x, the mean takeoff distance.) We can find the Z value for this point using equation (7–2).

$$Z = \frac{X - \mu_x}{\sigma_x} = \frac{4,000 - 4,000}{500}$$

$$= 0$$

Thus the Z value corresponding to the population mean, μ_x, is zero. This will be the case for all applications. Suppose we now select 4,500 feet as X. The Z value for this point is

$$Z = \frac{X - \mu_x}{\sigma_x} = \frac{4,500 - 4,000}{500} = \frac{500}{500}$$

$$= 1.00$$

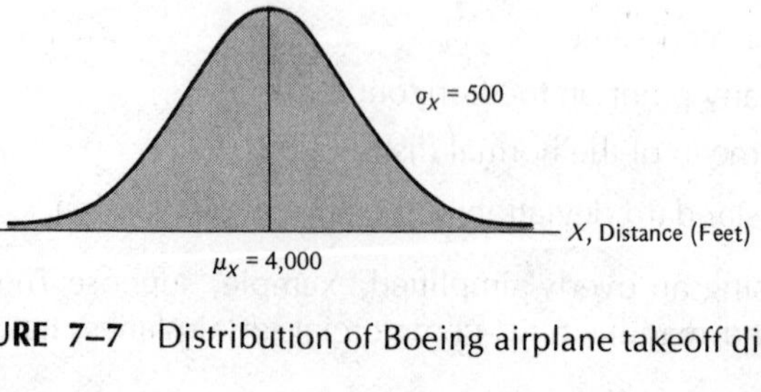

FIGURE 7–7   Distribution of Boeing airplane takeoff distances.

Verify for yourself that $X = 3,500$ feet corresponds to a Z value of -1.00.

If we examine equation (7–2) and the previous examples carefully, we see that **the Z value actually represents the number of standard deviations a point, X, is away from the population mean, μ_x.** In this Boeing jet example, σ_x, the standard deviation, is 500 feet. Therefore a takeoff distance of 4,500 feet is exactly 1.00 standard deviation from $\mu_x = 4,000$ feet. Likewise, a takeoff distance of 5,000 feet is 2.00 standard deviations from the mean.

Suppose The Boeing Company can make an engineering change that will reduce the standard deviation in takeoff distances to 250 feet without changing the mean takeoff distance. Now a takeoff distance of 4,500 feet is 2.00 standard deviations from μ_x, and 5,000 feet is 4.00 standard deviations from the mean of 4,000 feet.

Thus scaling an "actual" normal distribution to the standard normal distribution requires that we determine the Z values (number of standard deviations from μ_x) for each point on the horizontal axis. As we have shown, this is not a difficult process. Therefore, no matter what situation we are dealing with, if the random variable is normally distributed, we can use the standard normal distribution. This offers some specific advantages when it comes to finding probabilities for ranges of values under the normal curve because a table of standard normal distribution probabilities has been developed. This table, which appears in Appendix C, lists a series of Z values and the corresponding probability that a random variable value will fall between that Z and the mean of the distribution. Don't forget that the Z value is simply the number of standard deviations a point, X, is from the mean, μ_x. Therefore, if you want the probability a value will lie within 1.00 standard deviation of the population mean, use the normal table as follows:

1. Go down the left-hand column of the table to $Z = 1.0$.
2. Go across the top row of the table to 0.00 for the second decimal place in $Z = 1.00$.
3. Find the value where the row and column found in steps 1 and 2 intersect.

This value is 0.3413 and is the probability a value in a normal distribution will lie within 1.00 standard deviation *above* the population mean. Since the normal distribution is symmetrical, the probability a value will lie within 1.00 standard deviation *below* the population mean is also 0.3413. Therefore the probability that a value will lie within 1.00 standard deviation of the population mean in *either* direction is $0.3413 + 0.3413 = 0.6826$.

The same procedure is used to find the probability a value will be within 2.00 standard deviations of the mean and within 3.00 standard deviations of the mean. Figure 7–8 illustrates these probabilities. Make sure you can find the probability corresponding to any Z value in the standard normal table.

Suppose we return to the Boeing jet airplane example. Recall that the mean takeoff distance was 4,000 feet and the original standard deviation was 500 feet. The plane may be sold to developing nations where short runways (generally 5,000 feet) are common. Therefore the Boeing engineers are interested in the prob-

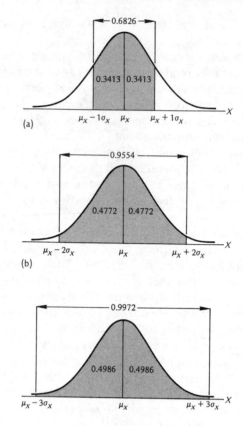

FIGURE 7–8 Standard normal distribution probabilities.

ability the plane will require more than 5,000 feet to take off. This probability corresponds to the area under the normal curve to the right of 5,000 feet.

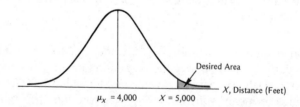

To use the standard normal table, we need to convert 5,000 feet to a Z value. This is equivalent to determining the number of standard deviations 5,000 feet is from the mean.

$$Z = \frac{X - \mu_X}{\sigma_X} = \frac{5,000 - 4,000}{500}$$

$$= 2.00$$

The area corresponding to $Z = 2.00$ is 0.4772, shown under the normal curve.

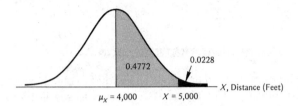

Since the normal curve is symmetrical and half the total area lies on each side of the mean, we can find $P(X > 5,000$ feet$)$ by subtracting 0.4772 from 0.5000.

$$0.5000 - 0.4772 = 0.0228$$

Thus the probability that the new jet will require at least a 5,000-foot runway is 0.0228.

As we stressed many times in the preceding chapters, variation in a process cannot be completely avoided. Because of ever-present variation in the business world, decision makers face uncertainty. The best we can hope for is that this variation, and thus uncertainty, can be reduced. For example, we just calculated the probability that a Boeing jet would not be able to take off from a runway 5,000 feet long as 0.0228. This means that, in the long run, slightly over 2 percent of the takeoff attempts would fail. The person who must decide whether to allow a jet to attempt a takeoff knows there is a high probability the plane will take off successfully, but yet is uncertain about what will actually occur on any single attempt. If the Boeing engineers could find a way to reduce the standard deviation of takeoff distances to 250 feet, the uncertainty would be reduced because the probability of an unsuccessful takeoff would be reduced. We can calculate this probability as follows:

$$Z = \frac{X - \mu_x}{\sigma_x} = \frac{5,000 - 4,000}{250}$$

$$= 4.00$$

The following normal curve shows the area corresponding to the probability of a takeoff taking more than 5,000 feet. Because $Z = 4.00$ exceeds the largest Z value in the standard normal table, the area between $Z = 4.00$ and μ_x is approximately 0.5000. Consequently the area to the right of $Z = 4.00$ is approximately zero. (Actually there is some very small area to the right of $Z = 4.00$, but for practical purposes we assume it to be zero.)

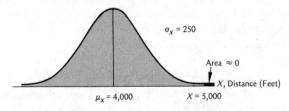

The probability of an unsuccessful takeoff from a runway 5,000 feet long is very small (essentially zero), and thus the uncertainty has been reduced. There is a greater chance of a successful takeoff because the airplane is more consistent in takeoff requirements.

7-5

OTHER APPLICATIONS OF THE NORMAL DISTRIBUTION

Suppose a consumer protection agency has decided to check packaging practices in the canned-food industry. This agency is checking to see if the canners are filling cans with an amount of food about equal to that stated on the can's label. For instance, if a can of corn is supposed to weigh 16 ounces, the agency requires it to weigh at least 15.2 ounces. If it does not, the canner is subject to heavy fines. Although variation is part of the filling process, each canner obviously is interested in the chances of a can weighing less than 15.2 ounces. One company has set its automatic fillers so that the *average* fill is 16 ounces for a 16-ounce can. The standard deviation in weights produced by the filling machines is 0.3 ounce.

To calculate the probability of a can weighing less than 15.2 ounces, we find the appropriate area under the normal curve as follows:

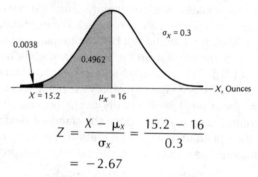

$$Z = \frac{X - \mu_x}{\sigma_x} = \frac{15.2 - 16}{0.3}$$

$$= -2.67$$

The negative 2.67 indicates that the value falls to the left of the mean. The probability corresponding to $Z = -2.67$ is 0.4962. This represents the probability a can will weigh between 15.2 and 16 ounces. To find the probability a can will weigh less than 15.2 ounces, subtract 0.4962 from 0.5000, which gives 0.0038. This means the canning company has less than 4 chances in 1,000 of selling a can weighing less than the acceptable limit.

The normal distribution table is not used exclusively to find probabilities. For instance, suppose the scores on a company's employment exam are normally distributed with a mean of 50 points and a standard deviation of 10 points. The personnel manager wants a minimum passing score such that only 15 percent of those who take the test pass. What should this minimum score be to insure that only the top 15 percent of the applicants pass? We can solve this problem using the normal distribution table in the following manner:

> *Step 1.* Rather than using the normal table to find the probability associated with a particular Z value, we find the Z value associated with the known probability. In this example, the probability of interest is 0.3500 (0.5000 − 0.1500). We look in the body of the normal table for the probability closest to 0.3500. This value is 0.3508, and the Z value corresponding to a probability of 0.3508 is 1.04.

Step 2. We use the Z value determined in step 1 and solve for the minimum passing score, X.

$$Z = \frac{X - \mu_X}{\sigma_X}$$

$$1.04 = \frac{X - 50}{10}$$

$$X = (10)(1.04) + 50$$

$$= 60.4$$

Thus the minimum passing score that will allow at most 15 percent of the applicants to pass the employment exam is 60.4 points. The following normal distribution illustrates these results:

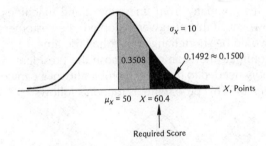

We can use the normal distribution in still another situation. We may know an X value and the standard deviation, but not μ_X. For example, suppose the production manager at the canning plant is faced with a problem. Because of the way the filling machines have been designed, the standard deviation of ounces per fill is fixed at 0.3 ounce, but the average ounces per fill can be adjusted. Under fear of reprisal by the consumer agency, the president of the canning company has ordered the production manager to set the filling machine so that there is a maximum 0.05 chance of a 16-ounce can weighing less than 15.2 ounces. To determine what the average fill should be, we perform the following analysis:

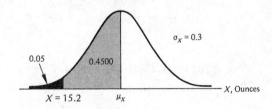

The area between 15.2 and μ_X is 0.4500. The Z value that corresponds to a probability of 0.4500 is approximately -1.64. Then, to solve for μ_X, we use

$$Z = \frac{X - \mu_X}{\sigma_X}$$

And substituting

$$Z = -1.64$$
$$X = 15.2 \text{ oz}$$
$$\sigma_X = 0.3 \text{ oz}$$

gives us

$$-1.64 = \frac{15.2 - \mu_X}{0.3}$$
$$\mu_X = 15.692 \text{ oz}$$

Therefore, to have a probability of 5 percent or less of filling a can with less than 15.2 ounces, the average fill should be 15.692 ounces. The president is faced with an ethical question here. So far, he has expressed total concern for getting caught with too little food in a can (that is, under 15.2 ounces in a 16-ounce can). To reasonably protect his company from fines, the filling machines should be set at 15.692 ounces. However, if this plan were adopted, the customer who expected 16 ounces in the can would be shortchanged over half the time.

To continue the example, suppose the president wants to know the average filling quantity needed to reduce the probability of a can containing less than 16 ounces to 10 percent. This probability is found as follows:

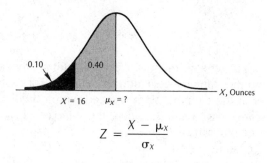

$$Z = \frac{X - \mu_X}{\sigma_X}$$

where

$$Z = -1.28$$
$$X = 16 \text{ oz}$$
$$\sigma_X = 0.3 \text{ oz}$$

Therefore

$$\mu_X = (1.28)(0.3) + 16$$
$$= 16.38 \text{ oz}$$

While the president is concerned about not filling the cans with enough food and not giving the customer the stated 16 ounces, he is also concerned about putting in more food than the customer is paying for. This last problem causes lost profits for the canning company. The president wants to know the average filling quantity such that no more than 15 percent of the 16-ounce cans actually weigh more than 16.2 ounces. To determine the setting to satisfy this requirement, we perform the following analysis:

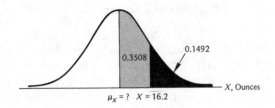

$$Z = \frac{X - \mu_x}{\sigma_x}$$

where

$Z = 1.04$ (1.04 corresponds to a probability of 0.3508)

$X = 16.2$ oz

$\sigma_x = 0.3$ oz

Now, solving for μ_x,

$$1.04 = \frac{16.2 - \mu_x}{0.3}$$

$$\mu_x = 15.888 \text{ oz}$$

Therefore, to satisfy the president's requirement of a maximum 15 percent of the cans having more than 16.2 ounces, the production manager will have to adjust the filling machine to fill at an average of 15.888 ounces.

Thus, with three different criteria, three different machine settings have been determined. The president must weigh the potential economic and social costs before deciding the fill setting.

7-6
NORMAL APPROXIMATION TO THE BINOMIAL DISTRIBUTION

The binomial distribution discussed in Chapter 6 is not easy to work with when the sample size is larger than those found in the binomial table, or when the p values are not given in the table. In these cases, we would have to apply the binomial formula, and it can be cumbersome. Fortunately, for large samples, an **approximation** can be used.

As a rule of thumb, **the normal distribution generally provides good approximations for binomial problems if the sample size is large enough to meet the following two conditions:**

$$np \geq 5$$

$$nq \geq 5$$

where: n = **sample size**

             p = **probability of a success**

             $q = 1 - p$ = **probability of a failure**

The best approximations occur when p approaches 0.5 and the sample size becomes large, because the binomial distribution approaches symmetry as p approaches 0.5. If $np \geq 5$ and $nq \geq 5$, the normal distribution can be used if we recall that for the binomial distribution, $\mu_x = np$ and $\sigma_x = \sqrt{npq}$. We can substitute these values into the Z formula and use the normal distribution table, as the following example illustrates.

    A large retail store has discovered that 5 percent of its sales receipts contain some form of human error. Out of the next 1,000 sales made, the credit manager wants to know the probability that more than 60 errors will be made.

    Since

$$np = (1,000)(0.05)$$
$$= 50 \geq 5$$

and

$$nq = (1,000)(0.95)$$
$$= 950 \geq 5$$

we will use the normal distribution to approximate the binomial. The first step is to determine the mean and standard deviation.

$$\mu_x = np = (1,000)(0.05)$$
$$= 50$$
$$\sigma_x = \sqrt{npq} = \sqrt{(1,000)(0.05)(0.95)}$$
$$= 6.89$$

Because the binomial distribution is a discrete distribution, we must make a slight modification when using the normal approximation. Specifically, we must treat 60 errors as the unit interval "59.5 to 60.5." Since we cannot find the probability of a point in the normal distribution, the probability of exactly 60 errors will correspond to the area between 59.5 and 60.5 errors. Thus the probability of *more* than 60 errors is found as follows:

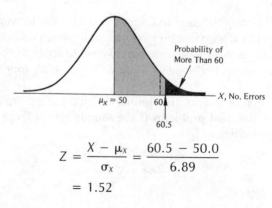

$$Z = \frac{X - \mu_x}{\sigma_x} = \frac{60.5 - 50.0}{6.89}$$
$$= 1.52$$

From the normal distribution table, the area between 60.5 and 50 ($Z = 1.52$) is 0.4357. Therefore the approximate probability of more than 60 incorrect sales slips is $0.5000 - 0.4357 = 0.0643$.

7-7
OTHER CONTINUOUS DISTRIBUTIONS

In subsequent chapters we will introduce other continuous probability distributions. Among these will be the *t* **distribution,** the **chi-square distribution,** and the *F* **distribution.** These additional distributions play important roles in statistical decision making. The basic concept that the area under a continuous curve is equivalent to probability is true for all continuous distributions.

7-8
CONCLUSIONS

In this chapter we have introduced continuous probability distributions. We have also shown that the normal distribution, with its special properties, is used extensively in statistical decision making. We discussed in some detail the standard normal distribution and showed how it can be adapted to any normal distribution application. We also showed that the normal distribution can be used to approximate the binomial distribution under certain circumstances.

The important statistical terms introduced in this chapter are summarized in the chapter glossary.

The main statistical formulas presented in this chapter are listed following the glossary.

chapter glossary

continuous probability distribution   The probability distribution of a variable that can assume an infinitely large number of values. The probability of any single value is theoretically zero.

normal distribution   A continuous distribution that is symmetrical (mean, median, and mode are all equal) and in theory has an infinite range. It is represented by

$$f(X) = \frac{1}{\sigma_x \sqrt{2\pi}}\, e^{-(X-\mu_X)^2/2\sigma_X^2}$$

standard normal distribution   A normal distribution with a mean equal to zero and a standard deviation equal to 1.0. All normal distributions can be standardized by forming a standard normal distribution of Z values.

$$Z = \frac{X - \mu_x}{\sigma_x}$$

where: Z = number of standard deviations X is from μ_x

chapter formulas

Normal distribution

$$f(X) = \frac{1}{\sigma_X \sqrt{2\pi}} e^{-(X-\mu_X)^2/2\sigma_X^2}$$

Standard Z value

$$Z = \frac{X - \mu_X}{\sigma_X}$$

solved problems

1. The length of steel I beams made by the Smokers City Steel Company is normally distributed with $\mu_X = 25.1$ feet and $\sigma_X = 0.25$ foot.
   a. What is the probability a steel beam will be less than 24.8 feet long?
   b. What is the probability a steel beam will be more than 25.25 feet long?
   c. What is the probability a steel beam will be between 24.9 and 25.7 feet long?
   d. What is the probability a steel beam will be between 24.6 and 24.9 feet long?
   e. For a particular applicaion, any beam less than 25 feet long must be scrapped. What is the probability a beam will have to be scrapped?

Solutions:

To answer all these questions, we will use the following procedure:

1. Draw a picture to show what area we are interested in.
2. Find the desired Z value.
3. Use the normal distribution table to find the desired area under the normal distribution curve (the needed probability).

a.

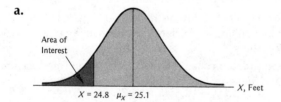

Area of Interest

$X = 24.8 \quad \mu_X = 25.1$ X, Feet

$$Z = \frac{X - \mu_X}{\sigma_X} = \frac{24.8 - 25.1}{0.25} = \frac{-0.3}{0.25}$$

$$= -1.20$$

The negative sign only means we are on the left half of the distribution.
From the normal distribution table, the area associated with a Z of -1.20 is 0.3849. However this is the area between 24.8 and 25.1 feet. We want the area less than 24.8 feet. We proceed as follows:

1. The total area under the curve (probability the beam will have some length) is 1.0.
2. The area under half the curve is 0.5.
3. The area we want is $0.5 - 0.3849 = 0.1151$.

b.

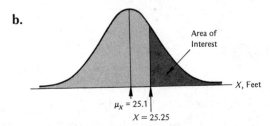

Area of Interest

X, Feet

$\mu_X = 25.1$

$X = 25.25$

$$Z = \frac{X - \mu_X}{\sigma_X} = \frac{25.25 - 25.1}{0.25} = \frac{0.15}{0.25}$$

$$= 0.60$$

The area between 25.1 and 25.25 from the table is 0.2257. The area greater than 25.25 is

$$0.5 - 0.2257 = 0.2743$$

c.

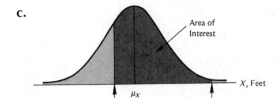

Area of Interest

X, Feet

μ_X

$X_1 = 24.9$ $X_2 = 25.7$

We do this problem in three steps:

Step 1. Find the area between 25.1 and 25.7 feet.
Step 2. Find the area between 24.9 and 25.1 feet.
Step 3. Add the probabilities found in steps 1 and 2.

Therefore

Step 1. $$Z = \frac{25.7 - 25.1}{0.25} = \frac{0.6}{0.25}$$

$$= 2.40$$

From the table, area $= 0.4918$.

Step 2. $$Z = \frac{24.9 - 25.1}{0.25} = \frac{-0.2}{0.25}$$

$$= -0.80$$

From the table, area $= 0.2881$.

Step 3. Total area $= 0.4918 + 0.2881 = 0.7799$.

d.

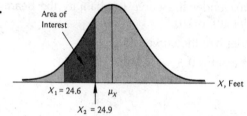

We also do this problem in three steps:

> *Step 1.* Find the area between 25.1 and 24.6.
>
> *Step 2.* Find the area between 25.1 and 24.9.
>
> *Step 3.* Subtract the area in step 2 from the area in step 1.

Therefore

Step 1.
$$Z = \frac{24.6 - 25.1}{0.25}$$
$$= -2.0$$

From the table, area = 0.4772.

Step 2.
$$Z = \frac{24.9 - 25.1}{0.25}$$
$$= -0.80$$

From the table, area = 0.2881.

Step 3. Desired area = 0.4772 − 0.2881 = 0.1891.

e.

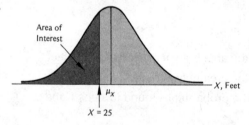

$$Z = \frac{25 - 25.1}{0.25} = \frac{-0.1}{0.25}$$
$$= -0.40$$

From the table, area = 0.1554. The probability we want is

$$0.5 - 0.1554 = 0.3446$$

Thus slightly more than 34 percent of the beams will have to be scrapped because they will not meet the required length.

---

2. The average absentee rate in large economics lecture sections is historically 15 percent. In a section of 200 students:

a. What is the probability that on a given day 160 or more students will attend class?

b. What is the probability that 180 or fewer students will attend?

c. What is the probability that between 165 and 185 students will attend?

Solutions:

Since

$$np = (200)(0.15) = 30 \geq 5$$
$$n(1 - p) = (200)(0.85) = 170 \geq 5$$

we can use the normal approximation to the binomial. For all parts of this problem,

$$\mu_X = np = (0.15)(200)$$
$$= 30$$
$$\sigma_X = \sqrt{np(1 - p)} = \sqrt{(200)(0.15)(0.85)}$$
$$= 5.05$$

a. This part asks for the probability of 40 or fewer absent. We must use $X = 40.5$ because we are using a continuous distribution to approximate a discrete distribution. The picture is as follows:

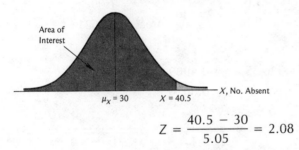

Area of Interest

$\mu_X = 30$ $X = 40.5$

X, No. Absent

$$Z = \frac{40.5 - 30}{5.05} = 2.08$$

From the table, $Z = 2.08$ gives an area of 0.4812. We have

$$P(X \leq 40.5) = P(30 \text{ to } 40.5) + P(\text{less than } 30)$$

but since $P(\text{less than } 30) = 0.5$,

$$P(X \leq 40.5) = 0.4812 + 0.5$$
$$= 0.9812$$

b. We want the probability 20 or more will be absent.

Area of Interest

$X = 19.5$ $\mu_X = 30$

X, No. Absent

$$Z = \frac{19.5 - 30}{5.05}$$
$$= -2.08$$

From the table,

$$P(19.5 \text{ to } 30) = 0.4812$$

so, therefore,

$$P(X \geq 19.5) = P(19.5 \text{ to } 30) + P(\text{more than } 30) = 0.4812 + 0.5$$
$$= 0.9812$$

c. We want the probability of from 15 to 35 absent.

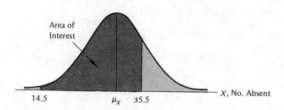

$$P(14.5 \text{ to } 35.5) = P(14.5 \text{ to } 30) + P(30 \text{ to } 35.5)$$

$$Z = \frac{14.5 - 30}{5.05} \qquad\qquad Z = \frac{35.5 - 30}{5.05}$$

$$= -3.07 \qquad\qquad = 1.09$$

$$\text{Area} = 0.4989 \qquad\qquad \text{Area} = 0.3621$$

So

$$P(14.5 \text{ to } 35.5) = 0.4989 + 0.3621$$
$$= 0.8610$$

problems

Problems marked with an asterisk (*) require extra effort.

1. Discuss three situations different from those presented in the chapter where you would use the normal distribution in decision making. How well do you think this distribution will approximate the true physical situations you describe?

2. Assuming that the distribution of family incomes in a west Boston neighborhood is normally distributed with a mean of $20,000 and a standard deviation of $4,000, find:
   a. The probability a family has an income under $12,000
   b. The probability a family's income is over $19,000
   c. The probability a family's income is between $18,000 and $26,000
   d. The probability of a family's income exceeding $40,000

3. Supppose the distribution of personal daily water usage in California is normally distributed with a mean of 18 gallons and a variance of 36 gallons.
   a. What percentage of the population uses more than 18 gallons?
   b. What percentage of the population uses between 10 and 20 gallons?
   c. What is the probability of finding a person who uses less than ten gallons?

4. Because of a perpetual water shortage in California, the governor wants to give a tax rebate to the 20 percent of the population who use the least amount of water. Referring to problem 3, what should the governor use as the maximum water limit for a person to qualify for a tax rebate?

5. Referring to problems 3 and 4, suppose the governor's proposed tax rebate causes a shift in the average water use from 18 gallons to 14 gallons per person per day but no change in the variance. What limit should be set on water use if 20 percent of the population are to receive a tax rebate?

6. Cattle are often fattened in a feedlot before being shipped to a slaughterhouse. Suppose the weight gain per steer at a feedlot averages 1.5 pounds per day with a standard deviation of 0.25 pound. Assume a normal distribution.
   a. What is the probability a steer will gain over 2 pounds on a given day?
   b. What is the probability a steer will gain between 1 and 2 pounds in any given day?
   c. What is the probability of selecting two steers that both gain less than 1.5 pounds assuming the two are independent?

7. Referring to problem 6, compute each probability under the assumption that the standard deviation is reduced to 0.20 pound. Discuss why the probabilities differ from those in problem 6.

8. Suppose light bulbs are packed in boxes of 1,000 bulbs. Although the packers try their best, 10 percent of the bulbs get broken before they reach the retail stores.
   a. Using the normal approximation to the binomial, find the probability that between 110 and 130 bulbs will be broken in a box.
   b. Using the normal approximation, find the probability that more than 130 bulbs will be broken.
   c. Suppose you have just found a box with 200 broken bulbs. Would you consider this unusual? Why or why not?

9. A market-research firm was hired to determine the percentage of people in a market area who would purchase *Playboy* magazine if a door-to-door sales campaign were undertaken. The firm stated that 40 percent would buy if contacted at home. Suppose the publisher has tried the sales campaign at 300 randomly selected homes.
   a. If the market research was accurate, what is the probability that fewer than 100 individuals will buy? Use the normal approximation to the binomial.
   b. Suppose the publisher actually sells *Playboy* to 70 people out of the 300 contacted. What would you conclude about the market research? About the sales campaign?
   c. Assuming the market research was done properly and the 40 percent is representative, how many sales are expected if the publisher attempts to sell to 5,000 homes?

10. Jamieson Airlines has a central office that takes reservations for all flights flown by the airline. The calls received during any week are normally distributed with a mean of 12,000 and a standard deviation of 2,500.

a. During what percentage of the weeks does the airline receive more than 11,000 calls?

b. During what percentage of the weeks does it receive fewer than 12,300 calls?

c. During what percentage of the weeks does it receive between 10,800 and 13,400 calls?

d. During what percentage of the weeks does the airline receive more than 12,800 or fewer than 11,100 calls?

*11. The Langley Pipe and Tile Company is replacing the sewer line in front of your house. Just before you leave for work in the morning, the supervisor warns you the crew may be in front of your house by day's end and that you may not, therefore, be able to move your car from the garage. Since you were planning on using your car that evening, you debate whether to move it now or leave it parked on the street. However, since it is a new car, you would rather leave it in the garage if possible.

Upon questioning, the supervisor states that on the average day, they are able to dig and lay 150 feet of sewer pipe. However some days are better than others, and so in 1 out of 20 days, they lay 175 feet or more, and in 1 out of 20 days, they lay 125 feet or less. The crew is presently 160 feet from your driveway.

a. Can you use the material presented in this chapter to help decide whether to move your car? If so, what assumption do you have to make?

b. If you decide *not* to move your car now, what is the probability you will not be able to move it from the garage at the end of the day?

*12. The supervisor in problem 11 also has a decision to make. A major water line crosses the street just in front of your house, 165 feet from where the crew is beginning to dig. Since settling dirt from the digging sometimes breaks the water line, the supervisor is required to have a plumbing crew stand by. However the plumbing crews have to be scheduled one-half day in advance and will charge for their time even if not used. Assuming the number of feet dug each day is normally distributed with the mean and standard deviation determined in problem 11, what is the probability the supervisor will need the plumbers today?

*13. The supervisor in problems 11 and 12 decides to consider costs before ordering the plumbing crew. If the crew is ordered and not used, the Langley Company will still be charged $150. However, if the water pipe is reached and the plumbing crew is not there, the digging crew will have to be sent home and paid for a full day. The supervisor estimates this would cost $250, counting the rent on the large backhoe. Given these costs, what should the supervisor do about ordering the plumbing crew?

14. According to a western timber expert, 10 percent of all trees cut cannot be used for construction lumber. If the necessary independence assumptions hold, and if 50 trees are cut down today, find:

a. The exact probability that five or more trees will not be usable for construction lumber

    b. The probability in part a using the normal approximation to the binomial. Explain any difference

    c. The answers to parts a and b if the number of trees cut is 20

15. The Dilmart Company is a large furniture dealer located in South Carolina. The company recently purchased a time-share computer system as an aid in handling its accounting and production data. The computer system is equipped with 32 terminals. The response time for computer systems depends on how many terminals are in use and what types of applications the terminals are employing. The computer manufacturer claims that under normal conditions, with all 32 terminals in operation, the response time will be normally distributed with a mean of 20 seconds and a standard deviation of 10 seconds. If the manufacturer's claim is true, what is the probability of two successive response times exceeding 45 seconds each?

16. Referring to problem 15, what would the mean response time have to be if the probability of a response time exceeding 50 seconds is to be, at most, 5 percent?

17. Referring to problem 15, what is the probability of observing a response time of five seconds or less?

18. Referring to problems 15 and 17, what would be the effect on the calculated probabilities if the standard deviation in response time was decreased from ten seconds to five seconds? Explain why the probabilities are changed.

19. The State National Bank has observed that the chance of a small business failing within two years after first opening is 60 percent. Suppose this percentage holds true in the future. What is the probability that of 10,000 businesses opening in 1981, 7,000 or more will be out of busines by the end of 1983? Assume the underlying probability distribution is binomial.

20. Referring to problem 19, what is the probability that between 5,000 and 7,200 small businesses will close by the end of 1983?

21. Airlines have a continuing problem with customers reserving space on flights and then not showing up. Currently there is no penalty for the customer, but the airlines incur opportunity costs and lost revenues. To offset this no-show problem, airlines overbook flights. That is, they sell more tickets than they have seats available.

    Assume an airline sells 115 tickets for a particular flight. If the distribution of customer arrivals can be approximated by a normal distribution with a mean of 96 and a standard deviation of 5, what is the probability that on a given day, more than 105 passengers will arrive for the flight?

22. Referring to problem 21, suppose the airline used on this flight has a capacity of 110 passengers. What is the probability of having more people show than seats available? Comment on this result looking from both the airline's and the customers' viewpoint.

*23. The Bryce Brothers Lumber Company is considering buying a machine that planes lumber to the correct thickness. The machine is advertised to produce

"6-inch lumber" having a thickness that is normally distributed with a mean of 6 inches and a standard deviation of 0.1 inch. If building standards in the industry require a 99 percent chance of a board being between 5.85 and 6.15 inches, should Bryce Brothers purchase this machine? Why or why not?

*24. Referring to problem 23, to what level would the company that manufactures the machine have to reduce the standard deviation for the machine to produce to industry standards?

25. The personnel manager for a large company is interested in the distribution of sick-leave hours for employees of her company. A recent study revealed the distribution to be approximately normal with a mean of 58 hours per year and a standard deviation of 14 hours.

An office manager in one division has reason to believe that during the past year, two of his employees have taken excessive sick leave relative to everyone else. The first employee used 74 hours of sick leave, and the second used 90 hours. What would you conclude about the office manager's claim, and why?

26. Suppose the company in problem 25 grants 40 hours of paid sick leave per year. Given the sick-leave distribution, what would you conclude about the adequacy of the company's sick-leave policy? Why?

27. Suppose the company in problems 25 and 26 is considering a change in its sick-leave policy for next year. The objective is to have the number of paid sick-leave hours at a level that will require 10 percent or fewer people to incur unpaid sick time. Assuming the sick-leave distribution described in problem 25 holds true next year, how many sick-leave hours should be paid by the company?

28. Considering problems 25–27, suppose a consultant has suggested the company hold its paid sick leave to 40 hours per year and hire a physician in an attempt to reduce the average number of sick-leave hours used to a more acceptable level. What would the mean have to be if the probability of an individual needing more than 40 hours of sick leave is to be, at most, 10 percent? Assume the standard deviation remains 14 hours and the distribution is normally distributed.

*29. The Fargo Box and Glue Company makes specialized products for the construction industry. One of these products is an electronic lock assembly used by many condominium builders. The lock provides added security for the homeowner and is recommended by several government safety agencies. The company maintains that the chance of a lock being defective is, at most, 5 percent. A few months ago a large hardware chain placed an order for 1,200 locks. Yesterday the Fargo Box and Glue Company received a letter from the hardware chain indicating it had found 74 defective locks in the 1,200.

What do you conclude about the 5 percent defective claim Fargo has been making?

30. The Hartco Investment Company is considering what rent to charge in its new office complex. Rental prices in this area are approximately normally distributed with a mean of $7.50 per square foot per year and a standard deviation of $2.00 per square foot.

Hartco wants to undercut 80 percent of the market. What price should it charge?

31. Referring to problem 30, what should the rental be if Hartco wishes to price its offices at a level exceeded by only 10 percent of other offices in the city?

32. A steel mill produces alloy sheets used for the body of airplanes. The mill produces sheets with an average thickness of 0.517 inch and a standard deviation of 0.037 inch. A new plane requires alloy sheets between 0.495 and 0.525 inch. What percentage of the sheets made by the mill will be suitable for the new plane?

33. A government agency has certified that the mileage rating for highway driving for a new-model diesel car is normally distributed with a mean of 37 m.p.g. and a standard deviation of 2 m.p.g. If you buy one of these cars:
a. What is the probability your mileage will be over 39 m.p.g.?
b. What is the probability your mileage will be more than 35.5 m.p.g.?
c. What is the probability your mileage will be between 35 and 38.5 m.p.g.?
d. Suppose you received only 24 m.p.g. What would be your conclusion?

=cases

7A EAST MERCY MEDICAL CENTER

Dorothy Jacobs was recently hired as assistant administrator of the East Mercy Medical Center. She is a new graduate of a well-regarded masters program in hospital administration and is expected to incorporate some advanced thinking into the apparently lax practices at East Mercy.

Hospitals have recently been under increasing pressure from both government and local sources because of escalating costs. Although members of the board of directors of East Mercy feel that cost considerations are secondary to quality of care, they, too, are sensitive to the increasing public pressure.

East Mercy is located in a rapidly growing area and is feeling capacity limitations. In particular, according to staff personnel, the obstetrics, adult medical/surgical, and pediatric wards are "bursting at the seams." East Mercy is considering an extensive expansion program, including expansion of the obstetric, adult medical/surgical, and pediatric wards. The board has allocated a total of $400,000 for new beds in these three wards. Dorothy is presently trying to determine how many beds current demand levels justify for each ward and how many beds to actually add given the $400,000 cost constraint.

Dorothy and her staff have computed statistics based upon the current year's patient census data in each of the three wards. These figures are as follows:

	Average No. Beds Used per Day	Standard Deviation
Obstetrics	24	6.1
Surgery	13	4.3
Pediatrics	19	4.7

Histogram plots of bed usage show a remarkably close approximation to a normal distribution for each department.

The present capacity of each of the wards is

$$Obstetrics = 30$$
$$Surgery \quad = 20$$
$$Pediatrics = 24$$

The hospital's architects have given the following estimates for the cost of adding one bed and all necessary supporting equipment to each of the wards:

$$Obstetrics = \$20,000$$
$$Surgery \quad = \$26,000$$
$$Pediatrics = \$15,500$$

It is possible for a ward to exceed its capacity, but according to state guidelines, this should not occur more than 5 percent of the time.

Dorothy is in the process of preparing a report to the administrator showing how many beds are to be added to each of the three wards.

references

LAPIN, LAWRENCE. *Statistics for Modern Business Decisions*. New York: Harcourt Brace Jovanovich, 1978.

NETER, JOHN, WILLIAM WASSERMAN, and G. A. WHITMORE. *Applied Statistics*. Boston: Allyn and Bacon, 1978.

RICHARDS, LARRY and JERRY LACAVA. *Business Statistics: Why and When*. New York: McGraw-Hill, 1978.

SPURR, WILLIAM A. and CHARLES P. BONINI. *Statistical Analysis for Business Decisions*. Homewood, Ill.: Irwin, 1973.

8

Sampling Techniques

why decision makers need to know

Data, transformed into useful information, are required before effective decisions can be made in the business world. Decision makers must decide what data to gather and how to go about getting them. Due to the many sources of *internal* and *external* data discussed in Chapter 2, decision makers are often faced with an "ocean" of data and potential information. Since decision makers cannot possibly use all the information that is potentially available, they must selectively gather only some. The process of selecting a subset of data from the population of all data is called **sampling.**

George Odiorne, a widely published writer on the subject of management, makes a very intuitive analogy about sampling. He says that managers will take little buckets and row their boats over this ocean of facts, dipping the buckets here and there. By examining what they find in the buckets, they are able to draw firm conclusions not only about what is in the buckets, but also about the entire ocean.*

This managerial process of rowing over an ocean of facts and occasionally dipping in a bucket should not be viewed as unscientific. Rather, operational

---

*George S. Odiorne, *Management Decisions by Objectives*, p. 196.

decision makers need both a procedure to tell them where to row their boats and dip their buckets and a method of scientifically drawing conclusions based on what they find in the sample buckets.

chapter objectives

To make effective decisions, managers need information. However, poorly gathered information may be worse than no information. Often, due to time and money considerations, decision makers are forced to gather information by sampling. To be useful, the sample information must be gathered correctly. In this chapter we shall consider some advantages of gathering information through sampling, and present some common sampling techniques.

student objectives

After studying the material in this chapter, you should be able to:

1. Recognize situations where information should be gathered by sampling and when information should be gathered by taking a census.
2. Discuss the common types of sampling techniques used to gather information, and the advantages and disadvantages of each of these techniques.
3. Choose the appropriate sampling technique to gather information in specific decision-making situations.

8-1
SAMPLING VERSUS A CENSUS

Sampling

Generally, managerial decisions can be separated into two major categories. First, there are those decisions involving situations where the physical event will occur in the future, and the needed information simply is not available. For instance, an electrical utility must decide whether to build a coal-fired power plant based on demand which may occur eight years in the future. In another case a company must decide whether to manufacture a product that will not hit the market for two years. The success or failure of these decisions will be determined in the future, and since no one can predict the future with certainty, most of the critical information needed to make these decisions is not available when they must be made.

In the second situation, complete information may be available, but it may require excessive time or money to gather. For instance, deciding whether to accept or reject a large shipment of parts is often based on the percentage of defective parts in the shipment. The only way to determine the true defective rate would be to test all parts in the shipment. Although some companies do test all incoming parts

(called 100 percent acceptance testing), other companies think this is too time consuming and expensive and base their accept/reject decision on a sample from the
incoming parts. Also, in many instances tests are destructive, and if all parts were
tested, there would be none left to use.

In Chapter 2 we defined a population as all items of interest to the data
gatherer. If we have access to the entire population, we could measure each item.
The set of measurements from the whole population is a *census*. Even if a decision
maker can take a census, there are often reasons to sample. These reasons fall into
the following categories:

1. **Time Constraints.**   Obviously, asking a question of 100 people
   will take much less time than asking the same question of 10,000.
   Thus a major advantage of sampling is that it is much *faster* than
   taking a census. An additional advantage, and one that is often
   overlooked, is that sample data can be processed much faster than
   census information. Although actual processing of data is now almost always done by computer, coding and putting the data into
   machine-readable form is time consuming. For instance, it takes
   the U.S. Census Bureau several years to make final ten-year census
   data available. Most managerial decisions have a time limit. The
   manager must make a good decision, but often this involves making
   a decision now instead of next year. A manager does not want to
   still be gathering data when the decisions must be made.

2. **Cost Constraints.**   The cost of taking a census may be tremendous.
   Consider, for example, the cost to General Motors of determining
   exactly what the U.S. population thinks of their latest automobiles.
   Gathering that data would be an astronomical undertaking. Even the
   federal government, with its large financial resources, attempts a
   nationwide census only once every ten years. Even then, much of
   the information supplied is based on a sample.

   This is not to say, however, that the cheapest approach
   is the best. In all cases the decision makers must balance the cost of
   obtaining information against the value of the information. Oftentimes decision makers can obtain more relevant and meaningful
   information from a sample than from a census. This generally occurs
   because more care can be taken with a sample since the data are
   gathered from fewer sources. Basically, if a certain amount of time
   is allotted to gather information, we can either gather a little information from a lot of observations, or a lot of information from a few
   observations. Often the value of the additional information from a
   few sources outweighs the advantage of having a census. Even the
   federal government, when taking the ten-year census, asks detailed
   questions of only a sample of the population.

3. **Improved Accuracy.**   The results of a sample may be more accurate than the results of a census. The reason is not hard to understand. A person gathering data from fewer sources tends to be more
   complete and thorough in both gathering and tabulating the data.

There are likely to be fewer human errors. However, if these measurement errors can be eliminated, a census will be more accurate than a sample.

4. **Impossibility of a Census.** Sometimes taking a complete census to gather information is economically impossible. In addition, obtaining information oftentimes requires a change in, or destruction of, the item from which information is being gathered. For instance, light-bulb manufacturers have to determine the average lifetime of their bulbs. The way to determine how long a bulb will last is to plug it in and wait until it burns out. Once the bulb fails, you have determined how long it would have lasted, but obviously are unable to sell the bulb. Many other information-gathering methods are destructive and hence cannot be used on a census basis.

      A slightly different problem occurs when gathering information involving legal restrictions or pesonal freedoms. An example is the testing of new drugs. Not too many people with an illness are willing to take a new drug until its effectiveness has been determined, and certainly individuals cannot be forced to test a drug. However you cannot determine a drug's effectiveness until it has been used. In fact, the Food and Drug Administration will not allow a drug to be sold until it has been extensively tested.

A Census

That a sample is often superior to a census for obtaining information does not mean a census should never be used. Usually the trade-off is whether the cost of taking a census is worth the extra information gathered by the census. In organizations where many census-type data are stored on computer files, the additional time and effort of gathering all information are not substantial. In other cases, technology in testing equipment has advanced to the point where many companies can run a 100 percent acceptance sample. Thus a census should not automatically be eliminated as an information-gathering technique.

      Generally, if the needed information has already been gathered as a by-product of another operation, or if technology allows information to be gathered rapidly from the population, and if the gathering process will not change or destroy the item being observed, a census should be considered. If this is not the case, a sample will most likely offer the "best" means of obtaining the needed information.

8-2
FUNDAMENTAL SAMPLING TECHNIQUES

Once the manager decides to gather information by sampling, there are many ways to select the sample. In this section we introduce several of the sampling techniques most often employed.

All sampling techniques fall into one of two categories: *statistical* or *nonstatistical. Statistical sampling* techniques include all those using **random** or *probability sampling.* With a random or probability sample, data items are selected by chance alone. Once the probabilities of selecting the items in the population are known, the individual selecting the sample does not change those probabilities. *Nonstatistical sampling,* or *nonprobability sampling,* **is** *nonrandom* **in nature.**

Both statistical and nonstatistical sampling techniques are commonly used by decision makers. However, nonstatistical sampling techniques have some widely recognized problems. For example, when a judgmental sample is selected, there is no objective means of evaluating the results. Because the results depend upon personal judgment, their reliability cannot be measured using probability theory.

We don't mean to imply that judgmental sampling is bad or should not be used. In fact, in many cases it represents the only feasible way to sample. For example, the J. R. Simplot Corporation, a manufacturer of frozen french fries, tests potatoes from each truck arriving at its manufacturing plant. The company uses the results of the sample to determine whether each truckload should be accepted. The quality control people judgmentally select a few potatoes from the top of each truck rather than from throughout the load as statistical sampling would require. It just is not feasible to take a statistical sample.

In the remainder of this chapter we discuss statistical and nonstatistical sampling techniques. While many techniques fall within each of these categories, we will cover only the most common.

Statistical (Probability) Sampling Techniques

Simple Random Sampling.   The most fundamental statistical sampling technique is a *simple random sample. Simple random sampling* **is a method of selecting items from a population such that every possible sample of a specific size has an equal chance of being selected.** Simple random samples can be selected from either *finite* or *infinite* populations. When the population being sampled is finite, or countable, selecting a simple random sample is straightforward.

As an example, suppose as a new insurance salesperson you wish to estimate the proportion of people in a local subdivision who already have life insurance policies. This would be important to you since the result would be an indication of your potential market in the subdivision.

For demonstration purposes, we greatly simplify this example. Suppose there are five families in the subdivision: Jones, Smith, Black, White, and Fitzpatrick. From the $N = 5$ families, we select $n = 3$ families. We can use the combinations counting technique to determine how many samples of $n = 3$ are possible from a population of $N = 5$.

$$_5C_3 = \frac{5!}{3!(5 - 3)!}$$

$$= 10 \text{ possible samples}$$

The sample space is

Jones, Smith, Black
Jones, Smith, White
Jones, Smith, Fitzpatrick
Jones, Black, White
Jones, Black, Fitzpatrick
Jones, White, Fitzpatrick
Smith, White, Fitzpatrick
Smith, Black, White
Smith, Black, Fitzpatrick
Black, White, Fitzpatrick

In a correctly performed simple random sample, each of these samples would have one chance in ten of being selected.

In this example we did not allow the same family to be selected more than once. This condition is **sampling without replacement** and is the most common sampling condition. However, sometimes after an item has been selected, it is placed back in the population and may be reselected. This is called **sampling with replacement.**

If we are sampling from an *infinite* population, we will not be able to count the number of possible samples that could be selected. Most production processes are considered infinite populations. We can obviously not insure that each possible sample has an equal chance of being selected since many items have not yet been made. Thus there is no way to categorically prove a sample selected from an infinite population is a true simple random sample. However, if the process that generates the population is stable over time and if the production process does not fluctuate (for example, from shift to shift, or from hour to hour), the population of items produced is stable, and a simple random sample is possible. From a practical point of view, if every effort is made to select the items in a random fashion without judgment, the decision maker can be fairly sure that the sample will possess the attributes of a simple random sample.

Simple random samples can be obtained in a variety of ways. We will present several examples to illustrate how simple random samples are selected in practice.

The Lottery.   Shortly before the military draft ended, the U.S. Selective Service Commission adopted a **lottery** system to select draftees. The purpose was to eliminate any possible favoritism in determining who would be and who would not be drafted. The lottery was held every year and the population each year consisted of all males who turned 19 since the last lottery. Each male in this population was assigned a lottery number corresponding to his birthdate. To assure a random selection of numbers, the following procedure was used:

1. A rotating drum was filled with 365 balls, each corresponding to a specific day of the year.

2. In front of a group of questioning members of Congress, news reporters, and television floodlights, the first ball (date) was selected

from the drum. All individuals with this birthdate were ranked number one in the draft priority system.

3. After the first date was removed, the drum was closed, turned, and the next date selected. The procedure was repeated for all 365 dates. However, constructing a fair lottery is difficult. As in this example, assuring that the balls in the drum are adequately mixed is difficult.

The Random Number Table.   Another way of selecting a random sample is to use a **random number table.** Appendix I contains a random number table. A page from Appendix I is reproduced as Table 8–1. Random number tables are constructed by recording values from a process that produces numbers which have the properties of randomness. Generally random numbers are generated by computer routines.

Suppose the personnel department at a large retail store in Seattle is considering changing the pay period from monthly to every two weeks. Since this decision will affect all 400 Seattle employees, the personnel manager has decided to select a simple random sample of 10 employees and ask their opinion about the change.

To select a random sample, each employee is assigned a number between 000 and 399. (For this application, no two employees will be assigned the same number.) The personnel manager decides on the following procedure for using the random number table:

1. Pick any two numbers between zero and 40.

2. Use the first number to locate the *row* for the starting random number and the second number to locate the *column.*

3. Beginning with this starting point, select ten three-digit numbers going down the column. If the bottom of the table is reached before ten numbers are selected, go to the top of the next column and continue. (*Note:* If a number exceeds 399, skip it and go to the next number.)

Assume that the personnel manager picks 12 and 17 as the numbers used to locate the starting point. Using Table 8–1, the starting point is the 2 found in row 12, column 17. The three-digit numbers are as follows:

216—1	026—7
459—Skip	831—Skip
134—2	531—Skip
834—Skip	441—Skip
519—Skip	107—8
378—3	697—Skip
031—4	530—Skip
966—Skip	230—9
815—Skip	952—Skip
096—5	344—10
192—6	

TABLE 8–1 Table of random numbers

1260	5529	9540	3569	8381	9742	2590	2516	4243	8130
8979	2446	7606	6948	4519	2636	6655	1166	2096	1137
5470	0061	1760	5993	4319	0825	6874	3753	8362	1237
9733	0297	0804	4942	7694	9340	2502	3597	7691	5000
7492	6719	6816	7567	0364	2306	2217	5626	6526	6166
3715	2248	2337	6530	1660	7441	1598	0477	6620	1250
9491	4842	6210	9140	0180	5935	7218	4966	0537	4416
5192	7719	0654	4428	9771	4677	3291	4459	7432	2054
1714	3725	9397	9648	2550	0704	1239	5263	1601	5177
1052	8415	3686	4239	3272	7135	5768	8718	7582	8366
6998	3891	9352	6056	3621	5395	4551	4017	2405	0831
4216	4724	2898	1050	2164	8020	5274	6688	8636	2438
7115	2637	3828	0810	4598	2329	7953	4913	0033	2661
6647	4252	1869	9634	1341	7958	9460	1712	6060	0638
3475	2925	5097	8258	8343	7264	3295	8021	6318	0454
0415	1533	7670	0618	5193	9291	2205	2046	0890	6997
7064	4946	2618	7116	3784	2007	2326	4361	4695	8612
9772	4445	0343	5238	0317	7531	5916	8229	3296	2321
5365	4306	4036	9873	9669	8505	8675	3116	1484	9975
6395	4681	9319	6908	8154	5415	5728	1593	9452	8213
1554	0411	3436	1101	0966	5188	0225	5615	8568	5169
6745	7372	6984	0228	1920	6710	3459	0663	1407	9211
2217	3801	5860	1673	0264	6911	7623	7137	0774	5898
9255	9297	4305	5060	8312	9192	6016	7238	5193	4908
1615	9761	5744	2733	5314	6985	6670	0975	5487	6107
6679	4951	5716	5889	4413	1513	5023	7313	0317	0517
5221	3207	0351	2452	1072	3830	5518	5972	6111	9352
7720	5131	4867	6501	6970	4075	4869	4798	7104	5342
2767	6055	2801	7033	5305	9382	2354	4135	5975	7830
9202	6815	8211	5274	2303	1437	8995	2514	6515	0049
9359	2754	8587	2790	9524	5068	1230	4165	7025	4365
1078	7661	0999	4413	3446	2971	7576	3385	3308	0557
5732	4853	2025	1145	9743	8646	4918	3674	8049	1622
1596	0578	1493	4681	3806	8837	4241	4123	6513	3083
0602	8144	8976	9195	9012	7700	1708	5724	0315	8032
3624	8592	7942	4289	0736	4986	4839	4507	4997	7407
9328	3001	1462	1101	0804	9724	4082	2384	9631	8334
1981	1295	1963	3391	5757	4403	5857	4329	3682	3823
6001	4295	1488	6702	9954	6980	4027	8492	8195	7934
3743	0097	4798	6390	6465	6449	7990	3774	3577	3895
3343	6936	1449	2915	6668	8543	3147	1442	6022	0056
9208	3820	5165	3445	2642	2910	8336	1244	6346	0487
4581	3768	1559	7558	6660	0116	7949	5609	2887	4156
3206	5146	7191	8420	2319	4650	3734	0501	0739	2025
5662	8315	4226	8395	2931	1812	3575	9341	5894	3691
4631	6278	9444	4058	0505	4449	5959	7483	8641	1311
1046	6653	8333	5813	6586	9820	0190	9214	1947	9677
5231	2788	7198	5904	8370	8347	6599	7304	6430	3495
5349	6641	3234	8692	4424	9179	2767	9517	1173	4160
8363	9625	3329	5262	5360	8181	9298	6629	2433	9414
5967	1261	2470	8867	1962	1630	8360	6024	6232	8386
6716	3916	8712	6673	1156	0001	6760	6287	4546	5743
0323	3514	8550	3709	6614	5764	0600	7444	2795	4426
1765	0918	8972	9924	5941	0331	6909	4872	5693	8957
4345	6886	2032	4817	2725	9471	2443	9532	4770	1271
9614	8571	2174	0071	7824	0504	9600	6414	5734	1371
1291	8246	5019	6559	5051	5265	1184	9030	6689	2776
3867	3915	8311	2430	1235	7283	6481	2012	8487	0226
1056	1880	3610	7796	7192	6663	5810	3512	1572	2921
1691	2237	6713	4048	8865	5794	3419	4372	6996	8342
7920	8490	2822	2647	1700	5335	0732	9987	7501	1223
1646	4251	7732	2136	4339	0331	9293	1061	2663	4821
7928	2575	2139	2825	3806	2082	9285	7640	6166	5758
3563	9078	1979	1141	7911	6981	0183	7479	1146	8949
6546	3459	3824	2151	3313	0178	9143	7854	3935	7300
8315	8778	1296	3434	9420	3622	3521	0807	5719	7764
8442	4933	8173	6427	4354	3523	3492	2816	9191	6261
7446	1576	2520	6120	8546	3146	6084	1260	3737	1333
2619	1261	2028	7505	0710	4589	9632	2347	1975	6839
0814	8542	1526	1202	8091	9441	0456	0603	8297	1412

SOURCE: Reprinted from Richard J. Hopeman, *Production and Operations Management*, 4th ed. (Columbus, Ohio: Charles E. Merrill, 1980), pp. 569–70. Used with permission.

The individuals whose numbers correspond to these three-digit numbers would be asked their opinions regarding the proposed change in pay periods.

Stratified Random Sampling.   Suppose we have been asked to estimate the average cash holdings of Oregon financial institutions on July 1. We will base our estimate on a sample. However not all financial institutions are the same size. A majority are classified as small, and there are some of medium size and only a few large ones. But the few large institutions have a substantial percentage of the total cash-on-hand in the state. By using a simple random sample, we might select proportionately either too few or too many large institutions. Although the simple random sample would be representative based on the number of institutions, it would not be representative based on cash-on-hand.

To avoid selecting a nonrepresentative sample, we might divide the institutions into three *classes*: small, medium, and large. We could then select a simple random sample of institutions from each group and estimate the average cash-on-hand from this combined sample.

What we have done is break the population (Oregon's financial institutions) into **subpopulations** called **strata. Selecting simple random samples from predetermined subpopulations, or strata, is called** *stratified random sampling*.

Populations can be stratified if they have a readily identifiable characteristic that can be used to separate the population members into subgroups. In our example the identifiable characteristic is institution size. In other cases the characteristic may be family size, location, or disposable personal income. In looking at company employees, we may stratify based on sex, age, education, or race.

If done correctly, stratified sampling can be used to estimate a population's characteristics with *less error* than the same size simple random sample. Since sample size is directly related to cost (in both time and money), this means a stratified sample can be more cost-effective than a simple random sample. Stratification is particularly useful when the population contains extreme points that can be grouped into separate strata.

Let us look at another example where stratified random sampling could be used. A manufacturer of automotive replacement parts recently landed a contract with several national retailers to manufacture parts to be sold under the retailers' brand names. Because of these large contracts, the company was forced to move from a single-shift to a three-shift operation. The manufacturing company has begun to receive complaints about the quality of its parts. Before operations were expanded, quality complaints were rare. The company owner is concerned about the increasing complaints and discusses the situation with her production manager. The production manager thinks the increase in complaints could be due to any one of several things. First, because the plant is now used more extensively, the machines may go out of adjustment faster than in the past. Consequently the solution may be to increase periodic maintenance. Second, most of the quality control programs could be occurring in the two added shifts, in which case additional training is needed. Third, the complaints may be increasing because the national retailers offer well-advertised money-back guarantees, in which case the complaints will likely continue.

The production manager has some historical quality information and so decides to sample the present output. He has decided to perform a stratified sample by shift. With results from the stratified sample, the production manager feels he will be able to identify the cause of the complaints and to recommend the correct course of action. Note, however, that once the population has been stratified, simple random sampling is used to select items from each stratum.

Systematic Sampling. The publishers of *Anvil* magazine are faced with a growing market. They are considering changing the magazine's information content to include more worldwide news. Assume *Anvil* has 150,000 subscribers and the publishers wish to question 1,500 subscribers about this proposed change.

Simple random sampling would involve assigning a unique number to each subscriber and using a random number table to select the sample of 1,500. (A computer could also be used if the subscriber lists are stored on disk or tape files. The computer could be programmed to generate random numbers similar to those in the table in Appendix I.)

An alternative to selecting a simple random sample would be to question every 100th subscriber from the list of subscribers. The procedure would involve using a random number table to select a number between 1 and 100 for the starting point. Using the random starting point, each 100th subscriber would be selected. If the starting point were 75, the sample would be

$$
\left.\begin{array}{c}
75 \\
175 \\
275 \\
375 \\
\cdot \\
\cdot \\
\cdot \\
149{,}975
\end{array}\right\} \; n = 1{,}500
$$

Selecting a sample based on this type of predetermined system is called *systematic sampling.* Systematic sampling is frequently used in business applications. **Systematic sampling should be used as an alternative to simple random sampling only when the population is randomly ordered.** For example, *Anvil* subscribers are no doubt listed alphabetically. To use systematic sampling in this case, we must assume opinion is randomly distributed through the alphabet.†

Systematic sampling, when applicable, has specific advantages. First, a systematic sample is easy to select. A highly trained expert is not needed to select, say, every 100th name from a subscription list or every third house in a subdivision. Also, depending upon the characteristics of the population, a systematic sample can be more evenly distributed across the population, and thus more representative.

Cluster Sampling. Taking a random sample requires not only that we randomly determine who from the population will be in the sample, but also that those selected be contacted and measured. This is not a problem when the population

---

†There are more statistically precise indications of when systematic sampling can be used instead of simple random sampling. See Leslie Kish, *Survey Sampling*, pp. 113–42.

is an incoming shipment of transistors. This *is* a problem when the population members selected for the sample are scattered across the country, or perhaps the world.

Suppose the Morrison-Knudsen Company, one of the largest construction companies in the world, wants to develop a new corporate bidding strategy. Upper management desires input on possible new strategies from its middle-level managers. Figure 8–1 illustrates the hypothetical distribution of middle-level managers throughout the world. For example, there are 25 middle-level managers in Algeria, 47 in Illinois, and so forth. The upper management decides to have personal interviews with a sample of these employees.

Algeria	Illinois	Scotland	Saudi Arabia	Alaska	New York	Florida	Idaho	Mexico	Australia
25	47	22	105	20	36	52	152	76	37

FIGURE 8–1   Clusters, Morrison-Knudsen geographical locations. (no. middle-level managers at each location)

One sampling technique is to select a simple random sample of size *n* from the population of middle managers. Unfortunately this technique would likely require that the interviewer(s) go to each state or country where Morrison-Knudsen has middle-level managers. This would be an expensive and time-consuming process. A systematic or stratified sampling procedure would probably also require visiting each location.

A sampling technique that overcomes the traveling (time and money) problem is **simple cluster sampling. Cluster sampling is a method by which the population is divided into groups, or clusters, and a sample of clusters is taken to represent the population.** Once the clusters are chosen, sampling of items from each cluster may be undertaken. In the Morrison-Knudsen example, the clusters are the geographical locations of the middle-level managers. As shown in Figure 8–1, this example has ten clusters.

The first step in simple cluster sampling is to randomly select *m* clusters from the total possible (*M*) clusters. Suppose Morrison-Knudsen randomly selects the following three clusters:

Mexico	Alaska	Algeria
76	20	25

The interview team can either question all middle managers in these three locations or select a random sample at each location. If they randomly sample from each primary cluster, they may select the following number of managers from the chosen clusters:

$$
\left.\begin{array}{ll} \text{Mexico} & 15 \\ \text{Alaska} & 8 \\ \text{Algeria} & 12 \end{array}\right\} \text{Ultimate clusters}
$$

Cluster sampling can be used in a variety of ways. Table 8–2 illustrates the relationship between population, population members, and clusters for this example and several other possible clusters.

TABLE 8–2 Examples of possible clusters

Population	Variable of Interest	Individual Elements	Clusters
Middle managers	Ideas on bidding strategies	Persons	Area locations
People in Boston	Political preference	Persons	City block
San Francisco workers	Commuting distance	Persons	Office building
Houses in Atlanta	Price	Houses	Subdivisions
Automobiles produced in Detroit	Mileage	Automobiles	State of current location
Harvard University	Parent's income	Students	Class standing

Two additional examples illustrate the potential of cluster sampling. A manufacturer of frozen bakery products, which has recently expanded into a new market area, prides itself in producing quality products and maintaining an extensive in-plant quality control system. In its original market area, the company handled its own distribution and maintained strict controls on its product. (If the frozen bakery goods begin to thaw, the quality is diminished.)

When expanding into the new market area, the firm decided to employ a regional distributor. Though the manufacturing firm has no reason to doubt that the distribution firm is adequately handling its product, it decides to take a sample from retail stores to check quality. To save time and money, the frozen-product firm decides to use cluster sampling. Management decides to randomly select four population centers in the new sales area and then randomly select stores from these four centers. Once a store is selected, the company representatives will randomly select a sample of frozen bakery products.

Another example concerns a manufacturer of electronics products. The company is considering producing a new marine radio that has a signal beacon and wants to make sure there is sufficient interest before going to the expense of producing and marketing the product. Since the product is technically advanced, specially trained interviewers are needed to explain the product to prospective buyers.

The company decides to use cluster sampling to determine its potential market. The clusters are the marinas on the West Coast. The company randomly selects five marinas. From each of these marinas, six registered boat owners are randomly selected and interviewed to determine the likelihood they would purchase this new radio. Based on the results of this sample, the company will decide whether to produce the new product.

A well-designed cluster sample will generally provide equal-quality information at less cost than a simple random sample. However cluster sampling has some potential disadvantages. Ideally each cluster should be representative of all the population, and thus heterogeneous with respect to the variable of interest. But in actual practice, this may not be the case. For instance, if the clusters are determined by geographical proximity, there will likely be some degree of homogeneity in the clusters. In this situation, cluster sampling may require a *larger* sample size than a simple random sample to provide the same level of information. Since similar people and items tend to group together, more clusters will have to be sampled than if the clusters were heterogeneous. This can reduce the cost advantage associated with cluster sampling. In addition, the costs and problems associated with statistically analyzing the sample data obtained from cluster sampling are generally increased.

Nonprobability Sampling Techniques

The simple random, stratified random, systematic, and cluster sampling techniques discussed previously are all examples of probability samples. When a probability sample is selected, there are a variety of statistical techniques available to analyze the sample data, as you will see in the remaining chapters of this text.

However sometimes a probability sample is either not possible or not desirable. In these instances *nonprobability sampling* is used. One example of nonprobability sampling is *judgment sampling.* **In *judgment sampling,* the person taking the sample has direct or indirect control over which items are selected for the sample.** Judgment sampling is appropriate when decision makers feel that some population members have more or better information than other members.

Judgment sampling is also used when the decision makers feel that some population members are more representative of the population than others. For instance, if you had to pick only 1 city in the United States in which to test-market a new product, you probably would not select that city randomly. You would likely pick the city based on a reason, such as, "Los Angeles is good for new products and it tends to lead the country," or, "Never introduce anything new in the Midwest." You would probably also use a judgment sample if you could select only 2 cities, or 3, or 4. However, if you could select 25 cities to test-market your product, you might very well choose these cities randomly.

Of course a judgment sample is only as good as the person(s) doing the sampling. If poor judgment is used, the information gathered may be nonrepresentative and misleading. For example, a decision maker can easily bias an opinion survey by sampling only individuals who share similar opinions.

Other nonprobability samples include *quota sampling* and certain *mail questionnaires.* **In *quota sampling,* the decision maker requires the sample to contain a certain number of items with a given characteristic.** For example, suppose an opinion survey is to be taken in a factory employing 60 percent males and 40 percent females. If the sample size is 1,000, the sampling quota might be 600 randomly selected males and 400 randomly selected females. Many political polls are, in part, quota samples.

Mail questionnaires have been used extensively since the 1920s to gather information. Mail questionnaires can belong to either the probability sampling category or the nonprobability sampling category. If the sample selection technique is not based on probability theory, the data from the mail questionnaire form a nonprobability sample. Even when random selection techniques are used, there is almost always a problem of *nonresponse bias.* By nonresponse bias we mean that those people who do not respond to the questionnaire may reflect a certain attitude or characteristic of the population which the questionnaire will therefore not detect or measure. Unless appropriate corrections are made to account for this nonresponse, the questionnaire data must be treated like a nonprobability sample.††

Although nonprobability sampling is often used in decision situations, the results cannot be statistically analyzed. Throughout the remainder of this text we shall stress statistical sampling applications that can be subjected to statistical analyses.

8-3
CONCLUSIONS

In this chapter we have presented an overview of some of the most commonly used sampling techniques. Sampling procedures are divided into two main categories: probability sampling and nonprobability sampling.

Probability sampling techniques include simple random sampling, stratified random sampling, systematic sampling, and cluster sampling. These techniques are all based on formal probability theory. Decision makers who properly use probability sampling can apply the statistical techniques discussed throughout this text to assist in the decision-making process.

Nonprobability sampling techniques include judgment sampling, quota sampling, and certain mail questionnaires. Oftentimes a nonprobability sample is preferred over a probability sample. However the possible statistical analysis is limited when such a sample is selected.

See the chapter glossary for definitions of terms introduced in this chapter.

chapter glossary ═══

census   A measurement of each item in the population.

cluster   A sampling unit containing several population elements. The cluster can be defined in any manner consistent with the study at hand.

cluster sampling   The selection procedure in which the population can be divided into M clusters and $n_1, n_2, n_3, \ldots, n_m$ samples are selected. In this procedure the unit of selection may contain more than one population element.

nonprobability sampling   A method of sampling whereby probabilities cannot be objectively assigned to the items. Consequently the reliability of the sample results cannot be measured in terms of probability. Examples are judgment sampling, quota sampling, and certain mail questionnaires.

population   A finite or infinite collection of measurements or items or individuals that make up the total of all possible measurements, items, or individuals within the scope of the study.

---

††See Leslie Kish, *Survey Sampling*, pp. 532–62 for a detailed discussion on nonresponse in sampling.

probability sampling   A sampling process where each unit selected has a known probability of selection. Examples are simple random sampling, systematic sampling, stratified random sampling, and cluster sampling.

random number table   A table that contains a series of numbers that satisfy the statistical properties of randomness.

sample   A subset of a population. The members of a sample may be selected using either probability sampling or nonprobability sampling.

simple random sampling   A selection procedure in which each possible sample of a given size n has the same chance of being selected.

statistical sampling   Another term for probability sampling. When statistical sampling is used, the reliability of the results can be measured in terms of probability.

strata   Subgroups of a population that contain homogeneous units with respect to the variable(s) under consideration.

stratified random sampling   The selection procedures whereby the population is divided into strata and simple random samples are selected from the strata.

systematic sampling   A method of selecting a sample based on a predetermined procedure, for instance selecting every fifth person on a class roster or every tenth television set manufactured.

=problems

1. Define in your own words what a simple random sample is and provide a business example where such a sample could be used.

2. Define in your own words what a stratified random sample is and give a business example where such a sample could be used.

3. Suppose a retail store is considering expanding into a new market area. As part of their study, the store's managers wish to find out more about people's spending habits for the kind of goods this new store would sell. Assuming the market area is about the size of your home town, how might the managers select a sample of individuals to talk to?

4. In problem 3 the market area was reasonably small (the size of your home town). How might your response change in terms of the type of sampling if the market area were about the size of your home state?

5. Discuss in your own words what is meant by *cluster sampling* and indicate why cluster sampling might offer an advantage over other types of statistical sampling.

6. The Basin Ski Resort is planning a telephone survey of the holders of its season lift ticket to determine the level of satisfaction with services this year. Management has a list of ticket holders in alphabetical order. Devise a sampling technique that might be used to select the individuals to call.

7. Under what circumstances might a decision maker use nonprobability sampling rather than one of the probability sampling techniques?

8. The maker of Creamy Good Ice Cream is concerned about the quality of ice cream being produced by its Illinois plant. Discuss a plan by which the Creamy Good managers might determine the percentage of defective cartons of ice cream. Would it be possible or feasible to take a census?

9. A random sample of skiers at the Aspen Ski Resort was selected by the makers of a particular brand of skiing equipment. Their method for selecting the sample required that individuals waiting in one of the lift lines be asked questions about various brands of skiing equipment. Comment on the method of sampling and indicate how you would design a sampling technique that might produce more statistically valid results.

10. A beer manufacturer is considering abandoning can containers and going exclusively to bottles because the sales manager feels that beer drinkers prefer drinking beer from bottles. However the vice-president in charge of marketing is not convinced the sales manager is correct. Describe a method by which a statistical sample of the company's customers could be selected.

11. With respect to problem 10, how might a cluster sampling approach be used if it is known that the beer is marketed at stores in all 50 states?

12. A recent Gallup poll for a political party was based upon responses of 1,856 people from across the country. Why do you suppose it is possible for a sample of this size to give a good indication of how 50 million voters will vote in a particular election?

13. If you were designing a stratified random sampling plan for a survey of city governments in your state to find out the amount of money they are spending on administrative salaries, what criteria might you use to form the strata? Provide several ideas and indicate the one that might give the best results.

14. Using the random numbers table in Appendix I, discuss how a simple random sample of service station operators could be selected.

15. We have indicated in this chapter that a cluster sampling plan might save you time and money, yet it might also cost you more money. Why is this? What effect do homogeneous clusters have on whether a cluster sample will be more or less costly than another form of statistical sampling?

16. Discuss the circumstances under which a systematic sample might be selected and provide an example of how such a sample might be obtained.

17. Student leaders often poll students to obtain opinions on topics of interest such as athletics, library hours, and so on. Concerning the last such poll on your campus, discuss the methods by which the sampling was performed, pointing out the strong and weak points of the process. If you need to, visit your student-body officers and find out their approach to sampling student opinion.

18. A questionnaire was recently sent out by a member of Congress. How might the statistical validity of the results be affected if many individuals don't fill out the questionnaire because they don't care for this person?

8A JOINT APPROPRIATIONS COMMITTEE

Vern Lyman is a student intern working with the state legislative Joint Appropriations Committee. The student intern program is reserved for only a few top students, and Vern feels very fortunate to have been selected. However, this will apparently be a tumultuous legislative year. The major cause of concern is a voter-passed initiative limiting the amount of property tax local government units in the state can collect. The state government has been running a surplus and is expected to "help out" the local units. Unfortunately the amount of the projected surplus will not offset the revenues lost to local units of government if the legislature enacts the initiative as passed by the voters.

While the Joint Appropriations Committee will not have any direct impact on implementing the initiative, it will be involved in determining which local governments receive state money. In particular, it will be involved in determining just how much state aid will go to the local units of government and how much revenue will be left for the state to operate with. If the state picks up a substantial portion of the lost local revenues, some state services will have to be cut. The Joint Appropriations Committee will then have to determine which state services will receive full funding and which will have to be cut back or eliminated.

The committee chairperson has called Vern to her office during a pre-legislative planning meeting. The chairperson is positive that the state will have to aid the local governments, but is unsure what will happen if the state is forced to cut back on its own services. She has asked Vern to come up with a sampling plan, including a questionnaire, to determine which state services the citizens would prefer be cut back if the need arises.

8B TRUCK SAFETY INSPECTION, PART 1

The Idaho Department of Law Enforcement, in conjunction with the federal government, recently began to formulate a truck inspection program in Idaho. The current truck safety program is limited to an inspection of only those trucks that appear (visually) to have some defect when they stop at one of the weigh stations in the state. The proposed inspection program will not be limited to the trucks with visible defects, but will potentially subject all trucks to a comprehensive safety inspection.

Mr. Lund of the Department of Law Enforcement is in charge of the new program. He has stated that the ultimate objective of the new truck inspection program is to reduce the number of trucks with safety defects operating in Idaho. Ideally all trucks passing through, or operating within, Idaho's borders would be inspected once a month, and substantial penalties applied to operators if safety defects were discovered. Mr. Lund is confident that such an inspection program would, without fail, reduce the number of defective trucks operating on Idaho's highways. However, each safety inspection takes about an hour, and because of limited money to hire inspectors, Mr. Lund realizes that all trucks cannot be inspected. He also knows it is unrealistic to have trucks wait to be inspected until trucks ahead of them have been checked. Such delays would cause problems with the drivers.

In meetings with his staff, Mr. Lund has suggested that before the inspection program begins, the number of defective trucks currently operating in Idaho needs to be estimated. This estimate can be compared with later estimates to see if the inspection program has been effective. To arrive at this initial estimate, Mr. Lund feels that some sort of sampling plan to select representative trucks from the population of all trucks in Idaho must be developed. He has suggested that this sampling be done at the eight weigh stations near Idaho's borders, but is unsure how to establish a statistically sound sampling plan that is practical to implement.

references

KISH, LESLIE. *Survey Sampling*. New York: Wiley, 1965.

MENDENHALL, W., L. OTT, and R. L. SCHEAFFER. *Elementary Survey Sampling*. Belmont, Calif.: Wadsworth, 1971.

ODIORNE, GEORGE S. *Management Decisions by Objectives*. Englewood Cliffs, N.J.: Prentice-Hall, 1969.

YAMANE, TARO. *Elementary Sampling Theory*. Englewood Cliffs, N.J.: Prentice-Hall, 1967.

9

Sampling Distribution of $\overline{X}$

Many business decisions involve determining the average, or mean, of a set of data. For example, when deciding what brand of tires to purchase, the manager of a rental-car company wants to know which brand will give the longest average wear. In another example, light-bulb manufacturers are now required to indicate the average life of each type of bulb they make. These companies must somehow determine the mean life of the bulbs they manufacture.

Calculating the mean of a set of data presents no particular problem as long as the data are available. However few business problems allow decision makers to measure the entire population of values. Either cost or time constraints limit the number of values they can use to calculate the mean. As we stated in Chapter 8, these constraints are the main reason for sampling.

When the mean is calculated from a sample, the value that results depends on which sample (of the many possible samples) is observed. Consequently decision makers who employ sampling to determine a mean value need to understand how those possible sample means are distributed. In this chapter we introduce the basic concepts of the *distribution of sample means*.

chapter objectives

In this chapter we introduce the sampling distribution of $\overline{X}$. We will show that the value of $\overline{X}$ obtained depends on the particular sample selected. Thus $\overline{X}$ comes from a distribution of possible sample means. We will also introduce one of the most important theorems in statistics—the central limit theorem—and will show how it is used in the decision-making process.

student objectives

After studying the material in this chapter, you should be able to:

1. Discuss the relationship between a population and the many samples that can be selected from it.
2. Explain why the central limit theorem is so important to statistical decision making.
3. Discuss the relationship between the sample size and the decisions that can be based upon sample information.
4. Discuss how variation in the population affects decisions that can be based upon sample information.

9-1
RELATIONSHIP BETWEEN SAMPLE DATA AND POPULATION VALUES

Decision makers are faced with a problem in addition to the problems associated with correctly gathering and arranging data in an understandable manner. This problem is that two samples from the same population will likely have different sample values and therefore possibly lead to different decisions. The following example will demonstrate what we mean.

Suppose the investment officer at SeaSide State Bank handles the retirement fund for all state government employees. Although most of the retirement money is invested in government bonds, the officer has been increasing the amount invested in corporate stocks.

The retirement committee, composed of state government employees, is naturally concerned with the rate of return being earned by the invested dollars. The greater this return, the larger will be the retirement fund and the greater the benefits to the employees when they retire. The committee has asked the investment officer to determine the average return for money invested in stocks only.

For the purposes of this example, assume the money has been invested in five stocks, with an equal amount in each stock. The returns on each stock last year were:

Stock	Return
A	7%
B	12
C	− 3
D	21
E	3

With only five stocks, the investment officer could easily report the population mean, μ_x, to the employee committee. The population mean is

$$\mu_x = \frac{\Sigma X}{N} \tag{9–1}$$

where: X = individual returns for the stocks

N = population size

Thus

$$\mu_x = \frac{7\% + 12\% + (-3\%) + 21\% + 3\%}{5}$$

$$= 8\%$$

To more fully describe the stock returns, the investment officer might also calculate the population standard deviation.

$$\sigma_x = \sqrt{\frac{\Sigma(X - \mu_x)^2}{N}} \tag{9–2}$$

Thus

$$\sigma_x = \sqrt{\frac{(7 - 8)^2 + (12 - 8)^2 + (-3 - 8)^2 + (21 - 8)^2 + (3 - 8)^2}{5}}$$

$$= 8.15\%$$

However, for this example, the investment officer has decided to select a simple random sample of three stocks and base her report on this sample. Although she will select only one sample, there are several possible samples from which to choose. We can determine exactly how many possible samples of three stocks can be selected without replacement from a population of size five by recognizing this as a combinations problem.

$$_NC_n = \frac{N!}{n!(N - n)!}$$

$$_5C_3 = \frac{5!}{3!(5 - 3)!} \tag{9–3}$$

$$= 10 \text{ possible samples}$$

Table 9–1 lists the ten possible samples and the sample mean, $\overline{X}$, of each. Since the $\overline{X}$ values range from 2.33 percent to 13.33 percent, the value of the sample mean reported to the employee committee will depend on the sample selected. The reported sample mean will also be different from the true population mean.

TABLE 9–1 Possible samples, SeaSide State Bank

Sample Stocks	Returns	$\bar{X}$
A, B, C	7%, 12%, −3%	5.33
A, B, D	7%, 12%, 21%	13.33
A, B, E	7%, 12%, 3%	7.33
A, C, D	7%, −3%, 21%	8.33
A, C, E	7%, −3%, 3%	2.33
A, D, E	7%, 21%, 3%	10.33
B, C, D	12%, −3%, 21%	10.00
B, C, E	12%, −3%, 3%	4.00
B, D, E	12%, 21%, 3%	12.00
C, D, E	−3%, 21%, 3%	7.00

where: $\bar{X} = \dfrac{\Sigma X}{n}$

n = sample size

9-2
SAMPLING ERROR

The investment officer at SeaSide State Bank is going to select only one of the ten possible samples. Notice in Table 9–1 that no sample mean equals the population mean of 8 percent. Thus her report to the employee committee will contain *sampling error.*

   Sampling error **is the difference between a population value and the corresponding sample value.** The amount of sampling error is determined by which $\bar{X}$ value is found. In the SeaSide example, if $\bar{X} = 8.33$ percent, the sampling error is fairly small. However, if $\bar{X} = 13.33$ percent, the sampling error is quite large. Because the investment officer cannot know how large the sampling error will be before selecting the sample, she should know how the possible sample means are distributed. **The distribution of possible sample means is called the sampling distribution of $\bar{X}$.**

9-3
SAMPLING DISTRIBUTION OF $\bar{X}$

We can use the SeaSide State Bank example to illustrate two important concepts for statistical decision making. The first concerns the relationship between the population mean, μ_x, and the average of the possible sample means, $\mu_{\overline{x}}$. We often call $\mu_{\overline{x}}$ the **mean of the means.** We find $\mu_{\overline{x}}$ as follows:

$$\mu_{\overline{x}} = \frac{\displaystyle\sum_{i=1}^{K} \overline{X}_i}{K} \tag{9–4}$$

where: $\overline{X}_i = i$th sample mean

$K =$ number of possible samples

Then

$$\mu_{\overline{x}} = \frac{5.33\% + 13.33\% + 7.33\% + \ldots + 7.00\%}{10}$$

$$= 8\%$$

We see that $\mu_{\overline{x}}$, **the average of all possible samples, equals the true population mean, μ_x.** This will always be true because $\overline{X}$ is an **unbiased estimator** of μ_x, the population mean. We will discuss the concept of unbiased estimates more fully in Chapter 10.

The second important concept concerns the relationship between the population standard deviation and the **standard deviation of the sample means.** The population returns ranged from -3 percent to 21 percent. However Table 9–1 illustrates that the sample means range from 2.33 percent to 13.33 percent. **The distribution of sample means is less variable than the population from which the samples were taken.**

Recall that in our example, the population standard deviation, σ_x, is 8.15 percent. Since there are only ten possible samples, we can calculate the standard deviation of the sample means, $\sigma_{\overline{x}}$, as follows:

$$\sigma_{\overline{x}} = \sqrt{\frac{\displaystyle\sum_{i=1}^{K} (\overline{X}_i - \mu_{\overline{x}})^2}{K}}$$

Therefore

$$\sigma_{\overline{x}} = \sqrt{\frac{(5.33 - 8.0)^2 + (13.33 - 8.0)^2 + \ldots + (7.0 - 8.0)^2}{10}} = \sqrt{11.07}$$

$$= 3.326$$

Note that $\sigma_{\overline{x}} = 3.326$ percent is less than $\sigma_x = 8.15$ percent. In fact, $\sigma_{\overline{x}}$ will be less than the population standard deviation, σ_x, for any application.

The value $\sigma_{\overline{x}}$, also called the *standard error of the mean*, indicates the spread in the distribution of all possible sample means.

9-4

SAMPLING FROM NORMAL DISTRIBUTIONS

In more realistic situations with larger populations, the number of possible $\overline{X}$ values can become very large. For example, if a sample of 5 is selected from a population of 100, the number of possible samples is

$$_{100}C_5 = \frac{100!}{5!95!}$$
$$= 75,287,520$$

In applications where the number of possible samples is very large, we cannot possibly calculate all the possible sample means to find $\sigma_{\overline{x}}$ and $\mu_{\overline{x}}$. However two important theorems allow decision makers to describe the distribution of sample means for any distribution. Theorem 9–1 is the first of these.

THEOREM 9–1

> If a population is normally distributed with mean μ_x and standard deviation σ_x, the sampling distribution of $\overline{X}$ values is also normally distributed with $\mu_{\overline{x}} = \mu_x$ and $\sigma_{\overline{x}} = \sigma_x/\sqrt{n}$.

Figure 9–1 shows a normal population distribution and two sampling distributions. As theorem 9–1 states, the average of the sample means, $\mu_{\overline{x}}$, equals the population mean, μ_x. Figure 9–1 also shows that the spread of the sampling distribution decreases as the sample size increases. From Theorem 9–1, the standard error of the mean is given by

$$\sigma_{\overline{x}} = \frac{\sigma_x}{\sqrt{n}} \qquad (9\text{–}5)$$

where:　σ_x = population standard deviation
　　　　n = sample size

Suppose scores for all students taking a standard college entrance examination are normally distributed with $\mu_x = 80$ and $\sigma_x = 10$. If a random sample of 100 scores is selected, the sampling distribution of possible $\overline{X}$ values will be normally distributed with

$$\mu_{\overline{x}} = \mu_x = 80$$

and

$$\sigma_{\overline{x}} = \frac{\sigma_x}{\sqrt{n}} = \frac{10}{\sqrt{100}} = 1.0$$

We see that the spread of the sampling distribution, $\sigma_{\overline{x}} = 1.0$, is considerably smaller than the spread of the population, $\sigma_x = 10$. Because the population

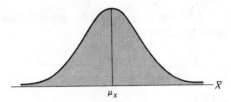

(a) Population Distribution

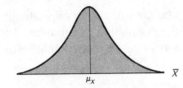

(b) Sampling Distribution, $n = 8$

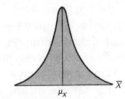

(c) Sampling Distribution, $n = 64$

FIGURE 9–1 Relationship between normal population and sampling distribution of $\overline{X}$.

standard deviation is divided by the square root of the sample size, as the sample size increases, the standard deviation of the sample means decreases. If the sample size were 400 instead of 100 scores, the mean of the sampling distribution would still be 80, but the standard deviation of the sampling distribution would be reduced to 0.50.

$$\sigma_{\overline{x}} = \frac{\sigma_x}{\sqrt{n}} = \frac{10}{\sqrt{400}} = 0.5$$

$\sigma_{\overline{x}}$ **is a measure of average sampling error. Therefore, increasing the sample size will reduce the average sampling error.**

9-5
SAMPLING FROM NONNORMAL POPULATIONS

Theorem 9–1 applies only if the population from which the sample is selected is normal. As we discussed in Chapter 7, there are many instances where the population of interest can be assumed to be normally distributed. However there are many other applications where the population of interest will not be normally distributed. If the population is nonnormal, theorem 9–1 cannot be used. In this case, however, theorem 9–2—the **central limit theorem**—does apply.

THEOREM 9–2

Central Limit Theorem

If random samples of n observations are taken from any population with mean μ_x and standard deviation σ_x, and if n is large, the distribution of possible $\bar{X}$ values will be approximately normal with $\mu_{\bar{x}} = \mu_x$ and $\sigma_{\bar{x}} = \sigma_x/\sqrt{n}$ regardless of the population distribution. The approximation becomes increasingly more accurate as the sample size, n, increases.

An important question when using the central limit theorem is, How large is a "large" sample size? Although there is no exact answer to this question, if the population is symmetric and unimodal about μ_x, sample sizes of 4 or 5 will produce approximately normal sampling distributions. In other cases where the population is extremely skewed, sample sizes of more than 25 are required to produce a normal sampling distribution. Many authors have adopted the rule of thumb that $n \geq 30$ will provide a distribution of sample means that is approximately normally distributed. We will also adopt this rule.

9-6
THE FINITE CORRECTION FACTOR

Both theorems 9–1 and 9–2 assume that either sampling is done *with replacement* or the population is large relative to the sample size. If sampling is done *without replacement* and the sample is large relative to the population, a modification must be made in calculating $\sigma_{\bar{x}}$. If n is greater than 5 percent of the population size and sampling is performed without replacement,

$$\sigma_{\bar{x}} = \frac{\sigma_x}{\sqrt{n}}\sqrt{\frac{N-n}{N-1}} \qquad (9\text{–}6)$$

where: σ_x = population standard deviation

            n = sample size

           N = population size

$\sqrt{\dfrac{N-n}{N-1}}$ = finite correction factor

In the SeaSide Bank example, the population mean, μ_x, was 8 percent, and the population standard deviation, σ_x, was 8.15 percent. Due to the small number of possible samples ($n = 10$), we were able to calculate $\sigma_{\overline{x}} = 3.326$. Because $n = 3$ is greater than 5 percent of $N = 5$ and sampling was without replacement, we can also use formula (9–6) to find $\sigma_{\overline{x}}$.

$$\sigma_{\overline{x}} = \frac{\sigma_x}{\sqrt{n}} \sqrt{\frac{N - n}{N - 1}} = \frac{8.15}{\sqrt{3}} \sqrt{\frac{5 - 3}{5 - 1}}$$

$$= 3.326$$

The term $\sqrt{(N - n)/(N - 1)}$ **is called the *finite correction factor* and is used if we are sampling without replacement and the sample size is large relative to the population.** Note that the finite correction factor always is less than 1.0, and that as the population gets large relative to the sample size, the factor approaches 1.0.

9-7

DECISION MAKING AND THE SAMPLING DISTRIBUTION OF $\overline{X}$

The concepts discussed in this chapter are extremely important to decision makers. The following example will demonstrate how the theorems are used in a decision environment. In this example, remember to distinguish between:

1. The population distribution
2. The distribution of all possible sample means
3. The mean of a single sample

These are three different factors in the decision-making situation and should not be confused.

Because of legislated mileage requirements, the major U.S. automobile makers are being forced to build smaller, lighter cars. But while building cars with smaller exteriors, the auto makers want to maintain interior dimensions that are comfortable for the majority of U.S. car buyers. One important dimension for riding comfort is the distance between the floorboard and the bottom of the dashboard. If this distance is too small, the rider's knees will hit the dash. Unfortunately for the auto makers, the distance from the foot to the knee is not the same for all car buyers.

Suppose the average foot-to-knee length for the population is $\mu_x = 20$ inches and the population standard deviation is $\sigma_x = 3$ inches. One maker has decided to select a random sample of potential customers to test the riding comfort of its latest "small" car. The quality control manager wants the test group to represent the population as a whole. The average foot-to-knee length for the test sample of 36 people is 21.5 inches. The quality control manager wishes to know whether this group is an unlikely selection from a population with $\mu_x = 20$ and $\sigma_x = 3$.

To help the quality control manager, we employ the central limit theorem. Thus the distribution of possible $\overline{X}$ values will be approximately normal with $\mu_{\overline{x}} = \mu_x$ and $\sigma_{\overline{x}} = \sigma_x/\sqrt{n}$. This sampling distribution is shown as follows:

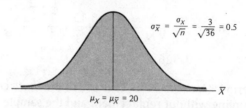

$$\sigma_{\overline{x}} = \frac{\sigma_x}{\sqrt{n}} = \frac{3}{\sqrt{36}} = 0.5$$

$$\mu_x = \mu_{\overline{x}} = 20$$

The sample mean selected is 21.5 inches. When the quality control manager wonders if this sample mean is an unlikely selection, he is really wondering about the chances of finding an $\overline{X} \geq 21.5$ inches. This probability is represented by the small black area in the following normal curve:

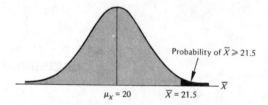

Probability of $\overline{X} \geqslant 21.5$

$\mu_x = 20$ $\overline{X} = 21.5$

Recall that to find areas under the normal curve, we first standardize the distribution so that we can work with Z values. Remember, Z represents the number of standard deviations a value is from the mean. When working with a sampling distribution, we find the Z value as follows:

$$Z = \frac{\overline{X} - \mu_x}{\sigma_{\overline{x}}} \qquad (9\text{--}7)$$

or

$$Z = \frac{\overline{X} - \mu_x}{\dfrac{\sigma_x}{\sqrt{n}}} \qquad (9\text{--}8)$$

If sampling is without replacement and $n > 5$ percent of the population,

$$Z = \frac{\overline{X} - \mu_x}{\dfrac{\sigma_x}{\sqrt{n}} \sqrt{\dfrac{N-n}{N-1}}} \qquad (9\text{--}9)$$

In the auto maker example, the sample size, $n = 36$, is certainly small relative to the population size, so we will use equation (9–8).

$$Z = \frac{\overline{X} - \mu_x}{\dfrac{\sigma_x}{\sqrt{n}}} = \frac{21.5 - 20}{\dfrac{3}{\sqrt{36}}}$$

$$= 3.00$$

Thus an $\bar{X}$ of 21.5 is 3.00 standard deviations away from μ_x of 20. From the standard normal distribution table in Appendix C, the area corresponding to $Z = 3.00$ is 0.4986.

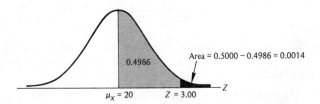

The probability of finding a sample mean equal to or greater than 21.5 inches is only 0.0014. Since this probability is so small, the quality control manager would most likely not want to use these 36 people to test the new car's comfort because they have foot-to-knee lengths that apparently do not represent the population.

9-8
CONCLUSIONS

When a manager selects a sample, it is only one of many samples that could have been selected. Consequently the sample mean, $\bar{X}$, is only one of the many possible sample means that could have been found. There is no reason to believe that the single $\bar{X}$ value will equal the population mean, μ_x. The difference between $\bar{X}$ and μ_x is called sampling error. Because sampling error exists, decision makers must be aware of how the sample means are distributed in order to discuss the potential sampling error.

In this chapter we have introduced two very important theorems. These theorems describe the distribution of sample means taken from any population. The most important of these theorems is the central limit theorem. Much of the material in Chapters 10, 11, and 12 is based on the central limit theorem.

We have also presented several new statistical terms, which are listed in the chapter glossary. Be sure you understand each one and how it applies to the material in this chapter. You will encounter these terms many times as you continue in this text.

A summary of the statistical equations used in this chapter follows the glossary.

chapter glossary

central limit theorem For random samples of n observations selected from any population with mean μ_x and standard deviation σ_x, and if n is large, the distribution of possible $\bar{X}$ values will be approximately normal with $\mu_{\bar{x}} = \mu_x$ and $\sigma_{\bar{x}} = \sigma_x/\sqrt{n}$. The approximation improves as n becomes larger.

finite correction factor   The factor used to adjust $\sigma_{\bar{x}}$ when sampling is done without replacement and the sample size is more than 5 percent of the population size. The formula is

$$\sqrt{\frac{N - n}{N - 1}}$$

sampling error   The difference between a population value and the corresponding sample value. Sampling error occurs when the sample does not perfectly represent the population from which it was selected.

standard error of the mean   A measure of the average sampling error. It is determined by dividing the population standard deviation by the square root of the sample size.

unbiased estimator   An unbiased estimator of a population value is an estimator whose expected value equals the population values. For example, $E(\bar{X}) = \mu_x$. The average of all possible sample means will equal the population mean.

chapter formulas

Population mean

$$\mu_x = \frac{\Sigma X}{N}$$

Population standard deviation

$$\sigma_x = \sqrt{\frac{\Sigma(X - \mu_x)^2}{N}}$$

Combinations (Number of possible samples of size n)

$$_NC_n = \frac{N!}{n!(N - n)!}$$

Sample mean

$$\bar{X} = \frac{\Sigma X}{n}$$

Standard error of the mean

If not all samples are known,

$$\sigma_{\bar{x}} = \frac{\sigma_x}{\sqrt{n}}$$

and

$$\sigma_{\bar{x}} = \frac{\sigma_x}{\sqrt{n}}\sqrt{\frac{N - n}{N - 1}}$$

Finite correction factor

$$\sqrt{\frac{N - n}{N - 1}}$$

SOLVED PROBLEMS

Z value

$$Z = \frac{\overline{X} - \mu_x}{\sigma_{\overline{X}}}$$

or

$$Z = \frac{\overline{X} - \mu_x}{\frac{\sigma_x}{\sqrt{n}}}$$

If sampling is without replacement and $n > 5$ percent of the population,

$$Z = \frac{\overline{X} - \mu_x}{\frac{\sigma_x}{\sqrt{n}}\sqrt{\frac{N - n}{N - 1}}}$$

====solved problems

1. In a local agriculture reporting area, the average wheat yield is known to be 60 bushels per acre with a standard deviation of 10 bushels.
   a. If a random sample of 64 acres is selected and the wheat yield recorded, what is the probability the sample mean will lie between 59 and 61 bushels?
   b. Suppose a sample size of 49 acres is selected. What is the probability the sample mean will lie between 59 and 61 bushels?
   c. Why is the probability found in part b different from that found in part a?

Solutions:

a. The sampling distribution will be normally distributed with $\mu_{\overline{X}} = 60$ and $\sigma_{\overline{X}} = 10/\sqrt{64}$, as follows:

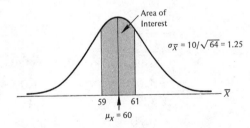

We standardize the sampling distribution by determining the Z values for $\overline{X} = 59$ and $\overline{X} = 61$.

$$Z = \frac{59 - 60}{\frac{10}{\sqrt{64}}} \quad \text{and} \quad Z = \frac{61 - 60}{\frac{10}{\sqrt{64}}}$$

$$= \frac{-1.0}{1.25} \qquad\qquad = \frac{1.0}{1.25}$$

$$= -0.80 \qquad\qquad = 0.80$$

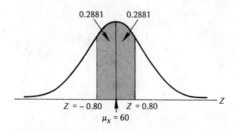

The probability can be determined by adding the areas corresponding to $Z = -0.80$ and $Z = 0.80$. From the normal table in Appendix C, the area corresponding to a Z of -0.80 is 0.2881. The area corresponding to a Z of 0.80 is also 0.2881. Therefore the probability an $\overline{X}$ value will lie between 59 and 61 bushels for a sample of 64 acres is 0.5762.

b. The solution here is the same as for part a except that $\sigma_{\overline{x}}$ is increased due to the smaller sample size. Therefore

$$\mu_{\overline{x}} = 60$$

$$\sigma_{\overline{x}} = \frac{10}{\sqrt{49}}$$

$$= 1.429$$

Thus

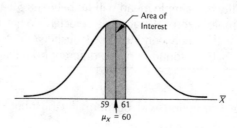

And

$$Z = \frac{59 - 60}{\dfrac{10}{\sqrt{49}}} \quad \text{and} \quad Z = \frac{61 - 60}{\dfrac{10}{\sqrt{49}}}$$

$$= \frac{-1.0}{1.429} \qquad\qquad = \frac{1.0}{1.429}$$

$$= -0.70 \qquad\qquad\quad = 0.70$$

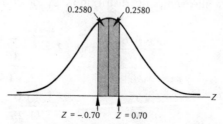

The probability, therefore, is $0.2580 + 0.2580 = 0.5160$.

c. When the sample size is reduced (64 to 49), the probability of obtaining a sample mean between 59 and 61 is reduced. With a smaller sample size, the sampling distribution is spread more, giving a greater opportunity for extreme $\overline{X}$ values.

---

2. A local insurance company has 240 employees who have an average annual salary of $21,000. The standard deviation of annual salaries is $5,000.

a. In a random sample of 100 employees, what is the probability the average salary will exceed $21,500?

b. What is the probability the sample mean found in part a will be less than $22,000?

c. Why does the finite correction factor need to be used in determining $\sigma_{\overline{x}}$?

Solutions:

a.

$$\mu_{\overline{x}} = \$21,000$$

$$\sigma_{\overline{x}} = \frac{\sigma_x}{n}\sqrt{\frac{N-n}{N-1}} = \frac{5,000}{\sqrt{100}}\sqrt{\frac{240-100}{239}}$$

$$= \$382.68$$

$$Z = \frac{\overline{X} - \mu_x}{\dfrac{\sigma_x}{\sqrt{n}}\sqrt{\dfrac{N-n}{N-1}}} = \frac{\$21,500 - \$21,000}{\$382.68}$$

$$= 1.31$$

From Appendix C, the area corresponding to a Z of 1.31 is 0.4049.

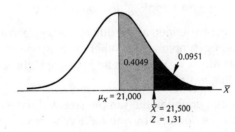

The probability of $\overline{X} \geq \$21,500$ is 0.0951.

b. To find the probability the mean salary of 100 employees will be less than $22,000, we must find the following area:

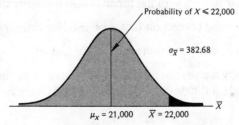

Then

$$Z = \frac{\$22,000 - \$21,000}{\$382.68}$$

$$= 2.61$$

From the normal distribution table in Appendix C, the area corresponding to $Z = 2.61$ is 0.4955. The probability of an $\overline{X}$ value below $22,000 is, therefore, $0.5000 + 0.4955 = 0.9955$.

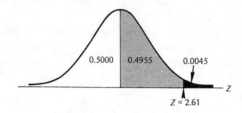

c. The finite correction factor must be used in determining $\sigma_{\overline{x}}$ because the sample size is quite large relative to the population size and the sampling is assumed to be without replacement.

problems

Problems marked with an asterisk (*) require extra effort.

1. Discuss in your own words the central limit theorem and indicate why it is important to decision makers.

2. What is meant by the term *sampling distribution of $\overline{X}$*?

3. Sampling error has been defined as the difference between the population value and the corresponding sample value. When dealing with the population mean, μ_x, and the sample mean, $\overline{X}$, why does an increase in the sample size decrease the average sampling error?

4. Would you agree that doubling the sample size will reduce the standard error of the sampling distribution of $\overline{X}$ by one-half? Why or why not?

5. Suppose a population is normally distributed. What is the probability of finding a sample mean, $\overline{X}$, that is greater than the population mean?

6. For a normally distributed population, what size sample is required to insure that the sampling distribution of the mean is also normally distributed? Explain.

7. When should the finite correction factor be used in sampling applications?

8. What is the effect of using the finite correction factor?

9. The Chair Company repairs old furniture and restores it to "better than original" condition. Records indicate that the time it takes to refinish and otherwise

restore a standard dining room set is normally distributed with a mean of 30 hours and a standard deviation of 5 hours. Recently a customer complained that he was charged too much for work performed by The Chair Company. To settle the argument, the manager of the company offered the customer the following option: "We will select a random sample of past work performed on tables similar to yours. If the sample mean based on five work times turns out less than yours, The Chair Company will refund your money. If the mean of this sample turns out to be greater than or equal to your billed time, you pay us half again the amount of the bill."

Taking into account the average and standard deviation of all work times on file, do you think the manager is wise to make such an offer if this customer's billed time was 32 hours? Discuss why or why not.

10. What would be your response if the customer's billed time in problem 9 was 34 hours?

11. Given The Chair Company problems just presented, what is the probability of finding a single time as great as or greater than the 32 hours billed to this customer? Discuss why this probability is different from the probability of the mean of a sample of five billed times being as great as or greater than 32 hours.

12. The Swim and Racquet Club is in the process of establishing a policy for how long a court may be reserved at any one time. To help make this decision, the club managers have selected a random sample of 100 tennis matches and determined that the mean time for completion is 75 minutes. What is the probability of finding a sample mean as small as or smaller than this if, in fact, the true average completion time is 90 minutes with a standard deviation of 10 minutes, as some members have claimed?

*13. The Environmental Protection Agency (EPA) requires all U.S. automobile makers to test their cars for mileage in the city and on the highway. One company has indicated that a certain model will get 25 m.p.g. in the city and 32 m.p.g. on the highway. However not all cars of a given model will get the same mileage; these mileage ratings are simply averages. Further, because there is variation among cars, the manufacturer has discovered that the standard deviation is 3 m.p.g. for city driving, and 2 m.p.g. for highway driving.

Given this information, suppose the San Francisco Police Department has purchased 64 cars from this company (a random sample). The police officers have driven these cars exclusively in the city and have recorded an average of 21 m.p.g. What would you conclude regarding the EPA mileage rating for this model car? Indicate the basis for your conclusion.

*14. Referring to problem 13, the police chief for the San Francisco Police Department has asked that his police officers drive the cars to Los Angeles and back to determine how these 64 cars perform in highway driving. These 64 cars averaged 34 m.p.g. Based upon the EPA rating, what should the police chief conclude? Explain.

15. The Mason Construction Company has built a total of 50 homes in the Seattle area in an average time of 35 days with a standard deviation of 10 days. A

prospective customer has interviewed 40 of the 50 homeowners about the quality of construction, and so forth. One of the questions asked the home-owners was how long it took the builder to construct their homes. The 40 responses averaged 46 days. What would you conclude about these responses? Explain.

16. The Sullivan Advertising Agency has determined that the average cost to de-velop a 30-second commercial is $20,000. The standard deviation is $3,000. Suppose a random sample of 50 commercials is selected and the average cost is $20,300. What are the chances of finding a sample mean this high or higher?

17. Suppose the Sullivan Advertising Agency is interested in establishing a pricing policy for prospective customers of 30-second commercials. Given that the mean cost is $20,000 with a standard deviation of $3,000, what are the chances of a given commercial costing between $19,500 and $22,000? What is the probability of a sample of 36 commercials having an average cost between $19,500 and $22,000? Explain why these probabilities are different.

18. The Baily Hill Bicycle Shop sells ten-speed bicycles and offers a maintenance program to its customers. The manager has found the average repair bill during the maintenance program's first year to be $15.30 with a standard deviation of $7.00. What is the probability that a random sample of 40 customers will have a mean repair cost exceeding $16.00?

19. Referring to problem 18, what is the probability that the mean repair cost for a sample of 100 customers will be between $15.10 and $15.80?

20. The central limit theorem indicates that the sampling distribution of $\overline{X}$ will have a standard deviation of $\sigma_x/\sqrt{n}$. Discuss in your own words why the sampling distribution of $\overline{X}$ should have less dispersion than the population.

21. As part of a marketing study, the Food King Supermarket chain has randomly sampled 150 customers. The average dollar volume purchased by the cus-tomers in this sample was $33.14.

Before sampling, the company assumed that the distribution of customer purchases had a mean of $30.00 and a standard deviation of $8.00. If these figures are correct, what is the probability of observing a sample mean of $33.14 or greater? What would this probability indicate to you?

22. The Bendbo Corporation has a total of 300 employees in its two manufacturing locations and the headquarters office. A study conducted five years ago showed that the average commuting distance to work for Bendbo employees was 6.2 miles with a standard deviation of 3 miles.

Recently a follow-up study based upon a random sample of 100 employees indicated an average travel distance of 5.9 miles. Assuming the mean and standard deviation of the original study hold, what is the probability of obtaining a sample mean of 5.9 miles or less? Based upon this probability, do you think the average travel distance may have decreased?

23. Referring to problem 22, a second random sample of size 40 was selected. This sample produced a mean travel distance of 5.9 miles. If the mean for all employees is 6.2 miles and the standard deviation is 3 miles, what is the probability of observing a sample mean of 5.9 miles or less?

24. Referring to problems 22 and 23, discuss why the probabilities differ. After all, the sample results were the same in each case!

*25. A marketing consultant has claimed that the average family income in a certain market area is at least $18,000 per year with a standard deviation of $4,000. A market research firm wishes to test this claim by selecting a random sample of 64 families. The process to test the claim begins with computing $\overline{X}$. If this sample mean is less than some specified value (called A), the claim will be rejected; otherwise, the claim will be accepted.

   Suppose the market-research company wishes to set the value A such that $P(\overline{X} \leq A) = 0.05$. Determine the appropriate value of A.

*26. Referring to problem 25, suppose the marketing research firm were to select a random sample of size 100 rather than 64 to test the claim. Determine A such that $P(\overline{X} \leq A) = 0.05$ given this change in sample sizes. Compare this answer to the one you found for problem 25.

*27. An automatic saw at a local lumber mill cuts four 2 x 4s to an average length of 120 inches. However, since the saw is a mechanical device, not all 2 x 4s are 120 inches. In fact, the distribution of lengths has a variance of 0.64. The saw operator just took a sample of 36 boards.
   a. If the saw is set correctly, what is the probability the average length of the sample boards is more than 120.6 inches?
   b. What is the probability the sample length is less than 119.3 inches?
   c. What should the saw operator conclude if she finds the sample to have an average length of 120.3 inches?

=references

LAPIN, LAWRENCE. *Statistics for Modern Business Decisions.* New York: Harcourt Brace Jovanovich, 1978.
MENDENHALL, WILLIAM and JAMES REINMUTH. *Statistics for Management and Economics.* North Scituate, Mass.: Duxbury, 1978.
NETER, JOHN, WILLIAM WASSERMAN, and G. A. WHITMORE. *Applied Statistics.* Boston: Allyn and Bacon, 1978.

10

Statistical Estimation— Large Samples

In Chapter 8 we emphasized that decision makers cannot always measure an entire population but often must rely on information gained by sampling the population. The decision made therefore often depends on the sample information, which, as we showed in Chapter 9, is subject to sampling error. Sampling error can cause problems for decision makers who are not familiar with statistical estimation. Suppose a market-research firm needs to know the average per capita income in a city before it can advise its client whether to open a new retail outlet in that city. This firm would likely select a statistical sample and compute the mean per capita income. However, as we showed in Chapter 9, the sample mean does not have to equal the population mean. Therefore the market-research firm has to include the possible sampling error in its estimate of the true population mean.

In this chapter we introduce the statistical techniques for estimating a population value with sample information if the sample is large.

chapter objectives

In this chapter we shall introduce the process of estimating population values based on samples from the population. Specifically, we will introduce point estimates and

221

interval estimates for such parameters as the population mean, the population proportion, the difference between two population means, and the difference between two population proportions. We will concentrate on estimation procedures for large sample sizes.

We shall also introduce the concepts of precision and confidence level for an estimate. We will show how to determine the sample size necessary to maintain a specified level of confidence and precision.

student objectives ════════════════════════════════

After studying the material in this chapter, you should be able to:

1. Discuss the difference between a point estimate and an interval estimate.
2. Discuss the advantages of a confidence interval estimate and recognize applications of such an estimate.
3. Calculate a confidence interval estimate for each of the following:
   a. Population mean
   b. Population proportion
   c. Difference between two population means
   d. Difference between two population proportions
4. Determine the impact of sample size on the confidence interval.
5. Discuss the importance of precision in a confidence interval estimate.
6. Determine the required sample size for specific estimation problems involving population means and proportions.
7. Identify business decision applications that require statistical estimation.

10-1
THE NEED FOR STATISTICAL ESTIMATION

A large motel chain with several hundred motels throughout the United States is considering changing the brand of television set in its motel rooms. The major consideration is to select the brand with the lowest combination of initial cost and maintenance cost. Although the motel management can get complete information on the initial purchase costs for the various brands, it is uncertain about the total maintenance costs. Management plans to keep the sets in the rooms for five years regardless of the brand it decides to purchase. The managers know that for any brand of television, some sets will require more maintenance than others. They also know that regardless of the brand they select, the actual total maintenance cost will not be known until the five years have passed. Obviously the managers cannot wait for five years of cost data to make their decision.

An electronics company in Delaware has developed a new testing simulator that uses heat to age the television parts. This electronics company can record all maintenance during a simulated five years for the television set being tested. The television industry has vouched for the simulator's accuracy, and the motel managers are confident of its results.

Since the motel chain will purchase 4,000 television sets, the only way the motel managers will be able to make a decision is to sample a few televisions of each brand. They will calculate the average maintenance cost for each brand, and use the sample means to estimate what the average maintenance cost will be for all televisions of each brand.

Like these motel managers, decision makers often need to use sample information to make estimates about a population. Market research relies heavily on statistical estimation. The market researchers select a sample of potential customers for a new product or service and, from that sample, estimate the proportion of people in the entire market area who will purchase the product or service. The list of business applications is endless. You will be introduced to a variety of applications of statistical estimation throughout this chapter.

10-2
POINT ESTIMATES

We have all seen the results of political polls taken during every election. These polls indicate the percentage of voters who favor a particular candidate or a particular issue. For example, suppose a poll indicates that 62 percent of the population favor a 55-mile-per-hour speed limit on the nation's highways. The pollsters have not contacted every person in the United States, but rather have sampled only a relatively few people to arrive at the 62 percent figure. In statistical terminology, the 62 percent is the **point estimate** for the true population percentage who favor a 55-mile-per-hour speed limit. **In general, a *point estimate* is a single number determined from a sample and is used to estimate the population value.**

The Environmental Protection Agency (EPA) tests automobile models sold in the United States to determine their mileage ratings. Following the testing, each model is assigned an EPA mileage rating based upon the test mileage. This rating is actually a point estimate for the true average of all cars of the given model.

Cost accountants make detailed studies of their company's production process to determine the costs of producing each product. These costs are often found by selecting a sample of items and following each item through the complete production process. The costs at each step in the process are determined, and the total cost is found when the process is completed. The accountants calculate the average cost for the items sampled and use this figure as the point estimate for the true average cost of all pieces produced. The point estimate becomes the basis for assigning a selling price to the finished good.

The federal government publishes many population estimates. Among these are estimates of the median family income and the proportion of unemployed persons. These values are calculated from samples and are point estimates of the population values.

Which point estimator the decision maker uses depends on the population characteristic the decision maker wishes to estimate. Often there is a choice of estimators for any population characteristic. For example, if we want to estimate the center of the population distribution, we might calculate the sample mean, the sample median, or the sample mode. Or, if we wish to estimate the spread of the population, we might calculate the sample variance or the range.

Given that decision makers are faced with a choice of estimators, how can they select the appropriate one? To answer this question, we shall discuss three criteria that are often used to judge point estimators. These criteria are **unbiasedness,** **consistency,** and **efficiency.**

Unbiasedness

We introduced the concept of an unbiased estimator in Chapter 9. We stated that **an estimator is *unbiased* if the average value of the estimator equals the population parameter.**

We also showed that the sample mean, $\overline{X}$, is unbiased since its expected value (or average) equals the population mean, μ_x. We could also show that the sample variance,

$$S_x^2 = \frac{\Sigma(X - \overline{X})^2}{n - 1} \qquad \textbf{(10–1)}$$

is an unbiased estimator of σ_x^2 if the population is infinite. This is why $n - 1$ rather than n is used in the denominator for the sample variance.

An unbiased estimator is correct on the average. However, this does not mean that each estimate will equal the population parameter. As we showed in Chapter 9, it is highly possible, and in fact likely, that $\overline{X}$ will not exactly equal μ_x. The difference is called *sampling error*, and can be quite large. Consequently, that any estimator is unbiased does not by itself mean that the estimator is a good estimator. Other criteria must be considered.

Consistency

Sampling error is the difference between the estimate and the parameter. Although unbiasedness tells us the average sampling error is zero, it says nothing about how close individual estimates will be to the parameter. The criterion of **consistency** recognizes the need for small sampling error. **If an estimator is *consistent*, in repeated sampling, the larger the sample size, the closer the estimator comes to the population parameter.** Take for example the sample mean, $\overline{X}$, as the estimator of the population mean, μ_x. As the sample size increases, the distribution of possible sample means becomes more concentrated around μ_x. Thus the potential for sampling error decreases as the sample size increases. The standard deviation of the sampling distribution, $\sigma_{\overline{x}}$, is $\sigma_x/\sqrt{n}$. Therefore $\sigma_{\overline{x}}$ decreases if n increases. Consequently $\overline{X}$ is said to be a consistent estimator of μ_x.

Efficiency

Suppose we have arrived at two *unbiased* estimators of the same parameter. Further, suppose that both of these are *consistent* estimators. **A third criterion for choosing** between the two is to select the one whose sampling distribution has the least variability for a given sample size. This would be the *most efficient estimator.* Figure 10–1 illustrates the concept of efficiency for two estimators, $\overline{X}$ and the median, of the population parameter μ_x. The sampling distribution of $\overline{X}$ is less spread out than the sampling distribution of the median. Therefore, when trying to etimate μ_x, the sample mean is preferred over the sample median because there is less chance of extreme sampling error for a given sample size.

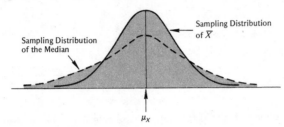

FIGURE 10–1 Efficiency in a statistical estimator.

Sufficiency

We have discussed three important evaluation criteria for point estimators. To be complete, we should mention at least one more—*sufficiency.* Although the mathematics necessary to fully explain this criterion are beyond the scope of this text, we shall provide a simple definition. **A sufficient estimator is an estimator that utilizes all information in a set of sample data.** For example, the sample mean, $\overline{X}$, is a sufficient estimator of μ_x.

Even if a point estimator is unbiased, consistent, efficient, and sufficient, you can be quite certain that the estimate will not equal the true population value. We always expect some sampling error. Thus, when cost accountants utilize $\overline{X}$, the average cost of a sample of pieces, to establish the total production cost, the point estimate will most likely be wrong. But they will have no way of determining how wrong it is.

To overcome this problem with point estimates, the most common procedure is to calculate an *interval* that the decision maker is confident contains the true population parameter. These intervals are called **confidence intervals** and are the subject of the next section.

10-3
CONFIDENCE INTERVAL ESTIMATION

The production manager of the Valley View Canning Company is responsible for monitoring the filling operations of cans. His company has recently installed a new machine that has been carefully tested and is known to fill cans of any size with a standard deviation, $\sigma_x = 0.2$ ounce. The manager's main problem is to adjust the

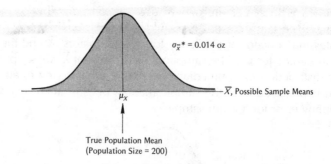

$$^*\sigma_{\bar{x}} = [\sigma_x/\sqrt{n}]\ [\sqrt{(N-n)/(N-1)}] = [0.2/\sqrt{100}]\ [\sqrt{(200-100)/(200-1)}]$$
$$= 0.014\ oz$$

The finite correction factor is used because the sample size is greater than 5% of the population size.

FIGURE 10-2   Sampling distribution of $\bar{X}$, Valley View Canning Company. (peach can fill in ounces)

average fill to the level specified on the can, for instance, 16 ounces, 24 ounces, and so forth.

Suppose he has just made a setting in an attempt to fill the peach cans to an average of 16 ounces. He starts the machine and fills 200 cans. From the 200 cans he selects a random sample of 100 and carefully weighs each one. The sample mean, $\bar{X}$, provides a point estimate of the population mean, μ_x. Suppose this value is

$$\bar{X} = 16.8\ oz$$

Because of the nature of any point estimate, the manager does not expect to find a sample mean that exactly equals 16 ounces. He also knows from the central limit theorem that the distribution of all possible samples means will be approximately normal around the true mean. This is illustrated in Figure 10-2.

Thus, although the manager does not expect a sample mean of 16 ounces, he will likely allow the process to continue if the sample mean is close to 16 ounces. To determine just what "close" is, the manager needs to determine a **confidence interval.**

10-4
CONFIDENCE INTERVAL ESTIMATION OF μ_x

Large Samples, σ_x Known

The Valley View Canning Company manager specifies that he wants a 95 percent confidence interval. This means that of all the intervals he might obtain, 95 percent will include the true mean. He figures that if the interval includes 16 ounces, the true mean might actually be 16 ounces and he will leave the machine setting as is.

The general format for the confidence interval is

Point estimate ± (interval coefficient)(standard error)

The point estimate for our canning example is $\overline{X}$. The standard error is $\sigma_{\overline{x}}$, which is determined by

$$\sigma_{\overline{x}} = \frac{\sigma_x}{\sqrt{n}} \sqrt{\frac{N - n}{N - 1}} \qquad (10\text{–}2)$$

Again, the finite correction factor is used because the sample is greater than 5 percent of the population.

The remaining factor, **the *interval coefficient,* is the number of standard errors on either side of the population mean necessary to include a percentage of the possible sample means equal to the confidence level.** When the sample size is *large* and the population standard deviation is *known*, the interval coefficient is a Z value from the standard normal distribution table in Appendix C. For example, if the desired confidence level is 95 percent, the Z value (interval coefficient) is 1.96, as shown in Figure 10–3.

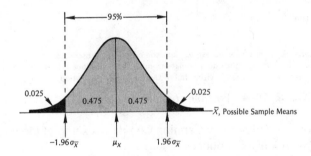

FIGURE 10–3   Interval coefficient, 95% confidence interval.

Thus the format for a confidence interval for estimating μ_x is

$$\overline{X} \pm Z\sigma_{\overline{x}} \qquad (10\text{–}3)$$

For this example, the confidence interval is

$$\overline{X} \pm 1.96 \frac{\sigma_x}{\sqrt{n}} \sqrt{\frac{N - n}{N - 1}} \qquad (10\text{–}4)$$

Any value of $\overline{X}$ between $\mu_x - 1.96\sigma_{\overline{x}}$ and $\mu_x + 1.96\sigma_{\overline{x}}$ will produce a confidence interval that contains μ_x. Using $Z = 1.96$ indicates that since 95 percent of the possible sample means come from this range, 95 percent of the potential confidence intervals will include the population mean. Figure 10–4 illustrates this important concept by showing a few of the possible intervals.

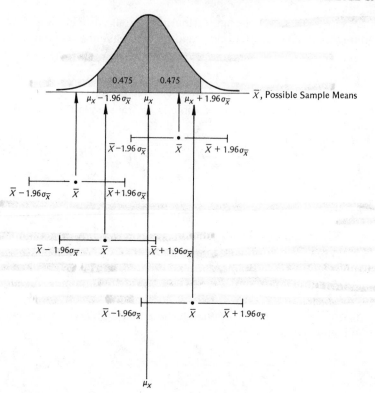

Note: Some intervals include μ_x, and some do not. Those intervals that do not contain the population mean are developed from sample means that fall in either tail of the sampling distribution.

FIGURE 10–4   Possible confidence intervals.

For the Valley View Canning Company example, the 95 percent confidence interval with $\bar{X}$ of 16.8 ounces is

$$\bar{X} \pm 1.96\, \frac{\sigma_x}{\sqrt{n}} \sqrt{\frac{N-n}{N-1}}$$

$$16.8 \pm 1.96\, \frac{0.2}{\sqrt{100}} \sqrt{\frac{200-100}{200-1}}$$

$$16.8 \pm 1.96(0.014)$$

$$16.8 \pm 0.027$$

Therefore the 95 percent confidence interval estimate for the true average fill is

$$16.773 \text{ oz} \underline{\hspace{2cm}} 16.827 \text{ oz}$$

Because this represents a 95 percent confidence interval, the manager is quite certain this interval *does* include the true mean. Since the desired average of 16 ounces is not contained within the limits, he will conclude (rightly or wrongly) that the true mean is not 16 ounces and will order further machine adjustment. The sampling and estimation process will be repeated.

Suppose after several adjustments the sample mean, $\overline{X}$, for a sample of size 100 from a run of 200 cans is 16.02 ounces. The 95 percent confidence interval developed from this point estimate is

$$\overline{X} \pm 1.96\sigma_{\overline{x}}$$

$$16.02 \pm 1.96(0.014)$$

$$15.99 \text{ oz} \underline{\hspace{2cm}} 16.04 \text{ oz}$$

The production manager knows that 95 percent of all possible confidence intervals should include μ_x, the true population mean. Therefore he feels confident that the true mean is between 15.99 ounces and 16.04 ounces. Because 16 ounces falls in this interval, the manager should not make any further adjustments in the machine setting at this time.

Suppose the filling process is allowed to continue and after eight hours, 10,000 cans of peaches have been filled. The production manager might then halt the run and select a random sample of 100 cans from the 10,000 to determine whether the filling machine has remained in adjustment. He will calculate a 95 percent confidence interval estimate of μ_x, the average content of the 10,000 cans. Suppose the sample mean is 16.01 ounces. Then the 95 percent confidence interval is

$$\overline{X} \pm 1.96 \frac{\sigma_x}{\sqrt{n}}$$

$$16.01 \pm 1.96 \frac{0.2}{\sqrt{100}}$$

$$16.01 \pm 0.0392$$

$$15.9708 \text{ oz} \underline{\hspace{2cm}} 16.0492 \text{ oz}$$

Note that we did not use the finite correction factor in calculating $\sigma_{\overline{x}}$, the standard error of the sampling distribution, because the sample is less than 5 percent of the population. The confidence interval includes 16 ounces. Therefore the production manager would have no reason to believe the true mean is not 16 ounces.

Precision of the Estimate.   The production manager for Valley View Canning is concerned about the width of his confidence interval. Though he is willing to believe, based on a sample of 100, that the true mean fill is 16 ounces because this value falls within the interval, he should also be willing to believe that the true mean fill is 16.03 ounces or 15.98 ounces because they also fall in the interval. The manager would like to decrease the width of the confidence interval estimate. **The width of a confidence interval is called the *precision* of the estimate.**

One way to increase the precision of an estimate is to decrease the confidence level. For example, if the manager is willing to accept a 90 percent confidence level, the interval coefficient can be reduced from 1.96 to 1.645 as shown

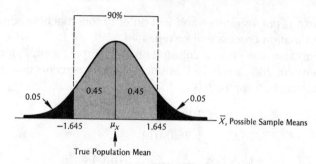

FIGURE 10–5 Interval coefficient, 90% confidence interval.

in Figure 10–5. For a sample of 100 selected from the production run of 10,000 cans, the 90 percent confidence interval from a sample mean of 16.01 is

$$\bar{X} \pm Z \frac{\sigma_x}{\sqrt{n}}$$

$$16.01 \pm 1.645 \frac{0.2}{\sqrt{100}}$$

$$16.01 \pm 0.0329$$

$$15.9771 \text{ oz} \underline{\hspace{2cm}} 16.0429 \text{ oz}$$

This represents a slight *increase* in precision (narrowing the interval) from the 95 percent confidence level. A confidence level of 80 percent would increase the precision even further.

$$\bar{X} \pm Z \frac{\sigma_x}{\sqrt{n}}$$

$$16.01 \pm 1.28 \frac{0.2}{\sqrt{100}}$$

$$16.01 \pm 0.0256$$

$$15.9844 \text{ oz} \underline{\hspace{2cm}} 16.0356 \text{ oz}$$

We have increased the precision by decreasing the confidence level, but what is the trade-off? Reducing the confidence from 95 percent to 80 percent means that now only 80 percent of all intervals are expected to include the true population mean. Thus the chances of obtaining an interval that includes the true mean have been decreased. The trade-off is between obtaining a narrow, more precise interval and increasing the chances of obtaining an interval estimate that does not contain the true population mean. The extent to which a decision maker is willing to reduce the confidence level depends upon the cost associated with an interval that does not contain μ_x. The higher this cost, the less willing the decision maker will be to reduce the confidence level.

A second alternative for increasing the precision of an interval estimate is to increase the sample size. Since an increase in sample size will reduce $\sigma_{\bar{x}}$, it will also reduce the confidence interval width. For example, suppose the production manager at Valley View Canning selects a random sample of 400 cans from the production run of 10,000 cans. Further, suppose the sample mean, $\bar{X}$, for this larger sample is also 16.01 ounces. The 95 percent confidence interval would be

$$\bar{X} \pm Z \frac{\sigma_x}{\sqrt{n}}$$

$$16.01 \pm 1.96 \frac{0.2}{\sqrt{400}}$$

$$16.01 \pm 0.0196$$

15.9904 oz _____ 16.0296 oz

This interval is more precise than the 95 percent interval for a sample of 100 cans, which was calculated previously to be

15.9708 oz _____ 16.0492 oz

If the sample size is increased to 800, precision can be increased further.

$$\bar{X} \pm Z \frac{\sigma_x}{\sqrt{n}}$$

$$16.01 \pm 1.96 \frac{0.2}{\sqrt{800}}$$

$$16.01 \pm 0.0138$$

15.9961 oz _____ 16.0238 oz

There are obvious trade-offs associated with selecting a larger sample size, generally in terms of time and money. Decision makers must decide the benefit to be gained from increased precision and balance this gain against the increased costs.

Sample Size Determination.   Often a decision maker has a desired confidence level and a specified precision level but is unsure about what sample size to select. For example, suppose an economist wishes to estimate the average annual family income of the residents in an agricultural county in Nebraska. Assume that the population standard deviation, σ_x, is known to be $1,000. The economist wants 95 percent confidence and a precision of $300. Therefore she needs a confidence interval with a total width of no more than $300. How large a sample size is required to satisfy the economist's confidence and precision requirements?

To answer this question, we must introduce a new concept, *tolerable error. Tolerable error* **is one-half the width of the confidence interval.** Recall that

$$\bar{X} \pm Z \frac{\sigma_x}{\sqrt{n}}$$

is the format of a confidence interval estimate of μ_x. Then

$$e = Z \frac{\sigma_x}{\sqrt{n}} \tag{10–5}$$

where: e = tolerable error

Z = interval coefficient

σ_x = population standard deviation

n = sample size

Since tolerable error is half the width of a confidence interval, the economist, who wishes an interval width of $300, will accept a tolerable error of $150. We can solve equation (10–5) for n, the sample size, as follows:

$$e = Z \frac{\sigma_x}{\sqrt{n}}$$

$$n = \frac{Z^2 \sigma_x^2}{e^2} \tag{10–6}$$

Equation (10–6) applies to problems where the population can be considered infinite or the sampling is performed with replacement.

For the economist's problem,

$$e = Z \frac{\sigma_x}{\sqrt{n}}$$

$$\$150 = 1.96 \frac{\$1,000}{\sqrt{n}}$$

$$n = \frac{(1.96)^2 (\$1,000)^2}{(\$150)^2} = 170.73$$

$$= 171 \text{ families}$$

Thus, to obtain a 95 percent confidence with a tolerable error of $150 (precision = $300), the economist would need to sample 171 families from the county.

What would be the impact on the sample size of increasing the required precision to $200? This means the tolerable error is reduced to $100. The required sample size would be calculated as follows:

$$e = Z \frac{\sigma_x}{\sqrt{n}}$$

$$\$100 = 1.96 \frac{\$1,000}{\sqrt{n}}$$

$$n = \frac{(1.96)^2 (\$1,000)^2}{(\$100)^2} = 384.16$$

$$= 385 \text{ families}$$

The economist may or may not be able to afford increasing the sample size from 171 to 385. As we indicated earlier, **decision makers will be forced, in practical applications, to strike a balance between required confidence and precision and the sample size they can afford.**

Large Samples, σ_x Unknown

In the previous section we discussed confidence interval estimates for *large* samples and a *known* population standard deviation. Although the population standard deviation may be known, we have found that in most business applications, if μ_x is not known, neither is σ_x. When σ_x is not known, it must also be estimated. The appropriate estimator is S_x, the **sample standard deviation,** calculated by

$$S_x = \sqrt{\frac{\Sigma(X - \overline{X})^2}{n - 1}} \qquad \text{(10–7)}$$

where: S_x = sample standard deviation

$\overline{X}$ = sample mean

n = sample size

If the sample size is *large*, the confidence interval estimate of μ_x is approximated by

$$\overline{X} \pm Z \frac{S_x}{\sqrt{n}} \qquad \text{(10–8)}$$

However, if the sample size is *large relative to the population,* and sampling is performed *without replacement,* the finite correction factor should be used. Thus the confidence interval is

$$\overline{X} \pm Z \frac{S_x}{\sqrt{n}} \sqrt{\frac{N - n}{N - 1}} \qquad \text{(10–9)}$$

Suppose the Ohio State Highway Department has undertaken a study of highway accidents and their causes. As one part of the study, the administrators need to estimate the average speed of vehicles around a dangerous curve on the interstate highway. They have selected a random sample of 500 vehicles and used radar to measure the vehicles' speed, finding

$$\overline{X} = 52.31 \text{ mi/h}$$

$$S_x = 6.30 \text{ mi/h}$$

The administrators want a 98 percent confidence interval. Therefore

$$\overline{X} \pm Z \frac{S_x}{\sqrt{n}}$$

$$52.31 \pm 2.33 \frac{6.30}{\sqrt{500}}$$

$$52.31 \pm 0.66$$

$$51.65 \text{ mi/h} \underline{\hspace{2cm}} 52.97 \text{ mi/h}$$

For the purposes of their report, the administrators are confident that the actual average speed for all cars traveling on this curve is between 51.65 and 52.97 miles per hour.

In another example, the postmaster at a large eastern post office was required to report the average weight per package mailed at her post office and the number of packages mailed during September. The post office auditors could then multiply the average weight by the number of packages to determine the total weight of all packages. Then they could multiply the total weight by the mailing rate to provide a rough estimate of what the income should have been from package mailing.

Because of the time it would take to record the weight of all the 49,007 packages mailed, the postmaster was granted permission to select a sample of packages. She reported the following information to the auditors:

$$N = 49,007$$
$$n = 100$$
$$\bar{X} = 3.12 \text{ lb}$$
$$S_x = 2.8 \text{ lb}$$

The auditors then used this information to construct a 95 percent confidence interval for estimating μ_x, the true average package weight.

$$\bar{X} \pm 1.96\sigma_{\bar{x}}$$

$$3.12 \pm 1.96\frac{2.8}{\sqrt{100}}$$

$$3.12 \pm 0.55$$

$$2.57 \text{ lb} \underline{\qquad\qquad} 3.67 \text{ lb}$$

The estimate for total weight is found by multiplying the lower limit and upper limit by 49,007 packages, which gives

$$125,949.99 \text{ lb} \underline{\qquad\qquad} 179,855.69 \text{ lb}$$

The auditors will multiply these limits by the appropriate rate to arrive at an interval estimate for income from package mailing. They will compare the post office's reported income with this estimate. If the reported income is within that interval, no formal audit will be conducted. Otherwise, the department will conduct a full-scale audit.

10-5
CONFIDENCE INTERVAL ESTIMATION OF A POPULATION PROPORTION

In Chapter 6 we introduced the binomial distribution. Recall that the binomial distribution applies when the following conditions are satisfied:

1. There are n identical trials, each of which results in one of two possible outcomes, success or failure.

2. The n trials are *independent* and the probability of a success remains constant from trial to trial.

Also recall that the *average of the binomial distribution* is

$$\mu_X = np \tag{10-10}$$

where: n = sample size (number of trials)

p = probability of a success

and that the *standard deviation of the binomial distribution* is

$$\sigma_X = \sqrt{npq} \tag{10-11}$$

where: $q = 1 - p$

The binomial distribution has many applications. However, rather than attempting to count the number of successes, decision makers will often want to determine the proportion or percentage of successes in the population. A Republican candidate for the U.S. Senate is concerned about the proportion of voters who will vote Republican. The quality control manager for a large toy manufacturer is concerned about the proportion of defective toys his company produces. The president of the United States and his economic advisors are concerned about the proportion of the labor force that are unemployed. These are but a few of the situations in which decision makers need to know the ***population proportion.*** The population proportion is determined as follows:

$$p = \frac{X}{N} \tag{10-12}$$

where: p = population proportion

X = total number of successes in the population

N = population size

If decision makers have access to the entire population, the calculated *p* value is a *parameter*. However, as is the case with any other population parameter, rarely can decision makers measure the entire population. Consequently a random sample must be selected and the population proportion estimated. The point estimate for the population proportion is

$$\hat{p} = \frac{X}{n} \tag{10-13}$$

where: $\hat{p}$ = sample proportion

X = number of successes in the sample

n = sample size

The sampling distribution of $\hat{p}$ values can be approximated by a normal distribution if the sample size is large and the population proportion, p, is not too close to zero or 1.0. The expected value and standard deviation of the sampling distribution are

$$E(\hat{p}) = \mu_{\hat{p}} = p \tag{10-14}$$

$$\sigma_{\hat{p}} = \sqrt{\frac{pq}{n}} \tag{10-15}$$

where: $\mu_{\hat{p}}$ = expected proportion of successes

$\sigma_{\hat{p}}$ = standard error of the sampling distribution of $\hat{p}$

$q = 1 - p$

n = sample size

There is a problem with equations (10–14) and (10–15). The mean and standard deviation of the sampling distribution, $\mu_{\hat{p}}$ and $\sigma_{\hat{p}}$, are determined by knowing p, the population proportion. But we want to estimate p in the first place. We overcome this problem by substituting $\hat{p}$ and $\hat{q}$ ($\hat{q} = 1 - \hat{p}$) for p and q in equation (10–15).

$$S_{\hat{p}} = \sqrt{\frac{\hat{p}\hat{q}}{n}} \qquad\qquad (10\text{–}16)$$

With this equation, we can construct a confidence interval estimate for the population proportion as follows:

$$\hat{p} \pm ZS_{\hat{p}} \qquad\qquad (10\text{–}17)$$

where: $\hat{p}$ = point estimate

Z = interval coefficient

$S_{\hat{p}}$ = estimated standard error of the sampling distribution

Wood Roofs, Inc., a growing business, operates in several northwestern states. This company sells cedar shake roofs to people who currently have a different type roof but want a more esthetically pleasing look. Wood Roofs, Inc. is considering opening an outlet in a northern California city but, before doing so, the manager wishes to estimate the proportion of homes in the city that already have cedar shake roofs. If this percentage is too high, the company will not operate in that city. The manager wants a 90 percent confidence interval and has selected a sample of 200 homes, with the following results:

$\hat{p} = 0.36$ (proportion of homes with cedar shakes)

$$S_{\hat{p}} = \sqrt{\frac{(0.36)(0.64)}{200}}$$

$$= 0.0339$$

The 90 percent confidence interval estimate for the true proportion of homes currently with cedar shake roofs is

$$\hat{p} \pm 1.645S_{\hat{p}}$$

$$0.36 \pm (1.645)(0.0339)$$

$$0.36 \pm 0.0557$$

$$0.3043 \underline{\hspace{2cm}} 0.4157$$

Based upon this interval, the manager is confident that the true proportion is between 0.3043 and 0.4157. From this estimate, she will have to decide whether to open an outlet in this city.

Sample Size Requirements

Suppose the manager of Wood Roofs, Inc. is unhappy with the precision of the confidence interval just calculated. She indicates that the precision (interval width) should be no greater than 0.05. Assuming the manager wishes to retain the 90 percent confidence level, how many houses should be included in the sample?

The appropriate sample size is found using the following equation:

$$e = Z\sqrt{\frac{pq}{n}} \qquad \text{(10–18)}$$

where: e = tolerable error = $\dfrac{\text{precision}}{2}$

Solving for n, we get

$$n = \frac{Z^2 pq}{e^2} \qquad \text{(10–19)}$$

Equation (10–19) assumes that sampling is with replacement or that the sample size is less than 5 percent of the population, and that the decision maker knows p. Of course, if p were already known, there would be no need to sample. There are two recommended ways of getting around this problem:

1. Use $p = 0.5$, since this will give the largest possible $\sigma_{\bar{p}}$, and thus a conservatively large sample size.
2. Select a *pilot* sample smaller than the expected sample size and calculate $\hat{p}$ for this sample.

Using the first approach, the required sample size for Wood Roofs, Inc. is

$$n = \frac{(1.645)^2(0.5)(0.5)}{(0.025)^2} = 1,082.41$$
$$= 1,083 \text{ houses}$$

Thus a sample of 1,083 houses will guarantee the desired precision and confidence.

Under the pilot sample approach, the Wood Roofs, Inc. manager may use the results of the sample of 200 homes to determine the required sample size. That is, she will use the $\hat{p} = 0.36$ value as a pilot estimate of p for determining the appropriate sample size as follows:

$$n = \frac{(1.645)^2(0.36)(0.64)}{(0.025)^2} = 997.55$$
$$= 998 \text{ houses}$$

Given the pilot sample information, the required sample size is 998 houses. Note that the 200 homes in the pilot sample can be included in the 998, meaning that only 798 additional homes need to be sampled.

10-6

CONFIDENCE INTERVALS FOR ESTIMATING THE DIFFERENCE BETWEEN TWO POPULATION PARAMETERS—LARGE SAMPLES

Managers often have to estimate either the mean of a single population or the proportion of successes in a population. The tools we have discussed up to now are very useful for these estimates. However the more common decision-making situation involves deciding between alternatives. Managers will therefore often have to estimate either the **difference between two population means** or the **difference between two population proportions.**

Difference between Two Population Means

The training director of a large industrial corporation is considering adopting one of two alternative training methods, referred to as "motive-goal" and "reward-based." The director has used each method on trial groups in the company and has maintained records on employee productivity after each training method. The manager would like to know the difference in average productivity scores for employees trained under the two methods.

The point estimator for the difference between population means, $(\mu_{x_1} - \mu_{x_2})$, is $(\overline{X}_1 - \overline{X}_2)$, the difference in sample means. As with a single sample mean, the sampling distribution of the difference, $(\overline{X}_1 - \overline{X}_2)$, will be normally distributed, with

$$E(\overline{X}_1 - \overline{X}_2) = \mu_{x_1} - \mu_{x_2} \qquad (10\text{--}20)$$

The standard deviation of the difference between two population means is

$$\sigma_{\overline{X}_1 - \overline{X}_2} = \sqrt{\frac{\sigma_{x_1}^2}{n_1} + \frac{\sigma_{x_2}^2}{n_2}} \qquad (10\text{--}21)$$

Suppose the training manager selects a random sample of 100 employee records for each training method and obtains the following results on productivity scores:

"Motive-Goal"	"Reward-Based"
$\overline{X}_1 = 89$ points	$\overline{X}_2 = 94$ points
$n_1 = 100$	$n_2 = 100$
$\sigma_{x_1} = 9$	$\sigma_{x_2} = 12$

From industry experience and federal government reports, the manager knows that the standard deviation for the "motive-goal" method, σ_{x_1}, is 9 and that the standard deviation for the "reward-based" method, σ_{x_2}, is 12.

The format for developing a confidence interval for estimating the difference between population means is the same as that for a single population mean. That is,

Point estimate $\pm$ (interval coefficient)(standard error of the estimate)

Thus the 95 percent confidence interval for this training method example is

$$(\bar{X}_1 - \bar{X}_2) \pm Z \sqrt{\frac{\sigma_{\bar{X}_1}^2}{n_1} + \frac{\sigma_{\bar{X}_2}^2}{n_2}}$$

$$(89 - 94) \pm 1.96 \sqrt{\frac{81}{100} + \frac{144}{100}} \qquad \text{(10–22)}$$

$$-5 \pm 2.94$$

$$-7.94 \text{ points} \underline{\hspace{2cm}} -2.06 \text{ points}$$

Therefore the training director is confident that the average difference in productivity scores from the two training techniques is between -7.94 and -2.06 points. The negative values indicate that the "reward-based" method tends to produce higher productivity scores on the average than does the "motive-goal" method. Based on these results, the training manager might decide to go with the "reward-based" system.

In the previous example, the standard deviations were known. While this will occasionally be the case, in most instances the standard deviations will have to be estimated. The confidence interval is now approximated by

$$(\bar{X}_1 - \bar{X}_2) \pm Z \sqrt{\frac{S_{\bar{X}_1}^2}{n_1} + \frac{S_{\bar{X}_2}^2}{n_2}} \qquad n_2 \geq 30 \qquad \text{(10–23)}$$

Equation (10–23) is applicable as long as the sample sizes are large; that is, n_1 and $n_2 \geq 30$.

Suppose a bus company has tested 90 tires from each of two brands and has found the following mileage information:

Brand A	Brand B
$\bar{X}_1 = 33,000$ mi	$\bar{X}_2 = 29,400$ mi
$S_{X_1} = 1,100$ mi	$S_{X_2} = 750$ mi
$n_1 = 90$	$n_2 = 90$

The 90 percent confidence interval for estimating the difference in tire mileage is

$$(\bar{X}_1 - \bar{X}_2) \pm 1.645 \sqrt{\frac{S_{\bar{X}_1}^2}{n_1} + \frac{S_{\bar{X}_2}^2}{n_2}}$$

$$(33,000 - 29,400) \pm 1.645 \sqrt{\frac{1,210,000}{90} + \frac{562,500}{90}}$$

$$3,600 \pm 230.85$$

$$3,369.15 \text{ mi} \underline{\hspace{2cm}} 3,830.85 \text{ mi}$$

The bus company might use this information to help decide which brand of tires to buy for its fleet of buses. Based on these sample data, brand A tires seem to last longer than brand B, on the average.

Difference between Two Population Proportions

An assistant quality control manager for the Clappenback Corporation has been experimenting with two quality control systems. System 1 has been used on one assembly line, which has produced 15,000 flashlight batteries. System 2 has been used on a second assembly line, which has produced 25,000 flashlight batteries. The manager is interested in determining the difference in the proportion of defectives produced under each quality control system.

Although the quality control manager wants to know $(p_1 - p_2)$, he cannot test every battery. Instead he will select a sample of batteries produced under each system and calculate $(\hat{p}_1 - \hat{p}_2)$, the point estimate for the difference in population proportions. If the sample sizes are sufficiently large, the sampling distribution of $(\hat{p}_1 - \hat{p}_2)$ will be approximately normally distributed, with

$$E(\hat{p}_1 - \hat{p}_2) = (p_1 - p_2) \tag{10-24}$$

and

$$\sigma_{\hat{p}_1 - \hat{p}_2} = \sqrt{\frac{p_1 q_1}{n_1} + \frac{p_2 q_2}{n_2}} = \text{standard error of the sampling distribution} \tag{10-25}$$

Equation (10–25), the standard error of the sampling distribution, contains p_1 and p_2, the population proportions. The quality control manager can substitute the estimates, $\hat{p}_1$ and $\hat{p}_2$, for these values. Therefore

$$S_{\hat{p}_1 - \hat{p}_2} = \sqrt{\frac{\hat{p}_1 \hat{q}_1}{n_1} + \frac{\hat{p}_2 \hat{q}_2}{n_2}} \tag{10-26}$$

Given equation (10–26), the quality control manager can develop a confidence interval estimate of $(p_1 - p_2)$.

$$(\hat{p}_1 - \hat{p}_2) \pm Z \sqrt{\frac{\hat{p}_1 \hat{q}_1}{n_1} + \frac{\hat{p}_2 \hat{q}_2}{n_2}} \tag{10-27}$$

As you can see, the format of equation (10–27) is the same as for all other confidence intervals. That is,

Point estimate ± (interval coefficient)(standard error of the sampling distribution)

Suppose the quality control manager has selected a sample of 400 batteries from each assembly line and tested them, with the following results:

Line 1	Line 2
$\hat{p}_1 = 0.08$	$\hat{p}_2 = 0.06$
$\hat{q}_1 = 0.92$	$\hat{q}_2 = 0.94$
$n_1 = 400$	$n_1 = 400$

The manager wants a 98 percent confidence interval estimate for the difference in the proportion of defectives produced under the two systems. He would find the estimate as follows:

$$(\hat{p}_1 - \hat{p}_2) \pm 2.33 \sqrt{\frac{\hat{p}_1 \hat{q}_1}{n_1} + \frac{\hat{p}_2 \hat{q}_2}{n_2}}$$

$$(0.08 - 0.06) \pm 2.33 \sqrt{\frac{(0.08)(0.92)}{400} + \frac{(0.06)(0.94)}{400}}$$

$$0.02 \pm 0.042$$

$$-0.022 \underline{\hspace{3cm}} +0.062$$

The manager is very confident that this interval includes the true difference between the proportion of defectives produced under the two quality control systems.

10-7
CONCLUSIONS

Many decision-making applications involve estimating a population value from a sample of the population. In this chapter we have introduced the concepts of statistical estimation for large samples. We discussed two types of estimates: point estimates and interval estimates.

Point estimates are values such as $\overline{X}$, S_X^2, $\hat{p}$, $(\overline{X}_1 - \overline{X}_2)$, and $(\hat{p}_1 - \hat{p}_2)$, which are calculated from a sample. Although there is no guarantee a point estimator will be close to the population value, the quality of a point estimator will be increased if it is unbiased, consistent, efficient, and sufficient.

Confidence interval estimates are recommended when a decision maker wants to estimate a population value and have an idea of the estimate's error. The decision maker can control the estimate's level of confidence and precision but constantly faces a trade-off between these two factors and the required sample size.

Even if a confidence interval that meets the decision maker's precision and confidence requirements is developed, there is no guarantee the calculated interval will actually include the population value of interest. The decision maker must recognize that there is always a chance of error anytime sampling and statistical estimation are involved. The best that can be said is that the decision maker has a measure of the chance the interval includes the true population value. If the confidence level is increased, the decision maker can be even more confident.

We have presented several new terms in this chapter. The chapter glossary presents a summary of these.

We have also covered several equations that can be used to develop interval estimates of various population parameters. A listing of these immediately follows the glossary.

chapter glossary

confidence interval   An interval developed from sample values such that if all possible intervals were calculated, a percentage equal to the confidence level would contain the population value of interest.

consistent estimator   An estimator is consistent if, as the sample size is increased, the expected sampling error decreases.

interval coefficient   The table value (Z value in this chapter) associated with a particular level of confidence. For example:

Confidence	Interval Coefficient
90	$Z = 1.645$
95	$Z = 1.96$
99	$Z = 2.58$

most efficient estimator   An estimator is most efficient if its sampling distribution is less variable for a given sample size than any other estimator.

parameter   A descriptive measure of the population that has a fixed value. Examples are the population mean, μ_x, and the population standard deviation, σ_x.

precision   The width of the confidence interval.

point estimator   A single number determined from a sample used to estimate a population parameter.

tolerable error   One-half the width of the confidence interval. Tolerable error is used in determining required sample size.

unbiased estimator   An estimator whose expected value equals the population parameter. If μ_x is the population parameter, $\overline{X}$ is an unbiased estimate if $E(\overline{X}) = \mu_x$.

chapter formulas

Finite correction factor

$$\sqrt{\frac{N - n}{N - 1}}$$

Standard error of the mean

σ_x *known*

$$\sigma_{\overline{x}} = \frac{\sigma_x}{\sqrt{n}}$$

or

$$S_{\overline{x}} = \frac{\sigma_x}{\sqrt{n}} \sqrt{\frac{N - n}{N - 1}}$$

σ_x *unknown*

$$S_{\overline{x}} = \frac{S_x}{\sqrt{n}}$$

or

$$S_{\overline{x}} = \frac{S_x}{\sqrt{n}} \sqrt{\frac{N - n}{N - 1}}$$

Sample mean

$$\bar{X} = \frac{\Sigma X}{n}$$

Sample variance

$$S_X^2 = \frac{\Sigma(X - \bar{X})^2}{n - 1}$$

Sample standard deviation

$$S_X = \sqrt{\frac{\Sigma(X - \bar{X})^2}{n - 1}}$$

Confidence interval for μ_X

$$\bar{X} \pm Z\sigma_{\bar{X}}$$

Standard error of the difference between two means

Population variances known

$$\sigma_{X_1 - X_2} = \sqrt{\frac{\sigma_{X_1}^2}{n_1} + \frac{\sigma_{X_2}^2}{n_2}}$$

Population variances unknown

$$S_{\bar{X}_1 - \bar{X}_2} = \sqrt{\frac{S_{X_1}^2}{n_1} + \frac{S_{X_2}^2}{n_2}}$$

Confidence interval for $\mu_{X_1} - \mu_{X_2}$

$$(\bar{X}_1 - \bar{X}_2) \pm Z\sqrt{\frac{S_{X_1}^2}{n_1} + \frac{S_{X_2}^2}{n_2}}$$

Sample proportion

$$\hat{p} = \frac{X}{n}$$

Standard error of proportions

$$\sigma_{\hat{p}} = \sqrt{\frac{pq}{n}}$$

and

$$S_{\hat{p}} = \sqrt{\frac{\hat{p}\hat{q}}{n}}$$

Confidence interval for p

$$\hat{p} \pm ZS_{\hat{p}}$$

Standard error of the difference between two proportions

p_1 and p_2 known

$$\sigma_{\hat{p}_1 - \hat{p}_2} = \sqrt{\frac{p_1 q_1}{n_1} + \frac{p_2 q_2}{n_2}}$$

p_1 and p_2 unknown

$$S_{\hat{p}_1 - \hat{p}_2} = \sqrt{\frac{\hat{p}_1 \hat{q}_1}{n_1} + \frac{\hat{p}_2 \hat{q}_2}{n_2}}$$

Confidence interval for ($p_1 - p_2$)

$$(\hat{p}_1 - \hat{p}_2) \pm Z S_{\hat{p}_1 - \hat{p}_2}$$

Sample size

One population mean

$$n = \frac{Z^2 \sigma_x^2}{e^2}$$

One population proportion

$$n = \frac{Z^2 pq}{e^2}$$

solved problems

1. The manager at a major U.S. airport wishes to estimate the proportion of flights that arrived late at the airport last year for a report she must submit to Civil Aeronautics Administration. She has indicated that a 95 percent confidence interval is required with precision of 4 percent.

   a. How large a sample should the airport manager select? (*Hint*: Use $\hat{p} = 0.5$ and explain why you can use this value.)

   b. Using the sample size determined in part a, suppose the sample proportion, $\hat{p}$, is 0.24. Develop the confidence interval and provide the appropriate interpretation for the airport manager.

Solutions:

a. To determine the appropriate sample size, we need the following information:

$$\text{Confidence level} = 95\%$$
$$\text{Interval coefficient} = 1.96$$
$$\text{Precision} = 0.04$$
$$\text{Tolerable error} = 0.02$$
$$\hat{p} = 0.5$$

(Note that using $\hat{p} = 0.5$ implies the greatest possible variance in the sampling distribution. Consequently the sample size we find will always be large enough to meet the confidence level and precision requirements.) Then

$$e = Z \sqrt{\frac{\hat{p}\hat{q}}{n}}$$

$$0.02 = 1.96 \sqrt{\frac{(0.5)(0.5)}{n}}$$

$$n = \frac{(1.96)^2(0.5)(0.5)}{(0.02)^2}$$

$$= 2,401$$

b. The 95 percent confidence interval is

$$\hat{p} \pm 1.96 \sqrt{\frac{(0.24)(0.76)}{2,401}}$$

$$0.24 \pm 0.017$$

$$0.223 \underline{\hspace{2cm}} 0.257$$

Note that this interval is more precise than that required by the airport manager. The reason for this is that $\hat{p}$ calculated is lower than the $\hat{p} = 0.5$ used to find the required sample size.

---

2. Suppose the airport manager in solved problem 1 wishes to estimate the difference in proportion of late flights for two airlines that use the airport. She has indicated a 95 percent confidence interval is required with precision equal to 0.03.
   a. What sample size is needed from each airline to provide an interval estimate with this confidence and precision? Let the sample sizes from the two airlines be equal. Also, use $\hat{p} = 0.24$ for both airlines since this value was calculated for a sample from all airlines using this airport.
   b. Using the sample size determined in part a, the airport manager found the following:

$$\hat{p}_1 = 0.22 \quad \text{and} \quad \hat{p}_2 = 0.27$$

Develop the 95 percent confidence interval estimate for the difference between population proportions.

Solutions:

a. To determine the required sample size, we must perform the following calculation:

$$e = 1.96 \sqrt{\frac{\hat{p}_1\hat{q}_1}{n_1} + \frac{\hat{p}_2\hat{q}_2}{n_2}}$$

Since $n_1 = n_2 = n$, we get

$$0.015 = 1.96\sqrt{\frac{(0.24)(0.76)}{n} + \frac{(0.24)(0.76)}{n}}$$

$$n = 6{,}228.5 = 6{,}229$$

Thus the airport manager needs to study 6,229 flights from each airline. This is quite a large sample. Chances are the manager will have to decrease the confidence level, the desired precision, or both.

b. The 95 percent confidence interval is

$$(\hat{p}_1 - \hat{p}_2) \pm 1.96\sqrt{\frac{\hat{p}_1\hat{q}_1}{n_1} + \frac{\hat{p}_2\hat{q}_2}{n_2}}$$

$$(0.22 - 0.27) \pm 1.96\sqrt{\frac{(0.22)(0.78)}{6{,}229} + \frac{(0.27)(0.73)}{6{,}229}}$$

$$(0.22 - 0.27) \pm 0.015$$

$$-0.065 \underline{\hspace{3cm}} -0.035$$

Thus the manager is confident that the true difference is between -0.065 and -0.035 (airline 2 has a higher proportion of late flights).

problems

Problems marked with an asterisk (*) require additional thought.

1. Discuss in your own words the difference between a point estimator and an interval estimator.

2. Give at least one business-related example for which a point estimate would be used rather than an interval estimate.

3. Give at least one business-related example for which an interval estimate would be preferred over a point estimate.

4. Discuss in your own words the general advantage of an interval estimate over a point estimate.

5. In this chapter we have emphasized that a confidence interval always takes the form

   Point estimate $\pm$ (interval coefficient)(standard error)

   Give the formula for a confidence interval estimating μ_x and identify the point estimate, interval coefficient, and standard error.

6. Referring to problem 5, list the formula for a confidence interval estimator for $(p_1 - p_2)$ and indicate the point estimate, interval coefficient, and standard error.

7. In this chapter we stated that an increase in sample size will improve the precision of a confidence interval estimate. What does the term *precision* mean? Discuss in your own words.

8. Would you agree that an increase in sample size will increase the confidence level of an interval estimate? Why or why not?

9. What is the effect on an estimate's precision of increasing the confidence level of the estimate? What is the reason for this change?

10. You have determined that the sample size necessary to estimate the average major league baseball attendance—with 90 percent confidence and tolerable error of 500 people—is $n = 40$ games. Suppose the baseball commissioner states the precision is too low. What are his alternatives for increasing precision? Discuss the advantages and disadvantages of each.

11. At a recent board meeting a district manager made the following statement: "We have selected a sample of potential customers and calculated a 95 percent confidence interval for the estimate of those who will purchase our new product. Based upon these results, I am 95 percent sure the true proportion is between 0.52 and 0.71." Comment on this manager's statement.

12. For each of the following confidence levels, determine the appropriate interval coefficient:
    a. 90 percent
    b. 95 percent
    c. 99 percent
    d. 94 percent
    e. 50 percent
    f. 89 percent
    g. 80 percent

13. The California Highway Department is studying traffic patterns on a busy highway near San Diego. As part of the study, the department needs to estimate the average number of vehicles that pass an off-ramp each day. A random sample of 64 days gives $\overline{X} = 14,205$ and $S_x = 1,010$.

    Develop the 90 percent confidence interval estimate for μ_x, the average number of cars per day. Be sure to interpret your results.

14. Referring to problem 13, suppose the Highway Department officials, after careful analysis, decide they really need 99 percent confidence. What is the 99 percent confidence interval estimate of μ_x, the average number of cars passing the off-ramp? Interpret your results.

15. After calculating the confidence interval in problem 14, the Highway Department officials feel the precision is too low for their needs. They feel the precision should be 300. Given this precision and the 99 percent confidence, what size sample is required?

16. A consumer organization has recently mounted a campaign to make seat-belt use mandatory in a southern state. As part of the argument, the organization has sampled 1,000 accident records and found that in 140 cases, people were thrown from their vehicles.

    Based upon this sample evidence, calculate and interpret a 90 percent confidence interval estimate for the true proportion of accidents in which people are thrown from their vehicles.

17. Suppose the following confidence intervals estimating the mean age in a southern retail market area have been constructed from three different simple random samples. Assume the age distribution is normal.

Sample	Lower Limit	Upper Limit
1	39.0	50.0
2	40.0	51.9
3	42.0	50.0

    If the three samples had the same value for $S_{\bar{x}}^2$ and all three intervals were constructed with a 95 confidence level, which sample was the largest? Explain how you know this. Which sample was the smallest? How do you know this?

18. Referring to the three samples in problem 17, suppose sample 2 was size 36 and had a sample variance of 256. Approximately what level of confidence should be associated with this interval?

19. Again referring to the three samples in problem 17, suppose each of the three samples contained 60 observations and all three intervals were constructed with an 80 percent confidence level. Which sample had the smallest standard deviation? Explain.

20. Referring to problems 17 and 19, compute the standard error for each sample. Then determine what change in sample size would be necessary to cut the tolerable error by one-half.

21. Referring to just the information provided in problem 17, which sample had the smallest mean? How do you know?

22. You have just completed a poll for a political candidate. You have estimated the percentage of voters who will vote for her at 0.51 ± 0.029. She is not happy with this estimate and would like a precision of 0.02. If each interview costs $1.75, how much will it cost to give her the interval width she wants? She will accept a 95 percent confidence level.

23. The manager at the Jolly Ralph Motor Inn has started receiving complaints about the time before being served in the dining room. The manager takes a sample of 120 diners and finds that the mean time from entrance to serving is 17.56 minutes with a standard deviation of 4.5 minutes.
    a. What is the confidence interval estimate of average serving time if the manager specifies a 94 percent confidence level?

b. What should the manager say to the next irate customer who enters his office claiming to have waited in the dining room for 45 minutes without being served?

c. Can you see any sampling problems that may affect the conclusions you can draw from the interval calculated in part a?

24. The Government Accounting Office (GAO) wants to estimate the average yearly repair cost for government copying machines. In some regions the government has bought a repair contract that costs $500 per year for each machine. The GAO wants to know whether this is a fair price. Fortunately government offices in several sections of the country do not have the repair contract and have been repairing the machines on an individual basis. The GAO has taken a random sample of 250 annual repair invoices and has found the average cost to be $426.78 with a sample standard deviation of $167.18.

Based on these sample data, would you recommend discontinuing the repair contract?

25. The manager of the North Saw Mill needs to estimate the percentage of clear and standard redwood contained in a recent shipment of logs. She samples a random selection of logs before they go to the drying kilns.

a. If out of a sample 350 boards she finds 89 clear, what is the 97 percent confidence level estimate of the overall proportion of clear boards?

b. Suppose, in an effort to increase the estimate's precision, the manager looks at an additional 350 boards and finds 249 of standard quality. What is her new estimate for the proportion of clear boards? (Standard boards have knots in them.)

26. The federal government has mandated that to qualify for federal highway funds, states must insure that at least 70 percent of the cars traveling on their highways are obeying the 55 mile-per-hour speed limit. A team of federal inspectors has gone to a southern state and randomly checked the speed of 180 cars. They found 117 obeying the limit.

Based on this evidence, and using a 96 percent confidence level, can the inspectors make a case for denying the state highway funds?

27. The district manager, upon reviewing the report submitted by the inspectors in problem 26, objects strongly. He claims the sample size was much too small to draw any meaningful conclusions. Do you agree or disagree? If the district manager wants to limit the tolerable error of all future estimates to 0.02, how large a sample size is needed?

28. Great Northern Chemical has developed a new, effective, insect repellent. The leading spray on the market now advertises an effective time of 8 hours. From past experience, Great Northern knows that the time a spray is effective is a function of personal skin chemistry, and the variability remains the same no matter what the type of spray. This variability gives a variance of 2.3 hours. Great Northern field-tests its spray on 65 volunteers and finds it to be effective for an average of 8.85 hours.

Can Great Northern state, based on a 90 percent confidence level estimate, that its spray is more effective than the leading brand?

29. The legal department of Great Northern Chemical from problem 28 is concerned about possible lawsuits if Great Northern advertises that its spray acts longer than the leading brand. At the same time, the lawyers recognize that brand-comparison advertising is very effective. They are worried about the precision of the estimate of effective time. They would be happy with a precision of 0.15 hour. What sample size will be needed to give this precision, again with a 90 percent confidence level?

　　The legal department is also concerned whether this advertising means that everyone will be protected for the stated length of time. They have asked your opinion.

30. The Army purchasing department is changing suppliers of black oxfords; the previous supplier has stopped making this type of shoe. Since these shoes are issued to new recruits and these recruits do a lot of marching, the Army is particularly concerned with the durability of the soles. They have devised a shoe tester that rubs the shoe bottom against a slab of concrete. The testers monitor the time needed to rub a hole in the sole.

　　The Army asks for a sample of 100 from each of two shoe manufacturers, but part of one shipment is lost. The following results are found:

	Manufacturer A	Manufacturer B
Sample size	100	80
Average time	3.2 h	2.9 h
Standard deviation	0.86 h	0.79 h

a. What is the 98 percent confidence level estimate for the difference between the average wear times for the two shoe manufacturers?
b. Based on this interval, should the Army prefer one manufacturer over the other?

*31. A small metal fabricator is developing a new material to use on stovetops in place of stainless steel. Although initial appearance is important, so is hardness. The fabricator has developed a measure of hardness based on resistance to scrubbing. Since the field has been limited to two materials based on previous tests, the fabricator has decided to market the material with the higher average hardness coefficient. The following test results are found:

	Material I	Material II
Sample size	64	64
Hardness coefficient average	0.66	0.64
Standard deviation	0.08	0.063

　　The manufacturing department has a slight preference for material II. Is there any reason this material should not be the one marketed? You must pick your own confidence level.

*32. A regional sales manager is very upset with the results of problem 31. In particular, she objects to the precision of the interval estimate. She states quite strongly that the company risks marketing an inferior product unless the estimate's precision is 0.02 or less. Do you agree? How large would the two sample sizes need to be to give this precision if the sizes are to be equal?

33. The research department of an appliance manufacturing firm has developed a solid-state switch for its blender, which the department claims will reduce the percentage of appliances being returned under the one-year, full warranty from 6 percent to 3 percent. To test this claim, the testing department selects a group of the blenders manufactured with the new switch and subjects them to a normal year's worth of wear.
    a. If out of 250 blenders tested with the new switch, 9 would have been returned, can you accept the research department's claim? Use a 95 percent confidence level.
    b. The testing department also tests 250 blenders with the old switch and finds that 16 fail. Is there reason to believe the new switch outperforms the old switch?

34. The personnel director at a large southern electronics plant is not satisfied with the present training program. He feels that although the present program is extensive, and expensive, too many people must be released because of inadequate work even after completing training. Unfortunately he really doesn't know what percentage are released for inadequate work or some other reason.

    In any event, he has been considering implementing a new training program and would like to compare the results of the new program with those of the old. He decides to randomly separate the next group of new trainees into two groups, train each group entirely by one of the two methods, and monitor the results. After several months, he has found the following:

	New Method	Old Method
Group size	214	215
No. released for poor work	63	72

    a. Using a 96 percent confidence level, is there any reason to change training methods?
    b. Suppose both methods are equivalent in cost, but trainees seem to like the new method much better than the old. Is there any reason not to change training methods?

35. A recent Chrysler advertisement states that 38 of 50 tested consumers preferred the Chrysler LeBaron to the Olds Cutlass. Assume the sample is a random sample.
    a. Develop a 98 percent confidence interval estimate for the true proportion of individuals who favor the LeBaron to the Olds Cutlass.
    b. Based upon these sample data, can we conclude that the majority of consumers would prefer the Chrysler car over the Oldsmobile? Discuss.

36. An initial survey indicates that 55 percent of Culver Height's residents are in favor of eliminating Saturday mail service if it will help balance the federal budget. However a local legislator wants to be able to estimate the true pop-

ulation value to within ± 2 percent at the 92 percent confidence level. How large will the total sample have to be to allow an estimate with these limits? Also, discuss what a 92 percent confidence level means.

*37. The manager whose department operates a fleet of company cars is trying to decide on which brand of tires to use. Her choices are brand F and brand M, and the price per tire is the same for both brands. The manager takes a sample of 64 tires from brand F and 55 tires from Brand M. She finds the following values:

	Brand F	Brand M
Average mileage	42,156	43,414
Sample standard deviation	3,455	2,981

Based on this sample information, is there any reason to prefer one brand of tires based on its average mileage? Use a confidence level of 95 percent.

38. The Early Dawn Egg Farm is trying to decide between two brands of egg containers. The major consideration is preventing breakage during shipment. Early Dawn Egg Farm ships 500 eggs in each type of container, with the following results:

	Brand 1	Brand 2
Percent broken	1.7	2.2

Using 95 percent confidence, can one brand be concluded to be better than the other in preventing breakage?

cases

10A ACE TRUCKING COMPANY

The Ace Trucking Company has just lost a labor arbitration suit in which it was accused of wrongly dismissing Al Farr, one of its drivers. The dismissal was ruled to be a violation of the union contract. In the current labor contract both parties (union and management) agreed that all such disputes that cannot be settled directly between the two parties be placed in binding arbitration. In the past few months there have been other cases arbitrated, and Ace has always been the "victor." However, in this case the ruling states that Al Farr must be rehired and given back pay equivalent to the amount he would have earned in the nine months since he was dismissed. The arbitrator stated, as is common in a case of this sort, that Al Farr must be "made whole."

Ace Trucking pays its drivers an hourly wage with time and a half for overtime and double time for holidays. The hours worked each week depend on the routes assigned. The assignment of routes is done pretty much randomly. Thus in some weeks certain drivers receive substantial overtime work and therefore more pay, while other drivers receive no overtime pay at all.

Dan Thomas, supervisor of Ace Trucking personnel relations, has been assigned the task of determining the amount of back pay that should be awarded to Al Farr. The arbitrator will determine whether the amount proposed by Ace Trucking is adequate, so Dan knows that he must be able to substantiate his figures.

The issue of back pay is complicated. The union has suggested a maximum-hours method in which Al Farr would receive pay equivalent to his pay rate times the maximum number of hours worked by any employee during the same nine-month period at Ace Trucking. Dan is resisting this plan because of the precedent it would set. At the same time, he has admitted that he does not know the exact amount that Al Farr should receive.

Because he must have a figure by Wednesday, Dan has decided to use personnel records and attempt to arrive at a dollar amount for Al that will be fair. During the first six months that Al Farr was out of work, he would have been paid $7.75 per hour, and $8.10 per hour during the final three months. Personnel department records show hourly work records for all Ace Trucking drivers. A sampling procedure Dan selected required him to pull a random sample of 50 employee records for each of the last three quarters (one quarter = three months). Table C–1 shows the results of the sampling after several calculations have been made.

Dan faces a long evening trying to make sense out of this information so he can make an effective presentation to the arbitrator.

TABLE C–1

	Quarter 1	Quarter 2	Quarter 3
Regular hours per week			
$\bar{X}$	39.4	39.7	39.6
S_x	1.6	2.1	1.3
Overtime hours per week			
$\bar{X}$	10.3	9.1	11.0
S_x	3.7	2.9	4.4
Holidays per quarter			
$\bar{X}$	12.7	6.2	8.8
S_x	3.1	4.7	2.6

10B   KRAMER TELEPHONE SOLICITATION COMPANY

Phyllis Martin is currently the operating manager of the Kramer Telephone Solicitation Company. Kramer contracts with both local and regional firms to perform telephone solicitations. Past work has included solicitation of magazine subscriptions, banquet attendance, and athletic season ticket sales, to name a few. Kramer uses phones with automatic logging devices and therefore is able to charge both for the number of calls made and the number of sales. Consequently the telephone pitch that Kramer operators use must be both quick and effective.

In recent work, Kramer has been using what the industry calls an "informative" pitch. However Phyllis Martin would like to change the pitch to what she terms the "informal" appeal. She has the authority to make this type of change, but will be under the gun to show that the new approach is more successful than the old.

If she makes an unsuccessful switch, she will almost certainly lose her position as manager. She has decided to try both methods on the next sales effort in an attempt to determine which will perform better.

Three of the most experienced callers have been trained in the new method. These three callers then randomly use the old method on one call, the new method on the next, and so on. At the end of the trial period Phyllis observed the following results:

	Caller A	Caller B	Caller C
Old method			
n	420 calls	385 calls	402 calls
No. sales	48	34	44
New method			
n	418 calls	390 calls	400 calls
No. sales	37	46	23

Phyllis wonders what interpretation she should make regarding these findings.

10C TRUCK SAFETY INSPECTION, PART 2

Mr. Lund of the Idaho Department of Law Enforcement has decided that sampling is necessary to carry out the first phase of the truck safety inspection program in Idaho. (See case 8B in Chapter 8.) The objective of the first phase of this program is to estimate the proportion of trucks operating in Idaho with some form of safety defect.

Mr. Lund has been informed that he has a free choice in determining his sampling plan, but that it must be statistically sound and practical to implement. Because he has limited funds for carrying out the inspections, he is very concerned about the sample size that is necessary to provide a "good" estimate of the proportion of defective trucks operating in Idaho. He knows that precision and confidence level selection are important in determining the required sample size but is unsure how to arrive at the appropriate value. He is in need of a good discussion of precision and confidence and their relationship to sample size determination. Additionally, he would like to know how large a sample to select.

references

HAYES, GLENN E. and HARRY G. ROMIG. *Modern Quality Control.* Encino, Calif.: Bruce, 1977.

PFAFFENBERGER, ROGER C. and JAMES H. PATTERSON. *Statistical Methods for Business and Economics.* Homewood, Ill.: Irwin, 1977.

WINKLER, ROBERT L. and WILLIAM L. HAYES. *Statistics: Probability, Inference, and Decision.* New York: Holt, Rinehart and Winston, 1975.

WONNACOTT, THOMAS H. and RONALD J. WONNACOTT. *Introductory Statistics for Business and Economics.* New York: Wiley, 1977.

11
Statistical Estimation— Small Samples

In Chapter 10 we introduced the fundamental concepts of statistical estimation. We emphasized the difference between point estimators and confidence interval estimators and showed how to calculate each. The discussion in Chapter 10 assumed that the decision makers were dealing with large samples of measurements from the populations of interest.

However, in many applications, obtaining measurements from large samples is not possible. Often the costs in time or money are prohibitive. In these instances decision makers must rely on small samples from the populations of interest, so the estimation techniques presented in Chapter 10 are inappropriate. Yet decision makers still need to be able to estimate various population values. Thus they need to understand the techniques for small-sample statistical estimation presented in this chapter.

chapter objectives

In this chapter we will introduce statistical estimation based upon small samples. We shall consider the Student t distribution and develop confidence intervals for estimates of the population mean, the difference between two population means for independent samples, and the difference between two population means for paired samples.

student objectives

After studying the material presented in this chapter, you should:

1. Know what the Student t distribution is.
2. Understand the fundamental differences between the t distribution and the normal distribution.
3. Be aware of the conditions under which the t distribution rather than the normal distribution should be used to determine the confidence interval estimate of a population value.
4. Know how to develop a small-sample confidence interval for estimating the population mean.
5. Know how to develop a small-sample confidence interval for estimating the difference between two population means for independent samples.
6. Know how to develop a small-sample confidence interval for estimating the difference between two population means for paired samples.
7. Understand the trade-offs between sample size, confidence level, and precision.

11-1
SMALL-SAMPLE ESTIMATION

In Chapter 10 we introduced statistical estimation for several population values. The common bond for all the confidence intervals developed in that chapter was that the sample sizes were large and the interval coefficient was a Z value from the standardized normal distribution. Although differentiating between a *large* and a *small* sample is to some extent arbitrary, a widespread convention is to consider the sample *large* if the sample size is greater than or equal to 30 ($n \geq 30$).

Many business applications do involve large sample sizes, but many others do not. The cost in terms of dollars and time often prevents a large sample from being used to estimate a population value. In other instances a large sample cannot be obtained because the population items are difficult to observe or few in number. And, in these cases, the large-sample procedures presented in Chapter 10 are simply not appropriate.

A major insurance company is conducting a study to determine if a downward change in its automobile collision rates can be justified. As part of the study, the company needs to estimate the average damage in dollars to a new car that hits a barricade head-on at 15 miles per hour. Because of the high cost of this type of sampling, the insurance company has decided to crash only ten cars.

Since 10 is less than 30, the insurance company does not have a large sample and, therefore, cannot employ the methods presented in Chapter 10. Rather, the company needs to use statistical techniques that are appropriate when dealing with small samples. Before we introduce these small-sample techniques, let's review four basic principles of interval estimation:

1. **The format for any confidence interval is**

    Point estimate ± (interval coefficient)(standard error of the estimate)

2. **The higher the confidence, the lower the precision for a given sample size.**

3. **An increase in sample size will increase the precision of the estimate for a given confidence level. This occurs because an increase in sample size decreases the standard error of the estimate.**

4. **A confidence interval carries with it the following interpretation: If all possible samples of a given size are selected, and all possible confidence intervals for a given confidence level are calculated, the percentage of intervals containing the true population value will equal the confidence level.**

These four principles were introduced in connection with large-sample estimation. They also apply to small-sample estimation. If you understand these four concepts, you should have little trouble understanding and applying the material presented in the remaining sections of this chapter.

11-2
THE STUDENT *t* DISTRIBUTION

The interval estimates in Chapter 10 were based on large samples. In developing the intervals, we expressed no concern about the shape of the population distribution because the central limit theorem tells us the sampling distribution will be approximately normal for large samples. Consequently the interval coefficient we used for each interval was a Z value from the standard normal distribution. For example, the confidence interval estimate for μ_x is given by

$$\bar{X} \pm Z \frac{\sigma_x}{\sqrt{n}} \qquad (11–1)$$

Even if σ_x is *not* known and must be estimated, the sampling distribution of $\bar{X}$ is still approximately normal if the sample size is large. The confidence interval estimate for μ_x with an unknown population standard deviation and a large sample size is

$$\bar{X} \pm Z \frac{S_x}{\sqrt{n}} \qquad (11–2)$$

However, when σ_x is unknown and both μ_x and σ_x must be estimated from a small sample, the interval coefficient in equation (11–2) cannot be determined using the standard normal distribution.

If the population from which the small sample is selected is normally distributed, an exact solution to the problem of small-sample interval estimation is available. W. S. Gosset, a brewmaster in Ireland, discovered the properties governing the sampling distribution of $\overline{X}$ for small samples. Because of fear of reprisal by his government, Gosset published his findings in 1908 under the pen name "Student" and called the sampling distribution the **t distribution.** He showed for small samples that the quantity

$$t = \frac{\overline{X} - \mu_x}{\frac{S_x}{\sqrt{n}}} \qquad\qquad (11\text{–}3)$$

follows a t distribution. "Student" (Gosset) set forth the following theorem:

If the population from which the sample is being selected is normally distributed with unknown standard deviation, the sampling distribution for $(\overline{X} - \mu_x)/(S_x/\sqrt{n})$ derived from the population will be described by a t distribution with $n - 1$ degrees of freedom.

The t distribution has two properties:

1. The t distribution is symmetrical, ranging from $-\infty$ to $+\infty$.
2. The standardized t distribution has a mean of zero and a variance equal to

$$\frac{n - 1}{(n - 1) - 2}, \qquad \text{for } n \geq 4$$

(Note that for $n \geq 4$, the variance exceeds 1.0.) These two properties indicate the similarities between the t distribution and the standard normal distribution and also point out the principal difference. Both the t and Z distributions are symmetrical and range from $-\infty$ to $+\infty$. Both standardized distributions have a mean of zero. The basic difference is in the spread of the two distributions. **The variance of the standard normal distribution is 1.0, whereas the variance of the t distribution is always greater than 1.0.** In fact, the exact variance of the t distribution depends on the sample size. If $n = 10$, the variance is 1.285. The variance for $n = 15$ is 1.167. The variance for $n = 25$ is 1.091.

Figure 11–1 illustrates the relationship between the t distribution and the Z distribution and shows how the variance of the t distribution changes as the sample size changes. When the sample size increases, the variance approaches 1.0.

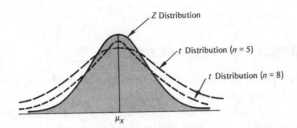

FIGURE 11–1 Comparison of t and Z distributions.

Thus, for large sample sizes, the t distribution approaches the standard normal distribution.

In Chapter 9 we showed that

$$Z = \frac{\bar{X} - \mu_x}{\frac{\sigma_x}{\sqrt{n}}}$$

and emphasized that Z represents the number of standard deviations $\bar{X}$ is from μ_x. Likewise,

$$t = \frac{\bar{X} - \mu_x}{\frac{S_x}{\sqrt{n}}}$$

where the t value is the number of standard deviations $\bar{X}$ is from μ_x.

As with the Z distribution, tables of t values have been developed. The standardized t-distribution table is contained in Appendix D. Across the top of the table are probabilities corresponding to areas in one tail, two tails, and for various confidence levels. Down the left side is a column headed "d.f." (degrees of freedom).* We will show you how to determine the degrees of freedom for all statistical procedures introduced in this text. For example, when we are developing a confidence interval for estimating μ_x, the number of degrees of freedom is $n - 1$.

The t value for a 90 percent confidence level with sample size 10 is 1.833. Note that we find this value by reading across the top to 0.90 and down the "d.f." column to 9. The value found at the row and column intersection is the appropriate t value. As another example, the t value corresponding to a 95 percent confidence level and sample size 19 is 2.101. The t value for 95 percent confidence and sample size 14 is 2.160. Make sure you can locate these values in the t-distribution table in Appendix D. You will be required to use the t-distribution table extensively in this chapter and in Chapter 13.

---

*The statistical concept of **degrees of freedom** is one of the most difficult for beginning students because of its many possible interpretations. The general expression for degrees of freedom is $N - K$, where N is the number of observations and K is the number of constants that must be calculated from the sample data to estimate the variance of the sampling distribution.

11-3

ESTIMATING THE POPULATION MEAN—SMALL SAMPLES

Let's return to the insurance company that needs to estimate the average damage to an automobile that crashes at 15 miles per hour.

The automobile industry in general has been attempting to build sturdier cars. Because of this, many insurance companies are examining their collision rates. The insurance company in our example needs to estimate each model's average damage for a 15-mile-per-hour crash to help the actuaries develop a premium rate for the model's owners.

Because of the costs involved, the company has selected a random sample of ten new cars and has crashed each one at 15 miles per hour. The resulting damage costs are shown in Table 11–1. Assuming the true distribution of damages is normal, the insurance actuaries can use these data to develop an interval estimate for the true average damage level as follows:

$$\bar{X} \pm t \frac{S_x}{\sqrt{n}} \tag{11-4}$$

where: $\bar{X}$ = point estimate

t = interval coefficient from the t distribution

S_x = estimate of σ_x

n = sample size

Assuming a 95 percent confidence interval is specified, and using the values for $\bar{X}$ and S_x shown in Table 11–1, the interval is

TABLE 11–1   Mean and standard deviation of automobile damage at 15 miles per hour

$1,954.00	$6,109.00
2,702.00	3,311.00
3,605.00	3,702.00
1,627.00	2,151.00
4,105.00	1,949.00

$$\bar{X} = \frac{\Sigma X}{n}$$

$$= \$3,121.50$$

$$S_x^2 = \frac{\Sigma(X - \bar{X})^2}{n - 1}$$

$$= 1,843,489.5$$

$$S_x = \sqrt{1,843,489.5}$$

$$= 1,357.75$$

$$\bar{X} \pm 2.262 \frac{S_x}{\sqrt{n}}$$

$$\$3,121.5 \pm 2.262 \frac{\$1,357.75}{\sqrt{10}}$$

$$\$3,121.5 \pm \$971.21$$

$$\$2,150.29 \underline{\hspace{2cm}} \$4,092.71$$

Therefore the actuaries are confident that the true average repair cost will be between $2,150.29 and $4,092.71.

Of course, the true mean will either fall in this interval, or it will not. If this precision is too low (interval is too wide), the insurance company has two ways of increasing it. The company can increase the sample size *or* decrease the confidence level.

Suppose because of the costs involved, the number of cars tested cannot be increased. If the confidence level is reduced to 90 percent, the interval coefficient, t, is reduced, giving the following confidence interval estimate for μ_x:

$$\bar{X} \pm 1.833 \frac{S_x}{\sqrt{n}}$$

$$\$3,121.50 \pm 1.883 \frac{\$1,357.75}{\sqrt{10}}$$

$$\$3,121.50 \pm \$787.01$$

$$\$2,334.49 \underline{\hspace{2cm}} \$3,908.51$$

Thus, decreasing the confidence level from 95 percent to 90 percent has decreased the interval width from $1,942.42 to $1,574.02.

As another example, consider the following situation. The Air Force has contracted for a new missile to bolster its worldwide defense position. The contractor has given the Air Force six missiles to test, as specified in the contract. Among other things, the Air Force wants to know the *average* range of the missiles. Suppose the Air Force fires these six missiles and finds the mean to be 647 miles with a standard deviation of 5 miles. Assuming that the distances for all such missiles are normally distributed, the Air Force can develop a 99 percent confidence interval by

$$\bar{X} \pm 4.032 \frac{S_x}{\sqrt{n}}$$

$$647 \pm 4.032 \frac{5}{\sqrt{6}}$$

$$647 \pm 8.23$$

$$638.77 \text{ mi} \underline{\hspace{2cm}} 655.23 \text{ mi}$$

Thus the Air Force is confident that the true mean is contained in the interval from 638.77 miles to 655.23 miles.

Although the Air Force may be satisfied with the average flight distance of the missiles, it may not be happy with the large standard deviation. If the manufacturer could reduce the standard deviation by making more consistent missiles, the estimate's precision would be increased. Do you understand why this would happen?

11-4
ESTIMATING THE DIFFERENCE BETWEEN TWO
POPULATION MEANS—SMALL SAMPLES

The Jacknee Corporation manufactures and distributes power hand tools in the Midwest. The plant managers at the two Jacknee locations have been responsible for their own quality control. They have always assumed that the quality of the tools coming from the two locations was the same. Recently Jacknee has received more customer complaints than usual. A corporate vice-president thinks that one plant may have lost some control on quality. Since the finished goods from the two locations are mixed at a central warehouse before they are distributed to the retailers, she cannot tell which plant is producing the inferior product. Consequently she has hired a testing company to examine 12 tools from each plant and assign a quality rating to each. The vice-president is interested in estimating the difference in the average ratings for tools produced at the two plants.

In Chapter 10 we discussed estimating the difference between two population means for large samples. We can estimate this difference for small samples using the t distribution, provided:

1. The two populations are normally distributed and independent
2. The two populations have equal standard deviations

If these two assumptions are satisfied, the small-sample confidence interval estimate of the difference between population means is

$$\left[(\bar{X}_1 - \bar{X}_2) \pm tS_{pooled} \sqrt{\frac{1}{n_1} + \frac{1}{n_2}} \right] \qquad (11\text{--}5)$$

where: $\bar{X}_1$ = sample mean from population 1

$\bar{X}_2$ = sample from population 2

t = interval coefficient from the t distribution with $n_1 + n_2 - 2$ d.f.

$$S_{pooled} = \sqrt{\frac{S_{X_1}^2(n_1 - 1) + S_{X_2}^2(n_2 - 1)}{n_1 + n_2 - 2}}$$

n_1 and n_2 = sample sizes from populations 1 and 2, respectively

Note that the small-sample confidence interval estimating the difference between two population means follows the usual format, that is,

Point estimate $\pm$ (interval coefficient)(standard error of the estimate)

As can be seen, the standard error of the estimate involves **pooling** the variances from each sample. The **pooled standard deviation, S_{pooled},** is the square root of a weighted average of the sample variances, where the weights are $n_1 - 1$ and $n_2 - 1$.

The interval coefficient is a t value from the Student t distribution. An important point about this t value is that the appropriate degrees of freedom are $n_1 + n_2 - 2$. Note that this is equal to the denominator in the equation for the pooled standard deviation.

Suppose the ratings for the 12 tools from each plant in our example give the following results:

Plant 1	Plant 2
$n_1 = 12$	$n_2 = 12$
$\overline{X}_1 = 49$	$\overline{X}_2 = 46$
$S_{x_1}^2 = 147$	$S_{x_2}^2 = 154$

The 90 percent confidence interval for estimating the difference between the mean ratings for tools from plants 1 and 2 is

$$(\overline{X}_1 - \overline{X}_2) \pm 1.717 S_{pooled} \sqrt{\frac{1}{n_1} + \frac{1}{n_2}}$$

$$(49 - 46) \pm 1.717 S_{pooled} \sqrt{\frac{1}{12} + \frac{1}{12}}$$

where:

$$S_{pooled} = \sqrt{\frac{(12 - 1)(147) + (12 - 1)(154)}{22}} = 12.26$$

Note that the interval coefficient, t, is found in the t-distribution table with $n_1 + n_2 - 2 = 22$ degrees of freedom.

The confidence interval is

$$(49 - 46) \pm (1.717)(12.26)(0.4082)$$

$$3 \pm 8.59$$

$$-5.59 \text{ points} \underline{\hspace{4cm}} 11.59 \text{ points}$$

The vice-president is confident, based on this interval estimate, that the true difference between the average quality ratings for the two plants is between 5.59 in favor of plant 2 and 11.59 in favor of plant 1. Since she believes that this interval contains the true mean difference, and since this interval includes zero, she would likely make no judgment about the quality control at either plant. However she may dislike the low precision (wide interval) and suggest a larger sample size or a decreased confidence level to increase the precision.

Note that the two sample sizes need not be equal when equation (11–5) is used to estimate the difference between two population means.

11-5
ESTIMATING THE DIFFERENCE BETWEEN TWO POPULATION MEANS—PAIRED SAMPLES

Comparing two population means often involves situations where the two samples are *not* independent. For example, the O.T. Herman Company makes tractor tires under two different processes. The quality assurance department wishes to estimate the difference in average wear of tires from the two processes. The test plan is to place a tire made from each process on the rear wheels of ten tractors and to measure tire wear after one month. The quality control people realize that tractor tire wear depends on the driver, the type of work performed, the amount of time the tractor is driven, and the tractor's age and type, so they place one tire from each process on each tractor to eliminate these factors. However, when they do this, the samples can no longer be considered independent. Instead they are paired samples, and the confidence interval estimate for the difference in population means is

$$\bar{d} \pm t \frac{S_d}{\sqrt{n}} \qquad\qquad (11\text{–}6)$$

where: $\bar{d}$ = average of the paired differences

t = interval coefficient

S_d = standard deviation of the paired differences

n = number of pairs

Table 11–2 presents the wear measurements for ten tires made under each process. By wear we mean the tire tread depth in inches.

TABLE 11–2   Tire wear measurements, O.T. Herman Tire Company (wear = tire tread depth in inches)

Tractor	Process 1	Process 2	d	$(d - \bar{d})^2$
1	3.25	3.14	0.11	$(0.11 - 0.309)^2 = 0.0396$
2	3.86	3.20	0.66	$(0.66 - 0.309)^2 = 0.1232$
3	7.15	6.93	0.22	$(0.22 - 0.309)^2 = 0.0079$
4	8.00	7.90	0.10	$(0.10 - 0.309)^2 = 0.0437$
5	1.14	1.01	0.13	$(0.13 - 0.309)^2 = 0.0320$
6	4.77	3.95	0.82	$(0.82 - 0.309)^2 = 0.2611$
7	5.24	4.93	0.31	$(0.31 - 0.309)^2 = 0.0000$
8	6.17	6.02	0.15	$(0.15 - 0.309)^2 = 0.0253$
9	8.32	8.13	0.19	$(0.19 - 0.309)^2 = 0.0142$
10	4.80	4.40	0.40	$(0.40 - 0.309)^2 = 0.0083$
			$\bar{d} = 0.309$	$\Sigma = 0.5553$

$$S_d = \sqrt{\frac{\Sigma(d - \bar{d})^2}{n - 1}}$$

$$= 0.2488$$

Using equation (11–6), we can develop a 95 percent confidence interval in the following manner. (Note that the degrees of freedom associated with the interval coefficient, t, are $n - 1$. Here, degrees of freedom = $10 - 1 = 9$.)

$$\bar{d} \pm 2.262 \frac{S_d}{\sqrt{n}}$$

$$0.309 \pm 2.262 \frac{0.2488}{\sqrt{10}}$$

0.1310 in. _____ 0.4869 in.

Thus the department management is confident that the true difference between the average wear for tires made under the two processes is between 0.1310 and 0.4869 inch, with process 1 showing less wear.

Note that this estimate was obtained by pairing observations *before* the measurements were taken. The reason was to control for differences that might result due to the tractor, the driver, or working conditions.

11-6

EFFECT OF VIOLATING THE ASSUMPTIONS FOR SMALL-SAMPLE ESTIMATION

When using the t distribution for statistical estimation with small samples, we assume that:

1. The populations are normally distributed
2. The populations have equal variances if two independent populations are involved

Although applications may exist for which these assumptions are strictly satisfied, our experience indicates this rarely happens. Fortunately, however, the estimation methods presented in this chapter are fairly robust. That is, for small departures from the assumptions of normality and equal variances, the t distribution is still appropriate. However, if the populations differ extensively from normality (the populations are extremely skewed) or the variances are far from equal, the decision maker will need to employ different statistical procedures or select a large sample size and use the large-sample estimation techniques presented in Chapter 10.

11-7

CONCLUSIONS

Very often decision makers need to estimate a population value based on a small sample from the population(s). Providing the population(s) are normally distributed (or at least not highly skewed), the t distribution forms the basis for statistical estimation with a small sample size.

The t distribution has a larger variance than the standard normal distribution, but approaches the normal distribution as the sample size becomes large. The t distribution is symmetrical and has degrees of freedom that depend on the application. We have indicated, for each use, how to calculate the degrees of freedom.

We have also introduced small-sample confidence interval estimation for μ_x, for $(\mu_{x_1} - \mu_{x_2})$ with independent samples, and for $(\mu_{x_1} - \mu_{x_2})$ with dependent samples.

chapter glossary

interval estimator An interval within which a population parameter may fall. The width of the interval is determined by a combination of sample size, confidence level, and variance.

large sample In this text, a sample of size $n \geq 30$.

parameter A measure of the population, such as μ_x, the population mean.

precision The width of the confidence interval. The narrower the interval, the greater the precision, and vise versa.

point estimator A single value calculated from a sample to be used to estimate a population parameter.

population A finite or infinite collection of measurements, items, or individuals of all possible measurements, items, or individuals within the scope of the study.

sample A subset of the population.

small sample In this text, a sample of size $n < 30$.

statistic A value computed from sample measures, such as $\overline{X}$, the sample mean. The statistic is usually a point estimator of the corresponding population parameter.

Student t distribution A sampling distribution of

$$t = \frac{\overline{X} - \mu_x}{\dfrac{S_x}{\sqrt{n}}}$$

for samples selected from a normally distributed population with σ_x estimated by S_x, and based upon a small sample. The standardized t distribution has a mean of zero and a variance equal to $(n - 1)/[(n - 1) - 2]$.

chapter formulas

Sample mean

$$\overline{X} = \frac{\Sigma X}{n}$$

Sample variance

$$S_x^2 = \frac{\Sigma(X - \overline{X})^2}{n - 1}$$

Sample standard deviation—Shortcut formula

$$S_x = \sqrt{\frac{\Sigma X^2 - \frac{(\Sigma X)^2}{n}}{n - 1}}$$

***t* statistic**

$$t = \frac{\overline{X} - \mu_x}{\frac{S_x}{\sqrt{n}}}$$

Confidence interval estimate for the population mean—Small sample

$$\overline{X} \pm t\frac{S_x}{\sqrt{n}}$$

where t has $n - 1$ d.f.

Confidence interval estimate for the difference between two population means—Independent small samples

$$(\overline{X}_1 - \overline{X}_2) \pm tS_{pooled}\sqrt{\frac{1}{n_1} + \frac{1}{n_2}}$$

where t has $n_1 + n_2 - 2$ d.f.

Pooled standard deviation

$$S_{pooled} = \sqrt{\frac{S_{x_1}^2(n_1 - 1) + S_{x_2}^2(n_2 - 1)}{n_1 + n_2 - 2}}$$

Paired-sample confidence interval—Small sample

$$\overline{d} \pm t\frac{S_d}{\sqrt{n}}$$

where t has $n - 1$ d.f., n = number of pairs.

solved problems

1. The ARTEE Corporation makes glass basketball backboards. The company has made ten boards using a new technique. Before it makes any more for commercial sale, ARTEE needs to estimate the average strength of the new boards. To determine the strength of a board, pressure is applied to each corner until the glass shatters. The pressure at which the glass shatters is recorded.

   Suppose that for the ten glass backboards tested, the mean was 1,526 pounds per square inch and the standard deviation was 15.25 pounds per square inch.

   a. Determine the 90 percent confidence interval for estimating μ_x assuming the population is normally distributed.

b. Determine the 99 percent confidence interval for estimating μ_x assuming the population is normally distributed.

c. Why are the precisions of the intervals calculated in parts a and b different?

Solutions:

a. The 90 percent confidence interval for μ_x is determined as follows:

$$\bar{X} \pm t \frac{S_x}{\sqrt{n}}$$

where t with 90 percent confidence and $n - 1 = 9$ degrees of freedom = 1.833. Then

$$1,526 \pm 1.833 \frac{15.25}{\sqrt{10}}$$

1,517.16 psi ——————— 1,534.84 psi

The ARTEE Corporation therefore can be confident that the average pressure at which a board will break is between 1,517.16 and 1,534.84 pounds per square inch.

b. The 99 percent confidence interval is

$$\bar{X} \pm 3.250 \frac{S_x}{\sqrt{n}}$$

$$1,526 \pm 3.250 \frac{15.25}{\sqrt{10}}$$

1,510.33 psi ——————— 1,541.67 psi

c. The interval calculated in part a is more precise than the interval calculated in part b because the confidence level is higher in part b. When confidence is increased, the interval width is increased, and thus precision is decreased.

✓**2.** Referring to solved problem 1, suppose the ARTEE Corporation wishes to estimate the difference in average strength between the original glass backboards and the new glass boards. Suppose the company has data for 20 original boards and the 10 new boards. The available information is as follows:

Original	New
$\bar{X}_1 = 1,500$ psi	$\bar{X}_2 = 1,526$ psi
$S_1 = 45.00$ psi	$S_2 = 15.25$ psi
$n_1 = 20$	$n_2 = 10$

a. Calculate the 95 percent confidence interval estimate of the difference between the two population means assuming each population is normally distributed and the population variances are equal.

b. Calculate the 90 percent confidence interval estimate for the difference between the two population means assuming each population is normally distributed and the population variances are equal.

Solutions:

a. The 95 percent confidence interval is

$$(\bar{X}_1 - \bar{X}_2) \pm tS_{pooled} \sqrt{\frac{1}{n_1} + \frac{1}{n_2}}$$

where

$$S_{pooled} = \sqrt{\frac{S_1^2(n_1 - 1) + S_2^2(n_2 - 1)}{n_1 + n_2 - 2}} = \sqrt{\frac{2{,}025(19) + 232.56(9)}{28}}$$

$$= 38.06$$

Note that the degrees of freedom for the interval coefficient are $n_1 + n_2 - 2 = 28$ for this problem. Then the interval is

$$(1{,}500 - 1{,}526) \pm (2.048)(38.06)\sqrt{\frac{1}{20} + \frac{1}{10}}$$

$$-56.18 \text{ psi} \underline{\hspace{2cm}} 4.19 \text{ psi}$$

The ARTEE Corporation could be confident that the true difference between the average strengths of the original and new boards is between -56.18 pounds per square inch (new board is stronger) and 4.19 pounds per square inch (old board is stronger). Since zero falls within the limits, the company could conclude that the true difference might be zero. This could then provide justification for the claim that the two types of boards are no different with respect to average strength.

b. The 90 percent confidence interval is

$$(\bar{X}_1 - \bar{X}_2) \pm 1.701 S_{pooled} \sqrt{\frac{1}{n_1} + \frac{1}{n_2}}$$

$$(1{,}500 - 1{,}526) \pm (1.701)(38.06)\sqrt{\frac{1}{20} + \frac{1}{10}}$$

$$-51.07 \text{ psi} \underline{\hspace{2cm}} -0.93 \text{ psi}$$

The ARTEE Corporation is confident that the true difference in average strength is between -51.07 (new is stronger) and -0.93 (new is stronger) pounds per square inch.

=**problems**

Problems marked with an asterisk (*) require extra effort.

1. Explain in your own words how to go about determining the degrees of freedom for two different estimation situations.

2. A drug manufacturer has asked you to help decide how to test a new drug designed to lower blood pressure. The drug has been tested on animals and is theoretically sound but has not yet been tested on humans.

a. Discuss some major considerations in using statistics is this situation. Do you expect to be able to use a large-sample statistical test or a small-sample test?

b. How important do you expect precision to be in your results and recommendations? Why?

c. Could you possibly use a two-sample test? If so, how would you go about employing such a test?

3. A congressional committee, concerned about highway safety, has assigned an investigator to check into complaints that long-distance truckers have been violating the federal law limiting the hours they can drive each day. The investigator has arrived at your state and randomly selects trucker log books from 22 long-distance haulers. Looking at the driving time for the last full day, he finds a mean of 11.3 hours with a standard deviation of 2.69 hours.

a. What is his 95 percent confidence level estimate of the true average driving time per day for truckers going through your state?

b. What is the 80 percent confidence level estimate?

c. What practical problems in using these data might make these conclusions invalid?

4. The Charles Athletic Shoe Company has a promotional agreement with the Trailmarkers professional basketball team. Charles will supply shoes and monetary consideration to each team member in exchange for rights to feature team members in its advertising. Charles has recently developed a new shoe sole material it thinks provides superior wear compared with the old material. The company gives each of the ten team members one shoe of each material and monitors wear until the shoe, for playing purposes, is worn out. The following times are found:

Player	New Material	Old Material
1	47.0 h	45.5 h
2	51.0	50.0
3	42.0	43.0
4	46.0	45.5
5	58.0	58.5
6	50.5	49.0
7	39.0	29.5
8	53.0	52.0
9	48.0	48.0
10	61.0	57.5

Based on these data, and assuming a 95 percent confidence interval, does the evidence indicate there is a difference between the two shoe materials?

5. The student union automatic coffee machine has been accused of filling cups with less than the advertised 6 ounces of coffee. A group of ten statistics students decide to test the machine after class. Each student buys a cup of coffee, and the results are a mean of 5.93 ounces and a standard deviation of 0.13 ounce.

a. Construct a 98 percent confidence level interval for the true machine-fill amount.

b. Does this interval support or refute the advertising on the machine?

c. Could the machine possibly be accurate and still give a sample average of 5.93 ounces?

*6. A company personnel director is interested in estimating the difference in absenteeism between the day and night shifts at her company's East Coast distribution center. She takes a random sample of 15 days from the day shift, but only 11 days from the night shift. She finds an average of 7.8 days for the day shift and an average of 9.3 days for the night shift.

Based on this evidence, the personnel director concludes that the night supervisor is being lax in hiring practices and recommends a possible change in supervisors. Comment on this action.

7. The night supervisor in problem 6, hearing about this possible action, comments strongly about the "evidence." In particular, she states that drawing a conclusion based on this small a sample is just plain wrong. She says, "Anyone who knows anything about statistics knows you have to have a sample size of at least 30 before you can draw any conclusions." Comment on this statement.

8. A worker in the personnel department finds that the data used by the director in problem 6 have the following standard deviations: 4.3 days for the day shift and 4.1 days for the night shift.

a. Should this additional information change the conclusion the personnel director has drawn?

b. What assumptions do you have to make before you are justified in using the procedures in this chapter to make conclusions based upon the information contained in these data?

9. The Heavenly Cake Mix Company has recently developed a new angel food cake mix. The company is certain that this mix is "lighter" than the mix of its leading competitor. To test the contention, Heavenly sends 22 numbered, but unidentified boxes of mix to its testing lab (11 boxes of its new mix and 11 of the competitor's mix). The cakes are baked, and the weights per square unit (in grams) are determined. The following values are found:

New	Competition
$\bar{X} = 2.14$ g	$\bar{X} = 2.27$ g
$S_X^2 = 0.29$ g	$S_X^2 = 0.32$ g

Based on these results, would Heavenly be justified in claiming that its cake is "lighter" than its leading competitor's?

10. Crash Halloway has developed a new engine to run in his Indianapolis-type race cars. While all race engines require extensive maintenance, endurance is a critical factor since Crash does not win those races in which his engine stops. However new engines are costly (over $50,000), so although Crash must test the new engine, he cannot test many of them.

Crash runs an endurance bench test on six engines to determine the expected running time before major failure. He finds the following times:

6.7 h	6.2 h
7.3	3.8
5.8	4.1

a. Estimate the average life of new engines, based on this sample information, at the 95 percent confidence level.

b. Crash feels an engine must be able to run four hours to have a chance in any major race. Comment on the chances of this new engine.

11. Suppose three confidence interval estimates of the mean monthly retirement income for people over 65 developed from three different small samples from the same normally distributed population are as follows:

Sample	Lower Limit	Upper Limit
1	$403.19	$596.81
2	413.31	586.88
3	285.35	714.65

Which of these confidence intervals is the most precise? Be sure to state your reasoning.

*12. Considering the confidence interval from sample 1 in problem 11, if the sample size is ten and the sample variance is $44,100, what level of confidence can be placed in the estimate? Explain how you arrived at your answer.

*13. Considering the confidence interval developed from sample 2 in problem 11, suppose the sample standard deviation is 210 and the confidence level is 95 percent. What size sample was used? Explain how you determined this.

*14. Referring to the confidence interval from sample 3 in problem 11, suppose the sample size is 25 and the confidence level is 95 percent. What was the value of the sample standard deviation? Show your calculations.

*15. Referring to problems 11–14, comment on the advantages of each interval and explain why the interval from sample 3 is so much wider than the other two. Would you expect such results from random samples from the same population?

*16. Referring to problems 11–14, which sample produced the largest sample mean? Explain your answer.

17. The regional administrator for the state Senior Citizen Assistance Fund has looked at the three intervals in problem 11. She wonders how there can be three different estimates and asks you to pick the most accurate interval. Which do you pick, and why?

18. A local educational television channel wants to estimate how many minutes each day preschool children watch educational programs. Workers contact a

random sample of 20 families and ask the parents to record the time each day for the next two weeks. (Only weekdays are considered.) Records show a daily average of 87 minutes and a sample standard deviation of 34 minutes. Construct a 95 percent confidence interval estimate of the average time spent watching educational programs.

19. Kent Billings, a well-known political science instructor, is experimenting with a different method of teaching introductory political science. He is teaching two sections this term and decides to use the old method in one section and the new, objective-based method in the other section. At the end of the term he gives a standardized test to each section and wants to estimate the difference in average scores between the two methods. The standardized test gives the following results:

	Old Method	New Method
Class size	37	33
Average score	78.9 points	81.3 points
Sample standard deviation	7.2 points	6.9 points

   a. Do you see any problems with the sampling techniques that would make the difference hard to estimate?
   b. Is Kent safe in assuming that the new method is superior to the old? Why or why not?

20. A manufacturer of signal flares has developed a new powder to use for its flares shot from flare guns. Muzzle velocity is critical since too high a velocity will destroy the flare and too low a velocity will not send the flare high enough. The manufacturer has determined that 500 feet per second is the optimal muzzle velocity. Eight flares are loaded with charge, and the muzzle velocities are measured. The following velocities are found:

505 ft/s	495 ft/s
425	510
483	475
465	515

Redesigning the flares to hold either more or less powder will be relatively expensive and should be avoided if possible.

   Based on these sample results, do you recommend the redesigning be done?

21. The Wild West Exploration Company has recently started drilling exploratory gas wells. The company expects to buy over one hundred quick-assembly drill platforms over the next several years, so it is receiving lots of attention from platform manufacturers. Wild West has narrowed the possible suppliers to two, but cannot decide between them. Wild West management has placed an initial order for ten platforms from each manufacturer and will choose the type of platform that takes less time to assemble. Management randomly ships the two

types of platforms to the next 20 sites and measures the time to assemble. The results are summarized as follows:

	Type A	Type B
Average time to assemble	5.2 days	4.9 days
Sample standard deviation	0.6 day	0.48 day

Based on this sample information, from which manufacturer should Wild West buy the other platforms?

*22. A local television station has added a consumer spot to its nightly news. The consumer reporter has recently bought ten bottles of aspirin from a local drugstore and has counted the aspirins in each bottle. Although the bottles advertised 500 aspirins, the reporter found the following numbers:

491	490
487	497
496	507
504	481
483	495

The consumer reporter claims this is an obvious case of the public being taken advantage of. Comment.

*23. Health Care, Inc. is a nonprofit company that provides health-care treatment to patients with heart conditions. Health Care, Inc. relies heavily on contributions from around the state to fund its operation. Recently the company initiated a fund-raising campaign in which it hoped for 15,000 contributions totaling $1.5 million.

In preparation for a board of directors meeting, the treasurer selected a random sample of 20 contributions that had already arrived and found the mean contribution to be $88.00 and the standard deviation to be $25.00.

Assuming the organization will receive 15,000 contributions, what is the 95 percent confidence interval estimate for the total dollars this campaign will generate? Based upon this estimate, what should the treasurer report with respect to the $1.5 million goal? Assume the population is normally distributed.

24. Referring to problem 23, how would the treasurer's report differ if the interval estimate were based upon a 99 percent confidence level?

*25. In recent years there has been a major shift in data processing from batch-oriented systems to on-line systems. One of the possible reasons for this is that programming is more efficient in the on-line environment.

A recent study by one large computer software company involved six senior programmers. Each programmer wrote the same two programs. The programs were judged to be of comparable difficulty. The first program was developed in a batch environment, and the second was developed in the on-line environment. The manager overseeing the study recorded the actual work hours from start to finish as follows:

Programmer	Batch	On-Line
1	9.8 h	8.5 h
2	11.7	10.0
3	5.6	5.8
4	10.5	8.6
5	14.3	13.0
6	9.2	11.0

Develop a 90 percent confidence interval estimate for the mean difference in programming time under the two approaches for these two programs. What conclusions can you make about batch versus on-line based on these data?

26. Referring to problem 25, develop a 95 percent confidence interval estimate for the mean difference in programming time for the two approaches. Interpret your results.

27. Referring to problems 25 and 26, explain how the two confidence intervals differ and why. Also, point out any similarities between the two interval estimates.

=cases

11A MIDCENTRAL WAREHOUSING

Midcentral Warehousing is a contract warehouse operation with locations in several large southern cities. It services many middle-size manufacturing and distributing companies that are not large enough to individually afford the extensive inventory control and security provided by Midcentral.

Midcentral has been using a large number of diesel-operated forklifts but recently has been considering changing to battery-operated lifts. The purchasing chief, Georgeanne Andrews, has been asked to consider the relative merits of battery versus diesel. Georgeanne feels that diesel has definite cost advantages for the moment, but that battery power may be more efficient in the future. However she is worried about how long battery-operated lifts can operate before they need to be recharged. The battery lifts require one hour of down time to charge. Naturally Georgeanne is concerned that there will be too much down time during the three shifts. Her question is whether to start replacing the older-model diesels with battery-operated lifts or with new diesels.

Georgeanne has decided that the critical factor is the time the lifts can operate between charges. The manufacturer specifications indicate the lifts will operate for at least two shifts, but Georgeanne, having been in the purchasing game for a long time, has learned to distrust specifications.

The lift manufacturer has agreed to supply Midcentral with records from 50 of its customers. The data consist of measurements on the number of hours between charges. Because of the large volume of data, Georgeanne has selected a random sample of 18 values from all of the data. These values (in hours) are:

8	12	7
10	11	9
13	10	8
3	9	10
12	9	12
11	10	11

Georgeanne reasons that she ought to be able to get an idea about the lasting power of the battery lifts from these sample data without looking at all the data.

11B   U-NEED-IT RENT ALL

Richard Fundt has operated the U-NEED-IT rental agency in a northern Wisconsin city for the past five years. One of the biggest rental items has always been chain saws, and lately the demand for these saws has increased dramatically. Richard buys chain saws at the industrial rate, and he then rents them for $10 per day. The average chain saw is used between 50 and 60 days per year. Although Richard makes money on any chain saw, he obviously makes more on those saws that last the longest.

Richard worked for a time as a repairperson and can make most repairs on the equipment he rents, including the chain saws. However he would also like to limit the time he spends making repairs. U-NEED-IT is presently stocking two types of saws—North Woods and Accu-Cut. Richard has a vague feeling that one of the models, Accu-Cut, doesn't seem to break down as much as the other. Richard presently has 8 North Woods saws and 11 Accu-Cut saws. He decides to keep track of the number of hours each saw is used between major repairs. He finds the following values:

Accu-Cut		North Woods	
48	46	48	78
39	88	44	94
84	29	72	59
76	52	19	52
41	57		
24			

The North Woods salesperson has stated that North Woods may be raising the price of its saws in the near future. This will make them slightly more expensive than the Accu-Cut models. However the prices have tended to move with each other in the past.

=====references

HAYES, GLEN E. and HARRY G. ROMIG. *Modern Quality Control*. Encino, Calif.: Bruce, 1977.

PFAFFENBERGER, ROGER C. and JAMES H. PATTERSON. *Statistical Methods for Business and Economics*. Homewood, Ill.: Irwin, 1977.

WINKLER, ROBERT L. and WILLIAM HAYES. *Statistics: Probability, Inference, and Decision*. New York: Holt, Rinehart and Winston, 1975.

WONNACOTT, THOMAS H. and RONALD J. WONNACOTT. *Introductory Statistics for Business and Economics*. New York: Wiley, 1977.

12

Introduction to Hypothesis Testing

=why decision makers need to know

In Chapters 10 and 11 we introduced the basic techniques of statistical estimation. These techniques have many business applications. All applications, however, have a common bond: sample information is used to help decision makers determine what the population is like. Although the marketing researcher, the accountant, and the operating manager all have their particular uses for statistical sampling, they all start out knowing very little about the population of interest. Through random sampling they are able to estimate population values such as the mean, variance, and proportion.

The estimation process of going from sample to population is useful and often applied. However there is another class of problems where sampling and statistical techniques are employed. In this class decision makers are faced with a claim, or **hypothesis,** about the population and must be able to substantiate or refute this claim. For example, a producer of electronic components claims that 3 percent or fewer of its components are defective. Before you buy 100,000 electronic components, you would want to test the manufacturer's claim. Since you could not feasibly test each component, you would likely select a sample and use the sample information to decide whether there really are 3 percent or fewer defectives in the shipment. **This process of using sample information to statistically decide on the state of a population is called** *statistical inference.*

In this chapter we introduce several statistical techniques used to test claims about population values. All decision makers need to have a solid understanding of these techniques if they are to be able to use the information available in a sample effectively.

chapter objectives

In this chapter we shall introduce statistical hypothesis testing. Hypothesis testing requires decision makers to first formulate a position or make a claim regarding the decision environment they are dealing with. Then they select a sample and, based on its contents, either affirm that this position is correct or conclude it is wrong. We will demonstrate how these predetermined positions or claims are formulated and how data are used to substantiate or refute the positions. The two types of errors that can be made in hypothesis testing will be discussed. The material in this chapter will also show how to establish decision-making rules in light of the chances of making each type of error.

student objectives

After studying the material in this chapter, you should be able to:

1. Discuss two forms of statistical hypotheses and know how to use sample information to test each.
2. Correctly identify and formulate a decision rule to accompany each statistical hypothesis.
3. Identify the two forms of potential error in hypothesis testing and discuss these errors and their consequences.
4. Develop the operating characteristic curve and power curve associated with each decision rule.
5. Recognize business applications in which statistical hypothesis testing can be performed.

12-1
REASONS FOR TESTING HYPOTHESES

Most managers would greatly increase their decision-making "batting average" if they knew the exact values of all the measures in their environment. Unfortunately they seldom do. We have spent considerable time in previous chapters discussing why managers need to gather data by sampling. Many times managers know what a population value should be because of company policy or contract specification, and must be able to use sample information to determine whether or not the policy or contract specification is being satisfied. For example, the Rock-Dee Corporation has a contract to produce a worm gear that will be used in Army helicopters. The contract specifies that the gears must have an average diameter of 3.4 inches. Further, 99 percent of all gears must be within 0.05 inch of this average. Certainly the Rock-Dee Corporation is interested in knowing whether or not it is satisfying the contract. A sample of gears could be selected, and based on the sample information, the production and quality control managers would decide whether or not to ship the last batch of gears.

Because information contained in a sample is always subject to sampling error, the decision regarding the gear production may be incorrect. Decision makers must recognize that the chance of making the wrong decision always exists.

Statistical hypothesis testing allows managers to use a structured analytical method to make decisions. It lets them make the decisions in such a way that the chances of decision errors can be controlled, or at least identified. Even though statistical hypothesis testing does not eliminate the uncertainty in the managerial environment, the techniques involved often allow managers to identify and control the level of uncertainty.

12-2
THE HYPOTHESIS-TESTING PROCESS

Denise Fitzgerald has been hired as head of production for the Cola Bottling Company. Some soft-drink bottlers have recently been under pressure from consumer groups, which claim that bottlers have been increasing the price of cola and filling the bottles with less than the advertised amount. Although Denise feels no manufacturer would purposely short-fill a bottle, she knows that filling machines sometimes fail to operate properly and fill the bottles less than full.

Since Denise is responsible for making sure the filling machines at her company operate correctly, she samples bottles every hour and, based on the sample results, decides whether to adjust the machines. If she is *not* interested in whether the bottles are filled with too much soft drink, she can identify two possible *states of nature* for 32-ounce bottles:

State 1. The bottles are filled with 32 or more ounces of soft drink on the average.

State 2. The bottles are filled with less than 32 ounces on the average.

Denise must base her decision about the filling process on the results of her hourly sample. As indicated earlier, when a decision is based on sample results, sampling error must be expected. Therefore, sometimes when the process is operating correctly, the sample will indicate that the average cola bottle contains less than 32 ounces. And sometimes the process will need adjustments when the sample results indicate it is operating correctly. To analyze such a situation and "best" select among the possible states of nature, decision makers need to formulate the problem such that a hypothesis test can be conducted. The first step is to restate the states of nature in hypothesis-test notation. For example, for our cola example,

$$\text{Null hypothesis—}H_0:\ \mu_x \geq 32\ oz$$

$$\text{Alternate hypothesis—}H_A:\ \mu_x < 32\ oz$$

The state of nature picked as the **null hypothesis** depends on the decision problem. **The *null hypothesis*, H_0, is the hypothesis, or claim, being tested and must contain the equality sign. The *alternative hypothesis*, H_A, represents population values other than those contained in the null hypothesis.**

Depending upon the sample information, the null hypothesis will either be supported or refuted. Careful thought should be given to establishing the null and alternative hypotheses since the conclusion reached may depend on the hypotheses being tested. Many examples throughout this and subsequent chapters will illustrate how to develop proper statistical hypotheses.

If Denise decides the filling process is operating correctly, she will allow it to continue, but if she decides it is operating incorrectly, she will adjust the filling machines. The decision will be based on the results of her sample. Thus, in addition to making the decision one way or the other, she can make two errors:

1. She may decide that the process is filling bottles with an average less than 32 ounces when, in fact, the average fill is 32 or more ounces. In this case she will *reject* a true null hypothesis. This error is a ***Type I statistical error.***

2. She may decide that the process is filling bottles with an average of 32 or more ounces when, in fact, it is not. Thus she might *accept* the null hypothesis when it is false. This error is a ***Type II statistical error.***

Figure 12–1 shows the possible actions and possible states of nature associated with all hypothesis-testing problems. As you can see, there are three possible outcomes: no error, Type I error, and Type II error. Only one of these outcomes will occur for every test of a null hypothesis. Of course, everyone would like to eliminate all chance of error; however the decision maker may, in fact, make

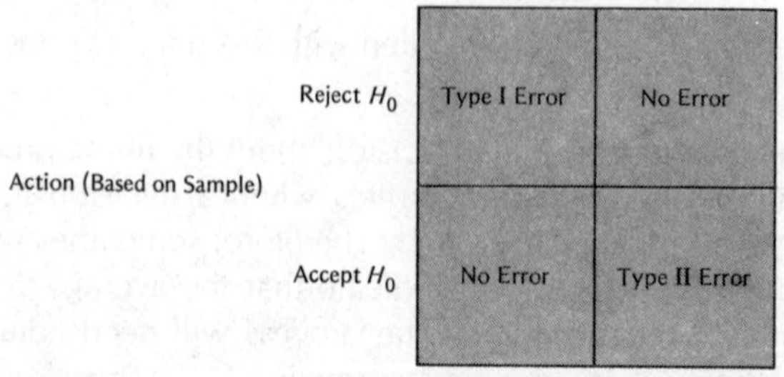

FIGURE 12–1   Hypothesis-testing outcome possibilities.

either a Type I or a Type II statistical error depending upon which decision is selected. Note, from Figure 12–1, that **if the null hypothesis is true and an error is made, it must be a Type I error.** On the other hand, **if the null hypothesis is false and an error is made, it must be a Type II error.**

Establishing the Decision Rule

The objective of a hypothesis test is to use sample information to decide whether to accept or reject the null hypothesis about a population value. But how do decision makers determine whether the sample information supports or refutes the null hypothesis? The answer is that they must compare the sample results with a predetermined decision rule.

Returning to the Cola Bottling Company example, the null and alternative hypotheses are as follows:

Hypotheses:

Null hypothesis—H_0: $\mu_x \geq 32$ oz
Alternative hypothesis—H_A: $\mu_x < 32$ oz

Denise Fitzgerald would establish the following decision rule:

Decision Rule:

Based on a sample,

If $\overline{X} \geq A$, conclude that the null hypothesis is correct and accept H_0.
If $\overline{X} < A$, conclude that the null hypothesis is false and reject H_0.

where: $\overline{X}$ = sample mean
            A = critical value

If the null hypothesis is rejected, Denise will halt production and have a maintenance crew adjust the filling machine to increase the average fill. On the other hand, if $\overline{X}$ is greater than or equal to A, she will accept the null hypothesis and conclude that the filling machines are working properly. Once the critical value is determined, the choice of actions is more or less mechanical and is not actually a decision. The real decision involves determining the critical value, A.

Selecting the Critical Value

Recall from the central limit theorem that for large samples, the distribution of possible sample means will be approximately normal with a center at the population mean, μ_x. The null hypothesis in the Cola Bottling Company example is $\mu_x \geq 32$ ounces, but even if the null hypothesis is true, we may get a sample mean less than 32 (sampling error). Assume for a minute that the population mean is exactly 32 ounces. Figure 12–2 shows the distribution of possible sample means for the cola company if $\mu_x = 32$. Selecting the critical value, A, for a hypothesis test requires the

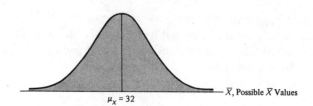

FIGURE 12–2   Sampling distribution of $\overline{X}$, Cola Bottling Company.

decision maker to answer the following question: What values of $\overline{X}$ will tend to refute the null hypothesis—values much smaller than μ_x, values much larger than μ_x, or values both much smaller *and* much larger than μ_x?

In our cola example, values of $\overline{X}$ greater than or equal to 32 ounces would tend to support the null hypothesis. However values of $\overline{X}$ below 32 ounces would tend to refute the null hypothesis. The smaller the value of $\overline{X}$, the greater the evidence that the null hypothesis should be rejected.

For example, the critical value in the cola example might be located as shown in Figure 12–3. If the null hypothesis is actually true ($\mu_x \geq 32$), the area under the normal curve to the left of A represents the *maximum* probability of rejecting a true null hypothesis, which is a Type I error. This probability is called **alpha (α).** The chances of committing a Type I error can be reduced if the critical value, A, is moved farther to the left of $\mu_x = 32$. This is illustrated in Figure 12–4.

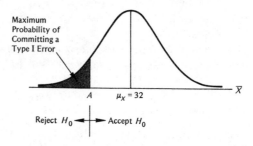

FIGURE 12–3   Critical value, Cola Bottling Company.

To determine the appropriate value for A, decision makers must determine how large an alpha they want. **Decision makers must select the value of alpha in light of the costs involved in committing a Type I error.** For example, if Denise Fitzgerald rejects the null hypothesis when it is true, she will needlessly shut down production and incur the cost of machine adjustment. In addition, the adjustment might be incorrect, and future production could be affected. Calculating these costs and determining the probability of incurring them is a subjective management decision. Any two managers might well arrive at different alpha levels. The important thing is that each would specify his or her alpha level for the hypothesis test.

Suppose Denise decides that she is willing to incur a 0.10 chance of committing a Type I error. Then, assuming the standard deviation of the production

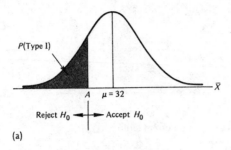

(a)

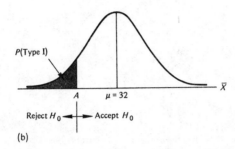

(b)

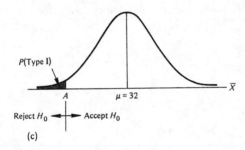

(c)

FIGURE 12–4 Type I error probabilities, Cola Bottling Company.

process is 0.5 ounce and the sample size is 64 bottles, the appropriate value of A is determined as in Figure 12–5. The decision rule is as follows:

Decision Rule:

Given $n = 64$.

If $\bar{X} \geq 31.92$ oz, accept H_0.
If $\bar{X} < 31.92$ oz, reject H_0.

Suppose the sample results in a sample mean of 31.50 ounces. As shown in Figure 12–6, the decision is to reject the null hypothesis and re-adjust the filling machines. However this decision must be made with the knowledge that the null hypothesis may, in fact, be true.

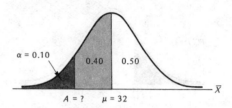

From the standard normal table,*

$$Z_{0.40} \approx -1.28$$

Then

$$Z = \frac{A - \mu_x}{\dfrac{\sigma_x}{\sqrt{n}}}$$

Solving for A,

$$A = \mu_x + Z\frac{\sigma_x}{\sqrt{n}} = 32 + (-1.28)\left(\frac{0.5}{\sqrt{64}}\right)$$

$$= 31.92 \text{ oz}$$

*The Z value from the standard normal table is used because the sampling distribution of $\overline{X}$ is approximately normal according to the central limit theorem when σ_x is known. Remember from Chapter 7, if A is to the left of the mean, Z will be negative.

FIGURE 12–5   Determining the critical value, Cola Bottling Company.

Hypotheses:

H_0: $\mu_x \geq 32$ oz
H_A: $\mu_x < 32$ oz

$\alpha = 0.10$

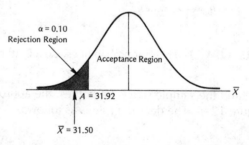

Since $31.50 < 31.92$, reject H_0.

FIGURE 12–6   Hypothesis test, Cola Bottling Company.

Summary

The hypothesis-testing process discussed in this section can be summarized in six steps:

1. **Determine the null hypothesis and the alternative hypothesis.**
2. **Determine the desired alpha level.**

3. **Choose a sample size.**
4. **Determine the critical value, A.**
5. **Establish the decision rule.**
6. **Select the sample and perform the test.**

These six steps can be followed to test any null hypothesis. Become well acquainted with the hypothesis-testing process, as it is a fundamental part of statistical decision making.

12-3
ONE-TAILED HYPOTHESIS TESTS

In the cola company example the null hypothesis could be refuted only if the sample mean was too small (that is, too far to the left of $\mu_x = 32$ ounces). Consequently the critical value, A, was placed in the left-hand (lower) tail of the normal curve. This example illustrates a *one-tailed hypothesis test.*

A one-tailed hypothesis test will assume one of two forms, which are determined by the way the null and alternative hypotheses are stated. Examples of these two forms are:

1	2
$H_0: \mu_x \le 50$	$H_0: \mu_x \ge 32$
$H_A: \mu_x > 50$	$H_A: \mu_x < 32$

If the first set of null and alternative hypotheses is used, we use a one-tailed hypothesis test with the critical value, A, in the right-hand (upper) tail of the normal curve, and we reject H_0 when $\overline{X} > A$. However, if the second set of null and alternative hypotheses is used (our cola example), the critical value, A, is placed in the left-hand tail, and we reject H_0 when $\overline{X} < A$.

The Ever-Stick Glue and Chemical Company manufactures many different chemicals in an eastern state. The firm uses vast amounts of water in its processing and has a conditional-use permit from the Environmental Protection Agency. Ever-Stick is allowed to return its waste water to the nearby river if it lowers the pollutant levels to allowable amounts. For a common toxic chemical, Ever-Stick is required to reduce the amount in the waste water to 20 parts per million or less.

Ever-Stick's pollution-abatement equipment will easily reduce the toxic chemical levels below the upper limit of 20 parts per million when working correctly, but like much pollution equipment it needs checking and adjustment to maintain the standards. The waste water is continually monitored, and based on the periodic sample results, the pollution equipment is either adjusted—a costly process—or allowed to continue as is. The appropriate hypotheses and decision rule for this situation are stated as follows:

Hypotheses:

H_0: $\mu_x \leq 20$ p.p.m., process can continue
H_A: $\mu_x > 20$ p.p.m., process requires adjustment

Decision Rule:

Take a random sample of n and calculate $\overline{X}$.

If $\overline{X} \leq A$, accept H_0.
If $\overline{X} > A$, reject H_0.

Ever-Stick decides to limit alpha to a maximum of 3 percent and base the decision on a sample of 40 trials. The decision rule is determined as shown in Figure 12–7. Note that the standard deviation, σ_x, is assumed to be five parts per million.

Hypotheses:

H_0: $\mu_x \leq 20$ p.p.m.
H_A: $\mu_x > 20$ p.p.m.

$\alpha = 0.03$

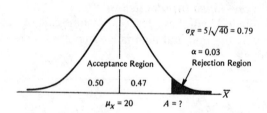

From the standard normal table,

$$Z_{0.47} \approx 1.88$$

Then

$$Z = \frac{\overline{X} - \mu_x}{\sigma_{\overline{x}}} = \frac{A - \mu_x}{\dfrac{\sigma_x}{\sqrt{n}}}$$

Solving for A,

$$A = \mu_x + Z\frac{\sigma_x}{\sqrt{n}} = 20 + 1.88(0.79)$$

$$= 21.485 \text{ p.p.m.}$$

Decision Rule:

If $\overline{X} \leq 21.485$ p.p.m., accept H_0.
If $\overline{X} > 21.485$ p.p.m., reject H_0.

FIGURE 12–7 Decision rule, Ever-Stick Glue and Chemical Company.

Suppose the Ever-Stick Glue and Chemical Company has selected a sample of 40 units of water that is scheduled to go into the river and has found an average of 23.4 parts per million of the toxic chemical. What should Ever-Stick do based upon this information? If you conclude the company should reject the null hypothesis and order the pollution-abatement equipment to be adjusted, you are consistent with the decision rule determined in Figure 12–7. However the possibility does exist that the null hypothesis is true and the adjustment is not needed.

12-4
TWO-TAILED HYPOTHESIS TESTS

The hypothesis-testing examples presented thus far have been one-tailed. The entire rejection region (alpha) has been located in one tail or the other. However sometimes decision makers face situations where a one-tailed hypothesis test is not appropriate. For example, Sam Stuart is a shirt buyer for a chain of southern department stores. Though shirts manufactured in foreign countries often provide a considerable cost savings over shirts made in the United States, Sam has noticed that the foreign manufacturers' quality control is often lacking. In particular, the shirt sizes tend to be inconsistent. A shirt labeled "16-33," which is supposed to have a 16-inch neck and a 33-inch sleeve, may not have the correct measurements.

Sam makes all purchases contingent on sizing being acceptable. However he cannot examine all the shirts since they arrive from the foreign manufacturers already wrapped. This means Sam must examine a sample of each shirt size. In the past Sam has made his accept/reject decision for shirts at each size based on a random sample of 45 shirts. For the sleeve length of the 16-33 shirt, the following hypotheses would apply:

Hypotheses:

H_0: $\mu_x = 33$ in.
H_A: $\mu_x \neq 33$ in.

Sam recognizes that even with the best random sampling techniques, he will occasionally make an error. Sometimes he will send back a shipment when he shouldn't (that is, reject the null hypothesis when it is true, a Type I error), and sometimes he will accept a shipment that should be sent back (that is, accept the null hypothesis when it is false, a Type II error).

In establishing the decision rule, Sam must decide which values of $\overline{X}$ will tend to refute the null hypothesis. In this case, the null hypothesis should be refuted for values of $\overline{X}$ either too much lower than $\mu_x = 33$ inches or values of $\overline{X}$ too much above $\mu_x = 33$ inches. This situation illustrates a ***two-tailed hypothesis test,*** meaning that the rejection region is divided between the two tails. (Although it is not required, in this text we will always divide the rejection region equally between the two tails of the sampling distribution.)

The correct decision rule for an equality hypothesis is as follows:

Decision Rule:

Select a sample of size n and determine $\overline{X}$.

If $A_L \leq \overline{X} \leq A_H$, accept H_0.
If $\overline{X} > A_H$, reject H_0.
If $\overline{X} < A_L$, reject H_0.

where: A_L = critical value in the lower tail

           A_H = critical value in the upper tail

Sam has specified an alpha level of 0.06 ($\alpha = 0.06$), meaning that he is willing to accept a 6 percent chance of rejecting a shipment of shirts of a given size if, in fact, they are satisfactory. Also, based on past experience with the manufacturers, Sam feels sure that the standard deviation, σ_x, is 0.25 inch. Figure 12–8 shows how the decision rule is developed. Note that the 6 percent rejection region is split between the two tails of the normal curve. Thus, if Sam selects a sample of 45 size 16-33 shirts, and the average sleeve length is less than 32.93 inches or greater than 33.07 inches, he will reject the null hypothesis and send back all shirts of this size in the batch.

Hypotheses:

H_0: $\mu_x = 33$ in.
H_A: $\mu_x \neq 33$ in.

$\alpha = 0.06$

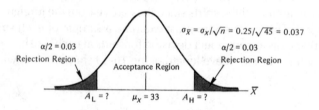

$$\sigma_{\bar{x}} = \sigma_x/\sqrt{n} = 0.25/\sqrt{45} = 0.037$$

From the standard normal table,

$$Z_{0.47} \approx \pm 1.88$$

Then, solving for A_L and A_H,

$$Z = \frac{A_L - \mu_x}{\sigma_{\bar{x}}} \qquad \text{and} \quad Z = \frac{A_H - \mu_x}{\sigma_{\bar{x}}}$$

$A_L = \mu_x + Z\sigma_{\bar{x}} = 33 + (-1.88)(0.037)$ $\qquad A_H = \mu_x + Z\sigma_{\bar{x}} = 33 + 1.88(0.037)$
     $= 32.93$ $\qquad\qquad\qquad\qquad\qquad\qquad = 33.07$

Decision Rule:

If $32.93 \leq \bar{X} \leq 33.07$, accept H_0.
If $\bar{X} < 32.93$, reject H_0.
If $\bar{X} > 33.07$, reject H_0.

FIGURE 12–8   Decision rule for a two-tailed test, Sam Stuart example.

12-5
HYPOTHESIS TESTING—POPULATION STANDARD DEVIATION UNKNOWN, LARGE SAMPLE

In our discussion of hypothesis testing we have assumed that the population standard deviation is known. When we discussed statistical estimation, we stated that in most cases, the population standard deviation is not known. In the estimation chapters our solution to an unknown standard deviation was to estimate the population standard deviation, σ_X, by the sample standard deviation, S_X. As you might suspect, we will also make this change for hypothesis testing.

Denise Fitzgerald, at the Cola Bottling Company, has installed a new filling machine. She wants to know whether the machine is filling bottles with 32 or more ounces, but has no past history for determining the population standard deviation. She can formulate the appropriate hypothesis in the same manner as before:

Hypotheses:

H_0: $\mu_x \geq 32$ oz
H_A: $\mu_x < 32$ oz

This null hypothesis will lead to the following decision rule:

Decision Rule:

Select a sample of size n and calculate $\overline{X}$.

If $\overline{X} \geq A$, accept H_0.
If $\overline{X} < A$, reject H_0.

Up to now the procedure has been the same as in the previous sections. However, since σ_X is not known, Denise will have to calculate the sample standard deviation in addition to the sample mean. If the sample size is large ($n \geq 30$), the sample standard deviation, S_X, can be substituted for σ_X. If Denise again specifies an alpha of 0.10, the critical value, A, becomes 31.89, as seen in Figure 12–9. Denise will compare the sample mean with the new critical value, and based on the decision rule, either accept or reject the null hypothesis.

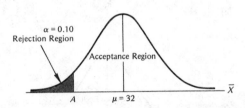

From the standard normal table,

$$Z_{0.40} = -1.28$$

Then

$$Z = \frac{A - \mu_x}{\frac{S_x}{\sqrt{n}}}$$

Denise takes a sample of 49 bottles and finds a sample standard deviation of 0.6. She can now solve for A.

$$A = \mu_x + Z\frac{S_x}{\sqrt{n}} = 32 + (-1.28)\left(\frac{0.6}{\sqrt{49}}\right)$$

$$= 31.89 \text{ oz}$$

FIGURE 12–9 Determining the critical value, population standard deviation unknown, Cola Bottling Company.

12-6

TYPE II ERRORS AND THE POWER OF THE HYPOTHESIS TEST

In the previous examples we have shown how decision rules for hypothesis tests of the population mean are determined and how the appropriate decision is based on the results of a sample from the population. In these examples we have determined the critical value(s) by first specifying alpha, the probability of committing a Type I error. As we indicated, **if the cost of committing a Type I error is high, the decision maker should specify a small alpha. A small alpha results in a small rejection region. If the rejection region is small, the sample mean is less likely to fall there, and the chances of rejecting a true null hypothesis are small.**

This logic provides a basis for establishing the critical value for the hypothesis test. However it completely ignores the possibility of committing a Type II error. Recall that accepting a false null hypothesis is a Type II decision error. If the rejection region is made small by selecting a small alpha level, the acceptance region will be large since the rejection region plus the acceptance region must add to 1.0. This is illustrated in Figure 12–10.

To complete the logic, **if the acceptance region is large, the probability of committing a Type II error [called *beta* (β)] will also tend to be large.** This is a

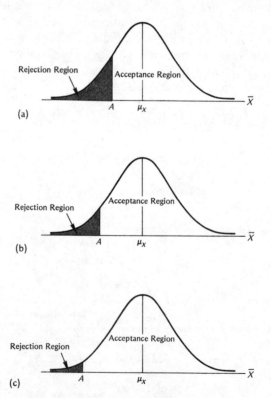

FIGURE 12–10   Acceptance and rejection regions.

point many decision makers often forget in their efforts to control Type I errors. A decrease in alpha will increase beta. However, as we will show, the increase in beta will not equal the decrease in alpha.

Decision makers must examine carefully the costs of committing Type I and Type II errors and, in light of these costs, attempt to establish acceptable values for alpha and beta. We will delay our discussion of how decision makers can simultaneously control the chances of making a Type I and a Type II error until Chapter 13. We will, however, now show you how to calculate the probability of committing a Type II error.

Calculating Beta (Probability of Committing a Type II Error)

Once alpha has been specified for a hypothesis test involving a particular sample size, beta cannot also be specified. Rather, the beta value is fixed, and all the decision maker can do is calculate it. We do not mean to imply that beta is a single value, because it is not. Since a Type II error occurs when a false null hypothesis is accepted (refer to Figure 12–1), there is a beta value for each possible population value for which the null hypothesis is false. For example, for the Ever-Stick Glue and Chemical Company, the null and alternative hypotheses were

$$H_0: \mu_x \leq 20 \text{ p.p.m.}$$

$$H_A: \mu_x > 20 \text{ p.p.m.}$$

Therefore the null hypothesis is false for all possible values of $\mu_x > 20$ parts per million. Thus, for each of these infinite number of possibilities, a value of beta can be determined. Suppose we assume that μ_x is actually 21 parts per million. If we use the alpha level of 0.03 and the decision rule determined in Figure 12–7, we can calculate beta using the following steps:

1. Draw a picture of the hypothesized sampling distribution showing the rejection region(s) and the acceptance region found by specifying an alpha level.
2. Immediately below the hypothesized sampling distribution, draw the "true" sampling distribution based upon the assumed new population mean. Note that the shape of the "true" distribution will be the same as the shape of the hypothesized distribution. Only the central location will be different.
3. Extend the critical value(s) from the hypothesized distribution down to the "true" distribution and shade the rejection region on the "true" distribution.
4. The unshaded area in the "true" distribution is beta, the probability of committing a Type II error.

Figure 12–11 shows how beta is determined if the "true" value of μ_x is 21 parts per million. Thus, by holding alpha to 0.03, the chance of committing a

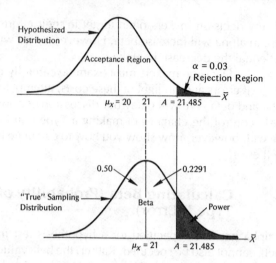

$$Z = \frac{\overline{X} - \mu_x}{\sigma_{\overline{x}}} \quad \text{(From the "true" distribution)}$$

$$= \frac{21.485 - 21}{0.79}$$

$$= 0.61$$

The area between $Z = 0.61$ and μ_x is 0.2291. Therefore

$$\text{Beta} = 0.5000 + 0.2291$$
$$= 0.7291$$
$$\text{Power} = 1 - \text{beta} = 1 - 0.7291$$
$$= 0.2709$$

FIGURE 12–11   Beta calculation, "true" $\mu_x = 21$.

Type II error is approximately 0.7291. However, what if the true mean is actually 22? Figure 12–12 shows how to calculate beta for this case.

The shaded area in the "true" distribution has also been calculated. This area is called the **power** of the hypothesis test. **Power is the probability of rejecting a false hypothesis.**[*] Naturally decision makers want power to be as large as possible, and this occurs if beta is made as small as possible since power and beta are inversely related. Power is determined as follows:

$$\text{Power} = 1 - \text{beta} \qquad\qquad (12–1)$$

Like beta, power changes depending on what the "true" value for μ_x is assumed to be.

Figures 12–11 and 12–12 show that as the "true" μ_x value is moved farther away from the hypothesized μ_x, beta becomes smaller. **The greater the dif-**

---

[*]Although this is a common definition of power, it is not universal. Some authors define power as the probability of rejecting the null hypothesis. For the purposes of an introductory text, this difference is not important.

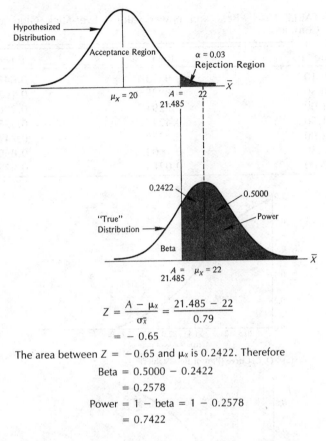

$$Z = \frac{A - \mu_x}{\sigma_{\bar{x}}} = \frac{21.485 - 22}{0.79}$$

$$= -0.65$$

The area between $Z = -0.65$ and μ_x is 0.2422. Therefore

$$\text{Beta} = 0.5000 - 0.2422$$
$$= 0.2578$$
$$\text{Power} = 1 - \text{beta} = 1 - 0.2578$$
$$= 0.7422$$

FIGURE 12–12 Beta calculation, "true" $\mu_x = 22$.

ference between the "true" mean and the hypothesized mean, the easier it is to tell the two apart, and the less likely we will accept the null hypothesis when it is actually false. Of course the opposite is also true. **As the "true" mean moves closer and closer to the hypothesized mean, the harder it is for the hypothesis test to recognize the difference between the two.**

Operating Characteristic Curves and Power Curves

As we indicated earlier, there are an infinite number of possible values for μ_x when the null hypothesis is false, and it would not be practical to calculate beta and power for each one. Instead we select several possible values for μ_x and find both beta and power. These values are plotted on separate graphs to form an **O.C. (operating characteristic) curve** for the beta values and a **power curve** for the power values. For example, the Ever-Stick Glue and Chemical Company might be interested in values for beta and power for the possible values of μ_x shown in Table 12–1. Make sure you can calculate these values also.

Figure 12–13 illustrates the O.C. curve and the power curve constructed from the values shown in Table 12–1. Notice that the several points have

TABLE 12–1 Beta and power values, Ever-Stick Glue and Chemical Company

"True" μ_x	Beta	Power
20.10	0.9573	0.0427
20.50	0.8925	0.1075
21.00	0.7291	0.2709
21.50	0.4247	0.5753
22.00	0.2578	0.7422
22.50	0.1003	0.8997
23.00	0.0274	0.9726

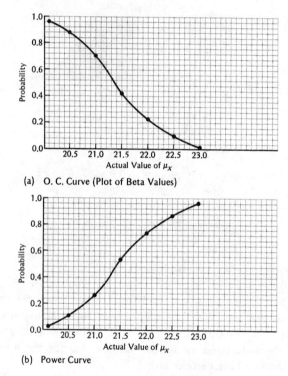

(a) O. C. Curve (Plot of Beta Values)

(b) Power Curve

FIGURE 12–13 O.C. curve and power curve, hypothesized $\mu_x = 20$.

been plotted and connected with a smooth curve. Thus, by using the O.C. curve and power curve in Figure 12–13, you can approximate both power and beta for any possible value of μ_x.

Impact of Sample Size on Power and Beta

In previous chapters we have emphasized that the size of the sample influences the spread in the sampling distribution of $\overline{X}$. As the sample size increases, $\sigma_{\overline{x}}$ decreases. The smaller $\sigma_{\overline{x}}$ becomes, the less the chance of extreme sampling error. This same concept applies in hypothesis testing and determining power and beta. As more information is available (larger sample size), with alpha held constant, the smaller beta (and the larger power) will be for any value of μ_x.

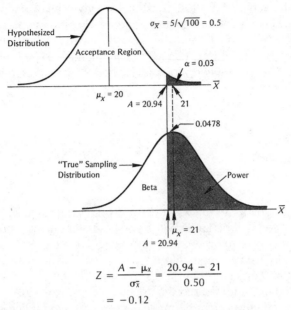

$$Z = \frac{A - \mu_x}{\sigma_{\bar{x}}} = \frac{20.94 - 21}{0.50}$$

$$= -0.12$$

The area between $Z = -0.12$ and μ_x is 0.0478. Therefore

$$Beta = 0.5000 - 0.0478$$

$$= 0.4522$$

$$Power = 1 - beta = 1 - 0.4522$$

$$= 0.5478$$

FIGURE 12–14 Beta and power calculations, Ever-Stick Glue and Chemical Company.

For example, the values of power and beta in Table 12–1 were determined assuming a sample size of $n = 40$. Suppose the sample size is increased to 100. Figure 12–14 shows how beta and power are determined for $\mu_x = 21$. Table 12–2 presents the values of beta and power for sample sizes of 40 and 100. Note that in all cases, a shift in sample size from 40 to 100 has reduced beta and increased power. Notice also that the increase in sample size has changed the critical value, A, from 21.485 to 20.94. The reduced critical value occurs because $\sigma_{\bar{x}}$ has decreased.

TABLE 12–2 Power and beta values, Ever-Stick Glue and Chemical Company
(H_0: $\mu_x = 20$)

"True" μ_x	Beta		Power	
	$n = 40$	$n = 100$	$n = 40$	$n = 100$
20.10	0.9573	0.9535	0.0427	0.0465
20.50	0.8925	0.8106	0.1075	0.1894
21.00	0.7291	0.4522	0.2709	0.5478
21.50	0.4247	0.1314	0.5753	0.8686
22.00	0.2578	0.0170	0.7422	0.9830
22.50	0.1003	0.00009	0.8997	0.9991
23.00	0.0274	0.00000*	0.9726	1.0000

*The true value here is small but positive.

This example has illustrated that **for a given level of alpha, the chances of making a Type II error can be reduced by increasing the sample size.** Depending upon the decision being made and the costs of making a Type II error, decision makers may want to control the chances of committing a Type II error. As we will show in Chapter 13, for a given alpha level, beta can be controlled by using an appropriate sample size.

12-7
CONCLUSIONS

This chapter has served as an introduction to hypothesis testing. As we have stated, decision makers often will be faced with hypotheses, or claims, about populations. The concepts introduced in this chapter, and extended in the next, provide decision makers with tools for using sample information to decide whether a given hypothesis should be accepted or rejected.

We have emphasized the importance of recognizing that when a hypothesis is tested, an error might be made. Type I and Type II decision errors have been discussed, and we have shown how to calculate the probability of both types of error. Decision makers must evaluate the costs associated with making a Type I error and weigh these against the costs of committing a Type II error. In the next chapter we illustrate how to control both alpha and beta in the same decision problem.

We have purposely limited our discussion to hypothesis tests of a single population mean. However the fundamentals introduced here will be applied in Chapters 13–17 of this text. Before proceeding to Chapter 13, make sure you have a firm grasp of the fundamentals of statistical inference presented in this chapter.

A summary of the important statistical terms introduced in this chapter are presented in the glossary. You should compare these definitions with the discussion in the body of the chapter.

chapter glossary

alpha   The maximum probability of committing a Type I error.

beta   The probability of committing a Type II error.

critical value   The value(s) in a hypothesis test that separate the rejection region from the acceptance region. The critical value can be in the same units as the population mean or it can be in standardized units.

hypothesis   A supposition about a true state of nature.

O.C. (operating characteristic) curve   The plot of the probability of accepting a false null hypothesis over the range of the alternative hypothesis.

one-tailed hypothesis test   A hypothesis test in which the entire rejection region is located in one tail of the sampling distribution.

power   The probability of rejecting a null hypothesis when it is false. Power is the complement of beta.

power curve   The plot of the probability of rejecting a false hypothesis.

states of nature   The uncertain events over which decision makers have no direct control.

statistical inference   The process by which decision makers reach conclusions about a population based on sample information collected from the population.

two-tailed hypothesis test   A hypothesis test in which the rejection region is split between the two tails of the sampling distribution.

Type I error   Rejecting the null hypothesis when it is, in fact, true.

Type II error   Accepting the null hypothesis when it is, in fact, false.

=solved problems

1. The personnel department of the East Coast Metal Fabrication Company needs help. A supervisor in the welding shop has repeatedly complained about one of her crew, a man named Robinson. The supervisor claims Robinson is a troublemaker, produces below-standard work, and is slow. East Coast Metal is bound by a strong union contract, and despite his "poor" work, Robinson is very active in union affairs. The personnel department feels that if Robinson is fired, the union will challenge the case and claim that Robinson was fired because of his union activities.

   Although the personnel department feels Robinson cannot be challenged on vague issues such as below-standard work or being a troublemaker, there are some rather strong clauses in the contract dealing with output. The personnel department feels that if Robinson produces below-contract output, he can be released.

   Robinson's crew is presently working on a prefabricated truck chassis unit, and Robinson has been assigned a job where the time standard is 11 minutes per piece. No one is expected to finish every welding job in 11 minutes, but the average time is to be 11 minutes. The supervisor assures the personnel department that Robinson is not producing at this standard. The personnel department decides to either take action against Robinson, or not, based on a sample of 40 work times.

   a. Formulate the null hypothesis for this decision.

   b. Discuss the ramifications of making a Type I error in this decision.

   c. If the standard deviation of times to perform this welding job is two minutes, what decision rule goes with your hypothesis if the personnel department is willing to commit a Type I error 5 percent of the time?

   d. If the sample of 40 welding operations showed an average completion time of 11.2 minutes, what decision would you make? Would your decision change if the sample average completion time were 12 minutes?

   e. If Robinson actually is doing his work in an average of 12 minutes, what is the chance of his being fired using the decision rule found in part c?

Solutions:

a. Hypotheses:

H_0: $\mu_x \leq 11$ min
H_A: $\mu_x > 11$ min

b. If a Type I error is made, the company will conclude that Robinson is performing below standard and will fire him when, in fact, his true average time, μ_x, is less than or equal to 11 minutes. The costs of this type of error would be social in nature, that is, firing a good employee. Also, Robinson may sue the company and obtain damages for being unjustly fired. In addition, the union may challenge the firing, demand a new sample, and win.

c. Decision Rule:

Take a sample of size n and determine $\overline{X}$.

If $\overline{X} \leq A$, accept H_0.
If $\overline{X} > A$, reject H_0.

The appropriate figure for a 5 percent Type I error is

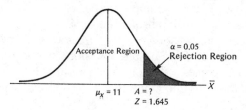

Calculating A,

$$A = 11 + Z\sigma_{\bar{x}} = 11 + 1.645 \frac{2}{\sqrt{40}}$$

$$= 11.520 \text{ min}$$

The rule becomes:

Decision Rule:

If $\overline{X} \leq 11.520$ min, accept H_0, keep Robinson.
If $\overline{X} > 11.520$ min, reject H_0, fire Robinson.

d. If $\overline{X} = 11.2$ minutes, from the decision rule the hypothesis is thought to be true, and Robinson would be retained. If $\overline{X} = 12$ minutes, from the decision rule the hypothesis is determined to be false, and Robinson would be fired.

e.

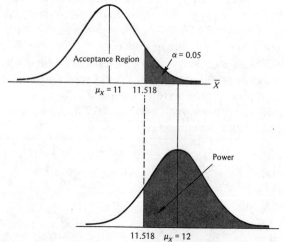

$$Z = \frac{11.520 - 12}{\dfrac{2}{\sqrt{40}}}$$

$$= -1.52$$

The area between $Z = -1.52$ and μ_x is 0.4357, so

$$P(\text{firing}) = 0.5000 + 0.4357$$
$$= 0.9357$$

This means there is a greater than 93 percent chance of firing Robinson when, in fact, he is actually producing the pieces in a time averaging 12 minutes. Note that beta = 1 − power is 1 − 0.9357 = 0.0643. This means there is slightly over a 6 percent chance this decision rule will lead to keeping Robinson if he is really producing at an average rate of 12 minutes per piece.

2. The Convoy Truck Company has been faced with steadily increasing fuel, labor, repair, and equipment costs. Although the company has raised rates as much as possible to reflect increased costs, many of the rates are subject to governmental regulation.

   Four years ago Convoy completed an extensive study of the costs and revenues associated with its operations. The average profit per truckload shipment was found to be $158.12. Convoy doesn't want to go through such an extensive study again because of the costs involved, and doesn't feel the need to because it hasn't altered operations. Rather than a complete study, the managers decide to select a random sample of invoices, determine the average profitability of this sample, and see if the per-run profitability has changed. The results of this sample are as follows:

   Sample size = 800 invoices
   Average sample profit = $149.76
   Sample standard deviation = $189.90

   a. Formulate the null hypothesis and decision rule for Convoy if the company wants to limit the chance of making a Type I error to 4 percent.
   b. Based on your decision rule and the sample data, what conclusions should the Convoy Truck Company reach?

Solutions:

a. Hypotheses:

H_0: $\mu_x = \$158.12$
H_A: $\mu_x \neq \$158.12$

Note that this is a two-tailed hypothesis test.

Decision Rule:

Take a sample of 800 and determine $\overline{X}$.

If $A_L \leq \overline{X} \leq A_H$, accept H_0.
If $\overline{X} < A_L$, reject H_0.
If $\overline{X} > A_H$, reject H_0.

The appropriate figure is

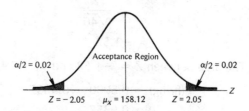

Note that we have split the rejection region into both tails.

$$A_L = \$158.12 + (-ZS_{\bar{x}}) \qquad \text{and} \qquad A_H = \$158.12 + ZS_{\bar{x}}$$

$$= \$158.12 - 2.05\frac{\$189.90}{\sqrt{800}} \qquad\qquad = \$158.12 + 2.05\frac{\$189.90}{\sqrt{800}}$$

$$= \$144.36 \qquad\qquad\qquad\qquad\qquad = \$171.88$$

The decision rule becomes:

Decision Rule:

If $\$144.36 \leq \bar{X} \leq \$171.88$, accept H_0.
If $\bar{X} < \$144.36$, reject H_0.
If $\bar{X} > \$171.88$, reject H_0.

b. Since $\bar{X} = \$149.76$, we should not reject H_0 and conclude, based on the sample data, that the average profit has not changed.

---

3. The manager of a Chicago meat-packing plant has established a standard which says that from a 900-pound steer (live weight), the company should obtain an average of 550 pounds of meat (cut and wrapped). Past experience indicates that even though the average may change, the standard deviation remains fairly constant at 40 pounds. To determine whether the standard is not being met, the manager selects a sample of 100 ($n = 100$) 900-pound steers and compares the sample average with the standard.
   a. Establish the null and alternative hypotheses.
   b. If the manager wishes to have no more than a 0.05 chance of committing a Type I error, specify the appropriate decision rule for this hypothesis test.
   c. Given an alpha level of 0.05 and the decision rule found in part b, what is the probability of committing a Type II error if an average of 547 pounds are cut and wrapped from the 900-pound steers?
   d. Calculate the power of this hypothesis test if the true mean is 547 pounds. Discuss what power means in this example.

Solutions:

a. Hypotheses:

H_0: $\mu_x = 550$ lb
H_A: $\mu_x \neq 550$ lb

Note that because the manager is concerned with detecting departures from the mean in either direction, the hypothesis is two-tailed.

b. The appropriate decision rule is determined as follows:

Hypotheses:

H_0: $\mu_x = 550$ lb
H_A: $\mu_x \neq 550$ lb

$\alpha = 0.05$

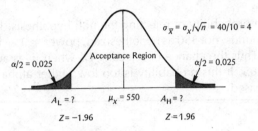

Solve for critical values.

$$A_L = \mu_x + (-Z)\sigma_{\bar{x}} \quad \text{and} \quad A_H = \mu_x + Z\sigma_{\bar{x}}$$
$$= 550 - 1.96(4) \qquad\qquad = 550 + 1.96(4)$$
$$= 542.16 \qquad\qquad\qquad = 557.84$$

Decision Rule:

If $542.16 \leq \bar{X} \leq 557.84$, accept H_0.
If $\bar{X} < 542.16$, reject H_0.
If $\bar{X} > 557.84$, reject H_0.

c. To determine the probability of committing a Type II error if μ_x is actually 547 pounds, we use the following procedure:

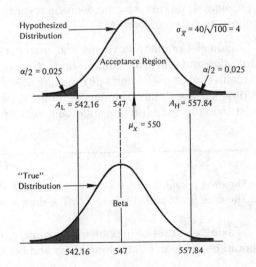

To find beta,

$$Z = \frac{A_L - \mu_x}{\sigma_{\bar{x}}} = \frac{542.16 - 547}{4} \quad \text{and} \quad Z = \frac{A_H - \mu_x}{\sigma_{\bar{x}}} = \frac{557.84 - 547}{4}$$

$$= -1.21 \qquad\qquad\qquad\qquad\qquad = 2.71$$

The area between $Z = -1.21$ and μ_x is 0.3869. The area between $Z = 2.71$ and μ_x is 0.4966. Therefore beta equals $0.3869 + 0.4966 = 0.8835$. Beta is the probability of accepting the null hypothesis when it is actually false.

d. Power is the probability of rejecting the null hypothesis. In this case, if μ_x is really 547 pounds, not 550 as hypothesized, power $= 1 - \text{beta} = 1 - 0.8835 = 0.1165$. Thus the chance of rejecting H_0 when μ_x is actually 547 pounds is only 0.1165. If this probability is too low, either alpha or the sample size can be increased.

problems

Problems marked with an asterisk (*) require extra effort.

1. The Balock Tree Trimming Company has a goal of trimming a tree in no more than 20 minutes on the average. Formulate the null and alternative hypotheses and explain why you set them up in this manner.

2. Discuss in your own words what is meant by a Type I decision error.

3. Define in your own words what is meant by a Type II decision error.

4. Suppose each morning you and a fellow executive have coffee. The ritual has always been that your friend flips a coin and if the coin comes up heads, he buys, and if it comes up tails, you buy. Suppose you have had to pay the last 20 times. What do you think about this situation? Why?

5. Referring to problem 4, discuss why the decision regarding the coin is actually a basic hypothesis-testing problem.

6. The Aldine Chemical Company develops and manufactures pharmaceutical drugs for distribution and sale in the United States. The pharmaceutical business can be very lucrative when useful and safe drugs are introduced into the market. Whenever Aldine's research lab considers putting a drug into production, the company must actually establish the following sets of null and alternative hypotheses:

Set 1	Set 2
H_0: The drug is safe.	H_0: The drug is effective.
H_A: The drug is not safe.	H_A: The drug is not effective.

Taking each set of hypotheses separately, discuss the considerations that should be made in establishing alpha and beta.

7. Referring to problem 6, without actually arriving at values for alpha and beta, indicate your choice for the relative levels of both for each set of hypotheses.

8. The Gladstone Insurance Company provides insurance coverage for automobile owners at a fixed premium. If customers are poor risks, Gladstone would like to not insure them. However, if the customers are good risks, Gladstone would like to have the business. Each time an individual applies for insurance, Gladstone is faced with the following null and alternative hypotheses:

H_0: The applicant is a good risk.

H_A: The applicant is a poor risk.

Given that Gladstone has recently been suffering decreasing profits, how would you go about assessing levels for alpha and beta? Discuss the factors that you consider important in arriving at this decision.

9. Referring to problem 8, how would your responses change if Gladstone were a new company anxious to grow and expand? Discuss your reasoning.

*10. If the chances of making a Type I error can be controlled by simply specifying the desired alpha level, why do we not set alpha equal to zero and, therefore, not worry about committing a Type I error?

11. The Jamison Secretarial Service claims that its typists average at least 3.4 pages per hour. Before placing this information in its advertising brochures, Jamison wishes to determine if this claim is true. A random sample of 400 records indicating typing productivity shows a mean of 3.2 pages per hour. Assuming the population standard deviation is known to be 0.5 page per hour and the desired alpha level is 0.05:
   a. Establish the appropriate null and alternative hypotheses.
   b. Develop the decision rule and test the null hypothesis. Indicate what the decision means to Jamison.

12. Referring to problem 11, would the decision change if the same sample results had been obtained from a sample of 36 records? Discuss why or why not.

13. Referring to problem 11, suppose the sample mean from a sample of size 300 is 3.34 pages per hour. What decision should Jamison make based upon these sample data? Explain what this means to Jamison.

*14. What part does the central limit theorem play in the test of a null hypothesis? Discuss fully.

15. A manufacturer of computer terminals claims that its product will last at least 50 hours without needing repairs. The Bo-Little Corporation is considering purchasing a great many of these computer terminals. Bo-Little's data-processing manager has determined that given the price of the terminal and the total dollars involved, Bo-Little should ask for some quality control records from the manufacturer.

Suppose the manufacturer produces records of a random sample of 100 terminals. The average time before the first breakdown was 48 hours with a sample standard deviation of 25 hours.

    a. Set up the appropriate null and alternative hypotheses.

    b. Determine the appropriate decision rule and indicate whether the sample information justifies rejecting the manufacturer's claim. Use an alpha level of 0.10.

    c. Discuss the ramifications of this decision and the potential costs of being wrong.

16. Referring to problem 15, suppose the sample mean of 100 terminals had been 45 hours rather than 48.

    a. What is the probability of finding a sample mean as small or smaller than 45 if the true mean is 50 hours?

    b. Now that you have determined the probability in part a, what does this mean to you with respect to whether or not the terminals can be expected to average at least 50 hours before the first breakdown?

    c. "If the probability of a sample result, given the null hypothesis, is too small, the null hypothesis should be rejected." Comment on this statement with respect to your answers in parts a and b.

    d. What relationship does alpha have with "too small" as discussed in part c?

17. The Back-Yard Picnic Table Company produces picnic tables in a variety of styles and sizes. The production manager is concerned that the output level during the past year has slipped below the standard average of 84 per day. To see if this is in fact the case, she has selected a random sample of $n = 100$ days' production records and has found a sample mean of 82 and a sample standard deviation of 10.

        Set up the appropriate null and alternative hypotheses. Then, using an alpha level of 0.05, perform the hypothesis test and discuss your conclusions.

18. Referring to problem 17, suppose the sample mean is 86 tables rather than 82. The sample standard deviation remains 10. What conclusion should the production manager at Back-Yard make now?

19. Referring to problem 17, what is the probability of accepting the null hypothesis that the mean is at least 84 tables per day if, in fact, the true mean is only 80 tables per day? (*Hint*: Use the sample standard deviation, $S_x = 10$.)

20. Referring to problems 17–19, what is the probability of a Type II error if the true mean is 83 tables per day?

*21. Referring to problems 17–20, what would be the impact on the values of beta if the sample size could be increased from 100 to 144? Let S_x remain 10. Indicate why the increased sample size affects the beta values this way.

*22. The managing partner of Brookings and Associates, a CPA firm, has a basic knowledge of hypothesis testing. One of his clients, a retail store, would like Brookings to perform an audit of the daily cash register tape against the actual dollar amount in the till.

        The client recognizes that occasionally an error is going to happen. As long as the error is in the store's favor, the store manager is not concerned. However, when the store comes up short, the store manager is very concerned.

The Brookings managing partner has indicated that he will perform the audit via sampling and hypothesis testing with the following hypotheses:

H_0: $\mu_x \geq \$0$ error, store at least comes out even on the average

H_A: $\mu_x < \$0$ error, the store loses some money on the average

From past experience in this kind of audit, the CPA feels that A, the critical value, should be set at $-\$2$. Therefore, if the average discrepancy between cash register tape and actual dollars is $2 or more at the store's expense, the null hypothesis will be rejected. If the null hypothesis is rejected, the clerk will be dismissed.

The CPA partner realizes that his client wants to be very sure the employee is fired if the firing is truly justified. Consequently he is concerned with knowing beta for various values of the "true," but unknown population mean. For example, if the sample size is 49 days and the standard deviation is known to be $4.00, what is beta if the true mean is actually $-\$1.50$?

*23. Referring to problem 22, calculate beta for a sample size of 64. The standard deviation remains $4, and the critical value, A, is held at $-\$2$.

*24. Referring to problem 23, why has an increase in sample size caused an increase rather than a decrease in beta? Discuss how this undesirable event happened and how you could have prevented it from happening in this case. [*Hint*: Consult the following article if you need some help: Herbert H. Tsang, "The Effects of Changing Sample Size on the Alpha and Beta Errors: A Pedagogic Note," *Decision Sciences* 8 (October 1977): 757–59.]

25. The Idaho State Tax Commission attempts to set up payroll tax withholding tables such that by the end of the year, an employee's income tax withholding is about $100 below his actual income tax owed to the state. The commission director claims that when all the Idaho returns are in, the average additional payment will be less than or equal to $100.

A random sample of 40 accounts revealed an average additional payment of $114 with a sample variance of $2,400. Testing at $\alpha = 0.10$, do the sample data refute the director's claim?

26. Referring to problem 25, suppose a division manager for the tax commission had claimed that the average additional payment will be exactly $100. Given the same sample information found, would you now support or refute the division manager's claim? Use $\alpha = 0.10$.

27. Referring to problem 25, suppose the "true" mean additional payment is $115 with a standard deviation of $50. What is the probability that a sample of size 40 will lead us to accept the hypothesis that the mean additional payment is $100 or less?

28. The director of financial services at Battlefield College claims that the average family income for students who have received grants-in-aid is less than or equal to $13,000 per year. During a recent audit of the financial services department, a sample of 36 students was selected. The mean family income for the students in this sample was $13,400 with a standard deviation of $6,000. Based upon these sample data, what should the auditors conclude about the director's claim? Test at $\alpha = 0.05$.

29. Referring to problem 28, what is the probability that a sample of size 36 will lead to a Type II error if the "true" mean is actually $13,100? Assume that the population standard deviation is $6,000, and use alpha equal to 0.05.

30. Referring to problems 28 and 29, what is the power of the hypothesis test with sample size 36 if the "true" mean is $13,200? Assume that the population standard deviation is $6,000 and the alpha level is 0.05. Define in your own words what power of the test means.

31. In problem 30 you were asked to find the power of the test when $n = 36$. Find the power of the hypothesis test when $\mu_x = \$13,200$ if the sample size is increased to 64. Assume $\alpha = 0.05$ and $\sigma_x = \$6,000$.

cases

12A   OVER-THE-HUMP TRANSPORT COMPANY, PART 1

Over-The-Hump Transport Company was started shortly after World War II by a group of Air Corps transport pilots. For the first few years the business operated primarily in Europe. However, in the 1960s the company expanded its operations into the U.S. market. This venture proved very successful, and by the late 1960s, Over-The-Hump entered the passenger business, providing primarily charter service in the Southwest.

Over-The-Hump was typical of many airlines of the period. It was owned and managed by individuals who knew a lot about flying and very little about business management. This lack of business knowledge was not a problem in the European operations because the company had very little competition. Nor was it a problem during the period of rapid expansion in the United States, since the extensive regulatory control exerted by government agencies limited direct competition. However, in current years many of the regulatory policies have been removed, allowing more direct competition between the smaller airlines.

The original owners of the company have for the most part retired, and a new group of managers is responsible for the current operations. These individuals have started to institute some basic cost control procedures. One of the first projects of the new managers was to try to determine which of Over-The-Hump's transport routes were profitable. Up to this point, all revenues were grouped together, as was the case with all the costs. The operational profit was simply the difference between total revenues and total costs. Individual routes were never analyzed.

For example, the airline has been running nightly cargo flights between Denver and Kansas City. The newly formed operations analysis department has determined that this flight must average 8,000 pounds of cargo per flight to break even. In the past, no records have been kept of the cargo weight by flight. However records have been maintained for the companies shipping on each flight. By going to the files, it is possible to compute the total weight on each flight, but this is a time-consuming process. Thus the operations analysis department cannot compute the average weight for the flights during the entire period that Over-The-Hump has operated this service. Instead the department has selected a random sample of 55 flights and, after much effort, computed the average shipping weight at 7,813 pounds with a standard deviation of 2,718 pounds.

The operations analysis department wonders how this information can be used to judge whether this route is profitable on the average.

12B OVER-THE-HUMP TRANSPORT COMPANY, PART 2

Although the results of the sample information proved very informative to Over-The-Hump management (see case 12A), there was some concern in the operations analysis department that a decision was going to be made with too little data. Specifically, those concerned felt that the sample size determination was based more on the time available than on the costs associated with making an incorrect decision.

Florence White, a senior systems analyst, was concerned that the sample might result in the route being accepted as profitable when, in fact, it was a money-loser. She proposed that a sample size as small as 55 would produce too large a chance of this happening.

Florence had an appointment with the manager of the operations analysis department that afternoon to justify her concern.

references

DUNCAN, ACHESON J. *Quality Control and Industrial Statistics*. 3rd ed. Homewood, Ill.: Irwin, 1965.

KAISER, HENRY F. "Directional Statistical Decisions." *Psychological Review* 67 (May 1960): 160–67.

LAPIN, LAWRENCE L. *Statistics for Modern Business Decisions*. New York: Harcourt Brace Jovanovich, 1978.

NETER, JOHN, WILLIAM WASSERMAN, and G. A. WHITMORE. *Applied Statistics*. Boston: Allyn and Bacon, 1978.

ROZEBOOM, WILLIAM W. "The Fallacy of the Null-Hypothesis Significance Test." *Psychological Bulletin* 57 (September 1960): 416–28.

SPURR, WILLIAM A. and CHARLES P. BONINI. *Statistical Analysis for Business Decisions*. Homewood, Ill.: Irwin, 1973.

WILSON, WARNER, HOWARD L. MILLER, and JEROLD S. LOWER. "Much Ado about the Null Hypothesis." *Psychological Bulletin* 67 (March 1967): 188–96.

WONNACOTT, THOMAS H. and RONALD J. WONNACOTT. *Introductory Statistics for Business and Economics*. New York: Wiley, 1977.

13

Additional Topics in Hypothesis Testing

why decision makers need to know

In Chapter 12 we introduced the fundamental concepts of testing hypotheses about a single population mean. We also discussed the types of errors decision makers can make when testing a hypothesis, and we showed how to calculate the probability of committing each error.

Whereas many business decisions involve testing hypotheses about a single population mean, many other applications require hypothesis tests of other population values such as variance. In still other applications, decision makers may be faced with testing hypotheses about the difference between values from two populations. For example, a financial investment advisor might need to determine whether there is a significant difference in the average performance of two mutual funds. In another case, a marketing manager may wish to test whether there is a difference in the proportion of individuals in a particular income level in two cities. The outcome of this test would provide useful information in helping decide, say, in which city to locate a new retail outlet. These are but a few of the possible examples where decision makers need to be familiar with techniques of hypothesis testing beyond those presented in Chapter 12.

chapter objectives

In this chapter we shall extend the hypothesis-testing techniques introduced in Chapter 12. We will cover such new topics as two-sample hypothesis testing, hypothesis testing about a single population proportion, and hypothesis testing about two population proportions. We will also discuss the statistical tests required for small-sample hypothesis testing and introduce tests about one and two population variances.

student objectives

After studying the material in this chapter, you should be able to:

1. Identify situations in which a small-sample hypothesis test must be employed.
2. Perform hypothesis tests of a population proportion.
3. Perform hypothesis tests for the difference between two population proportions.
4. Perform hypothesis tests for the difference between two population means for both large and small samples.
5. Perform hypothesis tests for one and two population variances.
6. Determine the sample size required to simultaneously control both Type I and Type II error probabilities in a hypothesis-testing problem.

13-1
CONTROLLING TYPE I AND TYPE II ERRORS

The Roller-Bed Company manufactures and sells waterbeds in several areas of the United States. The company has been purchasing the waterbed mattresses from a southern California firm but it manufactures its own wood frames at an Oregon plant.

To compete favorably in the waterbed market, Roller-Bed must guarantee its product. The chief competitor is currently offering a 36-month, money-back guarantee, and Roller-Bed feels that it must match or exceed this guarantee. To do this profitably, Roller-Bed's cost accountant and production manager have determined that the mattresses must have a strength rating that averages at least 1,200.

The production manager has suggested that a sample of mattresses be selected to determine whether or not Roller-Bed should introduce the 36-month guarantee for its product. Thus he wishes to test the hypothesis that the strength rating average is at least 1,200 versus the alternative that the rating average is less than 1,200. Although the mean strength rating may vary, the standard deviation of strength ratings is known to be 40.

The testing process destroys the mattress since pressure is applied until the rubber tears. Consequently the production manager wants to hold the sample size to a minimum. However he also knows that hypothesis tests involve the risk of committing either a Type I or a Type II error. If a Type I error is committed, Roller-Bed will incorrectly conclude that its mattresses are not strong enough and consequently will not offer the 36-month guarantee. The company will then be at a marketing disadvantage and may spend money to improve the quality when it is not necessary. Thus, if a Type I error is committed, Roller-Bed will needlessly lose sales and spend money. The potential costs are very high, so the company president wants a small probability of a Type I error. Specifically, she is willing to accept a probability of a Type I error no larger than 2 percent.

However, if the sample leads to accepting the null hypothesis, a Type II error might be made. The Type II error occurs if a false hypothesis is accepted. In this case, the company would incorrectly believe that the mattresses are strong enough to make the guarantee profitable. Depending upon how low the true average strength rating is, a Type II error could be very costly due to an excessive number of required replacements.

After a careful study of the costs involved, the cost accountant and the production manager have agreed that if the true average strength rating is equal to or less than 1,180, the company would be hit hard financially by making the 36-month guarantee. Because of this, the president is willing to accept only a 0.05 chance of committing a Type II error if the true average strength rating is 1,180.

The minimum sample size required to hold both Type I and Type II errors to the levels specified is determined by the following procedure:

Hypotheses:

H_0: $\mu_x \geq 1,200$
H_A: $\mu_x < 1,200$

Constraints:

1. The probability of a Type I error must be less than or equal to 0.02.
2. If the true mean strength rating is 1,180, the chances of a Type II error must not exceed 0.05.

Figure 13–1 illustrates these constraints.

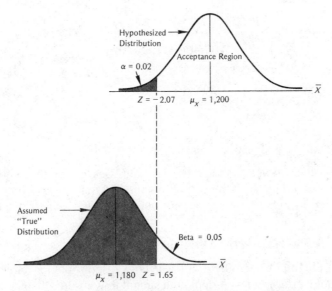

FIGURE 13–1   Controlling Type I and Type II errors, Roller-Bed Company.

The first step is to solve for the critical value, A. We can do this two ways.

For the hypothesized distribution:

$$Z = \frac{A - \mu_x}{\frac{\sigma_x}{\sqrt{n}}}$$

$$-2.07 = \frac{A - 1,200}{\frac{40}{\sqrt{n}}}$$

$$A = 1,200 - 2.07 \frac{40}{\sqrt{n}} \qquad \textbf{(13–1)}$$

For the assumed "true" distribution:

$$Z = \frac{A - \mu_x}{\frac{\sigma_x}{\sqrt{n}}}$$

$$1.65 = \frac{A - 1,180}{\frac{40}{\sqrt{n}}}$$

$$A = 1,180 + 1.65 \frac{40}{\sqrt{n}} \qquad \textbf{(13–2)}$$

Both equations (13–1) and (13–2) equal the same critical value, A. Thus we can set one equal to the other and solve for the optimal sample size n.

$$1,200 - 2.07 \frac{40}{\sqrt{n}} = 1,180 + 1.65 \frac{40}{\sqrt{n}}$$

$$20 - 2.07 \frac{40}{\sqrt{n}} = 1.65 \frac{40}{\sqrt{n}}$$

$$20\sqrt{n} - 2.07(40) = 1.65(40)$$

$$20\sqrt{n} - 82.8 = 66.0$$

$$20\sqrt{n} = 148.8$$

$$n = \left(\frac{148.8}{20}\right)^2 = 55.35$$

$$= 56$$

Thus the Roller-Bed Company should test 56 mattresses to limit the Type I and Type II errors to the desired levels. If the cost of sampling 56 mattresses is very high, either the Type I or the Type II error constraint, or both, will have to be relaxed.

A Two-Tailed Example

The Greater Columbia Water Division (GCWD) is responsible for minimum stream flows in several western states. A minimum stream flow insures that the fish have an adequate water supply. However the GCWD also has to make sure that water is not being wasted.

Water flow can be controlled if a dam is in the stream or river. The control is from the dam downstream. The flow can be measured very accurately as water passes over the dam.

In one river the required flow is 100,000 cubic feet per second. Paul Saulette of the GCWD has devised a plan for monitoring the stream flow. The plan consists of selecting a sample of readings at the dam and, depending upon the sample mean, making one of three decisions:

1. Do nothing; leave the flow alone.
2. Increase the flow.
3. Decrease the flow.

Paul recognizes that if the water flow is actually correct but he concludes that it is not, he will needlessly order an adjustment in the stream flow. On the other hand, if he leaves the flow alone when it is not 100,000 cubic feet per second, he will also be in error. In addition, not adjusting the flow when it is actually too low is more serious than not making the adjustment when the flow is too great.

If the standard deviation in stream flow is 5,000 cubic feet per second, given the following constraints, what is the minimum number of measurements that need to be taken?

Constraints:

1. The chance of committing a Type I error should be no more than 0.10.
2. If the true average stream flow is 97,000 cubic feet per second, the probability of a Type II error should be no more than 0.05.
3. If the true average stream flow is 110,000 cubic feet per second, the probability of a Type II error should be no more than 0.10.

To determine the required sample size, refer to Figure 13–2 and perform the following:

Hypotheses:

H_0: μ_x = 100,000 ft³/s
H_A: μ_x ≠ 100,000 ft³/s

For the hypothesized distribution:

$$A_L = \mu_x + (-1.65)\frac{\sigma_x}{\sqrt{n}}$$

$$= 100,000 + (-1.65)\left(\frac{5,000}{\sqrt{n}}\right)$$

For the assumed "true" distribution:

$$A_L = 97{,}000 + 1.65\frac{5{,}000}{\sqrt{n}}$$

Then set the two equations for A_L equal to each other and solve for n.

$$100{,}000 - 1.65\frac{5{,}000}{\sqrt{n}} = 97{,}000 + 1.65\frac{5{,}000}{\sqrt{n}}$$

$$3{,}000 - 1.65\frac{5{,}000}{\sqrt{n}} = 1.65\frac{5{,}000}{\sqrt{n}}$$

$$3{,}000\sqrt{n} - 1.65(5{,}000) = 1.65(5{,}000)$$

$$3{,}000\sqrt{n} = 16{,}500$$

$$\sqrt{n} = \frac{16{,}500}{3{,}000}$$

$$n = 30.25$$

$$= 31$$

This sample size of 31 satisfies constraints 1 and 2. However we must check to see whether constraints 1 and 3 require a larger sample size. We will use the larger of the two sample sizes since it will satisfy all constraints.

To determine the sample size necessary to satisfy constraints 1 and 3, we set the two equations for A_H equal to each other and solve for n.

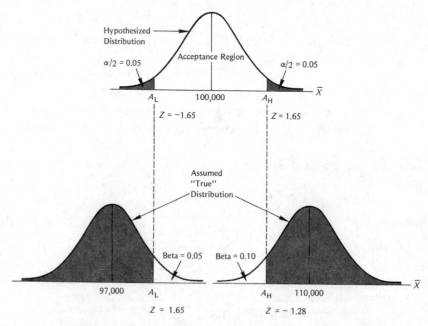

FIGURE 13–2 Controlling Type I and Type II errors, Greater Columbia Water Division.

For the hypothesized distribution:

$$A_H = 100,000 + 1.65 \frac{5,000}{\sqrt{n}}$$

For the assumed "true" distribution:

$$A_H = 110,000 + (-1.28)\left(\frac{5,000}{\sqrt{n}}\right)$$

Then

$$110,000 + (-1.28)\left(\frac{5,000}{\sqrt{n}}\right) = 100,000 + 1.65 \frac{5,000}{\sqrt{n}}$$

$$10,000 + (-1.28)\left(\frac{5,000}{\sqrt{n}}\right) = 1.65 \frac{5,000}{\sqrt{n}}$$

$$10,000\sqrt{n} + (-1.28)(5,000) = 1.65(5,000)$$

$$10,000\sqrt{n} = 14,650$$

$$\sqrt{n} = 1.465$$

$$n = 2.14$$

$$= 3$$

Therefore the larger sample size, $n = 31$, is required. The GCWD should take a minimum of 31 random measurements before rejecting or accepting the hypothesis that the average stream flow is 100,000 cubic feet per second.

13-2
HYPOTHESES TESTING ABOUT A POPULATION PROPORTION

The previous discussions in this chapter and in Chapter 12 involved hypothesis tests about a single production mean. While there are many decision problems that can be limited to a test of a population mean, there are many other cases where the value of interest is the **population proportion.** For example, the percentage of defective items produced on an assembly line might determine whether the assembly process should be restructured or left as it is. And, one measure of success for a life insurance salesperson might be the percentage of renewals generated from his or her existing customers.

Political polling is another example where a proportion or percentage is the value of interest. Senator Frances Chapel, a three-term legislator, has decided not to run for a fourth term unless she would be given a "mandate" by the people. Senator Chapel's staff assures her that at least 55 percent of the voters in the constituency will vote for her, should she decide to run for office.

Although Chapel has faith in her staff, she also recognizes that they may be overly optimistic. Consequently Senator Chapel has decided to hire the Band Wagon Opinion Company to select a sample of people in her district and test whether or not she can expect a "mandate" (at least 55 percent of the vote).

For various reasons, most notably that she wants to remain a senator, Frances Chapel will accept no more than a 2 percent chance of a Type I error. That is, she wants no more than a 2 percent chance of rejecting the hypothesis that she will receive a mandate when, in fact, the hypothesis is true.

Providing the sample size is relatively large, the normal distribution can be used to test hypotheses for proportions. Recall that in Chapter 10 we used the normal distribution to develop confidence intervals for population proportions. To use hypothesis testing in Chapel's situation, Band Wagon must follow the usual hypotheses statement-decision rule sequence. That is,

Hypotheses:

H_0: $p \geq 0.55$
H_A: $p < 0.55$

Decision Rule:

Select a sample of size n and calculate $\hat{p}$, where $\hat{p}$ is the sample proportion.

If $\hat{p} \geq A$, accept H_0.
If $\hat{p} < A$, reject H_0.

The critical value, A, is

$$A = p + Z\sigma_p \tag{13-3}$$

where: $\quad \sigma_p = \sqrt{\dfrac{p(1 - p)}{n}}$

$\qquad p$ = hypothesized proportion
$\qquad n$ = sample size

Note that the standard deviation, $\sqrt{p(1 - p)/n}$, depends upon the hypothesized value of p. As long as the population size is large relative to the sample size, there is no need to use the finite correction factor. However, when the sample size exceeds 5 percent of the population, the standard deviation should be calculated from

$$\sigma_p = \sqrt{\dfrac{p(1 - p)}{n}} \sqrt{\dfrac{N - n}{N - 1}}$$

Suppose the Band Wagon Opinion Company obtains opinions from 1,000 voters in Chapel's district. The appropriate decision rule is determined in Figure 13–3. From Figure 13–3, if a proportion smaller than 0.517 of the 1,000 voters in the sample favor Chapel, Chapel should reject the hypothesis of a "mandate."

Hypotheses:

$H_0: p \geq 0.55$
$H_A: p < 0.55$

$\alpha = 0.02$

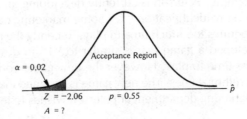

Decision Rule:

If $\hat{p} \geq A$, accept H_0.
If $\hat{p} < A$, reject H_0.

The Z value for a one-tailed test with $\alpha = 0.02$ is -2.06. Then

$$A = 0.55 + (-2.06)\left(\sqrt{\frac{0.55(0.45)}{1,000}}\right)$$

$$= 0.517$$

FIGURE 13–3 Decision rule, Senator Chapel example.

13-3
HYPOTHESIS TESTING ABOUT THE DIFFERENCE
BETWEEN TWO POPULATION MEANS

Many business decision problems require analyzing values from two or more populations. In this section we introduce methods to test hypotheses about the ***difference between two population means.*** These methods, however, are merely extensions of hypothesis tests for one population mean.

The Peterson Toy Company designs and manufactures games for children and adults. The company has several popular games currently in production but still finds itself with excess production capacity. Consequently the company introduces new games as fast as they are designed and at the same time eliminates poor sellers from production. The marketing people push for games with recognizable themes and attempt to time their introduction for a holiday season, especially Christmas.

The Peterson marketing people have performed extensive market research to determine what factors are most influential in a game's success or failure. A critical factor in children's games is the length of time needed to play the game. Games that are too complicated and too long generally are poor sellers, as are games that can be finished too quickly and are considered "simple."

After a new children's game has been designed, the company selects "typical" children to play it on a trial basis to determine the average time needed to finish. When the new game is similar to an existing game, the two are compared to determine whether they are equal with respect to average playing time.

For example, Peterson is currently developing an advertising plan for a game with the company identification "1." Some marketing department members fear that game "1" requires too much time to play. Recently the research and design department has developed a game that is similar to "1" in design but which they claim should take less time to play. However the new game contains other features that are important in selling games. The new game is assigned the identification "2."

The marketing department at Peterson wishes to test the following:

Hypotheses:

$H_0: \mu_1 \leq \mu_2$ or $\mu_1 - \mu_2 \leq 0$
$H_A: \mu_1 > \mu_2$ or $\mu_1 - \mu_2 > 0$

To simplify notation, we let μ_1 and μ_2 equal the population means, μ_{x_1} and μ_{x_2}, respectively. Also, we let σ_1^2 and σ_2^2 represent the population variances, $\sigma_{x_1}^2$ and $\sigma_{x_2}^2$, respectively. We shall follow this notation pattern throughout the remainder of the text whenever two or more populations are considered.

In words, the null hypothesis states that the average playing time for game "2" is *as long as or longer than* the average for game "1." If this is accepted, the company will continue with its marketing efforts for game "1." If the null hypothesis is rejected, the company will conclude that the average playing time for game "2" is *less than* the average time for game "1." In this case, Peterson will abandon game "1" in favor of game "2."

The decision rule for the Peterson hypothesis is as follows:

Decision Rule:

Select two samples, n_1 and n_2, and calculate $\overline{X}_1$ and $\overline{X}_2$.

If $\overline{X}_1 - \overline{X}_2 \leq A$, accept H_0.
If $\overline{X}_1 - \overline{X}_2 > A$, reject H_0.

In Chapter 9 we introduced the central limit theorem for the sampling distribution of $\overline{X}$. The central limit theorem also applies for the sampling distribution of the difference between two sample means. Thus, if the sample sizes are sufficiently large, the sampling distribution of $(\overline{X}_1 - \overline{X}_2)$ will be normally distributed, with

$$\mu_{\overline{x}_1 - \overline{x}_2} = \mu_1 - \mu_2 \tag{13-4}$$

and

$$\sigma_{\overline{x}_1 - \overline{x}_2} = \sqrt{\frac{\sigma_1^2}{n_1} + \frac{\sigma_2^2}{n_2}} \tag{13-5}$$

where: $\overline{X}_1, \overline{X}_2$ = sample means for population 1 and population 2, respectively

$\mu_1 - \mu_2$ = hypothesized difference between population means

σ_1^2, σ_2^2 = variances of population 1 and population 2, respectively

n_1, n_2 = sample sizes from population 1 and population 2, respectively

Thus the critical value for the large-sample statistical test of the difference between two population means is found by solving the following equation for A:

$$Z = \frac{A - (\mu_1 - \mu_2)}{\sqrt{\dfrac{\sigma_1^2}{n_1} + \dfrac{\sigma_2^2}{n_2}}} \qquad (13\text{–}6)$$

Then, solving for A,

$$A = (\mu_1 - \mu_2) + Z \sqrt{\frac{\sigma_1^2}{n_1} + \frac{\sigma_2^2}{n_2}}$$

Suppose the variances of the times needed to complete the games are $\sigma_1^2 = 900$ minutes and $\sigma_2^2 = 900$ minutes. The company is willing to accept a 0.07 chance of a Type I error ($\alpha = 0.07$) and selects a sample of 100 for each game. Figure 13–4 shows the results of the sampling and the decision rule.

The decision, based upon the sample results, is that game "2" takes as long or longer to play on the average than game "1." The company should continue to market game "1."

Before the hypothesis test is performed, the decision maker should consider the chances of committing both Type I and Type II errors. Calculating beta

Hypotheses:

$H_0: \mu_1 - \mu_2 \leq 0$
$H_A: \mu_1 - \mu_2 > 0$

$\alpha = 0.07$
$n_1 = 100, n_2 = 100$
$\sigma_1^2 = 900, \sigma_2^2 = 900$
$\bar{X}_1 = 50, \bar{X}_2 = 46$

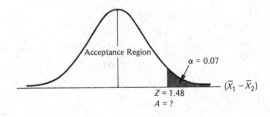

Decision Rule:

If $(\bar{X}_1 - \bar{X}_2) \leq A$, accept H_0.
If $(\bar{X}_1 - \bar{X}_2) > A$, reject H_0.

Solving for the critical level, A,

$$A = (\mu_1 - \mu_2) + Z \sqrt{\frac{\sigma_1^2}{n_1} + \frac{\sigma_2^2}{n_2}} = 0 + 1.48 \sqrt{\frac{900}{100} + \frac{900}{100}}$$

$$= 6.279$$

Since $(\bar{X}_1 - \bar{X}_2) = 4 < 6.279$, accept H_0.

FIGURE 13–4 Two-sample decision rule, Peterson Toy Company.

for the two-sample tests follows the same procedure as for the one-sample case discussed in Chapter 12. Further, alpha and beta are interpreted the same as for the one-sample test.

When the Population Variances Are Unknown

As is true with decisions involving one population, often the variances for a two-population hypothesis test are not known. However, if the sample sizes are reasonably large (generally over 30), the sample variances can be substituted for the population variances. Then, the critical value in the decision rule is found by solving for A in

$$Z = \frac{A - (\mu_1 - \mu_2)}{\sqrt{\dfrac{S_1^2}{n_1} + \dfrac{S_2^2}{n_2}}} \qquad (13\text{–}7)$$

The decision rule is found exactly as in the previous examples. For example, in the Peterson Toy Company problem, instead of assuming that $\sigma_1^2 = 900$ and $\sigma_2^2 = 900$, suppose we found the following sample variances: $S_2^2 = 808$ and $S_2^2 = 933$. Figure 13–5 shows this situation.

Hypotheses:

H_0: $\mu_1 - \mu_2 \leq 0$
H_A: $\mu_1 - \mu_2 > 0$

$\alpha = 0.07$
$Z = 1.48$
$n_1 = 100, n_2 = 100$
$\bar{X}_1 = 50, \bar{X}_2 = 46$
$S_1^2 = 808, S_2^2 = 933$

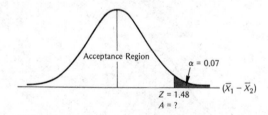

Decision Rule:

If $(\bar{X}_1 - \bar{X}_2) \leq A$, accept H_0.
If $(\bar{X}_1 - \bar{X}_2) > A$, reject H_0.

$$Z = \frac{A - (\mu_1 - \mu_2)}{\sqrt{\dfrac{S_1^2}{n_1} + \dfrac{S_2^2}{n_2}}}$$

$$A = (\mu_1 - \mu_2) + Z\sqrt{\frac{S_1^2}{n_1} + \frac{S_2^2}{n_2}} = 0 + 1.48\sqrt{\frac{808}{100} + \frac{933}{100}}$$

$$= 6.17$$

Since $(\bar{X}_1 - \bar{X}_2) \leq 6.17$, accept H_0.

FIGURE 13–5 Two-sample decision rule, variances unknown, Peterson Toy Company.

Other Considerations

The three basic forms for the null and alternative hypotheses for the two-sample test are

$$H_0: \mu_1 \leq \mu_2 \quad \text{or} \quad \mu_1 - \mu_2 \leq 0$$
$$H_A: \mu_1 > \mu_2 \quad \text{or} \quad \mu_1 - \mu_2 > 0$$

$$H_0: \mu_1 = \mu_2 \quad \text{or} \quad \mu_1 - \mu_2 = 0$$
$$H_A: \mu_1 \neq \mu_2 \quad \text{or} \quad \mu_1 - \mu_2 \neq 0$$

$$H_0: \mu_1 \geq \mu_2 \quad \text{or} \quad \mu_1 - \mu_2 \geq 0$$
$$H_A: \mu_1 < \mu_2 \quad \text{or} \quad \mu_1 - \mu_2 < 0$$

Also, the hypothesized difference between two population means need *not* be zero. For example, we might hypothesize that the average difference in attendance at two movie theaters is at least 120 people per day. In this case, the null and alternative hypotheses would be:

$$H_0: \mu_1 - \mu_2 \geq 120$$
$$H_A: \mu_1 - \mu_2 < 120$$

Another consideration is that the sample sizes selected from the two populations need *not* be equal. In the theater example, we might randomly sample $n_1 = 50$ days from theater 1 and $n_2 = 75$ days from theater 2. **Generally, if the population variances are not equal, the larger sample size should be selected from the population with the larger variance.**

13-4

HYPOTHESIS TESTING ABOUT THE DIFFERENCE BETWEEN TWO POPULATION PROPORTIONS

In Section 13–2 we introduced the methodology for testing hypotheses involving population proportions. In this section we extend the analysis to testing hypotheses about the *difference between two population proportions.*

Pomona Fabrication, Inc. produces hand-held hair dryers which several major retailers sell as their "house" brands. Pomona was an early entrant into this market and has developed substantial manufacturing and technological skills. However, in recent years the firm has faced increased competition from both domestic and foreign manufacturers. Pomona has been forced to reduce its prices, and this, coupled with ever-increasing production costs, has caused a substantial reduction in the company's profit margin.

A critical component of a hand-held hair dryer is the motor-heater unit. This component accounts for the majority of the dryer's cost and also for a majority of the product's reliability problems. Product reliability is extremely important to Pomona since the company currently offers a standard one-year warranty. Of course Pomona is also interested in reducing production costs.

Pomona's research and development department has recently developed a new motor-heater unit that will offer a 15 percent cost savings. However the company's vice-president of product development is unwilling to authorize the new component unless it is at least as reliable as the motor-heater currently being used.

The research and development department has decided to test samples of both units to see whether there is a difference in the proportions that will fail in one year. Two hundred and fifty units of each type will be tested under conditions that simulate one year's use.

Hypotheses:

H_0: $p_{new} \leq p_{old}$ or $p_{new} - p_{old} \leq 0$
H_A: $p_{new} > p_{old}$ or $p_{new} - p_{old} > 0$

where: p = proportion of dryers that fail before 1 yr

Decision Rule:

Select two samples, n_1 and n_2, and calculate $\hat{p}_1$ and $\hat{p}_2$, where $\hat{p}_1$ and $\hat{p}_2$ are estimates of the two population proportions, p_{new} and p_{old}, respectively.

If $\hat{p}_1 - \hat{p}_2 \leq A$, accept H_0.
If $\hat{p}_1 - \hat{p}_2 > A$, reject H_0.

If the null hypothesis is accepted, the company will continue to use the old motor-heater unit. Otherwise, the new motor-heater will be installed.

As with a hypothesis test of a single population proportion, the normal distribution can be used to test hypotheses about the difference between two population proportions, providing the sample sizes are sufficiently large. In this case, the critical value is found by solving the following equation for A:

$$Z = \frac{A - (p_1 - p_2)}{\sqrt{\dfrac{p_1(1 - p_1)}{n_1} + \dfrac{p_2(1 - p_2)}{n_2}}} \tag{13-8}$$

Note that the standard deviation of the sampling distribution for the difference between proportions is

$$\sigma_{p_1 - p_2} = \sqrt{\frac{p_1(1 - p_1)}{n_1} + \frac{p_2(1 - p_2)}{n_2}} \tag{13-9}$$

However, equation (13–9) requires us to know p_1 and p_2. Since these values are unknown and we have hypothesized zero difference between the two population proportions, we must calculate a **pooled estimator** by taking a weighted average of the observed sample proportions as follows:

$$\bar{p} = \frac{n_1\hat{p}_1 + n_2\hat{p}_2}{n_1 + n_2} \tag{13-10}$$

Note that the numerator is the total number of successes in the two samples and the denominator is the total sample size.

The standard deviation is

$$S_{\bar{p}_1 - \bar{p}_2} = \sqrt{(\bar{p})(1 - \bar{p})\left(\frac{1}{n_1} + \frac{1}{n_2}\right)}$$

(13–11)

Thus the Z formula is

$$Z = \frac{A - (p_1 - p_2)}{\sqrt{(\bar{p})(1 - \bar{p})\left(\frac{1}{n_1} + \frac{1}{n_2}\right)}}$$

(13–12)

Assume that Pomona is willing to accept an alpha level of 0.05, and that 75 of the new motor-heaters and 65 of the originals failed the one-year test. Figure 13–6 illustrates the decision rule development and the null hypothesis test. As you can see in the figure, Pomona should *not* reject the null hypothesis. Rather, based upon the sample information, the firm should conclude that the new motor-

Hypotheses:

$H_0: p_{new} - p_{old} \leq 0$
$H_A: p_{new} - p_{old} > 0$

$\alpha = 0.05$
$n_{new} = 250, n_{old} = 250$
$X_{new} = 75, X_{old} = 65$
$\hat{p}_{new} = 75/250 = 0.30, \hat{p}_{old} = 65/250 = 0.26$

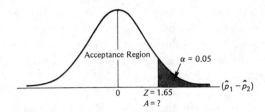

Decision Rule:

If $\hat{p}_{new} - \hat{p}_{old} \leq A$, accept H_0.
If $\hat{p}_{new} - \hat{p}_{old} > A$, reject H_0.

$$\bar{p} = \frac{250(0.30) + 250(0.26)}{250 + 250}$$

$$= 0.28$$

$$Z = \frac{A - (p_{new} - p_{old})}{\sqrt{\bar{p}(1 - \bar{p})\left(\frac{1}{n_{new}} + \frac{1}{n_{old}}\right)}}$$

Solving for the critical value, A,

$$A = (p_{new} - p_{old}) + Z\sqrt{(\bar{p})(1 - \bar{p})\left(\frac{1}{n_{new}} + \frac{1}{n_{old}}\right)} = 0 + 1.65\sqrt{(0.28)(0.72)\left(\frac{1}{250} + \frac{1}{250}\right)}$$

$$= 0.066$$

Since $\hat{p}_{new} - \hat{p}_{old} = 0.04 < 0.066$, accept H_0.

FIGURE 13–6 Decision rule for two-sample test of proportions, Pomona Fabrication, Inc.

heater is at least as reliable as the old one. Since the new one is less costly, it should be used.

13-5
HYPOTHESIS TESTING WITH SMALL SAMPLES, AND σ_X UNKNOWN

In many business applications involving sampling, the time and/or costs of collecting the sample data restrict the sample size. The hypothesis-testing techniques introduced thus far are appropriate if the sample size is large. In this section we consider hypothesis-testing techniques that can be used when the sample size is *small* and the population standard deviation is unknown. In particular, these techniques utilize the *t* distribution. If you need to review the underlying concepts of the *t* distribution, we recommend you reread Chapter 11. **You will find little difference between the small-sample hypothesis tests requiring the *t* distribution and the large-sample tests which utilize the standard normal distribution.**

Hypothesis Testing About the Population Mean— Small Samples

Samantha Edwards and Julie Adamson left high-level engineering design jobs several years ago to form their own company. They have recently been working on a new type of bumper system for automobiles. The federal government has specified a very strict crash standard for automobile bumpers. However the bumpers that currently meet the federal crash standard add weight to the car, which reduces gas mileage. Samantha and Julie think they have designed a light-weight bumper system that meets the federal crash standard.

The automobile manufacturers have expressed interest in the new bumper system. However the only way to make sure the bumper meets the standard is to perform crash tests with new automobiles. The damage from the tests is measured, and the average damage is compared with the standard of $150 average damage at 15 miles per hour.

The two entrepreneurs claim their bumper system will at least match the standard. Thus we have the following:

Hypotheses:

H_0: $\mu_x \leq \$150$
H_A: $\mu_x > \$150$

Decision Rule:

Take a sample of size n and determine $\overline{X}$.

If $\overline{X} \leq A$, accept H_0.
If $\overline{X} > A$, reject H_0.

Hypotheses:

H_0: $\mu_x \le \$150$
H_A: $\mu_x > \$150$

$\alpha = 0.05$

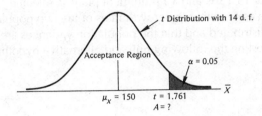

Decision Rule:

If $\overline{X} \le A$, accept H_0.
If $\overline{X} > A$, reject H_0.

     The critical value is

$$A = \mu_x + t\frac{S_x}{\sqrt{n}} = \$150 + 1.761\frac{\$80}{\sqrt{15}}$$

$$= \$186.37$$

Since $\overline{X} = \$155 < \$186.37$, accept H_0.

     FIGURE 13–7   Small-sample hypothesis test for μ_x, bumper example.

     One of the large auto makers has agreed to cosponsor the crash tests. It will furnish 15 cars for the test and will accept no more than a 0.05 chance of committing a Type I error.

     The first step is to perform the crash test for the 15 cars. Suppose the tests result in an average of $155 and a standard deviation of $80.

     The next step is to arrive at the appropriate critical value for deciding whether to accept or reject the null hypothesis. Because the sample size is small and the population standard deviation is unknown, we must use the t distribution in developing the decision rule, as shown in Figure 13–7. Note that the appropriate test statistic if the population of possible crash values is normally distributed is

$$t = \frac{A - \mu_x}{\dfrac{S_x}{\sqrt{n}}} \tag{13–13}$$

     Based upon the sample results, the conclusion is that the new bumper *does* meet or exceed the government standard.

Hypothesis Tests Involving Two Population Means— Small Samples

Among the fastest-growing investment alternatives in the United States are the tax-sheltered annuity (TSA) programs offered by large insurance companies. Certain people qualify to deposit part of their paychecks in the TSA and pay no federal income tax on this money until it is withdrawn. While the money is on deposit, the insurance companies invest it in stock or bond portfolios. If the portfolios perform well, the TSA accounts grow.

Some organizations with conventional retirement plans are concerned the TSAs may have an advantage in terms of growth. As part of a comparative study, a random sample of 15 TSAs and 15 retirement plans is selected.

If we assume that the growth rates of the two populations are approximately normally distributed and that the population variances are equal, a t statistic is appropriate for testing the following null and alternative hypotheses:

Hypotheses:

$H_0: \mu_1 - \mu_2 = 0$
$H_A: \mu_1 - \mu_2 \neq 0$

where: μ_1 = average growth rate for population 1 (TSAs)
μ_2 = average growth rate for population 2 (retirement plans)

Decision Rule:

Take two samples, n_1 and n_2, and determine $\bar{X}_1$ and $\bar{X}_2$.

If $A_L \leq (\bar{X}_1 - \bar{X}_2) \leq A_H$, accept H_0.
If $(\bar{X}_1 - \bar{X}_2) < A_L$, reject H_0.
If $(\bar{X}_1 - \bar{X}_2) > A_H$, reject H_0.

Drawing a parallel with the large-sample case, the two critical values are

$$A_L = (\mu_1 - \mu_2) + (-t)(S_{\bar{x}_1 - \bar{x}_2})$$
$$A_H = (\mu_1 - \mu_2) + t(S_{\bar{x}_1 - \bar{x}_2})$$

Determining the value of $S_{\bar{x}_1 - \bar{x}_2}$ requires that we estimate the pooled variance. In Chapter 11, when faced with the same situation, we estimated the pooled variance by finding a weighted average of the sample variances. The pooled variance here is

$$S_{pooled}^2 = \frac{S_1^2(n_1 - 1) + S_2^2(n_2 - 1)}{n_1 + n_2 - 2} \tag{13–14}$$

And the estimated standard deviation for the difference between two means is

$$S_{\bar{x}_1 - \bar{x}_2} = S_{pooled}\sqrt{\frac{1}{n_1} + \frac{1}{n_2}} \tag{13–15}$$

Thus the t statistic for the two-sample test is computed by

$$t = \frac{(\bar{X}_1 - \bar{X}_2) - (\mu_1 - \mu_2)}{S_{pooled}\sqrt{\frac{1}{n_1} + \frac{1}{n_2}}} \tag{13–16}$$

The critical values can be found using this t statistic, which will have $n_1 + n_2 - 2$ degrees of freedom.

Hypotheses:

$H_0: \mu_1 - \mu_2 = 0$
$H_A: \mu_1 - \mu_2 \neq 0$

$\alpha = 0.05$

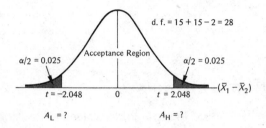

d. f. $= 15 + 15 - 2 = 28$

Acceptance Region

$\alpha/2 = 0.025$ $\alpha/2 = 0.025$

$t = -2.048$ 0 $t = 2.048$ $(\bar{X}_1 - \bar{X}_2)$

$A_L = ?$ $A_H = ?$

Decision Rule:

If $A_L \leq (\bar{X}_1 - \bar{X}_2) \leq A_H$, accept H_0.
If $(\bar{X}_1 - \bar{X}_2) < A_L$, reject H_0.
If $(\bar{X}_1 - \bar{X}_2) > A_H$, reject H_0.

The critical values are

$$A_L = (\mu_1 - \mu_2) + (-t)(S_{\bar{x}_1 - \bar{x}_2})$$

$$= (\mu_1 - \mu_2) + (-t)\left(S_{pooled}\sqrt{\frac{1}{n_1} + \frac{1}{n_2}}\right)$$

Then

$$S_{pooled} = \sqrt{\frac{(n_1 - 1)S_1^2 + (n_2 - 1)S_2^2}{n_1 + n_2 - 2}} = \sqrt{\frac{(15 - 1)58 + (15 - 1)61}{15 + 15 - 2}}$$

$$= 7.71$$

So

$$A_L = 0 + (-2.048)(7.71)\left(\sqrt{\frac{1}{15} + \frac{1}{15}}\right)$$

$$= -5.77$$

Performing similar calculations,

$$A_H = 5.77$$

Since $(\bar{X}_1 - \bar{X}_2) = 1.1$, accept H_0.

FIGURE 13–8   Hypothesis test of two population means, small sample, TSA versus retirement plans.

Figure 13–8 shows the procedure for testing the hypothesis that there is no difference in the average growth rates for the two populations. The sample information used to test the hypothesis with $\alpha = 0.05$ is

$\bar{X}_1 = 8.9\%$	$\bar{X}_2 = 7.8\%$
$S_1^2 = 58$	$S_2^2 = 61$
$n_1 = 15$	$n_2 = 15$

Summary

As you have seen, there are few differences in the methods employed for testing hypotheses for large and small samples. The primary difference is in the probability distribution used. If we have small samples with the standard deviation unknown, the t distribution can be used. Otherwise, the standard normal distribution generally applies.

13-6
AN ALTERNATIVE WAY OF TESTING HYPOTHESES

We have continually stressed the hypothesis statement-decision rule sequence for hypothesis-testing situations. Although this sequence will always apply, there is an often-used alternate way of stating the decision rule—one that does not involve finding a critical value.

The first example of this chapter involved the Roller-Bed Company testing mattress strength. If you remember, the hypotheses were

$$H_0: \mu_x \geq 1,200$$

$$H_A: \mu_x < 1,200$$

Whereas that problem involved finding the sample size, let's assume now that the sample size is given as 70 and that we want $\alpha = 0.04$. Using this null hypothesis, we would approach the problem by calculating a critical value, A. We would then compare the sample mean, $\overline{X}$, to this critical value to see if the hypothesis should be accepted or rejected.

Following a slightly different line of reasoning, we could state that if the sample mean, $\overline{X}$, is *enough less* than 1,200, the hypothesis should be rejected. The question is, How much less? We could measure this difference in terms of the number of standard deviations the sample mean is from 1,200. If this distance is large enough, the null hypothesis would be rejected.

Using this reasoning, an alternative form of the decision rule would be as follows:

Decision Rule:

Take a sample of size n and calculate $\overline{X}$ and Z where

$$Z = \frac{\overline{X} - \mu_x}{\frac{\sigma_x}{\sqrt{n}}} \quad \text{or} \quad Z = \frac{\overline{X} - \mu_x}{\frac{S_x}{\sqrt{n}}}$$

If $Z \geq Z_{critical}$, accept H_0.
If $Z < Z_{critical}$, reject H_0.

The value of $Z_{critical}$ is determined by the allowable level of Type I error. If we assume an allowable Type I error of 0.04, $Z_{critical}$ is -1.75. Again, the minus indicates that the rejection region is in the *left* tail of the the distribution.

If Roller-Bed samples 70 mattresses and finds a mean of 1,178 and a standard deviation of 53, then

$$Z = \frac{1,178 - 1,200}{\dfrac{53}{\sqrt{70}}}$$

$$= -3.417$$

Figure 13–9 illustrates $Z_{critical}$. Since $-3.417 < -1.75$, the null hypothesis would be rejected.

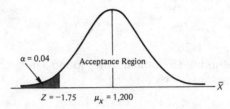

For this distribution, with $\mu_x = 1,200$, $\bar{X}$ will fall more than 1.75 standard deviations below the mean only 4% of the time. Hence $\alpha = 0.04$.

FIGURE 13–9   Hypothesis test for the alternative decision rule.

As a second example, reconsider the case of Senator Frances Chapel, where

Hypotheses:

H_0: $p \geq 0.55$, Chapel will receive at least 55% of the votes
H_A: $p < 0.55$

The new decision rule would be as follows:

Decision Rule:

Take a sample of size n and determine $\hat{p}$ and Z where

$$Z = \frac{\hat{p} - p}{\sqrt{\dfrac{p(1 - p)}{n}}}$$

If $Z \geq Z_{critical}$, accept H_0.
If $Z < Z_{critical}$, reject H_0.

Again, $Z_{critical}$ would be determined by the specified level of Type I error.

As a final example, consider the Edwards and Adamson bumper system just discussed. They hypothesized

$$H_0: \mu_x \leq \$150$$

$$H_A: \mu_x > \$150$$

and were willing to accept $\alpha = 0.05$. Since this is a small-sample problem, the appropriate decision rule is

Decision Rule:

Take a sample of size n and calculate $\overline{X}$ and t where

$$t = \frac{\overline{X} - \mu_x}{\frac{S_x}{\sqrt{n}}}$$

If $t \leq t_{\text{critical}}$, accept H_0.
If $t > t_{\text{critical}}$, reject H_0.

Since $\alpha = 0.05$, $t_{\text{critical}} = 1.265$.

After sampling 15 bumpers, they found a mean of \$155 and a standard deviation of \$80. Therefore

$$t = \frac{155 - 150}{\frac{80}{\sqrt{15}}}$$

$$= 0.242$$

Based on this decision rule, they should again conclude that the new bumper meets or exceeds the standard.

The two types of decision rules will always lead to the same decision (acceptance or rejection of the null hypothesis). As a decision maker, you should be aware of both procedures, and we suggest you solve problems using both methods.

13-7
SOME OTHER HYPOTHESIS TESTS

Our discussions in Chapter 12 and to this point in Chapter 13 have concentrated on the population mean as the value of interest. As we have shown, the mean is often the value upon which a decision is made. However, in some cases decision makers are more interested in the spread of a population than in its central location. For instance, military planes designed to penetrate enemy defenses have a ground-following radar system. The radar tells the pilot exactly how far the plane is above the ground. A radar unit that is correct *on the average* is useless if the readings are distributed widely around the true value. As a second example, many automatic transportation systems have stopping sensors to deposit passengers at the correct spot in a terminal. An automated stopping sensor that *on the average* lets passengers off at the correct point could leave many irritated people long distances up and down the track. Therefore many product specifications involve both an average value and some limit on the dispersion these values can have. These specifications are called **tolerances**. For example, the specification for a steel push-pin may be an average length of 1.78 inches $\pm$ 0.01 inch. The manufacturer using these pins would be interested in both the average length and how much these pins vary in length.

Hypotheses about a Single Population Variance

Signal Electronics, Inc. is developing a new radar system to be sold to the military. The contract is a performance contract, and its profitability depends on how well the unit works. While the Air Force team overseeing the contract has several performance criteria, one of the most important is how accurately the unit reads the distance off the ground. Before Signal's project engineer presents the trial units to the Air Force for evaluation, he will perform his own tests. One performance standard is that the variance in readings must be, at most, 15. Thus

$$H_0: \sigma_X^2 \leq 15$$
$$H_A: \sigma_X^2 > 15$$

The estimate of the population variance is S_X^2, where

$$S_X^2 = \frac{\Sigma(X - \bar{X})^2}{n - 1} \qquad (13\text{–}17)$$

To test a null hypothesis about a population variance, we compare S_X^2 with the hypothesized population variance, σ_X^2. To do this, we need to be able to standardize the distribution of sample variances in much the same way we used the Z distribution and the t distribution when hypothesizing about population means. **The standardized distribution for sample variances is a *chi-square distribution*.** The standardized chi-square variable is computed by the following equation:

$$\chi^2 = \frac{(n - 1)(S_X^2)}{\sigma_X^2} \qquad (13\text{–}18)$$

where: χ^2 = standardized chi-square variable

n = sample size

S_X^2 = sample variance

σ_X^2 = hypothesized variance

The shape of the standardized chi-square distribution depends upon σ_X^2 and the degrees of freedom, $n - 1$. Figure 13–10 illustrates chi-square distributions for various degrees of freedom. Note that **as the degrees of freedom increase, the chi-square distribution approaches a normal distribution.**

Returning to the Signal example, suppose the project engineer tested the radar system 20 times from a known height above the ground and found a variance of 16 feet squared. Figure 13–11 presents the hypothesis test at $\alpha = 0.10$.

Appendix E contains a table of chi-square values for various probabilities and degrees of freedom. The chi-square table is used in a manner similar to the t-distribution table. For example, to find χ^2_{critical} for the Signal Electronics example, we first determine that the degrees of freedom equal $n - 1$ (20 − 1 = 19) and that the desired alpha level is 0.10. Now we go to the chi-square table under the column

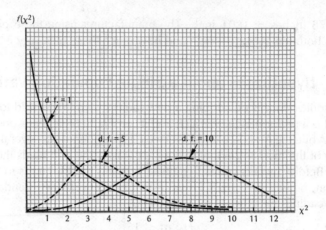

FIGURE 13–10   Chi-square distributions.

headed "0.10" (corresponding to the desired alpha) and find the χ^2 value in this column that intersects the row corresponding to the appropriate degrees of freedom. Table 13–1 illustrates how the chi-square table is used to find $\chi^2_{critical} = 27.204$. If the calculated χ^2 exceeds $\chi^2_{critical}$, the null hypothesis should be rejected.

As you can see in Figure 13–11, the sample variance is not *enough larger* than the hypothesized variance to cause the project engineer to reject the null hypothesis. He will conclude, based upon these results, that the radar system will meet the Air Force performance criterion under consideration.

Hypotheses:

$H_0: \sigma_x^2 \leq 15$
$H_A: \sigma_x^2 > 15$

$\alpha = 0.10$

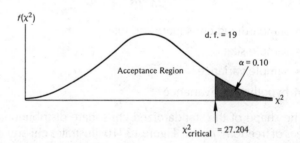

Decision Rule:

If $\chi^2 > \chi^2_{critical} = 27.204$, reject H_0.
Otherwise, do not reject H_0.

The chi-square test is as follows:

$$\chi^2 = \frac{(n-1)(S_x^2)}{\sigma_x^2} = \frac{19(16)}{15}$$

$$= 20.27$$

Since $\chi^2 = 20.27 < \chi^2_{critical} = 27.204$, accept H_0.

FIGURE 13–11   Chi-square test for one population variance, Signal Electronics, Inc.

TABLE 13–1 Finding critical values in the chi-square table

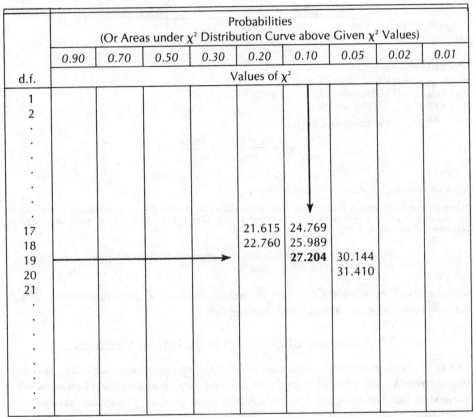

	Probabilities (Or Areas under χ^2 Distribution Curve above Given χ^2 Values)								
	0.90	*0.70*	*0.50*	*0.30*	*0.20*	*0.10*	*0.05*	*0.02*	*0.01*
d.f.	Values of χ^2								
1									
2									
.									
.									
.									
.									
.									
.									
17					21.615	24.769			
18					22.760	25.989			
19						**27.204**	30.144		
20							31.410		
21									
.									
.									
.									
.									
.									

In most applications, decision makers are concerned about the variance being too large. However there are instances in which decision makers are concerned with knowing whether the variance is *higher or lower* than a specified level. For instance, the ACME Assembly Company assembles clock radios. The number assembled from day to day will vary due to several factors, such as parts availability and employee attitudes. From past records, the production supervisor knows that the assembly variance should be 50. She has selected 25 days at random from this year's production reports and found a sample variance of 20. Figure 13–12 shows the statistical test using an alpha level of 0.20. As the test indicates, this sample information is so different from past information that the production supervisor should infer that the true variance has decreased. Based upon this conclusion, the supervisor would, no doubt, try to find out what caused the improved consistency in an effort

Hypotheses:

H_0: $\sigma_x^2 = 50$
H_A: $\sigma_x^2 \neq 50$

$\alpha = 0.20$

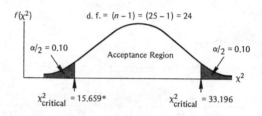

Decision Rule:

If $\chi^2 > \chi_{critical}^2 = 33.196$, reject H_0.
If $\chi^2 < \chi_{critical}^2 = 15.659$, reject H_0.
If $15.659 \leq \chi^2 \leq 33.196$, accept H_0.

The chi-square test is

$$\chi^2 = \frac{(n-1)(S_x^2)}{\sigma_x^2} = \frac{24(20)}{50}$$

$$= 9.6$$

Since calculated $\chi^2 < \chi_{critical}^2 = 15.659$, reject H_0.

*Since the hypothesis test is two-tailed, the lower critical value is found from the chi-square table under the column for an area $1 - \alpha/2 = 1 - 0.10 = 0.90$. The upper critical value is found from the chi-square table under the column for an area $\alpha/2 = 0.10$.

FIGURE 13–12   Chi-square test for one population variance, ACME Assembly Company.

to insure that it continues. Of course she would also want to determine if the assembly process is meeting the standard for average output.

Hypotheses about Two Population Variances

Just as decision makers are often interested in testing hypotheses regarding two population means, they are often faced with decision problems involving **two population variances.** For example, the Midlands Asphalt and Grading Company is considering purchasing a new paving machine. The company has two machines from which to choose. Midlands has decided to select the paver that provides the lesser variation in paving thickness. Variations in asphalt depth can have effects on both material cost and road strength. In addition, penalties are often assessed if testers find the road too thin.

One machine is considerably less expensive than the other, and although Midlands will buy the more expensive model if necessary, the company doesn't want to miss out on a good deal if the less expensive machine will spread asphalt with a comparable variation.

The purchasing agent at Midlands has arranged to sample 11 surfaces made by each machine to see whether he can detect a difference in thickness variation between the two. The hypotheses are

$$H_0: \sigma_1^2 = \sigma_2^2$$

$$H_A: \sigma_1^2 \neq \sigma_2^2$$

Since the population variances are not known, they must be estimated from the sample variances S_1^2 and S_2^2. Intuitively, you might reason that if the two population variances are actually equal, the sample variances should be nearly equal also.

If the samples are selected from two normally distributed populations with $\sigma_1^2 = \sigma_2^2$, the ratio of the sample variances, S_1^2/S_2^2, has a probability distribution known as an *F distribution*. The F statistic is

$$F = \frac{S_1^2}{S_2^2} \tag{13-19}$$

Because S_1^2 and S_2^2 have $n_1 - 1$ and $n_2 - 1$ degrees of freedom, respectively, the F distribution formed by the ratio of sample variances has $D_1 = n_1 - 1$ and $D_2 = n_2 - 1$ degrees of freedom.

The calculated F value gets larger if S_1^2 is greater than S_2^2. **Since we have control over which population we label as "1" and which we label as "2," we always select as population 1 the population with the larger sample variance. Therefore, if the calculated F gets too large, we conclude that the two populations have unequal variances.**

Suppose the samples for the two paving machines yield variances of $S_1^2 = 0.025$ and $S_2^2 = 0.017$. Figure 13-13 presents the test of the hypothesis that the two machines have equal variances. As was the case with the normal, t, and chi-square distributions, the F distribution has been tabulated. The F-distribution values for upper-tail areas of 0.05 and 0.01 are provided in Appendix F.

Hypotheses:

$H_0: \sigma_1^2 = \sigma_2^2$
$H_A: \sigma_1^2 \neq \sigma_2^2$

$\alpha = 0.10$

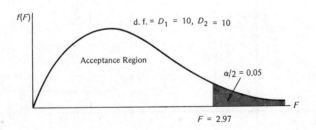

Decision Rule:

If $F \leq F_{critical} = 2.97$, accept H_0.
If $F > F_{critical}$, reject H_0.

The F test is

$$F = \frac{S_1^2}{S_2^2} = \frac{0.025}{0.017}$$

$$= 1.47$$

Since $F = 1.47 < F_{critical}$ with $D_1 = 10$ and $D_2 = 10$ degrees of freedom $= 2.97$, accept H_0.

Note: The right-hand tail of the F distribution always contains an area of $\alpha/2$ if the hypothesis test is two-tailed.

FIGURE 13-13 F test for two population variances, Midlands Asphalt and Grading Company.

The relationship between the specified alpha level and the appropriate F table is important. Two situations can occur:

1. **If the hypothesis test is two-tailed (that is, H_0: $\sigma_1^2 = \sigma_2^2$), the appropriate F table is the one with an upper-tail area (area above the critical value) equal to one-half alpha.** For example, if you have a two-tailed test with $\alpha = 0.10$, you should use the F-distribution table containing values of F for the upper 5 percent of the distribution. If $\alpha = 0.02$, you should use the F table containing values of F for the upper 1 percent.

2. **If the hypothesis test is one-tailed (that is, H_0: $\sigma_1^2 \leq \sigma_2^2$), the appropriate F table is the one with the upper-tail area equal to alpha.**

TABLE 13–2   Finding critical values in the F table (values of F for upper 5 percent)

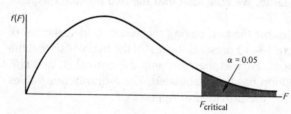

D_2 \ D_1	1	2	3	4	5	6	7	8	10	12
1										
2										
3										
4										
5										
6										
7										
8										
10								3.07	2.97	
12								2.85	2.76	
14									2.60	2.53
16										
20										
24										
.										
.										
.										
.										
.										
.										

For example, if $\alpha = 0.05$, use the F table containing values of F for the upper 5 percent of the distribution.

Thus the keys to which F table to use are the alpha level and the type of hypothesis test. We can illustrate this concept with two examples. First, take the case of a two-tailed test with stated alpha = 0.10 and degrees of freedom $D_1 = 10$ and $D_2 = 12$. We can use Table 13–2 to illustrate how $F_{critical}$ is determined. First, because it is a two-tailed test, go to the F table having an upper-tail area equal to one-half alpha (in this case, 0.10/2 = 0.05). Next, go to the column corresponding to $D_1 = 10$ and read down to the row corresponding to $D_2 = 12$. The value at the intersection of the row and column is the desired $F_{critical}$ (2.76).

As a second example, take the case of a one-tailed test with stated alpha = 0.01 and degrees of freedom $D_1 = 6$ and $D_2 = 10$. Because the test is one-tailed, we use the F table having an upper-tail area equal to alpha. Table 13–3 illustrates how we find the appropriate critical value. Figure 13–13 illustrates the

TABLE 13–3   Finding critical values in the F table (values of F for upper 1 percent)

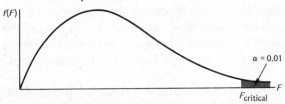

D_2 \ D_1	1	2	3	4	5	6	7	8	10	12
1										
2										
3										
4										
5										
6										
7										
8						6.37				
10					5.64	5.39	5.21			
12					5.06	4.82				
14										
.										
.										
.										
.										
.										
.										
.										
.										

hypothesis test for the Midlands Asphalt and Grading Company example. Since the F test leads to accepting the null hypothesis, Midlands should conclude, at the alpha = 0.10 level, that there is no difference between the two machines with respect to variability in paving thickness.

13-8
CONCLUSIONS

In this chapter we have extended our discussion of statistical hypothesis testing beyond the basic concepts introduced in Chapter 12. We have discussed the methods involved in simultaneously controlling the chances of committing both Type I and Type II errors. We have introduced the methods necessary to test hypotheses involving two population means, as well as one and two population proportions. Also, tests involving small samples using the t distribution have been presented.

We introduced two new probability distributions in this chapter: the chi-square distribution and the F distribution. The hypothesis tests using these distributions are tests of one and two population variances, respectively. You will use the F distribution extensively in Chapter 14, where we introduce analysis of variance. The chi-square distribution will be used extensively in Chapter 15 for nonparametric statistical tests.

You have probably noticed that statistical estimation, discussed in Chapters 10 and 11, and hypothesis testing, discussed in the last two chapters, have a lot in common. Some decision makers, and statistical authors, seem to favor one technique over the other. We feel both have their place and both should be learned. Our experience indicates that estimation procedures are most useful when decision makers have little idea about the population values and are primarily concerned with determining these values. On the other hand, if the decision makers have some idea about the parameter's value, the hypothesis-testing format is most useful.

You should make sure you are familiar with the terminology and equations at the end of this chapter before moving to Chapter 14.

chapter glossary

alpha The probability of committing a Type I error.

beta The probability of committing a Type II error.

one-tailed test A hypothesis test in which the entire rejection region is placed in one tail of the sampling distribution.

pooled estimator of population proportions The weighted average of two sample proportions. The pooled estimator, $\bar{p}$, is used in tests of two population proportions.

pooled variance A weighted average of two sample variances where the weights are the degrees of freedom associated with each variance.

proportion   The ratio of the number of successes to the number of observations.

two-tailed test   A hypothesis test in which the rejection region is split equally between the two tails of the sampling distribution.

Type I error   Rejecting a true null hypothesis.

Type II error   Accepting a false null hypothesis.

chapter formulas

Critical value

$$A = \mu_x + Z\frac{\sigma_x}{\sqrt{n}}$$

Z value

One population proportion

$$Z = \frac{\hat{p} - p}{\sqrt{\dfrac{p(1 - p)}{n}}}$$

Two population proportions

$$Z = \frac{(\hat{p}_1 - \hat{p}_2) - (p_1 - p_2)}{S_{\hat{p}_1 - \hat{p}_2}}$$

One population mean

$$Z = \frac{\overline{X} - \mu_x}{\dfrac{\sigma_x}{\sqrt{n}}}$$

Two population means

$$Z = \frac{(\overline{X}_1 - \overline{X}_2) - (\mu_1 - \mu_2)}{\sqrt{\dfrac{\sigma_1^2}{n_1} + \dfrac{\sigma_2^2}{n_2}}}$$

Two population means, variances unknown

$$Z = \frac{(\overline{X}_1 - \overline{X}_2) - (\mu_1 - \mu_2)}{\sqrt{\dfrac{S_1^2}{n_1} + \dfrac{S_2^2}{n_2}}}$$

t value

One population mean

$$t = \frac{\overline{X} - \mu_x}{\dfrac{S_x}{\sqrt{n}}}$$

Two population means

$$t = \frac{(\bar{X}_1 - \bar{X}_2) - (\mu_1 - \mu_2)}{S_{pooled}\sqrt{\frac{1}{n_1} + \frac{1}{n_2}}}$$

Pooled estimator of *p*

$$\bar{p} = \frac{n_1\hat{p}_1 + n_2\hat{p}_2}{n_1 + n_2}$$

Estimated variance of the difference between two population proportions

$$S_{\hat{p}_1 - \hat{p}_2} = \sqrt{(\bar{p})(1 - \bar{p})\left(\frac{1}{n_1} + \frac{1}{n_2}\right)}$$

Pooled standard deviation

$$S_{pooled} = \sqrt{\frac{S_1^2(n_1 - 1) + S_2^2(n_2 - 1)}{n_1 + n_2 - 2}}$$

Chi-square value (One population variance)

$$\chi^2 = \frac{(n - 1)(S_X^2)}{\sigma_X^2}$$

***F* value** (Two population variances)

$$F = \frac{S_1^2}{S_2^2}$$

solved problems

1. A major television manufacturer claims that its average set will be defect-free for more than two years' of use. A consumer reporting service has decided to test this claim. The service finds a random sample of 20 set owners and determines the time to the sets' first repair. The sample results, in years, are

1.97	2.87	3.01	2.75
2.09	1.34	1.62	1.10
2.24	1.79	2.81	0.57
3.17	3.89	3.10	2.05
1.01	4.16	2.59	1.67

Based on these sample data, is the manufacturer's claim supported, assuming the burden of proof is on the manufacturer to justify its advertising claim?

Solution:

Hypotheses:

H_0: $\mu_x \leq 2$ yr, claim is not true
H_A: $\mu_x > 2$ yr, claim is true

Decision Rule:

Take a sample of size n and determine $\overline{X}$.

If $\overline{X} \leq A$, accept H_0.

If $\overline{X} > A$, reject H_0.

The alpha level is 0.01.

Since σ_x is unknown and the sample size is small, we use the sample standard deviation and the t distribution to determine the A value if we assume the distribution of time to the first breakdown is normally distributed. Then

$$A = \mu_x + t\frac{S_x}{\sqrt{n}}$$

Next, we calculate S_x.

$$S_x = \sqrt{\frac{\Sigma(X - \overline{X})^2}{n - 1}}$$

$$\overline{X} = \frac{\Sigma X}{n}$$

$$= 2.29$$

$$S_x = \sqrt{\frac{17.115}{19}}$$

$$= 0.949$$

The t value for $\alpha = 0.01$ for a one-tailed test with $n - 1 = 19$ degrees of freedom is 1.729. Thus

$$A = 2 + 1.729\frac{0.949}{\sqrt{20}}$$

$$= 2.3669$$

Decision Rule:

If $\overline{X} > 2.3669$ yr, reject H_0.

If $\overline{X} \leq 2.3669$ yr, accept H_0.

Since $\overline{X} = 2.29 < 2.3669$, we should accept the null hypothesis and conclude that the manufacturer's claim cannot be supported. Note that the manufacturer would no doubt argue for a larger alpha level or that the burden of proof be shifted to the reporting service.

2. Property taxes are based upon the assessed valuation of real estate. The higher the valuation the greater the tax on that property. In one southern county, a controversy is taking place between some citizens and the county tax appraiser. The citizens claim that during a recent reappraisal, the residential property was increased in value by a greater average percentage than commercial property. If the citizens' claim is true, they will end up paying a greater relative share of property taxes than owners of commercial property.

An outside consulting firm has been hired to study the situation. As part of their study, the consultants have selected a sample of 400 residential properties and 300 commercial properties and determined the average percent increase in assessed valuation for each class of property. The sample means along with the sample variances are

Residential	Commercial
$\bar{X}_R = 108\%$	$\bar{X}_C = 102\%$
$S_R^2 = 1,400$	$S_C^2 = 1,650$
$n_R = 400$	$n_C = 300$

Does this sample evidence support the citizens' claim? Test at the 0.10 alpha level.

Solution:

Even though this problem deals with percentages, the problem is still one of hypothesis testing about two population means.

Hypotheses:

$H_0: \mu_R - \mu_C \leq 0$
$H_A: \mu_R - \mu_C > 0$

The sampling distribution and the acceptance and rejection regions are as follows:

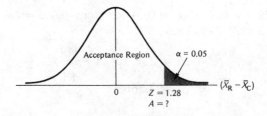

We can test the null hypothesis two ways. First, we can solve for Z using

$$Z = \frac{(\bar{X}_R - \bar{X}_C) - 0}{\sqrt{\dfrac{S_R^2}{n_R} + \dfrac{S_C^2}{n_C}}}$$

Note that we use the normal distribution even though the population variances are unknown since the samples are large. Therefore

$$Z = \frac{(108 - 102) - 0}{\sqrt{\dfrac{1,400}{400} + \dfrac{1,650}{300}}}$$

$$= 2.00$$

Since $Z = 2.00 > Z_{critical} = 1.28$, we reject the null hypothesis and conclude that the citizens' claim is true.

A second approach is to solve for A and then compare the difference between the two sample means to the decision rule.

$$A = 0 + 1.28 \sqrt{\dfrac{1,400}{400} + \dfrac{1,650}{300}}$$

$$= 3.84$$

Decision Rule:

If $(\bar{X}_R - \bar{X}_C) \le 3.84\%$, accept H_0.
If $(\bar{X}_R - \bar{X}_C) > 3.84\%$, reject H_0.

Since $(\bar{X}_R - \bar{X}_C) = 6$ percent, we reject H_0. Rejecting the null hypothesis does not mean that the assessments are in error. They may actually reflect the change in market values or an adjustment from previous inequities against commercial property.

3. AGI Food Stores, Inc. operates two stores in an Ohio city. Each store uses a different marketing approach. Store A caters to the higher-income shopper by carrying specialty items and gourmet foods. The store's physical layout is spacious, and much money is spent keeping the floors and shelves clean. Store B is directed at lower-income customers and provides good selections of the basic food products.

AGI management has recently performed an extensive study of shoppers in the two stores. One manager has claimed that the dollar volume of purchases at store B is more consistent than the dollar volume of purchases at store A. To test this claim, the following data have been collected:

Store A	Store B
$n_A = 41$	$n_B = 61$
$S_A^2 = \$100.00$	$S_B^2 = \$87.50$

Establish the appropriate null and alternative hypotheses and test at the $\alpha = 0.05$ level.

Solution:

Since the claim is that store B's sales amounts are more consistent than store A's, a test of variances is appropriate.

Hypotheses:

$H_0: \sigma_B^2 \geq \sigma_A^2$, claim is not true
$H_A: \sigma_B^2 < \sigma_A^2$, claim is true

The appropriate test is the F test for two variances.

$$F = \frac{S_A^2}{S_B^2}$$

Decision Rule:

If $F \leq F_{critical}$, accept H_0.
If $F > F_{critical}$, reject H_0.

The following figure shows the F distribution and the rejection region:

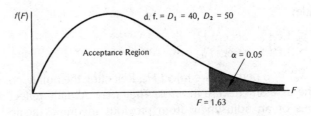

$f(F)$ d. f. = $D_1 = 40$, $D_2 = 50$

Acceptance Region

$\alpha = 0.05$

$F = 1.63$

Now

$$F = \frac{S_A^2}{S_B^2} = \frac{100}{87.50}$$

Since $F = 1.14 < 1.63$, we conclude the claim is not true—that the purchase volume for store B is not more consistent than that for store A.

problems

Problems marked with an asterisk (*) require extra effort.

1. Discuss in your own words what a Type I error is and give an example of your own where committing a Type I error would be very costly.

2. Discuss in your own words what a Type II error is. Provide a business-oriented example of your own where committing a Type II error would be very costly.

3. Discuss your own example of a business situation where committing either a Type I or Type II error would be very costly.

4. Referring to your answer to problem 3, how can the chances of the Type I and Type II errors be controlled?

5. What factors should decision makers take into account when considering how large a chance they are willing to accept of making a Type I or Type II error?

6. Suppose a fast-food chain is considering expanding into a new community but management knows from past experience that the chances of success are limited if the income level in the community is too low. To justify expansion, management feels that the average per capita income must be at least $8,100 per year.

To determine the feasibility of expansion, the chain wishes to select a sample to test the hypothesis that the income is at least $8,100. Suppose another study has indicated that the standard deviation of income in this community is $1,900. What decision rule should the company adopt if it will accept the following:

   a. No more than a 7.35 percent chance of a Type I error when $\mu_x = \$8,100$
   b. No more than a 10 percent chance of committing a Type II error if $\mu_x = \$8,000$

Remember, the decision rule includes the required sample size and the appropriate critical value.

7. Referring to problem 6, show the impact on the required sample size if the probability of the Type II error is held to 0.05 when $\mu_x = \$8,000$. State why the sample size changed in this manner.

8. Suppose the cost of the sample sizes determined in problems 6 and 7 is too great. What alternatives do the decision makers have? Discuss why each will likely lower the required sample size.

9. The Green Thumb Nursery is considering selling a new species of shrub that is supposed to be very hardy. The manager would like to place an unlimited guarantee on the shrub which states that if one does not survive, it will be replaced free of charge. To do this, he figures that at least 90 percent must survive. He decides to select a sample of 100 shrubs and test the hypothesis that at least 90 percent of all shrubs will survive.

Using an alpha level of 0.10, what should the manager conclude if 85 survive?

10. Referring to problem 9, suppose before the sample had actually been selected, the manager wished to know the chances of accepting the null hypothesis when, in fact, the true proportion of shrubs that would survive is 0.88. Find beta in this case, assuming the manager holds alpha at 0.10.

11. The manager of the Green Thumb Nursery in problems 9 and 10 wishes to reduce beta to half the level calculated in problem 10 while holding alpha at 0.10. What is the sample size required to meet these constraints?

12. Discuss the meaning of a Type I error as it refers to the Green Thumb Nursery problems 9–11.

13. What does a Type II error mean to the manager of the Green Thumb Nursery referred to in problems 9–11? Which error, Type I or Type II, do you think is most serious in this case, and why?

14. The agricultural extension agent in a rural Idaho county wishes to see whether there is a difference between the production levels for two different types of hay. She has hypothesized that there is no difference.

A random sample of production from 100 acres for each brand of hay has been selected, with the following results:

Brand 1	Brand 2
$\bar{X}_1 = 3.6$ tons/acre	$\bar{X}_2 = 4.2$ tons/acre
$S_1^2 = 9$	$S_2^2 = 12$
$n_1 = 100$	$n_2 = 100$

Based upon these sample results, what should the extension agent conclude if alpha is set at 0.04?

*15. Referring to problem 14, assume that the population variances are known to be $\sigma_1^2 = 11.2$ and $\sigma_2^2 = 9.6$. Before the samples are actually selected, the extension agent wants to know the chance of a Type II error if the true difference between the average production is 0.5 ton per acre. Leaving alpha at 0.04, find beta.

*16. Referring to problems 14 and 15, find the required sample sizes if alpha is 0.04 and the beta is half that found in problem 15 when the true difference is 0.5 ton per acre. (*Hint:* The sample sizes must be equal.)

*17. With respect to problem 16, what would the required sample size be if alpha were allowed to increase to 0.10? Why would the required sample size decrease?

18. The Allentown Fire Department has come under pressure from the city council because many citizens have complained about poor response rates. Last year the town had 1,525 fires. The required average response time is 5 minutes.

A random sample of 25 fires is selected and the results show a mean of 7 minutes and a variance of 36. Testing at the alpha = 0.05 level, should the council chastize the fire department, or should it conclude that the standard is being satisfied?

19. The Altus Park and Recreation Department has claimed that the city parks are being used extensively by adults rather than children. In fact, a recent report claims that the average age of park users is at least 30. If this claim is true, the city council plans to institute a research study to determine whether a children's park is needed.

In an effort to verify whether the recreation department's claim is true, a sample of park users has been selected and their ages recorded. A sequential random sample of people entering the park was used, giving the following results:

11 yr	5 yr
17	66
31	59
59	14
18	18

Based upon this sample evidence, should the department's claim be supported? Assume the sample was random. Test at the alpha = 0.05 level.

20. Comment on the way in which the sampling was conducted in problem 19. What problems might result from this type of sampling in this case?

21. Suppose the registrar at the local technical school claims that the academic scholarships given at the school have been biased toward females. He maintains that the grade-point average (GPA) of males is no different from that of females but that females get 60 percent of the scholarships. To test his claim, a random sample of 12 male and 12 female students has been selected from the eligible students, with the following results:

Male GPA		Female GPA	
2.25	1.99	3.25	2.87
3.16	3.25	3.00	3.47
3.00	3.00	3.16	3.62
3.87	2.97	3.04	4.00
2.62	1.72	2.98	2.19
2.30	3.88	2.93	2.45

Based on these data, what conclusion should the registrar reach if he tests the claim at the 0.05 alpha level? Be sure to indicate the assumptions necessary to perform this hypothesis test.

*22. Referring to problem 21, where a small-sample t test was required, test the assumption that the population variances for male and female grade-point averages are equal. Use an alpha level of 0.10. Based upon your results, comment on the test conducted in problem 21.

23. The J. R. Reindeer Corporation grows Christmas trees commercially. The company has many acres planted in various species of trees.
Because retailers prefer uniform-sized trees, the manager in charge of tree harvesting is concerned with the variation in tree growth. A five-year-old tree is generally considered to be of cutting age, assuming it has grown to the proper size. The trees in a particular field are all planted at the same time. However the manager recognizes that not all trees will grow at the same rate. Therefore she expects some variation in tree size. However, if there is too much variation in a particular species, that species will be phased out.
Out of one field that has just been harvested, a sample of 30 trees showed a standard deviation of 4.2 inches. If the acceptable standard deviation is 3 inches, what should the manager conclude about the trees in this field? Test the hypothesis at the 0.05 alpha level.

24. Referring to problem 23, suppose the manager has just selected a sample of 30 trees of a different species from a second field and found the variance to be 15. What should the manager conclude about the two samples in terms of relative variation in growth? Use alpha equal to 0.02 to test the hypothesis.

25. The State University registrar claims that fewer than 20 percent of the students who enroll at State graduate in four years. To test this claim, a random sample of 100 students was selected, and 18 were found to have graduated in four years. Using alpha equal to 0.10, are the sample data sufficient to refute the registrar's claim?

26. Referring to problem 25, suppose the Board of Education wishes to test the claim that there is no difference in the proportions of students from State University and City College who graduate in four years. A sample of 100 students was selected from each, and 18 State students and 21 City College students were found to have graduated in four years. Using an alpha equal to 0.05, do these data support or refute the claim?

27. The Utah Department of Transportation conducted a study of bridges in the state. To receive a federal grant, the department had to show that at least 40 percent of the bridges need repair, as claimed in its grant proposal. A random sample of 49 bridges revealed 18 bridges in need of repair. Do these data support or refute the claim made in the grant proposal? Use alpha equal to 0.01.

28. In problem 27, an alpha level of 0.01 was used to conduct the test. Assume that this alpha was selected by the Utah Department of Transportation. Discuss why the department might have selected such a small alpha in this case.

*29. Referring to problem 27, suppose the "true" proportion of defective bridges is 0.38. What is the probability that a sample of size 49 will lead to a Type II error when the hypotheses are

$$H_0: p \geq 0.40$$
$$H_A: p < 0.40$$

and the alpha level is 0.01? Comment on this result from both the Utah point of view and the federal government point of view.

*30. Referring to problems 27–29, what size sample is required to meet the following constraints?
a. No more than a 5 percent chance of committing a Type I error
b. No more than a 10 percent chance of committing a Type II error when the true proportion of defective bridges, p, is 0.38.

*31. The Consumer Information Company is interested in performing a test to validate an advertising claim made by the Tremco Drill Company that the average useful life of its electronic drill equipment is at least 400 hours. A prior study has shown the standard deviation to be 20 hours. What size sample should Consumer Information select if it wishes to satisfy the following constraints?
a. No more than a 5 percent chance of committing a Type I error
b. No more than a 10 percent chance of committing a Type II error when $\mu_x = 390$ hours
c. No more than a 14 percent chance of committing a Type II error when $\mu_x = 395$ hours

32. Using the sample size determined in problem 31, establish the appropriate decision rule for the Consumer Information Company to follow in conducting its test.

33. The Bertflo Investment Company claims that the variance in dollar profits per $1,000 invested for its clients is less than or equal to 400. To test this claim, another investment advisory service selects a sample of 20 Bertflo clients and determines that the sample variance is 600. Is this result sufficient to refute Bertflo's claim at an alpha level of 0.10?

===cases

13A ESSEX CHEMICAL COMPANY

The Essex Chemical Company of New Jersey has recently been indicted by a federal grand jury for knowingly dumping pollutants into the West River. This indictment resulted in the firing of the plant manager and three of his chief assistants. At the home office, the vice-president of production operations has suddenly taken early retirement. The shake-up in the organization has been substantial, and the actions taken by the Essex board of directors have reflected its disapproval of the pollution violations.

Essex was required by court order to sign a decree stating that the company would never knowingly pollute the West River again. Additionally, Essex has agreed to provide the state and federal regulatory agencies all existing records of its waste output volumes on demand.

The president of Essex has hired a new plant manager, Paula Douglas, to take over the New Jersey plant and has made it clear to Paula that complying with the pollution-control standards is an absolute requirement. Essex is not about to go through such negative publicity a second time.

One of the first actions that Paula Douglas took after becoming plant manager was to install a new pollution-control device which is a combination of electrostatic and chemical filters. Scientists and company engineers have informed her that under normal plant operations, the pollution levels will vary somewhat. The federal standards allow an average waste output into the West River of 4 parts per million. However the government will not impose heavy fines unless the average output reaches 5 parts per million. From all indications, the new pollution-control device would meet these standards. In fact, the device is supposed to hold pollution to an average of 3 parts per million.

The pollution-control device demands periodic regeneration, particularly the chemical filter. Since there is really no way to accurately predict when the filters will need regeneration, the plant has to rely on a series of monitoring devices in the canal leading to the West River.

Paula Douglas is trying to determine how much to budget for these monitoring devices. She estimates that the first three will cost $20,000 each and that each additional can be purchased for $10,000.

Tests on the filtered output from the chemical plant have indicated a variance of 7 parts per million regardless of the average level. Paula recognizes that since variation exists in the monitoring process, there is always the possibility of sampling error. If the new machine will actually allow pollution to reach an average of 5 parts per million, she is willing to incorrectly accept that it will hold the average to 3 parts per million only 1 time in 1,000. However, because the regeneration process is expensive and time consuming, she wants to regenerate needlessly only 1 time in 20.

Paula has to turn in her budget request this afternoon and needs to know how many of the monitoring devices to buy. (Note that each monitoring device will produce the equivalent of one sample measure. The average of the sample measures will be used to determine if the control device needs to be regenerated.)

13B   GREEN VALLEY ASSEMBLY COMPANY

The Green Valley Assembly Company assembles consumer electronics products for manufacturers that need temporary extra production capacity. As such, they have periodic product change. Since the products Green Valley assembles are marketed under the label of well-known manufacturers, high quality is a must.

Tom Bradley of the Green Valley personnel department has been very impressed by recent research concerning job-enrichment programs. In particular, he has been impressed with the increases in quality that seem to be associated with these programs. However some studies have shown no significant increase in quality and imply that the money spent on such programs has not been worthwhile.

Tom has talked to Sandra Hansen, the production manager, about instituting a job-enrichment program in the assembly operation at Green Valley. Sandra was somewhat pessimistic about the potential but agreed to introduce the program. The plan was to implement the program in one wing of the plant and continue with the current methods in the other wing. The procedure was to be in effect for six months. Following that period, a test would be made to determine the effectiveness of the job-enrichment program.

After the six-month trial period, a random sample of employees from each wing produced the following output measures:

Old	Job-Enriched
n_1 = 50 employees	n_2 = 50 employees
$\overline{X}_1$ = 11/h	$\overline{X}_2$ = 9.7/h
S_1 = 1.2/h	S_2 = 0.9/h

Both Sandra and Tom wonder whether the job-enrichment program has affected production output. They would like to use these sample results to determine if the average output has been changed and to determine if the consistency of the employees was affected by the new program.

A second sample from each wing was selected. The measure was the quality of the products assembled. In the "old" wing, 79 products were tested and 12 percent were found to be defectively assembled. In the "job-enriched" wing, 123 products were examined and 9 percent were judged defectively assembled.

With all these data, Sandra and Tom are beginning to get a little confused, but they realize that they must be able to use the information somehow in order to make a judgment about the effectiveness of the job-enrichment program.

13C MILLER MANUFACTURING COMPANY

Miller Manufacturing has its main plant near the Gulf of Mexico. In 1979 the plant was severely damaged by a hurricane. While some damage is covered by insurance, the bulk of the damage has been classified as an act of God and will not be covered. Consequently the salvage operation is very important.

Harry Adams, who is in charge of salvage operations, has successfully supervised temporary repairs of the buildings and some key machinery. He feels Miller is almost ready to resume regular operations. Although the company is taking advantage of low-cost federal loans, its extensive fixed costs have continued during the down time. Consequently, the sooner Miller begins manufacturing the better.

Miller makes several power tools, many of which contain an electromechanical servo control mechanism. These control units were stored in a warehouse that received extensive water and dirt damage. The control mechanism supplier has informed Miller management of a shortage. The supplier will not be able to ship new units until at least the following month. If the control mechanisms Miller has are damaged by dirt, they will fail after a short period of use, and Miller's reputation for high-quality power tools will be damaged. However, if the units are good, Miller will be able to start generating a cash flow rapidly.

After looking at the fixed to variable cost ratios and the cost of replacing units, Harry has estimated that if 10 percent or fewer of the control units are damaged, the optimal procedure will be to make the power tools and offer free replacement for any equipment that fails in the first year due to electromechanical control mechanism failure. However, if it is concluded that the damaged rate exceeds 10 percent, Miller will wait until its supplier can ship new parts.

Harry randomly selects 300 switches and finds that 33 have been damaged by dirt and water.

13D DOWNTOWN DEVELOPMENT: BAYVIEW, NORTH DAKOTA

Bayview, North Dakota, is a rapidly growing city with a high number of corporate headquarters. In addition, Bayview has a substantial amount of light industry and an increasing number of service businesses to support the population.

The city government has been involved in a controversy for the past five years over the issue of whether to build a regional shopping center in the downtown area or in the suburbs. The city council has gone on record as favoring the downtown site but has received heavy opposition from a citizens' group called KNOW (Keep Nice Our World) which feels that the best site is in the suburbs. KNOW's argument is that less energy will be needed to reach the shopping center if it is placed closer to the people. The city council, on the other hand, has passed a "Metro" plan which calls for downtown development and mass transit of people to the downtown area, which it argues would save energy.

Endless meetings and hearings have been held on the issue, but never has there been a vote or even a legitimate poll of the people. The city council and the KNOW representatives have agreed to hire a marketing consultant from Rock Springs, Wyoming, to select a random sample of citizens and ask them which location they favor for the shopping center.

The city council claims that at least 50 percent of the population favor the downtown site and feels quite confident that the sample will bear this out. However the KNOW group claims that 50 percent or fewer prefer the downtown location and is equally confident in its claim.

The Rock Springs consultant selected a random sample of 384 persons and found 188 in favor of the downtown site and 196 opposed to the downtown site.

The day following the tabulation of the sample results, the *Bayview Gazette* ran a headline story relating the sample results and statements from the city council and KNOW, both of which stated that the results proved their claims. The story mentioned something about both parties testing their hypotheses at the 0.05 alpha level.

One citizen was heard that morning to say, "What does this mean? How can they both be right? This just doesn't make sense. It can't be."

references

DUNCAN, ACHESON J. *Quality Control and Industrial Statistics*. 3rd ed. Homewood, Ill.: Irwin, 1965.

KAISER, HENRY F. "Directional Statistical Decisions." *Psychological Review* (May 1960): 160–67.

LAPIN, LAWRENCE L. *Statistics for Modern Business Decisions*. New York: Harcourt Brace Jovanovich, 1978.

NETER, JOHN, WILLIAM WASSERMAN, and G. A. WHITMORE. *Applied Statistics*. Boston: Allyn and Bacon, 1978.

ROZEBOOM, WILLIAM W. "The Fallacy of the Null-Hypothesis Significance Test." *Psychological Bulletin* 57 (September 1960): 416–28.

SPURR, WILLIAM A. and CHARLES P. BONINI. *Statistical Analysis for Business Decisions*. Homewood, Ill.: Irwin, 1973.

WILSON, WARNER, HOWARD L. MILLER, and JEROLD S. LOWER. "Much Ado about the Null Hypothesis." *Psychological Bulletin* 67 (March 1967): 188–96.

WONNACOTT, THOMAS H. and RONALD J. WONNACOTT. *Introductory Statistics for Business and Economics*. New York: Wiley, 1977.

14

Analysis of Variance

Hypothesis testing is a widely used application of statistical tools. Chapters 12 and 13 served as an introduction to hypothesis testing. The main objective in those chapters was to provide the essentials of decision making using statistical hypotheses. We discussed at some length the two types of errors decision makers can make and indicated how the chances of committing these errors can be controlled.

The examples of hypothesis testing presented in Chapters 12 and 13 dealt with one or, at most, two populations. However decision makers will often find applications where more than two populations are involved. For instance, a personnel manager may be administering a different piece-rate pay scale at each of four divisions in his company. He might hypothesize that the average employee productivity levels are equal at all four divisions. Or, a study may be conducted to determine whether students graduating from the six universities in a single state have different average starting salaries. The statistical tests presented in Chapters 12 and 13 cannot handle hypothesis tests involving multiple populations, but there are procedures for doing so. One such procedure is *analysis of variance.*

chapter objectives

In this chapter we shall introduce analysis of variance and show how this tool can be used in business decision making. We will indicate when analysis of variance should be used and demonstrate how to test whether the means of two or more populations are equal. Sometimes analysis of variance leads to the conclusion that the population means are not all equal. For these cases, we present two methods to help you decide which population means are different.

student objectives

After reading the material in this chapter, you should be able to:

1. Discuss why using multiple two-sample t tests is not an appropriate alternative to analysis of variance.
2. Describe what is meant by partitioning the sum of squares.
3. Perform an analysis of variance for a one-way experimental design.
4. List the assumptions necessary to use analysis of variance.
5. Apply Tukey's method for pairwise comparisons of population means.
6. Apply Scheffé's method for pairwise comparisons of population means.
7. Recognize business applications for which analysis of variance is the appropriate statistical tool.

14-1
ANALYSIS OF VARIANCE—ONE-WAY DESIGN

The national sales manager for Ambell, Inc. was recently asked by the company's president whether there is a difference in the average weekly sales per salesperson between the four regions covered by the sales force. The president suspects that average sales productivity has not been equal in the four regions. If his suspicions are true, he will recommend that the national sales manager direct more effort to the regions with lower average sales. In addition, the company will increase advertising in the regions with significantly lower average sales.

To answer this question, Pam Burke, the national sales manager, selected a random sample of eight salespeople from each region and calculated each person's sales level for the past year. Pam has formulated the following null and alternative hypotheses:

$$H_0: \mu_1 = \mu_2 = \mu_3 = \mu_4$$
$$H_A: \text{Not all means are equal.}$$

She has also established an alpha level of 0.05 for this test.

Why Two-Sample t Tests Won't Work

One method to test the null hypothesis involving four population means would be to use the two-sample t test discussed in Chapter 13. Pam Burke could set up a series of null and alternative hypotheses involving *all* possible pairs of sales regions. The two-sample t test could be used to test each null hypothesis. With four populations, there are

$$_4C_2 = \frac{4!}{2!(4-2)!}$$

$$= 6$$

separate pairs of regions. Thus, to test the null hypothesis that all four population means are equal would require six separate t tests of the form

$$H_0: \mu_1 = \mu_2$$
$$H_A: \mu_1 \neq \mu_2$$

with the test statistic

$$t = \frac{(\overline{X}_1 - \overline{X}_2) - 0}{S_{pooled}\sqrt{\dfrac{1}{n_1} + \dfrac{1}{n_2}}} \qquad (14\text{–}1)$$

If the six separate t tests are performed, and the null hypothesis is accepted in each case, we could conclude that all four population means are equal. However using multiple t tests is acceptable only if Pam is *not* concerned with holding alpha to the 0.05 level. The problem with using a series of two-sample t tests is that although each test has an alpha level of 0.05, the true alpha level for all tests combined is greater than 0.05. The actual alpha level is

$$\alpha_{actual} \leq [1 - (1-\alpha_1)(1-\alpha_2)(1-\alpha_3)\ldots(1-\alpha_6)]$$
$$\leq [1 - (0.95)(0.95)(0.95)(0.95)(0.95)(0.95)]$$
$$\leq 0.2649$$

Thus the maximum probability of committing one or more Type I errors using a series of six t tests, each with $\alpha = 0.05$, is, at most, 0.2649. **The logic behind this increase in alpha is that as more tests are performed, the risk of rejecting a true hypothesis is increased.**

Due to this problem, a series of two-sample t tests is not generally considered adequate to test hypotheses involving more than two populations. **However a statistical tool known as *analysis of variance (ANOVA)* can be used without "compounding" the probability of committing a Type I error.**

The Rationale of One-Way Analysis of Variance

In the Ambell example, Pam Burke needs to determine whether the average sales output per salesperson is equal between regions. *Analysis of variance* is a statistical procedure which, as its name implies, is used to examine population variances to determine whether the population means are equal.

Three assumptions (the same ones as for a two-sample t test) must be satisfied before analysis of variance can be applied:

1. The samples must be independent random samples.
2. The samples must be selected from populations with normal distributions.
3. The populations must have equal variances (that is, $\sigma_1^2 = \sigma_2^2 = \ldots = \sigma_k^2 = \sigma^2$).

The rationale behind analysis of variance might best be understood by studying Table 14–1, which presents the sales data Pam Burke has collected. You should notice several things about these data. First, the salespeople in the sample did *not* sell exactly the same number of units. Thus variation exists in the units sold by the 32 people. This is called the **total variation** in the data. Second, within any particular region, the salespeople did *not* all sell an equal number of units. Thus variation exists within regions. This variation is called **within-sample variation**. Finally, the sample means for the four regions are *not* all equal. Thus variation exists between the regions. As you might have guessed, this is called **between-sample variation**.

TABLE 14–1   Sales data, Ambell, Inc.
(units sold by each salesperson)

Salesperson	Region 1	Region 2	Region 3	Region 4	
1	3	9	7	12	
2	5	8	5	8	
3	9	8	4	7	
4	10	7	9	7	
5	7	10	9	8	
6	3	9	6	10	Grand
7	5	9	8	5	total
8	6	4	8	7	↓
Total	48	64	56	64	232

$$\overline{X}_1 = 6 \qquad \overline{X}_2 = 8 \qquad \overline{X}_3 = 7 \qquad \overline{X}_4 = 8$$

$$\text{Grand mean } (\overline{\overline{X}}) = \frac{232}{32}$$

$$= 7.25$$

The basic principle of one-way analysis of variance is that

Total sample variation = between-sample variation + within-sample variation

From our discussion of variance in Chapter 4, we know that the variability in a set of measurements is proportional to the sum of the squared deviations of the measurements from the mean as given by

$$\Sigma(X - \overline{X})^2 \qquad\qquad \textbf{(14–2)}$$

Therefore the total sample variation in the data shown in Table 14–1 is proportional to the sum of the squared deviations of the 32 sales figures around the **grand mean.** This is called the **total sum of squares** and is calculated as follows:*

$$TSS = \sum_{i=1}^{K} \sum_{j=1}^{n_i} (X_{ij} - \overline{\overline{X}})^2 \tag{14–3}$$

where: TSS = total sum of squares

K = number of populations (columns)

n_i = sample size from population i

X_{ij} = jth measurement from population i (in the present example, the 32 different measurements)

$\overline{\overline{X}}$ = grand mean (here, the mean of all 32 measurements)

The total sum of squares for the Ambell example is

$$TSS = (3 - 7.25)^2 + (5 - 7.25)^2 + (9 - 7.25)^2 + \ldots + (7 - 7.25)^2$$
$$= 148$$

We stated earlier that total variation equals the between-sample variation plus the within-sample variation. Likewise, it can be shown that

$$\underbrace{\sum_{i=1}^{K} \sum_{j=1}^{n_i} (X_{ij} - \overline{\overline{X}})^2}_{TSS} = \underbrace{\sum_{i=1}^{K} n_i(\overline{X}_i - \overline{\overline{X}})^2}_{SSB} + \underbrace{\sum_{i=1}^{K} \sum_{j=1}^{n_i} (X_{ij} - \overline{X}_i)^2}_{SSW} \tag{14–4}$$

Equation (14–4) shows that the total sum of squares (TSS) can be **partitioned** into two parts: the **sum of squares between (SSB)** and the **sum of squares within (SSW).** For the Ambell example, we get

$$TSS = SSB + SSW$$
$$148 = 22 + 126$$

where

$$SSB = 8(6 - 7.25)^2 + 8(8 - 7.25)^2 + 8(7 - 7.25)^2 + 8(8 - 7.25)^2$$
$$= 22$$
$$SSW = TSS - SSB = 148 - 22$$
$$= 126$$

Pam Burke wants to determine whether the mean sales in the four regions are equal. However she must decide this based on a sample from each region. Pam wants to know if the four distributions shown in Figure 14–1 best describe the sales distributions, or whether Figure 14–2 applies. If Figure 14–1 applies, a null hypothesis that all means are equal should be accepted. If Figure 14–2 is the case,

---

*The chapter glossary contains calculation formulas for each of the sum of squares formulas contained in this section.

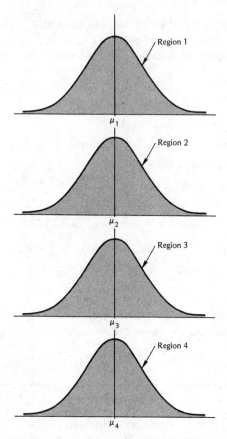

FIGURE 14–1 Normal populations with equal means and equal variances.

the null hypothesis should be rejected. Note that these figures illustrate the assumptions of normal populations and equal population variances.

To determine whether to accept or reject the null hypothesis using analysis of variance, we perform the following:

1. Establish the null hypothesis to be tested:

$$H_0: \mu_1 = \mu_2 = \mu_3 = \ldots$$

H_A: Not all means are equal.

2. Make two estimates of the population variance, one based on individual regions, one based on differences in regional averages.

3. If the two estimates are about the same, we conclude that Figure 14–1 applies, and that the means are equal.

The computational procedures necessary to make the variance estimates are not complicated. The first estimate is

$$\text{MSW} = \frac{\text{SSW}}{N - K} = \text{unbiased estimate of } \sigma^2 \qquad \textbf{(14–5)}$$

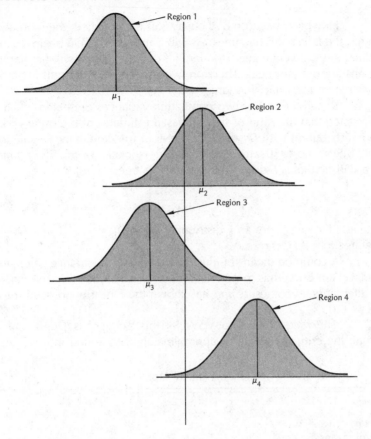

FIGURE 14–2 Normal populations with unequal means and equal variances.

where: MSW = mean square within

SSW = sum of squares within

N = total number of measurements from all samples

K = number of groups (here, 4 regions)

Also, if the region means are truly equal (the null hypothesis is true), the second estimate is

$$\text{MSB} = \frac{\text{SSB}}{K - 1} = \text{unbiased estimate of } \sigma^2 \qquad \textbf{(14–6)}$$

where: MSB = mean square between

SSB = sum of squares between

K = number of groups

Thus, **if the null hypothesis is true (that is, all means are equal), MSW and MSB are both estimates of the population variance, σ^2, and we would expect MSW and MSB to be nearly equal.**

However, because of the way SSB is calculated, the more the sample means differ, the *larger* SSB becomes. As SSB is increased, MSB begins to differ from MSW. When this difference gets "too large," our conclusion will be that the population means are not all equal. Therefore we would reject the null hypothesis. But how do we determine what "too large" is?

Recall the test of two population variances presented in Chapter 13. There, we stated that **the ratio of two unbiased estimates of the same variance, σ^2, forms an *F* distribution with D_1 and D_2 degrees of freedom.** If the population means are equal, MSW and MSB are both unbiased estimates of σ^2. Their ratio should produce a statistic that is *F*-distributed when H_0 is true.

$$F = \frac{\text{MSB}}{\text{MSW}} \qquad (14\text{–}7)$$

The calculated *F* comes from an *F* distribution with $D_1 = K - 1$ and $D_2 = N - K$ degrees of freedom if H_0 is true.

A common means of illustrating analysis of variance calculations is in table format. Pam Burke has formulated Table 14–2. Notice that she has listed the null and alternative hypotheses, and has shown the *F* distribution and the decision rule for $\alpha = 0.05$.

The ratio of MSB to MSW for this example is 1.628. This value is smaller than the critical *F* value of approximately 2.92 found in the *F*-distribution

TABLE 14–2 ANOVA, Ambell, Inc.

Hypotheses:

$H_0: \mu_1 = \mu_2 = \mu_3 = \mu_4$
H_A: Not all means are equal.

$\alpha = 0.05$

Source of Variation	SS	d.f.	MS	F Ratio
Between samples	22	$K - 1 = 3$	MSB $= 7.33$	MSB/MSW $= 1.628$
Within samples	126	$N - K = 28$	MSW $= 4.5$	
Total	148	$N - 1 = 31$		

d. f. $= D_1 = 3, D_2 = 28$

Acceptance Region

$\alpha = 0.05$

$CV \approx 2.92$

$F = 1.628$

Decision Rule:

If $F > F_{\text{critical}} \approx 2.92$, reject H_0.
Otherwise, do not reject H_0.
Since $1.628 < 2.92$, do not reject H_0.

table in Appendix F. Note that we have used degrees of freedom $D_1 = 3$ and $D_2 = 30$ to obtain the critical F value. The exact F value could be found from an F table that contains entries for degrees of freedom $D_1 = 3$ and $D_2 = 28$. Since the calculated F value is smaller than the critical F value, Pam should *not* reject the null hypothesis. She would conclude that the average sales per salesperson in the four regions could, in fact, be equal. There is no justification for individual treatment in any one region based upon these results. The president's concern seems unfounded.

The analysis of variance has allowed Pam to test a null hypothesis involving four population means without "compounding" alpha above the 0.05 level.

One-Way Analysis of Variance—A Second Example

A city in Arizona has recently received a federal grant to purchase two new fire trucks and hire six additional firefighters. The city manager, in compliance with the grant, will assign the trucks and firefighters to the two busiest fire districts in the city. However the city manager doesn't know which districts are the busiest or, for that matter, if there is a difference between the city's three districts.

TABLE 14–3 Response data by fire district
(no. responses)

Day	District 1	District 2	District 3
1	3	4	2
2	2	4	2
3	1	4	1
4	0	4	0
5	0	3	0
6	0	5	0
7	2	4	3
8	2	4	0
9	1	4	2
10	4	4	0
Total	15	40	10

$$\overline{X}_1 = 1.5 \qquad \overline{X}_2 = 4 \qquad \overline{X}_3 = 1$$

$$\text{TSS} = \sum_{i=1}^{K} \sum_{j=1}^{n_i} (X_{ij} - \overline{\overline{X}})^2 = (3 - 2.16)^2 + (2 - 2.16)^2 + \ldots + (0 - 2.16)^2$$

$$= 82.16$$

$$\text{SSB} = \sum_{i=1}^{K} n_i(\overline{X}_i - \overline{\overline{X}})^2 = 10(1.5 - 2.16)^2 + 10(4 - 2.16)^2 + 10(1 - 2.16)^2$$

$$= 51.76$$

$$\text{SSW} = \text{TSS} - \text{SSB}$$

$$= 30.40$$

The city manager develops the following null hypothesis:

H_0: There is no difference in the average number of responses per day between the three fire districts.

To test this hypothesis (and help determine which districts will get the firefighters and equipment), the city manager selected a random sample of ten days during the current year. He asked the district fire marshals to supply the number of responses made in their districts on each day. These data are shown in Table 14–3.

Assuming the populations are normally distributed with equal variances, analysis of variance can be used to test the hypothesis of equal means. Table 14–4 shows the null and alternative hypotheses, as well as the analysis of variance and decision rule for this example.

Based upon the analysis of variance shown in Table 14–4, the city manager should conclude that the average number of responses is not equal for the three districts.

TABLE 14–4 ANOVA, Arizona fire districts

Hypotheses:

H_0: $\mu_1 = \mu_2 = \mu_3$
H_A: Not all means are equal.
$\alpha = 0.01$

Source of Variation	SS	d.f.	MS	F Ratio
Between districts	51.76	2	25.88	23.10
Within districts	30.40	27	1.12	
Total	82.16	29		

Decision Rule:

If $F > F_{critical} = 5.43$, reject H_0.
Otherwise, do not reject H_0.

Since $F = 23.10 > 5.43$, reject H_0.

*The critical F value denoted CV is found by interpolating the F distribution for values of F between degrees of freedom $D_1 = 2$, $D_2 = 24$ and $D_1 = 2$, $D_2 = 30$.

Given this conclusion, the next logical step is to determine which means are different. Several statistical procedures exist to help the city manager accomplish this and decide which districts get the trucks and firefighters. We will introduce two of these techniques: **Tukey's method** and **Scheffé's method**.

14-2
TUKEY'S METHOD OF MULTIPLE COMPARISONS

Once the analysis of variance leads to rejecting the null hypothesis of equal means, decision makers need a method to determine which means are not equal. (Remember, t tests are not acceptable; they "compound" the probability of Type I errors.) One method, developed by John W. Tukey, can be used when the samples from the populations are the same size. *Tukey's method* involves establishing a *T range* in the following manner:

$$T \text{ range} = T\sqrt{\text{MSW}} \qquad\qquad \textbf{(14–8)}$$

where: $\quad T = \dfrac{1}{\sqrt{n}} q$

$\qquad q$ = value from the Studentized range table (Appendix G) given
$\qquad\qquad \alpha$ and $D_1 = K$ and $D_2 = N - K$ d.f.

$\qquad n$ = common sample size

If any pair of sample means has an **absolute difference**, $|\bar{X}_1 - \bar{X}_2|$, greater than the T range, we can conclude that the population means are not equal. For example, our Arizona city has three fire districts. The possible **pairwise** comparisons (formally referred to as **contrasts**) are

$$|\bar{X}_1 - \bar{X}_2| = |1.5 - 4.0| = 2.5$$
$$|\bar{X}_1 - \bar{X}_3| = |1.5 - 1.0| = 0.5$$
$$|\bar{X}_2 - \bar{X}_3| = |4.0 - 1.0| = 3.0$$

The calculated T range for this example with $\alpha = 0.05$ is

$$T \text{ range} = T\sqrt{\text{MSW}}$$

where

$$T = \frac{1}{\sqrt{n}} q_{(1-\alpha),\ D_1=K=3,\ D_2=N-K=27}$$

$$= \frac{1}{\sqrt{10}} 3.51 \qquad \text{[Note that } q \text{ is from the Studentized range table in Appendix G}$$
$$\text{(Value interpolated)]}$$

$$= 1.109$$

Then

$$T \text{ range} = 1.109\sqrt{1.12}$$
$$= 1.173$$

And thus

$$|\bar{X}_1 - \bar{X}_2| = |1.5 - 4.0| = 2.5 > 1.173$$
$$|\bar{X}_2 - \bar{X}_3| = |4.0 - 1.0| = 3.0 > 1.173$$
$$|\bar{X}_1 - \bar{X}_3| = |1.5 - 1.0| = 0.5 < 1.173$$

Therefore, based on Tukey's method, the city manager can conclude that the mean number of responses for district 1 is *not* equal to the mean for district 2, and that the mean for district 2 is *not* equal to the mean for district 3. However he doesn't have enough statistical evidence to conclude that the means of district 1 and district 3 are different.

By examining the sample means, the city manager can infer that district 2 has more responses on the average than district 1, and more than district 3. Thus district 2 should receive some new firefighters, and at least one truck. However, based on this study, the city manager has no statistical basis for assigning a truck or crew to a second district. He cannot conclude that districts 1 and 2 differ with respect to average responses. Some other factor(s) will have to be used to make the final decision.

Tukey's method of multiple comparisons allows decision makers to determine which population means are not equal once analysis of variance leads to rejecting the null hypothesis of equal population means. This method does not "compound" the alpha level, but only applies when the sample sizes are equal.

14-3
SCHEFFÉ'S METHOD OF MULTIPLE COMPARISONS

The main restriction of Tukey's method is that it requires equal sample sizes. In some controlled applications managers can guarantee equal sample sizes from each population. However, in many other applications equal sample sizes are not possible. When decision makers have unequal sample sizes, the analysis of variance equations still apply, as does the interpretation of results. But, since Tukey's method does not apply, decision makers need a method of multiple comparisons that allows unequal sample sizes. Henry Scheffé has developed such a procedure.

The Bend River Corporation assembles wristwatches for several watch manufacturers. The assembly process is completely manual. The company has become successful by steadily improving its productivity level. A large part of the improvement is due to the production supervisors, who are always looking for ways to improve the work flow.

Recently the supervisors identified five possible sequences by which new-model watches could be assembled. To determine which sequence, or sequences, are best, they selected a random sample of 50 workers. Ten workers were randomly assigned to each sequence. At the end of a two-week training period, the workers began assembling watches for 20 consecutive days using their assigned sequences. Table 14–5 shows the number of watches assembled by each worker for each sequence. As might be expected in a study of this kind, several workers missed days of work and had to be eliminated from the study. Therefore unequal sample sizes resulted, as shown in Table 14–5.

Table 14–6 presents the analysis of variance results, the null and alternative hypotheses, and the decision rule. The null hypothesis that the five assembly sequences are equal with respect to the average number of watches produced is

TABLE 14–5 Watch-assembly data, Bend River Corporation (no. watches assembled)

Sequence 1	Sequence 2	Sequence 3	Sequence 4	Sequence 5
47	86	65	58	49
52	85	63	62	51
46	81	67	63	47
49	90	64	64	55
53	84	59	59	54
39	88	67	55	42
51	77	70	70	53
	79	59	59	57
	82		60	59
			61	48

$n_1 = 7$ $n_2 = 9$ $n_3 = 8$ $n_4 = 10$ $n_5 = 10$ $N = 44$

$\bar{X}_1 = 48.14$ $\bar{X}_2 = 83.55$ $\bar{X}_3 = 64.25$ $\bar{X}_4 = 61.10$ $\bar{X}_5 = 51.50$ $\bar{\bar{X}} = 62.02$

$$\text{TSS} = \sum_{i=1}^{K} \sum_{j=1}^{n_i} (X_{ij} - \bar{\bar{X}})^2 = (47 - 62.02)^2 + (52 - 62.02)^2 + \ldots + (48 - 62.02)^2$$

$$= 7{,}446.97$$

$$\text{SSB} = \sum_{i=1}^{K} n_i (\bar{X}_i - \bar{\bar{X}})^2 = (48.14 - 62.02)^2 + \ldots + (51.50 - 62.02)^2$$

$$= 6{,}675.40$$

$$\text{SSW} = \text{TSS} - \text{SSB}$$

$$= 771.57$$

clearly rejected. The production supervisors should conclude that not all assembly sequences will produce equal average outputs.

Once the analysis of variance indicates that the null hypothesis should be rejected, **Scheffé's method of multiple comparisons** can determine which means are different. **Like Tukey's method, Scheffé's method produces a range to which the absolute differences in pairs of sample means (called *contrasts*) can be compared.** This range is

$$S \text{ range} = S\hat{\sigma}$$

where:
$$S = \sqrt{(K - 1)(F_{\alpha, D_1 = K-1, \ D_2 = N-k})}$$

$$\hat{\sigma} = \sqrt{\left(\frac{1}{n_i} + \frac{1}{n_j}\right)(\text{MSW})}$$

One contrast for an alpha level of 0.05 is

$$|\bar{X}_1 - \bar{X}_2| = |48.14 - 83.55| = 35.41$$

If $35.41 > S$ range, the production supervisors should conclude that sequences 1 and 2 differ significantly with respect to average output. In this case the S range is

$$S \text{ range} = S\hat{\sigma}$$

TABLE 14–6 ANOVA, Bend River Corporation

Hypotheses:

$H_0: \mu_1 = \mu_2 = \mu_3 = \mu_4 = \mu_5$
$H_A:$ Not all means are equal.

$\alpha = 0.05$

Source of Variation	SS	d.f.	MS	F Ratio
Between methods	6,675.40	4	1,668.85	84.35
Within methods	771.57	39	19.78	
Total	7,446.97	43		

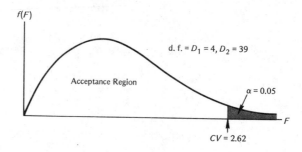

Decision Rule:

If $F > F_{\text{critical}} = 2.62$, reject H_0.
Otherwise, do not reject H_0.

Since $84.35 > 2.62$, reject H_0.

where

$$S = \sqrt{(K - 1)(F_{\alpha, D_1, D_2})} = \sqrt{4(2.62)}$$

$$= 3.23$$

$$\hat{\sigma} = \sqrt{\left(\frac{1}{n_1} + \frac{1}{n_2}\right)(\text{MSW})} = \sqrt{\left(\frac{1}{7} + \frac{1}{9}\right)(19.78)}$$

$$= 2.24$$

Therefore

$$S \text{ range} = 3.23(2.24)$$

$$= 7.235$$

Since $|\bar{X}_1 - \bar{X}_2| = 35.41 > 7.235$, the supervisors will conclude that the mean levels for these two sequences do differ. Based on the sample means, they infer that asembly sequence 2 will give significantly higher average output than sequence 1.

The Scheffé method can be applied to any or all possible pairwise contrasts. The S range will differ slightly from contrast to contrast if the sample sizes differ. Table 14–7 shows all possible pairwise contrasts and their associated S ranges for our watch-making example. The only contrasts for which we can conclude that

TABLE 14–7   Scheffé test, Bend River Corporation

Contrast	S Range	Significant?
$|\bar{X}_1 - \bar{X}_2| = |48.14 - 83.55| = 35.41$	7.235	yes
$|\bar{X}_1 - \bar{X}_3| = |48.14 - 64.25| = 16.11$	7.434	yes
$|\bar{X}_1 - \bar{X}_4| = |48.14 - 61.10| = 12.96$	7.079	yes
$|\bar{X}_1 - \bar{X}_5| = |48.14 - 51.50| = 3.36$	7.079	no
$|\bar{X}_2 - \bar{X}_3| = |83.55 - 64.25| = 19.30$	6.980	yes
$|\bar{X}_2 - \bar{X}_4| = |83.55 - 61.10| = 22.45$	6.600	yes
$|\bar{X}_2 - \bar{X}_5| = |83.55 - 51.50| = 32.05$	6.600	yes
$|\bar{X}_3 - \bar{X}_4| = |64.25 - 61.10| = 3.15$	6.814	no
$|\bar{X}_3 - \bar{X}_5| = |64.25 - 51.50| = 12.75$	6.814	yes
$|\bar{X}_4 - \bar{X}_5| = |61.10 - 51.50| = 9.60$	6.424	yes

the population means are not different are μ_1 and μ_5, and μ_3 and μ_4. Based on the sample means, the production supervisors infer that sequence 2 will produce a higher average output than the other four sequences.

14-4
OTHER ANALYSIS OF VARIANCE DESIGNS

In the previous sections we have introduced one-way analysis of variance and presented the rationale for using analysis of variance to test the null hypothesis that the means of two or more populations are equal. One-way analysis of variance has many applications. However it is just one of many analysis of variance designs available to decision makers. The purpose of this section is to discuss, in general terms, a few other commonly used analysis of variance designs. If one of these designs looks appropriate for a particular decision problem you should consult texts by Guenther (1964), Mendenhall (1968), and Neter and Wasserman (1974) for detailed explanations of the procedures.

Randomized Block Design

The *randomized block* analysis of variance design is an extension of the paired-sample *t* test discussed in Chapter 11. The main difference between the randomized block design and the one-way design is the way the experimental units are assigned to the populations. For example, suppose a commercial fertilizer manufacturer is concerned with the relative effectiveness of four different brands. If the manufacturer were to randomly select land parcels to test the four different fertilizer brands, the best land might always receive one brand, or the worst land might always receive a different brand. If this were to happen, any differences in the brands might be confused by the differences in land quality. In the randomized block design, the goal is to eliminate a variable (in this case, land quality) of no interest to allow better conclusions about the variable that is of interest. This is done by forming homogeneous blocks as shown in Table 14–8. As you can see, we make sure that each class (*block*) of land receives each type of fertilizer.

TABLE 14–8 Randomized block design

Land Quality	Fertilizer 1	Fertilizer 2	Fertilizer 3	Fertilizer 4
Excellent	X_{11}	X_{12}	X_{13}	X_{14}
Good	X_{21}	X_{22}	X_{23}	X_{24}
Fair	X_{31}	X_{32}	X_{33}	X_{34}
Poor	X_{41}	X_{42}	X_{43}	X_{44}

Latin Squares Design

Whereas the randomized block design is used to eliminate one source of variation, the *Latin squares* design can control two sources of variation. For example, suppose our fertilizer company wants to control for the amount of irrigation, as well as for the land quality, in comparing the four fertilizers. Say the company has identified four irrigation levels—heavy, medium, light, and dry—and four land-quality levels—excellent, good, fair, and poor. The goal is to test each brand of fertilizer under each combination of land and irrigation level one time. Table 14–9 illustrates a possible Latin squares design for the fertilizer example. Note that the ordering shown in Table 14–9 is only one of several possible designs. The main concern is that the rows, columns, and treatments are randomly assigned. The advantage of the Latin squares design over the randomized block design is that it reduces the required sample size.

TABLE 14–9 Latin squares design

Land Quality		Irrigation Level			
		Heavy	Medium	Light	Dry
	Excellent	F_1	F_4	F_2	F_3
Land Quality	Good	F_2	F_3	F_4	F_1
	Fair	F_4	F_1	F_3	F_2
	Poor	F_3	F_2	F_1	F_4

where: F_1 = fertilizer brand 1
F_2 = fertilizer brand 2
F_3 = fertilizer brand 3   } Treatments
F_4 = fertilizer brand 4

Factorial Design

The variable in which the manager is interested is commonly called a *factor.* The one-way, randomized block, and Latin squares designs can all be applied to only one factor. However decision makers often are interested in looking at two or more factors at the same time. For example, the manager of a plywood mill might be interested in studying how glue type affects plywood strength, and at the same time how wood type affects plywood strength.

A *factorial* analysis of variance design is used when two or more factors (here, glue and wood type) are considered simultaneously. The mill manager may hypothesize that the average plywood strength is equal regardless of which of three types of glue is selected. At the same time, she may hypothesize that the average plywood strength is equal regardless of which of four types of wood is used. The mill manager can also use the factorial design to determine whether there is any interaction between factors. For example, do certain wood types work better with certain glues?

Table 14–10 shows how the factorial analysis of variance design could be set up for the plywood example. The advantage of performing the two-factor study rather than two one-factor studies is that a smaller sample size is required to reach the same level of statistical significance.

TABLE 14–10 Factorial ANOVA, plywood mill example

		Glue		
		1	2	3
Wood	Pine	X_{111} X_{112} X_{113} . . . X_{11n}	X_{121} X_{122} X_{123} . . . X_{12n}	X_{131} X_{132} X_{133} . . . X_{13n}
	Fir	X_{211} . . . X_{21n}	X_{221} . . . X_{22n}	X_{231} . . . X_{23n}
	Birch	X_{311} . . . X_{31n}	X_{321} . . . X_{32n}	X_{331} . . . X_{33n}
	Larch	X_{411} . . . X_{41n}	X_{421} . . . X_{42n}	X_{431} . . . X_{43n}

where: X_{111} = first strength measure for pine wood and glue 1

14-5
CONCLUSIONS

Managers are often faced with comparing alternatives involving more than two populations. Because hypothesis testing is such a powerful decision-making tool, managers need to be able to extend the hypothesis-testing procedure to more than two populations. Analysis of variance allows this extension.

By using analysis of variance, managers are able to determine whether the sample observations come from the same population or from different populations. Analysis of variance allows managers to separate observations into categories and see whether this separation explains some of the variation in the sample observations. The ability to test for significant relationships between sample observations falling into different categories makes analysis of variance a powerful decision-making tool.

chapter glossary

between-sample variation (SSB)   The variation that exists between categories of observations. This variation exists due to partitioning the data into separate groups.

factorial design   A method of classifying interval (or ratio) data into different categories defined by two or more cardinal or ordinal variables.

one-way design   A method of classifying interval (or ratio) data into different categories of a single variable.

total sample variation (TSS)   The variation that exists in the observations before the data are partitioned into groups.

within-sample variation (SSW)   The variation in the observations that exists within each category. This variation exists after the data have been partitioned into separate groups.

chapter formulas

Total sum of squares

$$TSS = \sum_{i=1}^{K} \sum_{j=1}^{n_i} (X_{ij} - \overline{\overline{X}})^2$$

Sum of squares between

$$SSB = \sum_{i=1}^{K} n_i(\overline{X}_i - \overline{\overline{X}})^2$$

Sum of squares within

$$SSW = \sum_{i=1}^{K} \sum_{j=1}^{n_i} (X_{ij} - \overline{X}_i)^2$$

Mean square within

$$MSW = \frac{SSW}{N - K}$$

Mean square between

$$MSB = \frac{SSB}{K - 1}$$

F statistic for one-way analysis of variance calculation formulas

$$F = \frac{MSB}{MSW}$$

One-way analysis of variance

$$TSS = \sum_{i=1}^{K} \sum_{j=1}^{n_i} X_{ij}^2 - \frac{\left(\sum_{i=1}^{K} \sum_{j=1}^{n_i} X_{ij} \right)^2}{N}$$

$$SSB = \sum_{i=1}^{K} \frac{\left(\sum_{j=1}^{n_i} X_{ij} \right)^2}{n_i} - \frac{\left(\sum_{i=1}^{K} \sum_{j=1}^{n_i} X_{ij} \right)^2}{N}$$

$$SSW = TSS - SSB$$

solved problems

1. The plant manager of a small Midwest fishing-equipment manufacturing plant is convinced that he can come up with a method of determining why his workers are sometimes very accurate and at other times very careless. Although he hasn't had much luck up to this point in determining causes, he has recently completed a book dealing with bio-rhythms. This seems to be the final explanation, so he has bio-rhythm charts prepared for all assembly employees. He divides the employees into four groups according to their positions on the chart: high, low, going up, and going down. He monitors the "error per hundred" rate for each employee and finds the following results:

High	Low	Going Up	Going Down
2.3	3.2	3.6	1.7
3.1	1.8	2.7	2.5
1.9	2.2	2.9	2.0
2.7	2.5	3.0	1.8
2.1	1.9	3.8	2.1
3.3	2.0	3.4	2.6

a. Formulate the null hypothesis and decision rule necessary to test this situation if you want to control Type I error probability at the 1 percent level.
b. How many degrees of freedom are associated with the between-group estimate? How do you determine this value?
c. How many degrees of freedom are associated with the within-group estimate?
d. Set up the appropriate analysis of variance table and formulate a conclusion about the possible relationship between error rate and bio-rhythm level.
e. What is meant by MSW and MSB? What do they estimate?

Solutions:

a. Hypotheses:

H_0: $\mu_{high} = \mu_{low} = \mu_{up} = \mu_{down}$
H_A: Not all means are equal.

Decision Rule:

If $F > F_{critical}$, reject H_0.
If $F \leq F_{critical}$, accept H_0.

where $F_{critical} = 4.94$ (from Appendix F).

b. The degrees of freedom for the between-group estimate equal the number of groups minus 1. In this problem,

$$d.f. = K - 1 = 4 - 1$$
$$= 3$$

c. For the within-group estimate, the degrees of freedom equal the total sample size minus the number of groups. Here,

$$d.f. = N - K = 24 - 4$$
$$= 20$$

d.

Source of Variation	SS	d.f.	MS	F Ratio
Between groups	4.411	3	1.470	6.504
Within groups	4.528	20	0.226	
Total	8.939	23		

Since $6.504 > F_{critical} = 4.94$, the null hypothesis is rejected.

e. MSW is the population variance estimate based on the variation found within the groups after partitioning. MSB is the population variance estimate based on the average values found for each group.

2. An equipment rental firm is trying out three new types of grease in the transmissions of its rental front-end loaders. The maintenance manager is interested in whether any of the greases reduces the time before the transmissions have to be repaired. The greases were randomly distributed among a set of new loaders; however the numbers with each type of grease are not equal. The following data show the hours of use until repair for each front-end loader:

Grease 1	Grease 2	Grease 3
314	401	426
423	307	377
298	267	450
267	217	479
298		503
		523

a. Formulate the appropriate null hypothesis and decision rule to test this situation. Assume $\alpha = 0.01$.

b. Construct the analysis of variance table for this situation and either accept or reject your hypothesis.

c. If you conclude one or more greases are associated with a longer average time to repair, determine which greases result in different means.

Solutions:

a. Hypotheses:

H_0: $\mu_1 = \mu_2 = \mu_3$
H_A: Not all means are equal.

Decision Rule:

If $F \leq F_{critical}$, accept H_0.
If $F > F_{critical}$, reject H_0.

where $F_{critical} = 6.93$ (from Appendix F).

b.

Source of Variation	SS	d.f.	MS	F Ratio
Between groups	81,477	2	40,738.50	10.41
Within groups	46,957	12	3,913.08	
Total	128,434	14		

Since $F = 10.41 > 6.93$, the null hypothesis should be rejected.

c. Since the number of observations differ between groups, we have to use Scheffé's method, where

$$S \text{ range} = S\hat{\sigma}$$
$$S = \sqrt{(K - 1)F_{critical}}$$
$$\hat{\sigma} = \sqrt{\left(\frac{1}{n_1} + \frac{1}{n_2}\right)(MSW)}$$

We have

$$F_{critical} = 6.93$$

so

$$S = \sqrt{(3 - 1)(6.93)} = 3.723$$

For groups 1 and 2:

$$\hat{\sigma} = \sqrt{\left(\frac{1}{5} + \frac{1}{4}\right)(3,913.08)} = 41.96$$
$$S \text{ range} = 3.723(41.96) = 156.23$$

For groups 1 and 3:

$$\hat{\sigma} = \sqrt{\left(\frac{1}{5} + \frac{1}{6}\right)(3{,}913.08)} = 37.88$$

S range $= 3.723(37.88) = 141.02$

For groups 2 and 3:

$$\hat{\sigma} = \sqrt{\left(\frac{1}{4} + \frac{1}{6}\right)(3{,}913.08)} = 40.38$$

S range $= 3.723(40.38) = 150.33$

Also, $\bar{X}_1 = 320$, $\bar{X}_2 = 298$, and $\bar{X}_3 = 459.667$.

Contrasts			S Range	Significant?
$\lvert \bar{X}_1 - \bar{X}_2 \rvert = \lvert 320 - 298$	$\rvert =$	22	156.23	no
$\lvert \bar{X}_1 - \bar{X}_3 \rvert = \lvert 320 - 459.667 \rvert =$		139.667	141.02	no
$\lvert \bar{X}_2 - \bar{X}_3 \rvert = \lvert 298 - 459.667 \rvert =$		161.667	150.33	yes

problems

Problems marked with an asterisk (*) require extra effort.

1. Discuss in your own words each of the following:
   a. Within-group variation
   b. Between-group variation
   c. Total sum of squares
   d. Degrees of freedom

2. You are responsible for installing emergency lighting in a series of state office buildings. You have bids from four manufacturers of battery-operated emergency lights. The costs are about equal, so you decide to base your buy decision on which type lasts the longest. You receive a sample of four lights from each manufacturer, turn on the lights, and record the time until each light fails. You find the following values:

Type A	Type B	Type C	Type D
24 h	27 h	21 h	23 h
21	25	20	23
25	26	25	21
22	22	22	21

   Based on this evidence, can you conclude that the mean times to failure of the four types are equal? You will have to specify an alpha level.

3. A large metropolitan police force is considering changing from full-size cars to intermediates. The force purchases 5 cars from each of three manufacturers.

Each car is run for 5,000 miles, and the operating cost per mile computed. Unfortunately 1 car is involved in an accident, 1 is run into a river during a high-speed chase, and 1 is "lost" due to a paperwork mix-up. The operating costs for the remaining 12 cars are distributed as follows:

Car A	Car B	Car C
13.3¢/mi	12.4¢/mi	13.9¢/mi
14.3	13.4	15.5
13.6	13.1	14.7
12.8		14.5
14.0		

a. Perform an analysis of variance on these data. Assume $\alpha = 0.01$.

b. Do the experimental data provide evidence that the operating costs per mile for the three types of police cars are different?

c. Discuss the advantages associated with having all sample sizes equal when using analysis of variance.

4. The police department discussed in problem 3 is presently using three different makes of police cars (primarily to spread governmental purchases among different car dealers). The department is interested in determining whether one of four possible brands of gas gets better mileage and whether any interaction exists between brand of gas and type of car. Discuss how to formulate an experiment to test these questions.

5. A nationwide moving company is considering three different types of nylon tie-down straps. The purchasing department randomly selects straps of each type and determines their breaking strengths. The following values are found:

Type 1	Type 2	Type 3
1,950 lb	2,210 lb	1,820 lb
1,870	2,300	1,730
1,900	1,990	1,760
1,880	2,190	1,700
2,010	2,250	1,810

a. Construct the analysis of variance table for this set of data.

b. Based on your analysis, with a Type I error of 0.05, can you conclude that a difference exists between the types of nylon ropes?

*6. A leading manufacturer of beer is considering five different types of advertising displays for a new low-calorie beer. The displays are each tested in 5 different randomly selected stores. A total of 25 stores are in the sample. The average monthly sales and variances for the first three months for each type of display are as follows:

Display Type	Mean	Variance
A	98 cases	10
B	77	8
C	84	8
D	103	11
E	91	9

Can the manufacturer conclude that it really doesn't matter which type of display is used? Assume an alpha of 0.01.

7. Referring to problem 6, the test performed was one-way analysis of variance. Suppose the manufacturer is concerned that store size might be an additional important factor in determining beer sales. Illustrate an appropriate experimental design to allow this manufacturer to investigate the significance of both factors.

8. Referring to problem 3, suppose the alpha level had been set at 0.05 rather than 0.01. What would the conclusion have been regarding the null hypothesis that the three car makers have equal average operating costs?

*9. Given that your conclusion in problem 8 was that the average operating costs differ between the three car types, between what pair or pairs of car types does the significant difference exist?

*10. Referring to problem 6, examine all possible pairs of beer brands and determine which pairs have means that are significantly different at the alpha = 0.01 level.

*11. Referring to problem 5, perform the Tukey procedure to see which pairs are different.

*12. Is it possible that Tukey's procedure will not detect a difference when, in fact, the analysis of variance led to rejecting the null hypothesis of equal means? Why would this occur? Why did this not occur in this case?

13. Channel 9 Television in Bextfort, Washington, recently conducted a study of television news viewers. One item of interest to Channel 9 management was whether the average age of viewers watching Channel 9 was the same as for the other two stations in Bextfort. The sample included 24 viewers of each station, with the following results: TSS = 2,900 and SSW = 700.

Using an alpha level of 0.05, what should Channel 9 conclude about the average ages of news viewers of the three stations? Why?

14. Referring to problem 13, develop the T range and indicate how it can be used to determine which pair or pairs of means are significantly different.

15. Referring to problem 14, develop the S range for each pairwise contrast.

16. Referring to problems 13–15, indicate whether the T range or S range should be used to make the pairwise comparisons. Why?

*17. Examine the T range and the S range you determined in problems 13 and 14 and discuss why, when the sample sizes are equal, the T range is preferred over the S range.

18. Discuss in your own words why decision makers should use analysis of variance rather than multiple two-sample t tests when testing hypotheses involving more than two populations.

19. Ajax Mountain Ski Company operates a small snow-skiing operation with two chair lifts. Recently some customers have complained that the lines at chair 1 are too long. The Ajax manager has collected the following data, which represent the number of people in line at the two chair lifts at randomly observed times during a week:

Chair 1	Chair 2
10	14
3	13
14	19
7	7
19	11
33	9
28	12
11	13
26	15

Use a two-sample t test to determine whether there is a significant difference in the average numbers of people waiting at the two lifts. Test this at the alpha = 0.05 level.

20. Using the information contained in problem 19, use analysis of variance to test the null hypothesis that the average numbers waiting at the two chair lifts are equal.

*21. Referring to problems 19 and 20, what observations can you make about the relationship between the two-sample t test and two-sample analysis of variance?

22. The Savouy Corporation recently purchased a bicycle-manufacturing plant formerly owned by the American Traveling Company. American had been outfitting its bikes with tires produced by the Leach Corporation. Savouy management is considering whether to stay with Leach tires or to change to another brand. Three other brands are being considered, all of which cost about the same as the Leach tire. The criterion for tire selection will be average tread life.

Samples of 20 have been selected from the Leach tires and from brands A, B, and C. The following results were found:

$$\bar{X}_{Leach} = 111\,h \qquad \bar{X}_A = 126\,h \qquad \bar{X}_B = 100\,h \qquad \bar{X}_C = 105\,h$$

$$TSS = 19,620$$

(Note that $\bar{X}$ indicates the hours of use until the tread was reduced to a specified level.)

Using an alpha of 0.05, test that there is no difference in average tread life for the four brands.

23. Referring to problem 22, use the appropriate method of multiple comparisons to determine which pairs of means are significantly different. Based upon your analysis, make a recommendation as to which brand of tire Savouy should use. Let $\alpha = 0.05$.

24. A ski resort in Idaho has been charged with discriminating against some nationalities of ski instructors in the amount they are allowed to earn from private lessons. Three nationalities teach at this resort: Canadians, Austrians, and Germans. Random samples of six Canadians, eight Austrians, and seven Germans were selected. The following statistics were recorded:

$$\bar{X}_{Canadian} = \$3,111 \qquad \bar{X}_{Austrian} = \$2,005 \qquad \bar{X}_{German} = \$3,511$$
$$TSS = 18,328,128$$

Based upon these values, what do you conclude about average salaries for ski instructors at this ski resort? Use $\alpha = 0.05$.

25. Referring to problem 24, apply the appropriate method of multiple comparisons to determine between which pairs of means significant differences exist. Use $\alpha = 0.05$.

cases

14A   CONSUMER INFORMATION ASSOCIATION

Yolanda Carson is a newly hired research assistant for the Consumer Information Association. The association is a nonprofit group whose major purpose is to supply information necessary to help consumers make better, informed decisions. Yolanda has been assigned to work with the group studying consumer practices in the banking industry.

Yolanda is aware of studies that seem to indicate that services provided by banks and the interest banks charge for loans are related to demographic factors such as the size of the city in which the banks are located and whether the state allows branch banks. She has been asked to determine whether there is a difference in consumer loan charges between major sections of the country.

Yolanda has been assured the cooperation of the American Banking Institute and has been given access to any data the institute has. However she knows that loan charges may depend on many factors and feels compelled to study banks firsthand. In particular, she has decided to randomly select banks in all parts of the country and apply for an automobile loan at each bank selected. She has decided to make the test during two time periods six months apart.

Since consumer interest rates have been changing rapidly lately, Yolanda has recorded all rates in terms of "prime plus ____%." (The prime rate is the rate large banks charge their largest corporate customers.) In the first test, Yolanda found the following values charged:

Northeast	Southeast	Midwest	West
prime + 3.2%	prime + 2.7%	prime + 3.4%	prime + 3.7%
2.9	2.9	3.5	3.6
2.8	3.0	2.9	3.6
3.5	2.9	3.7	3.9
3.4	2.8	3.4	4.0
4.0	2.5	3.5	3.8
3.2	2.7	3.0	3.4
	2.9		3.8

The executive director of the Consumer Information Association is going to be holding a news conference in a few days to discuss the work the organization has been doing. He would like to be able to cite Yolanda's study as an example of its services.

14B   SINGLEAF DEPARTMENT STORE, PART 1

The Singleaf Department Store operates in several large eastern cities. Singleaf was one of the first department stores to offer customers its own credit card, and since then the profits on its credit operations have approached those of its sales operations.

Since credit sales are such an important part of operations, the Singleaf board of directors monitors the credit area very closely. In particular, the board demands a quarterly report on the "age" of the account receivables. ("Aging" accounts is a common practice. The accounts are arranged into a distribution according to the time since a payment has been made to the accounts.)

The board has noticed a trend toward "older" accounts. Although outstanding balances mean extra interest on the accounts, they also increase the chance of having to write off the account as a bad debt. Beth Hansen, staff assistant to the president in charge of customer relations, has been considering methods to reduce the age of the accounts while at the same time not alienating the credit customers and losing future sales.

Many financial concerns turn old accounts over to a professional collection agency, but Singleaf does not. Singleaf's management feels the goodwill lost when a customer is called on by a professional bill collector may be more than the potential loss if the account is written off. While Singleaf makes an effort to collect on delinquent accounts, this effort has been handled by each of its stores individually.

Beth Hansen has been asked to supervise a study of collecting on delinquent accounts. She has decided to start by determining if the type of collection letter sent to the customer makes a difference in the response received. She has devised the following letter system:

1. Reminder letter
2. Reminder letter with return envelope included
3. "Tough" letter
4. "Tough" letter with return envelope included

Beth has had 40 letters (10 of each type) sent to randomly selected credit customers with "old" accounts of approximately the same balance. The following amounts of money were collected from these customers in the two weeks following the date the collection letter was mailed:

	Letter Type		
1	2	3	4
$45	$48	$56	$59
0	0	18	28
39	43	58	63
0	0	0	15
33	44	63	57
0	12	0	9
35	36	11	72
5	0	0	41
41	44	0	73
0	0	20	0

The president is anxious to report the findings to the board, along with any conclusions that can be drawn.

14C   AMERICAN TESTING SERVICES

When P.T. Miller formed American Testing Services, he anticipated that there was a market for consulting services in the market-research area. By the amount of work that his firm has had during the past three years, he is sure he was correct.

Recently P.T. was approached by the Convestal Corporation to perform an analysis of the five running shoes Convestal makes. The basic research question is whether people who own Convestal running shoes share an equal opinion (on the average) of the shoes without regard to the particular type or style of Convestal shoe they own. If not, then which styles seem to be most liked, and which seem to be most disliked, by the customers?

P.T. suggested that Convestal ask a random sample of shoe customers to rate their shoes on a scale of 1 to 100, with 100 being the best. The following data represent the results of the sampling:

Style 1	Style 2	Style 3	Style 4	Style 5
84	67	78	60	88
54	56	79	64	79
88	70	89	60	84
90	67	84	72	78
88	59	78	77	68
76	66	84	70	70
	70	90	66	80
	56	85	70	80
	70		59	78
	67		70	84
				75

What should P.T. Miller conclude?

references

GUENTHER, W. C. *Analysis of Variance.* Englewood Cliffs, N.J.: Prentice-Hall, 1964.

MENDENHALL, W. *An Introduction to Linear Models and the Design of Experiments.* Belmont, Calif.: Wadsworth, 1968.

NETER, JOHN and WILLIAM WASSERMAN. *Applied Linear Statistical Models.* Homewood, Ill.: Irwin, 1974.

15

Hypothesis Testing Using Nonparametric Tests

================why decision makers need to know

In the previous three chapters we presented an introduction to hypothesis testing. We discussed tests for one sample, two samples, and more than two samples. We also constructed tests for large and small samples. However all of these tests have a common bond: they all assume that the data are at least interval-scaled. Additionally, for the *t* test and analysis of variance, the population distribution(s) are assumed to be normal.

For many applications, the data are at least interval-scaled, and the populations can be assumed normally distributed. Production and financial situations often correspond to these assumptions. However there are many other applications where these conditions do not hold, for instance in many marketing and personnel situations. For situations where the assumptions do not hold, decision makers still need a set of statistical techniques to assist in decision making. The techniques that can be employed without the strict data and distribution assumptions are called ***nonparametric statistics***.

chapter objectives

In this chapter we shall introduce several useful nonparametric statistical tests, including the chi-square goodness of fit test, the Mann-Whitney U test, the Kolmogorov-Smirnov two-sample test, contingency analysis, and the Kruskal-Wallis one-way analysis of variance. We will emphasize how each test can be applied to assist the decision-making process but will *not* develop these tests from a mathematical standpoint. However we *do* expect this chapter to show you how to recognize when a nonparametric test is required, as well as how to use several popular nonparametric tests.

student objectives

After studying the material in this chapter, you should be able to:

1. Apply the chi-square goodness of fit test and the Mann-Whitney U test in a decision application.
2. Apply the Kolmogorov-Smirnov two-sample test.
3. Use the chi-square test in a contingency analysis application.
4. Employ the Kruskal-Wallis one-way analysis of variance in a decision-making setting.
5. Determine when a nonparametric statistical test is required rather than one of the statistical tests presented in Chapters 12, 13, or 14.

15-1
CHI-SQUARE GOODNESS OF FIT TEST

Rebecca Sweetlittle is the managing partner of Sweetlittle and Associates Dry Cleaners. Sweetlittle is open six days a week and performs virtually every available dry-cleaning service for its customers. Rebecca has always had the policy that "service brings customers back." Because of this policy, she has always had 15 employees working each day the cleaners is open. Her assumption has been that the day of the week does *not* influence business volume. However a few customers have recently complained that on some days, the service seems slower than on other days.

Over the past 24 weeks, Rebecca collected the data shown in Table 15–1. The frequencies indicate the total number of customers served during each day of this 24-week period.

Under her assumption that the day of the week should make no difference in customer volume, Rebecca would expect an equal number of customers on each of the six days. In this case, since there were 9,792 customers in all, she would expect 1,632 customers each day. As seen in Table 15–1, the observed number of customers was not 1,632 on each day. However, Rebecca wonders if the difference between what she has observed and what she would expect is great enough to offset her assumption that the day of the week makes no difference in customer volume.

TABLE 15–1 Customer frequency by day of week, Sweetlittle and Associates Dry Cleaners

Day	Observed Frequency
Monday	1,525
Tuesday	1,711
Wednesday	1,655
Thursday	1,497
Friday	1,603
Saturday	1,801
Total	9,792

A statistical test known as the *chi-square goodness of fit test* has been developed to help answer this type of question. As the name implies, **the chi-square goodness of fit test measures how well observed data fit what would be expected under specified conditions.** Suppose Rebecca establishes the following null and alternative hypotheses:

H_0: The number of customers is evenly spread over the six working days.

H_A: The number of customers is not evenly spread over the six working days.

The chi-square goodness of fit test statistic is

$$\chi^2 = \Sigma \frac{(f_o - f_e)^2}{f_e} \tag{15–1}$$

where: f_o = observed frequency

f_e = expected frequency

The chi-square test statistic is distributed as a chi-square variable with $K - 1$ degrees of freedom, where K is the number of categories, or **cells,** specified in the null hypothesis. Figure 15–1 illustrates a chi-square probability distribution and shows the acceptance and rejection regions. The chi-square goodness of fit test is one-tailed, and the rejection region is determined by the alpha level selected.

As you can see from Figure 15–1, if the calculated χ^2 gets large, the null hypothesis should be rejected. This makes sense when we examine equation (15–1). **When the expected frequencies differ from the observed frequencies by a large amount, χ^2 will become large.**

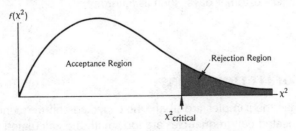

FIGURE 15–1 Chi-square probability distribution.

TABLE 15–2 Chi-square goodness of fit test, Sweetlittle and Associates Dry Cleaners

Hypotheses:

H_0: The number of customers is evenly spread over six working days.
H_A: The number of customers is not evenly spread over six working days.

$\alpha = 0.10$

Day	f_o	f_e	$(f_o - f_e)^2$	$(f_o - f_e)^2/f_e$
Monday	1,525	1,632	11,449	7.015
Tuesday	1,711	1,632	6,241	3.824
Wednesday	1,655	1,632	529	0.324
Thursday	1,497	1,632	18,225	11.167
Friday	1,603	1,632	841	0.515
Saturday	1,801	1,632	28,561	17.500
				$\chi^2 = 40.345$

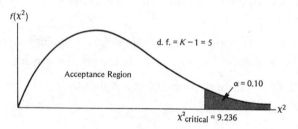

Decision Rule:

If $\chi^2 \leq \chi^2_{\text{critical}}$, accept H_0.
If $\chi^2 > \chi^2_{\text{critical}}$, reject H_0.

Since $40.345 > 9.236$, reject H_0. Conclude that the customers are not spread evenly over the six working days.

Table 15–2 presents the chi-square goodness of fit test for Sweetlittle. Calculated χ^2 is 40.345, and the critical value for five degrees of freedom and $\alpha = 0.10$ is 9.236 from the chi-square table in Appendix E. Based on these results, Rebecca Sweetlittle should conclude that her company's customer volume is not spread evenly over the six working days. She will no doubt want to explore the possibility of adding more employees on some days or shifting the work schedule to accommodate the heavier customer days, such as Saturday.

15-2
CHI-SQUARE GOODNESS OF FIT LIMITATIONS

In instances when the samples are small, the *expected* cell frequencies can be very small. If the expected cell frequencies are *too* small, the calculated χ^2 may be overstated, which can lead to rejecting the null hypothesis more often than the data justify. Two generally accepted rules of thumb can be used to decide whether the expected cell frequencies are too small:

1. **When the degrees of freedom equal 1 (cells = 2), the expected cell frequencies should be at least 5.0.**

2. **When the degrees of freedom exceed 1, at least 80 percent of the cells should have expected frequencies greater than 5.0, and all cells should have expected cell frequencies greater than 1.0.**

If neither of these two conditions is satisfied, the chi-square goodness of fit test should not be used, or the expected frequencies can be increased by combining cells. However, combining should only be done if the combined cells are meaningful. For example, the regional safety manager for the U.S. Soil Conservation Service has hypothesized that the number of workers having accidents would be spread evenly among workers' grade levels. Table 15–3 presents the accident data for the previous year. If the safety manager's claim is true, we would expect 4.5 accidents in each cell, or job grade. Since the expected frequency is below 5.0 in all cells, we should conclude either that the chi-square goodness of fit test should not be used or that some cells (job grades) must be combined to increase the expected cell frequencies. Table 15–4 illustrates the grouping performed by the safety manager, and also the statistical test. Now the expected cell frequencies are all greater than 5.0, and the chi-square goodness of fit test is applicable.

TABLE 15–3 Accident records, U.S. Soil Conservation Service

Job Grade	Observed Accidents
GS-9	3
GS-7	7
GS-5	4
GS-3	4
Total	18

As shown in Table 15–4, the safety manager, using this information, must conclude that accidents are spread evenly over the grade levels.

Other limitations of the chi-square goodness of fit test center around the requirement that the data be grouped in meaningful categories. As such, when this test is applied to continuous distributions, the results are only approximations and are sensitive to the manner in which the data have been grouped. Therefore two decision makers might apply the chi-square goodness of fit test to the same data, but because of a difference in the way they grouped the data, arrive at different conclusions.

TABLE 15–4   Chi-square goodness of fit test, Soil Conservation Service

Hypotheses:

H_0: The frequency of accidents is spread evenly between the GS-9, GS-7, and the combination of GS-5 and GS-3 employees.

H_A: The frequency of accidents is not spread evenly between the job grades.

$\alpha = 0.05$

Job Grade	f_o	f_e	$(f_o - f_e)^2$	$(f_o - f_e)^2/f_e$
GS-9	3	6	9	1.500
GS-7	7	6	1	0.167
GS-5 and below	8	6	4	0.677
				$\chi^2 = 2.344$

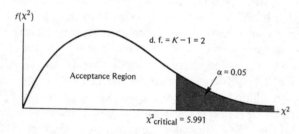

Decision Rule:

If $\chi^2 \leq \chi^2_{critical}$, accept H_0.

If $\chi^2 > \chi^2_{critical}$, reject H_0.

Since $\chi^2 = 2.344 < \chi^2_{critical} = 5.991$, do not reject H_0.

15-3
MANN-WHITNEY *U* TEST

When managers are concerned with determining whether two *independent* samples come from populations with the same distributions, and the data they have are of at least *ordinal* scale, the most commonly used nonparametric statistical procedure is the **Mann-Whitney U test.** This test is generally employed when the available data are in the form of **rankings,** or the decision makers do *not* want to make the assumption of normal populations as required by the two-sample *t* test discussed in Chapter 13.

The Ada County Highway District (ACHD) was recently formed to consolidate the street and highway maintenance and construction activities in Ada County. Formerly, the Urban division took care of all the streets and roads in the county's largest city, and the Rural division handled all street and road work outside that city. The Urban and Rural divisions had separate managements, and because of the perceived duplication of managerial activities, the consolidation was mandated by the Ada County commissioners. However the working force of the ACHD presently remains divided in the divisions, Rural and Urban.

A few months following the consolidation, several Rural division supervisors began claiming that the Urban division employees waste gravel from the county gravel pit. They claimed that the Urban division uses more gravel per mile of road maintenance than the Rural division.

In response to these claims, Dennis Millier, the ACHD materials manager, decided to perform a test. He selected a random sample of weeks from the district's job cost records from work performed by the Urban (U) division and a random sample of work performed by the Rural (R) division. The data in Table 15–5 represent the yards of gravel used per mile of road for each week sampled.

TABLE 15–5   Yards of gravel per mile, Ada County Highway District

Rural	Urban
460	600
830	652
720	603
930	594
500	1,402
620	1,111
703	902
407	700
1,521	827
900	490
750	904
800	1,400

The data are of a ratio-level measurement, but Dennis Millier is not willing to make the normality assumptions necessary to employ the two-sample *t* test. Thus he has decided to use the Mann-Whitney *U* test with following hypotheses:

$$H_0: \mu_U \leq \mu_R$$
$$H_A: \mu_U > \mu_R$$
$$\alpha = 0.05$$

The first step in testing this hypothesis using the Mann-Whitney *U* test is to combine the raw data from the two samples into one unit and assign the values rankings as shown in Table 15–6. **The logic of the Mann-Whitney *U* test centers around the idea that if the sum of the rankings of one group differs greatly from the sum of the rankings of the second group, we should conclude that there is a difference in central locations of the populations.**

We calculate a *U* value for each sample as follows:

$$U_R = n_1 n_2 + \frac{n_1(n_1 + 1)}{2} - \Sigma \text{ ranks}_1 \qquad (15\text{–}2)$$

$$U_U = n_1 n_2 + \frac{n_2(n_2 + 1)}{2} - \Sigma \text{ ranks}_2 \qquad (15\text{–}3)$$

TABLE 15–6   Ranking of yards of gravel per mile, Ada County Highway District

Rural $(n_1 = 12)$		Urban $(n_2 = 12)$	
	Rank		Rank
460	2	600	6
830	16	652	9
720	12	603	7
930	20	594	5
500	4	1,402	23
620	8	1,111	21
703	11	902	18
407	1	700	10
1,521	24	827	15
900	17	490	3
750	13	904	19
800	14	1,400	22
Σ ranks$_1$ = 142		Σ ranks$_2$ = 158	

Thus, for our example using the ranks in Table 15–6,

$$U_R = 12(12) + \frac{12(13)}{2} - 142$$

$$= 80$$

$$U_U = 12(12) + \frac{12(13)}{2} - 158$$

$$= 64$$

Note that $U_R + U_U = n_1 n_2$. This is always the case and provides a good check on the correctness of the rankings in Table 15–6.

     We select either U_R or U_U and call this U_{test}. When the samples are "large" (n_1 and n_2 both at least ten), the distribution of U_{test} will be approximately normally distributed, with

$$\text{Mean} = \frac{n_1 n_2}{2} \tag{15–4}$$

and

$$\text{Standard deviation} = \sqrt{\frac{n_1 n_2 (n_1 + n_2 + 1)}{12}} \tag{15–5}$$

Therefore the Mann-Whitney U test can utilize the normal distribution in a manner similar to the hypothesis tests presented in Chapters 12 and 13.

     If the hypothesis test is *one-tailed,* and the alternate hypothesis indicates the rejection region is in the *lower* tail of the distribution, U_{test} should be set equal to whichever of U_R or U_U will be *smaller* if the alternate hypothesis is true. Conversely, if the test is *one-tailed,* and the alternate hypothesis says **the rejection** region is in the *upper* tail of the distribution, U_{test} should be set equal to whichever of U_R or U_U will be *larger* if the alternate hypothesis is true. Finally, if the test is *two-tailed*, U_{test} can be set equal to *either* U_R or U_U.

The Z value is

$$Z = \frac{U_{\text{test}} - \dfrac{n_1 n_2}{2}}{\sqrt{\dfrac{n_1 n_2 (n_1 + n_2 + 1)}{12}}} \qquad (15\text{-}6)$$

Table 15-7 presents the test for the ACHD example. We see that the sample evidence does not lead to rejecting the null hypothesis, so Dennis Millier must conclude that the Urban division is no more wasteful than the Rural division in the use of gravel on construction jobs. However this test does not indicate whether or not both divisions are wasteful.

TABLE 15-7 Mann-Whitney U test, large samples

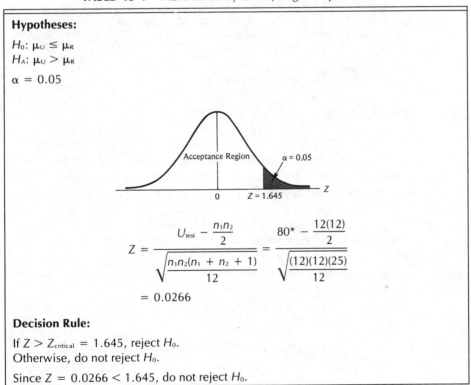

Hypotheses:

$H_0: \mu_U \le \mu_R$
$H_A: \mu_U > \mu_R$

$\alpha = 0.05$

Acceptance Region

$\alpha = 0.05$

0 $Z = 1.645$ Z

$$Z = \frac{U_{\text{test}} - \dfrac{n_1 n_2}{2}}{\sqrt{\dfrac{n_1 n_2 (n_1 + n_2 + 1)}{12}}} = \frac{80^* - \dfrac{12(12)}{2}}{\sqrt{\dfrac{(12)(12)(25)}{12}}}$$

$$= 0.0266$$

Decision Rule:

If $Z > Z_{\text{critical}} = 1.645$, reject H_0.
Otherwise, do not reject H_0.

Since $Z = 0.0266 < 1.645$, do not reject H_0.

*One-tailed upper tail; U_{test} is the larger of U_R and U_U.

The normal approximation for the Mann-Whitney U test improves as the sample sizes increase. For sample sizes smaller than ten, a small-sample Mann-Whitney U test must be employed. You should consult Conover (1971) or Marascuilo and McSweeney (1977) for the details of these tests. You should also consult these sources for a discussion of the procedures necessary for adjusting the Mann-Whitney U test when there are tied rankings in the data.

15-4

KOLMOGOROV-SMIRNOV TWO-SAMPLE TEST— LARGE SAMPLES

Foodway is a retail food chain with several hundred stores spread throughout the United States. Some top-level managers favor closing stores in towns with populations under 10,000. These managers propose placing all the company's efforts in the larger population centers. One of their arguments is that stores in small towns are not as productive as stores in larger cities.

Before making a decision, the board of directors decided to test the productivity argument by applying a store rating system developed by a regional sales manager. Basically, the rating system assigns each store a numerical value between 1 and 100 as a measure of many factors of productivity. The board randomly selected 50 stores from "small" cities and 50 stores from "large" cities. The distribution of store ratings is shown in Table 15–8.

TABLE 15–8 Store productivity ratings, Foodway

Productivity Rating		"Small" Frequency	"Large" Frequency
0–20	poor	2	3
21–30	fair	11	17
31–50	good	15	18
51–70	very good	19	8
Over 70	excellent	3	4
Total		50	50

The Foodway board of directors needs to determine whether there is, in fact, a difference in productivity ratings between the two groups of stores. The **Kolmogorov-Smirnov two-sample test** will test for this difference as long as the measurements are *at least ordinal* and the two samples are *independent*. Like the Mann-Whitney U test, the Kolmogorov-Smirnov two-sample test can be used for either one- or two-tailed tests. In the Foodway example, the board might make a one-tailed test by predicting that the stores in "large" cities will have significantly higher productivity ratings than stores in "small" cities. The one-tailed test statistic, when both sample sizes are greater than 40, is

$$\chi^2 = 4D^2 \frac{n_1 n_2}{n_1 + n_2} \quad \text{with 2 d.f.} \quad (15\text{–}7)$$

where: D = maximum difference between cumulative relative frequencies
= maximum $S_1(X) - S_2(X)$

$S_1(X)$ = cumulative relative frequency of the population hypothesized to have the largest values

$S_2(X)$ = cumulative relative frequency of the other group

n_1 = sample size of the first population

n_2 = sample size of the second population

The χ^2 calculated by using equation (15–7) is approximately chi-square-distributed with two degrees of freedom when the sample sizes are larger than 40.

The **Kolmogorov-Smirnov two-sample test deals exclusively with the maximum difference in cumulative relative frequency.** If the maximum difference, D, is *not* large, then the two groups (in this case, classes of stores) are assumed not to be significantly different. Table 15–9 illustrates how the Kolmogorov-Smirnov two-sample test is performed in the Foodway example.

TABLE 15–9 Kolmogorov-Smirnov two-sample test, one-tailed, Foodway

Hypotheses:

H_0: Stores in "large" and "small" cities do not differ with respect to productivity ratings.
H_A: Stores in "large" cities will have generally higher productivity scores than stores in "small" cities.

$\alpha = 0.05$

		Productivity Rating					
		0–20	21–30	31–50	51–70	Over 70	Total
	"Large"-city frequency	3	17	18	8	4	50
	Cumulative relative frequency $= S_1(X)$	3/50	20/50	38/50	46/50	50/50	
Store Location	"Small"-city frequency	2	11	15	19	3	50
	Cumulative relative frequency $= S_2(X)$	2/50	13/50	28/50	47/50	50/50	
	$S_1(X) - S_2(X)$	1/50	7/50	**10/50**	− 1/50	0	

$$\uparrow$$
$$D$$

Decision Rule:

If $\chi^2 \leq \chi^2_{critical}$, accept H_0.
If $\chi^2 > \chi^2_{critical}$, reject H_0.

Then

$$\chi^2 = 4D^2 \frac{n_1 n_2}{n_1 + n_2} = 4(0.2)^2 \frac{50(50)}{100}$$

$$= 4$$

Since $\chi^2 = 4 < \chi^2_{critical} = 5.991$, H_0 should *not* be rejected.

The maximum difference in cumulative frequency is 10/50, which occurs in the 31–50 rating level. However this difference is not large enough to allow the conclusion that stores in "large" cities have significantly higher productivity levels than stores in "small" cities at the alpha = 0.05 level. This information might well be used to support the continuation of stores in the smaller cities.

Two-Tailed Kolmogorov-Smirnov Test

If decision makers wish to test a two-tailed hypothesis using the Kolmogorov-Smirnov test, a slight modification in the procedure is required. Rather than using equation (15–7) as we did for the one-tailed test, we compare the maximum absolute difference in cumulative relative frequency to a critical value for the specified alpha level found in the table of D values in Appendix H. **If the calculated D is greater than the critical value, the decision makers can conclude that there is a significant difference between the two groups being compared.**

To demonstrate this, suppose Ace Electronics is concerned about what its customers think of its two regional service centers. The corporate service manager believes the two service centers have been performing equally well and is confident customer responses to a questionnaire will bear this out. She has asked 200 customers

TABLE 15–10   Kolmogorov-Smirnov two-sample test, two-tailed, Ace Electronics

Service Rating	Northern $S_1(X)$		Southern $S_2(X)$		Difference $S_1(X) - S_2(X)$
	Frequency	Cumulative Relative Frequency	Frequency	Cumulative Relative Frequency	
0–5	0	0/200 = 0	8	8/400 = 0.020	0.02
6–10	2	2/200 = 0.010	52	60/400 = 0.150	0.14
11–15	5	7/200 = 0.035	90	150/400 = 0.735	**D = 0.340**
16–20	93	100/200 = 0.500	180	330/400 = 0.825	0.325
21–25	42	142/200 = 0.710	50	380/400 = 0.950	0.240
26–30	58	200/200 = 1.000	20	400/400 = 1.000	0
	$\Sigma = 200$		$\Sigma = 400$		

$D = 0.340$

Hypotheses:

H_0: The ratings for the northern service center will be no different than the ratings for the southern service center.

H_A: There is a statistical difference in service ratings for the two service centers.

$\alpha = 0.05$

The critical value from Appendix H is

$$D_{critical} = 1.36 \sqrt{\frac{200 + 400}{80,000}}$$

$$= 0.117$$

Since $D = 0.340 > D_{critical} = 0.117$, we reject the null hypothesis and conclude that there is a difference in ratings between the two regions.

from the northern region to rate the Ace service center on a scale of 1 to 30, with 30 the highest rating. Likewise, she has asked 400 customers in the southern region to rate the southern service center. Table 15–10 illustrates the survey results along with the Kolmogorov-Smirnov two-sample test. This test clearly indicates that service ratings for the two regional service centers differ significantly. Based upon the direction of the difference, the corporate service manager should infer that customers in the northern region rate their service center higher than do customers in the southern region. The service manager might now try to find out why a difference exists between the two regions.

The Kolmogorov-Smirnov two-sample test can be used for both grouped and ungrouped data. However, if the data are grouped in an arbitrary manner, the choice of intervals can influence the test results. We recommend you use as many intervals as possible for the data.

15-5
CONTINGENCY ANALYSIS

Wilt Roderick, the loan manager at State Bank, is interested in developing a set of criteria that he and other loan officers can use in determining whether a customer should be granted an automobile loan. During the past two years, State Bank has made 400 automobile loans. Upon careful analysis of these loans, Wilt has classified them as follows:

Class	Frequency
Good loan (payments made on time)	300
Fair loan (payments made but consistently late)	60
Poor loan (repossession required)	40

For each loan, Wilt has data on a number of variables. One variable is whether the borrower is buying or renting a home. Wilt is interested in determining whether a relationship exists between the buy-rent variable and the loan class variable. **Contingency analysis,** also called the **chi-square test of independence,** offers a means by which the loan manager can test to see whether the two variables are statistically *independent.* We begin by developing a contingency table from the available data as follows:

		Rent	Buy	
	Good	140	160	300
Loan Class	Fair	20	40	60
	Poor	20	20	40
		180	220	400 = N

The row and column totals in the contingency table correspond to the number of individuals in each category. For example, 180 borrowers are renting their homes, and the remaining 220 are buying. The values inside the table represent the *joint* occurrence of two variables. For instance, 140 borrowers are renting and also have a good loan.

The null and alternative hypotheses are

H_0: The row and column variables are independent.

H_A: The row and column variables are *not* independent.

If the null hypothesis is true, the probability of a good loan for a person renting a house should equal the probability of good loan for a person buying. These two probabilities should also equal the probability of a good loan without considering the buy-rent variable.

To illustrate, we can find the probability of a good loan as follows:

$$P_{good} = \frac{\text{number of good loans}}{\text{number of loans granted}} = \frac{300}{400}$$

$$= 0.75$$

Then, if the null hypothesis is true,

$$P(good|rent) = 0.75$$

$$P(good|buy) = 0.75$$

Thus we would expect 75 percent of the 180 renters, or 135 people, to have good loans, and 75 percent of the 220 buyers, or 165 people, to have good loans. We can use this same reasoning to determine the expected number of borrowers in each cell in the contingency table.

		Rent	*Buy*
	Good	Actual = 140 Expected = 135	Actual = 160 Expected = 165
Loan Class	Fair	Actual = 20 Expected = 27	Actual = 40 Expected = 33
	Poor	Actual = 20 Expected = 18	Actual = 20 Expected = 22

If the null hypothesis of independence is true, we would expect to find the observed frequencies in the cells equal to the corresponding expected frequencies. **The greater the difference between the observed and expected, the more likely the null hypothesis of independence is false and should be rejected.** If this sounds to you like a chi-square goodness of fit problem, you are right. The appropriate test statistic is

$$\chi^2 = \sum_{i=1}^{r} \sum_{j=1}^{c} \frac{(f_{o_{ij}} - f_{e_{ij}})^2}{f_{e_{ij}}} \quad \text{with d.f.} = (r - 1)(c - 1) \qquad \textbf{(15–8)}$$

where: f_o = observed cell frequency

f_e = expected cell frequency

r = number of rows

c = number of columns

Don't be confused by the double summation in equation (15–8); it merely indicates that all rows and columns must be used in calculating x^2.

Table 15–11 presents the hypotheses and the test results. We see that the calculated x^2 value is less than the critical value from the chi-square probability table. Therefore Wilt Roderick *cannot* conclude, based on these data, that buying or

TABLE 15–11 Chi-square test of independence, State Bank

Hypotheses:

H_0: Loan class is independent of buying or renting.
H_A: Loan class is not independent of buying or renting.

$\alpha = 0.01$

		Rent	Buy
	Good	$f_o = 140$ $f_e = 135$	$f_o = 160$ $f_e = 165$
Loan Class	Fair	$f_o = 20$ $f_e = 27$	$f_o = 40$ $f_e = 33$
	Poor	$f_o = 20$ $f_e = 18$	$f_o = 20$ $f_e = 22$

$$x^2 = \sum_{i=1}^{r} \sum_{j=1}^{c} \frac{(f_{o_{ij}} - f_{e_{ij}})^2}{f_{e_{ij}}}$$

$$= \frac{(140 - 135)^2}{135} + \frac{(160 - 165)^2}{165} + \frac{(20 - 27)^2}{27} + \ldots + \frac{(20 - 22)^2}{22}$$

$$= 4.04$$

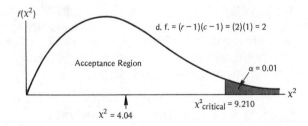

d. f. $= (r-1)(c-1) = (2)(1) = 2$

$x^2_{critical} = 9.210$

$x^2 = 4.04$

Decision Rule:

If $x^2 \leq x^2_{critical}$, accept H_0.
If $x^2 > x^2_{critical}$, reject H_0.

Since $x^2 = 4.04 < x^2_{critical} = 9.210$, we cannot reject the null hypothesis of independence.

renting makes a difference in the resulting loan classification, and he should not attempt to use this variable as a screening criterion.

Another variable for which Wilt has data is family income. The contingency table for income crossed with loan class is as follows:

		Income Level				
		$0–$10,000	$10,000–$20,000	$20,000–$30,000	Over $30,000	
	Good	12	43	87	158	300
Loan Class	Fair	18	20	18	4	60
	Poor	20	10	7	3	40
		50	73	112	165	400 = N

As is always the case with contingency analysis, the null hypothesis is that the rows and columns are independent. To test this hypothesis, we must compare the observed cell frequencies to the expected cell frequencies. A shortcut formula for determining the expected frequencies is

$$f_{e_{ij}} = \frac{\Sigma \text{ row } i \ \Sigma \text{ column } j}{\text{total no. of observations}}$$

For example,

$$f_{e_{11}} = \frac{300(50)}{400} = 37.50$$

$$f_{e_{12}} = \frac{300(73)}{400} = 54.75$$

The completed contingency table and the null hypothesis test for the State Bank example are shown in Table 15–12. Clearly, the hypothesis of independence between income level and loan classification must be rejected. Based upon the data, Wilt might infer that individuals with higher incomes tend to end up with loans in a higher classification.

Certainly, Wilt will want to explore other variables before arriving at his loan-screening criteria. When large numbers of data are involved, and when many cross-tabulations are required, computer programs can be of great assistance. One such program that we recommend for its ease of use and output format is SPSS (Statistical Package for the Social Sciences) [Nie et al. (1975)].

Limitations of Contingency Analysis

The chi-square test of independence requires that the expected frequencies in all cells be at least 5.0. To insure this, either the rows or columns may have to be combined, as was illustrated for the chi-square goodness of fit test. However, remember to make sure that there is a basis for grouping categories, and that the meaning of the results is not lost when the grouping is performed. **Note that the degrees of freedom are reduced when rows and/or columns are combined.**

TABLE 15–12 Chi-square test of independence, State Bank

Hypotheses:

H_0: Loan classification is independent of income level.
H_A: Loan classification is not independent of income level.

$\alpha = 0.01$

		Income Level			
		$0–$10,000	$10,000–$20,000	$20,000–$30,000	Over $30,000
	Good	$f_o = 12.00$ $f_e = 37.50$	$f_o = 43.00$ $f_e = 54.75$	$f_o = 87.00$ $f_e = 84.00$	$f_o = 158.00$ $f_e = 123.75$
Loan Class	Fair	$f_o = 18.00$ $f_e = 7.50$	$f_o = 20.00$ $f_e = 10.95$	$f_o = 18.00$ $f_e = 16.80$	$f_o = 4.00$ $f_e = 24.75$
	Poor	$f_o = 20.00$ $f_e = 5.00$	$f_o = 10.00$ $f_e = 7.30$	$f_o = 7.00$ $f_e = 11.20$	$f_o = 3.00$ $f_e = 16.50$

$$\chi^2 = \sum_{i=1}^{r} \sum_{j=1}^{c} \frac{(f_{o_{ij}} - f_{e_{ij}})^2}{f_{e_{ij}}} = \frac{(12.00 - 37.50)^2}{37.50} + \frac{(43.00 - 54.75)^2}{54.75} + \cdots + \frac{(3.00 - 16.50)^2}{16.50}$$

$$= 127.72$$

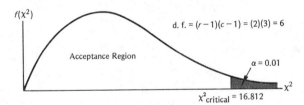

d. f. $= (r-1)(c-1) = (2)(3) = 6$

Acceptance Region

$\alpha = 0.01$

$\chi^2_{critical} = 16.812$

Decision Rule:

If $\chi^2 \leq \chi^2_{critical}$, accept H_0.
If $\chi^2 > \chi^2_{critical}$, reject H_0.

Since $\chi^2 = 127.72 > \chi^2_{critical} = 16.812$, we reject the hypothesis of independence and conclude that a relationship exists between the two variables.

15-6
KRUSKAL-WALLIS ONE-WAY ANALYSIS OF VARIANCE

We showed that the Kolmogorov-Smirnov two-sample test is a useful nonparametric procedure for determining whether two samples are from populations with the same characteristics. However, as we discussed in Chapter 14, many decisions involve comparing more than two populations. In that chapter we introduced one-way analysis of variance and showed how, if certain assumptions are satisfied, the F distribution can be used to test the hypothesis of equal population means. However, what if the decision makers are not willing to assume normally distributed populations? In that case, they must turn to a nonparametric procedure to compare the populations. *Kruskal-Wallis one-way analysis of variance* is the nonparametric counterpart to the analysis of variance procedure presented in Chapter 14. It is applicable any time the variable in question has a continuous distribution, the data are least ordinal, and the samples are independent.

The Bartholomew Company is considering acquiring a new computer system to handle its on-line data-processing activities, including inventory management, production scheduling, and general accounting and billing applications. Based upon cost and performance standards, Gladys Coil, Bartholomew's data-processing manager, has reduced the possible suppliers to three. One critical factor for Gladys is down time (how often the computer is nonoperational). When the computer goes down, the on-line applications are halted and normal business activities are interrupted. Gladys has received, from each supplier, a list of firms that are using the computer system Bartholomew is considering. From these lists, Gladys selected random samples of nine users of each computer system. In a telephone interview, she found the number of hours of down time in the previous month for each system. The down times are shown in Table 15–13.

TABLE 15–13 Computer down times, Bartholomew Company (hours/month)

System A	System B	System C
4.0	6.9	0.5
3.7	11.3	1.4
5.1	21.7	1.0
2.0	9.2	1.7
4.6	6.5	3.6
9.3	4.9	5.2
2.7	12.2	1.3
2.5	11.7	6.8
4.8	10.5	14.1

To use the Kruskal-Wallis analysis of variance here, we first replace each down-time measurement by its *relative ranking* within all groups combined. The smallest down time is given a rank of 1, the next smallest a rank of 2, and so forth, until all down times for the three systems have been replaced by their relative rankings. Table 15–14 shows these rankings for the 30 observations. Notice that the rankings are summed for each computer system. **The Kruskal-Wallis test will determine whether these sums are so different that it is not likely they came from populations with equal means.**

If the samples actually do come from populations with equal means (that is, the three systems have the same per-month average down time), then the H statistic calculated as follows will be distributed as a chi-square variable with $K - 1$ degrees of freedom, where K equals the number of samples under study:

$$H = \frac{12}{N(N + 1)} \sum_{i=1}^{K} \frac{R_i^2}{n_i} - 3(N + 1) \qquad \textbf{(15–9)}$$

TABLE 15–14 Rankings of computer down times, Bartholomew Company

System A	System B	System C
11	19	1
10	23	4
15	27	2
6	20	5
12	17	9
21	14	16
8	25	3
7	24	18
13	22	26
Σ ranks = 103	Σ ranks = 191	Σ ranks = 84

where: N = total of all observations

   K = number of samples

   R_i = sum of the ranks in the ith sample

   n_i = size of the ith sample

   If H is larger than χ^2_{critical}, the hypothesis of equal means is rejected, and Gladys would conclude that the populations from which the samples were selected have different means. Table 15–15 presents the hypotheses and statistical test for the Bartholomew example. The Kruskal-Wallis one-way analysis of variance shows that Gladys Coil should conclude that *not all* three computer systems have equal average down times. From this analysis, the supplier of computer system B would most likely be eliminated from consideration unless other factors such as price or service offset the apparent longer down times.

Limitations and Other Considerations

The Kruskal-Wallis one-way analysis of variance does *not* require the assumption of normality and is therefore often used instead of the analysis of variance technique discussed in Chapter 14. **However the Kruskal-Wallis test as discussed here applies only if the sample sizes from each population are at least five and the samples are independently selected.**

   When ranking observations, decision makers will sometimes encounter *ties.* When ties occur, each observation is given the mean rank for which it is tied. The H statistic is influenced by ties and should be corrected by dividing equation (15–9) by

$$1 - \frac{\sum\limits_{i=1}^{g} (t_i^3 - t_i)}{N^3 - N} \qquad \textbf{(15–10)}$$

where: g = number of different groups of ties

   t = number of tied observations in the tied group of scores

   N = total number of observations

TABLE 15–15 Kruskal-Wallis one-way analysis of variance test, Bartholomew Company

Hypotheses:

H_0: The mean down times are equal for all three computer systems.
H_A: The mean down times are not equal for all three computer systems.

$\alpha = 0.10$

Using the rankings and sums in Table 15–14,

$$H = \frac{12}{N(N + 1)} \sum_{i=1}^{K} \frac{R_i^2}{n_i} - 3(N + 1)$$

$$= \frac{12}{27(27 + 1)} \left[\frac{(103)^2}{9} + \frac{(191)^2}{9} + \frac{(84)^2}{9} \right] - 3(27 + 1)$$

$$= 11.49$$

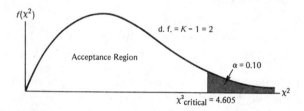

d. f. $= K - 1 = 2$

Acceptance Region

$\alpha = 0.10$

$\chi^2_{critical} = 4.605$

Decision Rule:

If $H \leq \chi^2_{critical}$, accept H_0.
If $H > \chi^2_{critical}$, reject H_0.

Since $H = 11.49 > \chi^2_{critical} = 4.605$, reject H_0.

Thus the correct formula for calculating the Kruskal-Wallis H statistic when ties are present is

$$H = \frac{\dfrac{12}{N(N + 1)} \sum_{i=1}^{K} \dfrac{R_i^2}{n_i} - 3(N + 1)}{1 - \dfrac{\sum_{i=1}^{g} (t_i^3 - t_i)}{N^3 - N}} \qquad (15\text{--}11)$$

Correcting for ties increases H and thus makes rejecting the null hypothesis more likely than if the correction is not used. **A rule of thumb is that if no more than 25 percent of the observations are involved in ties, the correction factor is not required.** Solved problem 3 in this chapter illustrates the use of the correction for ties for the Kruskal-Wallis test.

15-7
CONCLUSIONS

Many people make the mistake of learning one or two statistical techniques and then using these techniques in all situations. Surprisingly, some people even get emotional about being able to analyze a particular problem using their favorite technique. As future managerial decision makers, you cannot afford the luxury of defining a problem situation so you can apply your favorite technique. To you, statistics should be an aid to decision making, not an end in itself.

Many powerful statistical tools discussed in this book rest on the assumptions that the data being analyzed can be measured by at least an interval scale, and that the underlying populations being analyzed are normal. If these assumptions come close to being satisfied, many of the tools discussed prior to this chapter apply and are useful. However, in many practical situations these assumptions just do not apply, in which case the tools discussed in this chapter may be appropriate. In any case, nonparametric statistical tests should be part of every decision maker's "bag of tools."

chapter glossary

contingency table   A table used to classify sample observations according to two or more identifiable characteristics.

nonparametric statistics   Statistical techniques that do not depend on the population conforming to a predetermined distribution.

parametric statistics   Statistical techniques that require the underlying population distribution to conform to a predetermined distribution.

chapter formulas

Chi-square goodness of fit statistic

$$\chi^2 = \Sigma \frac{(f_o - f_e)^2}{f_e}$$

Kolmogorov-Smirnov test statistic

$$\chi^2 = 4D^2 \frac{n_1 n_2}{n_1 + n_2}$$

Kruskal-Wallis H statistic

$$H = \frac{12}{N(N + 1)} \sum_{i=1}^{K} \frac{R_i^2}{n_i} - 3(N + 1)$$

Kruskal-Wallis adjustment factor

If 25 percent of the ranked observation are ties, divide H by

$$1 - \frac{\sum_{i=1}^{g} (t_i^3 - t_i)}{N^3 - N}$$

Mann-Whitney U mean

$$\text{Mean} = \frac{n_1 n_2}{2}$$

Mann-Whitney U standard deviation (No ties)

$$\text{Standard deviation} = \sqrt{\frac{n_1 n_2 (n_1 + n_2 + 1)}{12}}$$

solved problems

1. Burtco Incorporated designs and manufactures gears for heavy-duty construction equipment. One such gear, #9973, has the following specifications:

    1. Mean diameter 3 inches
    2. Standard deviation 0.001 inch
    3. Output normally distributed around the mean

    The production control manager has selected a random sample of 500 gears from the inventory and found the following distribution:

Gear Diameter (inches)	Frequency
Under 2.995	3
2.995 and under 2.996	4
2.996 and under 2.997	5
2.997 and under 2.998	19
2.998 and under 2.999	98
2.999 and under 3.000	146
3.000 and under 3.001	124
3.001 and under 3.002	83
3.002 and under 3.003	11
3.003 and over	7
Total	500

    Based upon this sample information, does gear #9973 meet specifications? Use a significance level of 0.05.

Solution:

This is an example of a one-sample goodness of fit problem. We can hypothesize as follows:

H_0: Gear #9973 is normally distributed with mean diameter 3 inches and standard deviation 0.001 inch.

H_A: Gear #9973 is not within specifications.

$\alpha = 0.05$

An appropriate statistical procedure is the chi-square goodness of fit test. This test calculates the expected frequencies for each interval and compares the expected to the observed. If the calculated χ^2 gets too large, we will conclude that the fit is not good and reject the null hypothesis.

To determine the expected frequencies, we calculate the probability of a gear having a diameter in each of the intervals, assuming the gear meets the required specifications. Then we multiply the probability by 500 to obtain the expected frequency. The procedure is as follows:

$P(\text{less than } 2.995) = $ area under normal curve to the left of 2.995

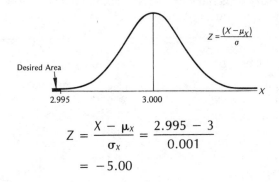

$$Z = \frac{X - \mu_x}{\sigma_x} = \frac{2.995 - 3}{0.001}$$

$$= -5.00$$

The area to the left of $Z = -5.00$ is essentially zero. Therefore the expected frequency is $(0)(500) = 0$.

As another example,

$P(2.997 \leq X \leq 2.998) = $ area under the normal curve between 2.997 and 2.998

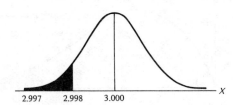

$$Z = \frac{2.997 - 3}{0.001} = -3; \text{ area to the right} = 0.4986$$

$$Z = \frac{2.998 - 3}{0.001} = -2; \text{ area to the right} = 0.4772$$

Therefore the desired probability is $0.4986 - 0.4772 = 0.0214$, and the expected frequency is $(0.0214)(500) = 10.7$.

In a like manner, we find the expected frequencies for each interval.

Gear Diameter (inches)	f_o	f_e
Under 2.995	3	0
2.995 and under 2.996	4	0.015
2.996 and under 2.997	5	0.635
2.997 and under 2.998	19	10.75
2.998 and under 2.999	98	67.95
2.999 and under 3.000	146	170.65
3.000 and under 3.001	124	170.65
3.001 and under 3.002	83	67.95
3.002 and under 3.003	11	10.75
3.003 and over	7	0.650

Because the chi-square goodness of fit test does not work well when more than 20 percent of the cells have expected cell frequencies below 5.0, as is the case in this example, we must combine the cells.

Gear Diameter (inches)	f_o	f_e	$(f_o - f_e)^2$	$(f_o - f_e)^2/f_e$
Under 2.998	31	11.4	384.16	33.70
2.998 and under 2.999	98	67.95	903.00	13.29
2.999 and under 3.000	146	170.65	607.62	3.56
3.000 and under 3.001	124	170.65	2,176.22	12.75
3.001 and under 3.002	83	67.95	226.50	3.33
3.002 and over	18	11.4	43.56	3.82
				$\chi^2 = 70.45$

$$\chi^2 = \sum_{i=1}^{K} \frac{(f_{oi} - f_{ei})^2}{f_{ei}}$$

$$= 70.45$$

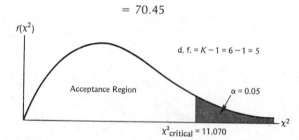

Since $\chi^2 = 70.45 > \chi^2_{critical} = 11.070$, we reject the null hypothesis and conclude that the gears are *not* within specifications. The production manager would most likely want to halt production and attempt to adjust the process to bring the gears within specifications. However, as often happens, he may discover that the specifications are unreachable with the machinery he has to work with, in which case either new machinery will be required or the specifications will have to be renegotiated.

2. Automobile insurance companies have for a number of years used age, sex, and marital status for determining rates. For example, single males under 25 are considered the highest risk and are charged the highest premiums. Single females under 25 are charged somewhat lower premiums.

Recently the National Ranch Insurance Company studied 1,000 of its policyholders. The purpose of the study was to determine whether such factors as age, sex, and marital status are independent of whether or not the policyholder has filed an accident claim. With regard to age for all drivers, both male and female, National Ranch found the following:

	Age				
	Under 25	25–40	40–55	Over 55	
Reported Claim	93	72	53	63	281
No Claim	115	155	265	184	719
	208	227	318	247	1,000 = N

Based upon these data, what should National Ranch conclude about age and claim status?

Solution:

The basic question facing National Ranch is whether age is independent of claim status. Therefore we set up our null and alternative hypotheses as follows:

H_0: Whether or not an insurance claim has been filed by a policyholder is independent of the policyholder's age.
H_A: Age and claim status are not independent.
$\alpha = 0.05$

An appropriate test is the chi-square test of independence.

$$\chi^2 = \sum_{i=1}^{r} \sum_{j=1}^{c} \frac{(f_{o_{ij}} - f_{e_{ij}})^2}{f_{e_{ij}}}$$

The expected frequencies are

$$f_{e_{11}} = \frac{281(208)}{1,000} = 58.44$$

$$f_{e_{12}} = \frac{281(227)}{1,000} = 63.78$$

$$f_{e_{13}} = \frac{281(318)}{1,000} = 89.35$$

.

.

.

$$f_{e_{24}} = \frac{719(247)}{1,000} = 177.59$$

The completed contingency table is as follows:

Age

	Under 25	25–40	40–55	Over 55
Reported Claim	$f_o = 93.00$ $f_e = 58.44$	$f_o = 72.00$ $f_e = 63.78$	$f_o = 53.00$ $f_e = 89.35$	$f_o = 63.00$ $f_e = 69.40$
No Claim	$f_o = 115.00$ $f_e = 149.55$	$f_o = 155.00$ $f_e = 163.21$	$f_o = 265.00$ $f_e = 228.64$	$f_o = 184.00$ $f_e = 177.59$

Then

$$\chi^2 = \frac{(93.00 - 58.44)^2}{58.44} + \frac{(72.00 - 63.78)^2}{63.78} + \ldots + \frac{(184.00 - 177.59)^2}{177.59}$$

$$= 51.28$$

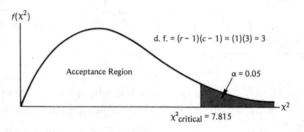

d. f. $= (r - 1)(c - 1) = (1)(3) = 3$

Acceptance Region

$\alpha = 0.05$

$\chi^2_{critical} = 7.815$

Since $\chi^2 = 51.28 > \chi^2_{critical} = 7.815$, National Ranch should reject the hypothesis of independence. Based upon the data, the insurance company might infer that young drivers have more accident claims than expected, and therefore higher rates might be justified.

---

3. The Bixby Company operates in three states. Its primary business is to give review seminars for individuals who plan to sit for the CPA (certified public accountant) examination. Bixby has three teams that put on the seminars, one for each state. Michelle Bixby is quite concerned that there be consistent performance by the three teams. She feels she can measure the success of her teams by the CPA exam scores received by persons who have taken the seminars.

   Michelle wishes to determine whether the average score received by those who took the seminar is the same in the three states. Because of the way the test is scored, she is *not* willing to assume the populations are normally distributed. A sample of 12 scores from each state provided the following data:

Maine	New York	New Jersey
90	90	55
65	88	55
72	83	90
83	83	83

65	65	65
50	65	65
65	83	60
83	83	55
83	90	65
55	90	72
72	65	90
65	72	65

Based on these data, what should Michelle conclude about the average scores of the three states? Use $\alpha = 0.10$. Assume the population variances are equal.

Solution:

This problem involves three independent samples and can be analyzed using Kruskal-Wallis one-way analysis of variance. We begin by establishing the null and alternative hypotheses as

H_0: The average CPA exam score for seminar students will be the same in the three states.

H_A: The average CPA exam score for seminar students will not be the same in all three states.

$\alpha = 0.10$

The first step in the Kruskal-Wallis procedure is to change the raw scores to ranks. Note that there are many ties in these data. Thus we will give each tied value the average of the ranks for which it is tied. Also, we will calculate the Kruskal-Wallis H statistic using the correction factor.

The rankings are as follows:

Maine	New York	New Jersey
33.5	33.5	3.5
12	30	3.5
19.5	25.5	33.5
25.5	25.5	25.5
12	12	12
1	12	12
12	25.5	6
25.5	25.5	3.5
25.5	33.5	12
3.5	33.5	19.5
19.5	12	33.5
12	19.5	12
Σ ranks = 201.5	Σ ranks = 288.0	Σ ranks = 176.5

Then

$$H = \cfrac{\cfrac{12}{N(N+1)} \sum \cfrac{R^2}{n} - 3(N+1)}{1 - \cfrac{\Sigma(t^3 - t)}{N^3 - N}}$$

$$= \cfrac{\cfrac{12}{(36)(37)}\left[\cfrac{(201.5)^2}{12} + \cfrac{(288.0)^2}{12} + \cfrac{(176.5)^2}{12}\right] - 3(37)}{1 - \cfrac{[(4^3 - 4) + (11^3 - 11) + (4^3 - 4) + (8^3 - 8) + (6^3 - 6)]}{36^3 - 36}}$$

$$= \frac{5.1400}{0.9538}$$

$$= 5.389$$

Note that the denominator in the Kruskal-Wallis H equation is the correction factor for the ties in the rankings. The closer this correction factor is to zero, the greater impact it has on the calculated H value.

The null hypothesis test is a chi-square test with $K - 1 = 2$ degrees of freedom. If H exceeds the critical value from the chi-square table, we will reject the null hypothesis of equal means.

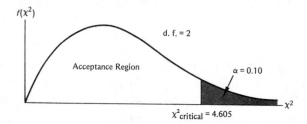

Since $H = 5.389 > \chi^2_{critical} = 4.605$, we reject the null hypothesis and conclude that the mean scores in the three states are not equal. Michelle Bixby would infer that the seminar team in New York is turning out students who do better on the average than New Jersey or Maine students. Of course this does not automatically mean the New York team is superior to the others.

problems

Problems marked with an asterisk (*) require extra effort.

1. Explain in your own words the difference between a parametric statistical test and a nonparametric statistical test. Think of at least three new decision-making situations where each type of test would likely be used.

2. Discuss the concepts behind contingency tables and how to construct them.

3. Ramona Lane, the manager of Rapid Way Super Discount, always keeps four checkout stands open. However she frequently notices lines for registers 1 and 2. She isn't sure whether the layout of the store channels customers into these registers or whether these checkers are simply slower than the other two.

Ramona kept a record of which stands 1,000 shoppers checked out through. The shoppers checked out of the four stands according to the following pattern:

Stand 1	Stand 2	Stand 3	Stand 4
292	281	240	187

Based on these data, can Ramona conclude that an equal number of shoppers are likely to use each of the stands?

4. Jack O'Connell, chief of officials for the National Basketball League, reviews films of all games to evaluate calls made by the referees. Jack rates each call "good" or "bad." During a week's worth of games, Jack found the following distribution of calls for two officials:

Call	Official A	Official B
Good	463	518
Bad	51	38

a. Do these data indicate that the proportion of bad calls is the same for each official?
b. Test the data using a chi-square statistic with alpha = 0.05.
c. Compare the test in part b with the Z test for the difference between two population proportions.

5. The Rock Company, an income tax assistance firm, has decided to aim its new advertising campaign at taxpayers who earn less than $25,000 per year. The company has been using four separate advertising themes, one in each of four sections of the country. The president has asked the four regional managers to randomly select returns from their regions and indicate how many fall into the above- and below-$25,000 categories. Unfortunately the president forgot to specify the number of returns each manager was to sample. The main office of the Rock Company gathers the following data:

	Northeast	South	Midwest	West
Less than $25,000	57	112	109	71
More than $25,000	162	197	148	153

Do these sample results indicate that one of the advertising campaigns has been more effective in luring in taxpayers earning less than $25,000? You will have to pick your own error level and justify it. What mistake were you trying to guard against?

6. A manufacturer of packaged food has decided to market a new quick brownie mix. This venture is rather risky since the new mix will compete with three established brands. The production manager argues that what will sell the new mix is quality. The marketing manager states that the average consumer can't tell the difference between brands anyway, and that the critical factor will be the advertising campaign.

To test the marketing manager's contention, the four types of brownies were baked under equal conditions. Shoppers were randomly stopped in several supermarkets, and asked to sample each brownie and indicate which of the four tasted the best. Which brownie was tasted first was randomly determined, and the production manager and the marketing manager felt the process yielded representative results. The following data show how many shoppers rated each brand the tastiest:

		Brand		
	A	B	C	New
	37	23	27	41

Which contention do these data support? Use a Type I error of 0.01.

7. Ralph Rogers has developed a highly successful practice as a black-market acupuncture specialist. Ralph's success is built on his money-back guarantee. If his treatment wears off, he'll treat you again. His accountant, in trying to set up an allowance for future visits account, hypothesizes that whether or not a patient will demand a retreatment is related to the price of the original treatment. The following data show the relationship between price and return treatment:

		Price		
		High	Medium	Low
	In Less Than 2 Years	26	13	16
Retreatment	In 2–5 Years	43	35	52
	None in 5 Years	87	79	103

a. What should the accountant conclude regarding the hypothesis?
b. What factors that the accountant apparently has not considered might be important to the analysis?

*8. A regional cancer treatment center has had success treating localized cancers with a linear accelerator. While admissions for further treatment nationally average 1.1 per patient per year, the center's director feels that readmissions with the new treatment are Poisson-distributed with a mean of 0.5 per patient per year. He has collected the following data on a random sample of 400 patients:

Readmissions Last Year	Patients
0	139
1	87
2	48
3	14
4	8
5	1
6	1
7	0
8	2

a. Formulate the null hypothesis to test the director's claim.
b. Calculate the test statistic to test this claim.
c. Assume the Type I error is to be controlled at 0.05. Do you agree with the director's claim? Why?

9. A major car manufacturer is experimenting with three new methods of pollution control. The testing lab must determine whether the three methods produce equal pollution reductions. Readings from a calibrated carbon monoxide meter are taken from groups of engines randomly equipped with one of the three control units. The following data are found:

Method 1	Method 2	Method 3
45	39	26
49	37	31
48	31	32
50	42	40
33	43	44
46	47	41
46	46	34
47	27	28
46	29	30
39	31	34
	36	35
		38

Use the Kruskal-Wallis test to determine whether the three pollution-control methods will produce equal results.

10. The Miltmore Corporation specializes in performing market-research studies for companies that plan to enter well-established markets with new products. Miltmore's method is to select a random sample of potential customers and have half of them rate the new brand and the other half rate the market's leading brand. If Miltmore concludes that the new brand rates at least as high as the leading brand, it recommends that the company perform a complete market-research program under Miltmore's direction.

The Raineer corporation is considering entering the electric type-writer market and has approached Miltmore about testing its typewriter against the leader in the industry. Miltmore has collected ratings on a 1–40 scale from 200 randomly selected people. Half of these rated the Raineer machine, and half rated the leading machine, with the following results:

		Frequency	
Satisfaction Rating		Raineer	Leader
1–10	poor	36	10
11–20	fair	40	16
21–30	good	10	40
31–40	very good	14	34
		$\Sigma = 100$	$\Sigma = 100$

Based upon these sample results, what should the Miltmore Corporation conclude regarding the Raineer typewriter and the desirability of more extensive market research? Select an alpha level and justify the level you have selected.

11. The Miltmore Corporation referred to in problem 10 also performs consulting services for companies that think they have image problems. Recently Miltmore was approached by the Bluedot Beer Company. Bluedot executives were concerned that the company's image relative to its two closest competitors' had diminished. The Federal Food and Drug Administration ordered that 50,000 cases of Bluedot beer be recalled due to a potential health problem with the water used in making the beer.

In response to Bluedot's problem, Miltmore conducted an image study in which a random sample of eight people were asked to rate Bluedot's image. Five people were asked to rate competitor A's image, and ten people were asked to rate competitor B's image. The image ratings were made on a 100-point scale, with 100 the best possible rating. The results of the sampling were:

Bluedot	Competitor A	Competitor B
40	95	50
60	53	80
70	55	82
40	92	87
55	90	93
90		51
20		63
20		72
		96
		88

Based upon these sample results, should Bluedot conclude that there is an image difference between the three companies? Select your own alpha level and justify it.

*12. Referring to problem 11, should Bluedot infer that its image has been damaged by the federal government recall of its product? Discuss why or why not.

13. Referring to problems 11 and 12, suppose Bluedot contracted with Miltmore to obtain image ratings before the recall from 100 randomly selected people, and another 100 image ratings after the recall. The following information reflects the result of Miltmore's work:

	Frequency	
Rating	Before	After
0– 19	3	8
20– 39	28	36
40– 59	30	40
60– 79	30	14
80–100	9	2
	$\Sigma = 100$	$\Sigma = 100$

Based on these results, what should Bluedot conclude? Test this at an appropriate alpha level and justify the alpha you select.

14. Why might the decision maker in problem 11 wish to use parametric analysis of variance rather than the corresponding nonparametric test? Discuss.

*15. National Bank is contemplating purchasing one of two small banks in nearby towns. One thing that concerns National management is the stability of the customers at these two banks. Specifically, management wonders if the two banks differ with respect to customer longevity.

Each bank was asked to furnish National with a breakdown of the age of savings accounts. The following data were collected and represent a random sample from the accounts:

	Frequency	
Account Age	Bank A	Bank B
0–1 yr	140	221
1–4	76	80
4–10	100	55
Over 10 yr	70	40
	$\Sigma = 386$	$\Sigma = 396$

Based upon these sample data, what should National managers conclude? Use alpha equal to 0.05.

16. McDougals is a nationwide fast-food company with corporate headquarters in Bellview, Washington. For the past few months the company has undertaken a new advertising study. Initially, company executives selected 22 of the retail outlets that were similar with respect to sales volume, profitability, location climate, economic status of customers, and experience of store management. Each of the outlets was randomly assigned one of two new advertising plans promoting a new sandwich product. The following data represent the number of the new sandwiches sold during the specific test period at each retail outlet:

Advertising Plan 1	Advertising Plan 2
1,711	2,100
1,915	2,210
1,905	1,950
2,153	3,004
1,504	2,725
1,195	2,619
2,103	2,483
1,601	2,520
1,580	1,904
1,475	1,875
1,588	1,943

McDougals executives are interested in determining whether these data indicate that the two advertising plans lead to significantly different average sales levels for the new product. They wish to test this at the 0.05 alpha level but are not willing to make the assumptions necessary to use the t test.

17. Two small regional life insurance companies are being studied by the Triangle Life Insurance Company as candidates for a possible merger. Triangle can merge with only one of these regional companies at this time, and as part of its study, wishes to determine if there is a difference in the average annual premiums received by the two regional companies. The following data represent sample policy premiums for each company:

Company 1	Company 2
$246	$300
211	305
235	308
270	325
411	340
310	295
450	320
502	330
311	240
200	360

Do these data indicate a difference in mean annual premiums for the two companies? Apply the Mann-Whitney U test with $\alpha = 0.10$.

18. Referring to problem 17, apply the t test to determine whether the data indicate a difference between annual premiums for the two regional companies. Use $\alpha = 0.10$. Also, indicate what assumptions must be made to apply the t test.

19. Referring to problem 16, McDougals executives wish to apply the Kolmogorov-Smirnov two-sample test to determine whether the data indicate a difference

between the mean sales for the two advertising plans. Use the following group-ings and test at the 0.05 alpha level:

1,100–1,500
1,501–1,900
1,901–2,300
2,301–2,700
2,701–3,100

20. Referring to problem 19, how would the results of the test change if the follow-ing groupings were used?

1,100–1,800
1,801–2,500
2,501–3,200

Perform the test using alpha equal to 0.05 and discuss how the selection of groupings can affect the Kolmogorov-Smirnov two-sample test.

21. The Triangle Life Insurance Company in problem 17 has expanded the study of the two regional life insurance companies it is considering for a possible merger. Triangle has collected the following data on annual premiums:

		Premium			
		$100–$200	$201–$300	$301–$400	Over $400
Frequency	Company 1	7	21	40	2
	Company 2	3	33	25	9

Based upon these data, what should Triangle conclude about the average annual premiums collected by these two companies? Use $\alpha = 0.10$.

*22. The J. Scholten CPA firm performed a study of last year's income-tax business. In one part of the study the accountants collected data on their clients' gross taxable incomes and the associated tax payments. These data are shown in the following table, where, for example, there are 25 clients whose gross incomes were below $10,000 and who paid $3,000 or less in taxes.

		Taxes			
		$0–$3,000	$3,001–$5,000	$5,001–$10,000	Over $10,000
Gross Income	$0–$10,000	25	0	0	0
	$10,001–$20,000	17	5	0	0
	$20,001–$40,000	15	40	8	3
	Over $40,000	3	27	22	14

Based upon these data, can Scholten conclude that its clients' gross incomes are independent of the income taxes paid? Test at the alpha 0.05 level. What comment would you have made had independence been con-cluded?

23. Referring to problem 22, Scholten also studied the time it took its accountants to complete each client's tax return, and related this time to the taxes paid by the client. Scholten managers were interested in determining whether a relationship exists between these two variables or whether they could consider the two variables independent. The following data are available:

		Taxes			
		$0–$3,000	$3,001–$5,000	$5,001–$10,000	Over $10,000
No. Work Hours	0–2	27	30	5	2
	2–4	22	30	5	6
	Over 4	10	12	20	10

Based upon these data, what should the Scholten firm conclude? Use an alpha of 0.10.

cases

15A BENTFORD ELECTRONICS, PART 1

On Saturday morning Jennifer Bentford received a call at her home from the production supervisor at Bentford Electronics Plant #1. The supervisor indicated that she and the supervisors from Plants #2, #3, and #4 had agreed that something must be done to improve company morale and, thereby, increase the production output of their plants. Jennifer Bentford, president of Bentford Electronics, agreed to set up a Monday morning meeting with the supervisors to see if together they could arrive at a plan for accomplishing these objectives.

By Monday each supervisor had compiled a list of several ideas, including a four-day work week and interplant competition of various kinds.

After listening to the discussion for some time, Jennifer Bentford asked if anyone knew if there was a difference in average daily output for the four plants. When she heard no positive response, she told the supervisors to select a random sample of daily production reports from each plant and test whether there was a difference. They were to meet again on Wednesday afternoon with test results.

By Wednesday morning the supervisors had collected the following data on units produced:

Plant #1	Plant #2	Plant #3	Plant #4
4,306	1,853	2,700	1,704
2,852	1,948	2,705	2,320
1,900	2,702	2,721	4,150
4,711	4,110	2,900	3,300
2,933	3,950	2,650	3,200
3,627	2,300	2,480	2,975

The supervisors had little trouble collecting the data, but were at a loss how to determine if there was a difference in the output of the four plants. Jerry Gibson, the company's research analyst, told the supervisors that there were statistical procedures that could be used to test hypotheses regarding multiple samples if the daily output was distributed in a bell shape (normal distribution) at each plant. The supervisors expressed dismay because none thought his or her output was normal. Jerry Gibson indicated that there were techniques that didn't require the normality assumption, but he didn't know what they were.

The meeting with Jennifer Bentford was scheduled to begin in three hours.

15B BENTFORD ELECTRONICS, PART 2

Following the Wednesday afternoon meeting (see case 15A), Jennifer Bentford and her plant supervisors agreed to implement a weekly contest called the NBE Game of the Week. The plant turning out the most production each week would be considered the NBE Game of the Week winner and would receive ten points. The second-place plant would receive seven points, and the third- and fourth-place plants would receive three points and one point, respectively. The contest would last 26 weeks. At the end of the 26-week period, a $200,000 bonus would be divided among the employees in the four plants proportional to the total points accumulated by each plant.

The announcement of the contest created a lot of excitement and enthusiasm at the four plants. No one complained about the rules since the four plants were designed and staffed to produce equally.

At the close of the contest, Jennifer Bentford called the supervisors into a meeting, at which time she asked for data to determine whether the contest had significantly improved productivity. She indicated that she had to know this before she could authorize a second contest. The supervisors, expecting this response, had put together the following data:

Units Produced (4 plants combined)	Before-Contest Frequency	During-Contest Frequency
0– 2,500	11	0
2,501– 8,000	23	20
8,001–15,000	56	83
15,001–20,000	15	52
	$\Sigma = 105$ days	$\Sigma = 155$ days

Jennifer examined the data and indicated that it looked like the contest was a success, but she wanted to base her decision to continue the contest on more than just an observation of the data. "Surely there must be some way to statistically test the worthiness of this contest," Jennifer stated. "I have to see the results before I will authorize the second contest."

15C SINGLEAF DEPARTMENT STORE, PART 2

Beth Hansen has completed her study of different types of collection letters (see case 14B). However she is also concerned about the effect of the collection efforts on the goodwill of Singleaf stores. As mentioned previously, Singleaf does not sell the past-due accounts to a collection agency because of the potential loss of goodwill.

Beth has decided to use a telephone interview system to determine customer attitudes toward the Singleaf stores. She has devised a series of questions to measure customer attitudes on many factors connected with the stores: clerk helpfulness, billing procedures, merchandise selection, and so on. The scores on each question are combined to give an overall measure of the customer attitude toward the Singleaf organization (ten points maximum). In an effort to determine what customers in general think about the store and the effect of the collection letters, Beth has decided to randomly select customers from the following categories:

1. Frequent customers with nondelinquent accounts
2. Infrequent customers with nondelinquent accounts
3. Delinquent customers who have received reminder collection letters
4. Delinquent customers who have received "tough" collection letters
5. Delinquent customers who have not received collection letters

Since the telephone questionnaire will be expensive, Beth would like to have an idea about the potential findings before asking the president in charge of customer relations to authorize spending the money. The sample telephone survey gave the following overall ratings for Singleaf stores:

		Customer Category		
1	*2*	*3*	*4*	*5*
9	3	5	4	5
8	5	9	9	4
8	5	6	9	4
9	7	9	6	3
8	8	7	1	7
8	9	10	3	3

Beth has an appointment with the Singleaf president tomorrow morning.

references

CONOVER, W. J. *Practical Nonparametric Statistics*. New York: Wiley, 1971.

DANIEL, WAYNE W. *Applied Nonparametric Statistics*. Boston: Houghton Mifflin, 1978.

MARASCUILO, LEONARD A. and MARYELLEN MCSWEENEY. *Nonparametric and Distribution-free Methods for Social Sciences*. Monterey, Calif.: Brooks/Cole, 1977.

NIE, NORMAN H., C. HADLAI HULL, JEAN G. JENKINS, KAREN STEINBRENNER, and DALE H. BENT. *SPSS: Statistical Package for the Social Sciences*. New York: McGraw-Hill, 1975.

16

Simple Linear Regression and Correlation Analysis

The statistical techniques discussed thus far have all dealt with a single variable. For example, in Chapters 10 and 11, where we introduced the fundamentals of statistical estimation, decision makers were interested in estimating such values as the average weight of a can of corn or the average income of families in a certain geographical area. In Chapters 12–14 we introduced the basic concepts of hypothesis testing. Again the hypothesis involved a single variable from one or more populations.

Although many business applications involve only one variable, in other instances decision makers need to consider the relationship between two or more variables. For example, the sales manager for a tool company may notice his sales are not the same each month. He also knows that the company's advertising expenditures vary from month to month. This manager would likely be interested in whether a relationship exists between tool sales and advertising. If he could successfully define the relationship, he might use this information to improve predictions of monthly sales and, therefore, do a better job of planning for his company.

In this chapter we introduce simple linear regression and correlation analysis. These techniques are important to decision makers who need to determine the relationship between two variables. Our experience indicates that regression and correlation analysis are two of the most often applied statistical tools for decision making.

chapter objectives

In this chapter we shall introduce simple linear regression and correlation techniques. We will use examples to demonstrate the three main uses of regression analysis: prediction, description, and control.

In this chapter we will also show how decision makers can determine whether a significant linear relationship exists between two variables. We show how to determine whether regression analysis is actually useful in a practical decision-making situation. In addition, we show how to develop confidence intervals for the estimates made using regression analysis.

Finally, in this chapter we shall introduce the assumptions behind regression analysis and discuss some problems that might occur if regression analysis is incorrectly used.

student objectives

After studying the material in this chapter, you should be able to:

1. Calculate the simple correlation between two variables.
2. Determine if the correlation is significant.
3. Calculate the simple linear regression equation for a set of data and know the basic assumptions behind regression analysis.
4. Determine whether a regression model is significant.
5. Develop confidence intervals for the regression coefficients.
6. Interpret the confidence intervals for the regression coefficients.
7. Recognize regression analysis applications for purposes of prediction, description, and control.
8. Recognize some potential problems if regression analysis is used incorrectly.

16-1
STATISTICAL RELATIONSHIPS BETWEEN TWO VARIABLES

Harry Aims has just taken over as the Bato Tool Company general manager. The owners have been concerned about the company's varying quarterly sales. Variation in sales causes cash-flow and inventory-stocking problems, to name a few. To get a handle on the extent of sales variation, Harry has tabulated sales data for the past 12 quarters. As can be seen in Table 16–1, quarterly sales do vary. Y. R. McNeese, the company's sales manager, claims the sales variability occurs because the marketing department constantly changes its advertising expenditures. McNeese is quite certain there is a relationship between sales and advertising but doesn't know what the relationship is.

TABLE 16–1   Sales, Bato Tool Company
($\times$ $1,000)

Quarter	Sales
1	22
2	28
3	22
4	26
5	34
6	18
7	30
8	38
9	30
10	40
11	50
12	46

In Figure 16–1 are *scatter plots* which depict several potential relationships between a dependent variable, Y, and an independent variable, X. **A *dependent variable* is the variable whose variation we wish to explain. An *independent variable* is a variable used to explain variation in the dependent variable.** Figures 16–1(a) and (b) are examples of *linear* relationships between X and Y. This means that as the independent variable, X, changes, the dependent variable, Y, tends to

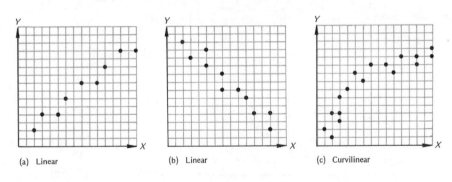

(a) Linear   (b) Linear   (c) Curvilinear

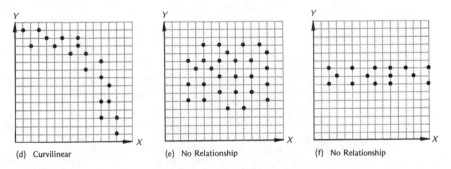

(d) Curvilinear   (e) No Relationship   (f) No Relationship

FIGURE 16–1   Two-variable relationships.

change systematically in a straight-line manner. Note that this systematic change can be positive (Y increases as X increases) or negative (Y decreases as X increases).

Figures 16–1(c) and (d) are examples of **curvilinear** statistical relationships between variables X and Y. And, Figures 16–1(e) and (f) are examples where there is **no relationship** between X and Y. That is, when X increases, sometimes Y decreases, and other times Y increases.

The three situations shown in Figure 16–1 are all possibilities for describing the relationship between sales and advertising for the Bato Tool Company. The first step in determining the appropriate relationship is to collect advertising expenditure data for each quarter for which sales data are available. These values are shown in Table 16–2. Next, the scatter plot shown in Figure 16–2 should be constructed.

Figure 16–2 indicates that advertising and sales are somewhat linearly related. However the "strength" of this relationship is questionable. Harry Aims needs a measure of the strength of the linear relationship between sales and advertising.

TABLE 16–2   Sales and advertising data, Bato Tool Company
(× $1,000)

Quarter	Sales Y	Advertising X
1	22	0.8
2	28	1.0
3	22	1.6
4	26	2.0
5	34	2.2
6	18	2.6
7	30	3.0
8	38	3.0
9	30	4.0
10	40	4.0
11	50	4.0
12	46	4.6

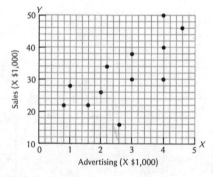

FIGURE 16–2   Plot of sales versus advertising expense, Bato Tool Company.

16-2
CORRELATION ANALYSIS

The quantitative measure of strength in the linear relationship between two variables is called the *correlation coefficient*. The correlation coefficient for two variables (here, sales and advertising expenditures) can be estimated from sample data using

$$r = \frac{n\Sigma XY - \Sigma X \Sigma Y}{\sqrt{[n(\Sigma X^2) - (\Sigma X)^2][n(\Sigma Y^2) - (\Sigma Y)^2]}} \qquad \text{(16–1)}$$

where: r = sample correlation coefficient

n = sample size

X = value of the independent variable

Y = value of the dependent variable

The correlation coefficient can range from a perfect positive correlation, +1.0, to a perfect negative correlation, −1.0. If two variables have no linear relationship, the correlation between them is zero. Consequently, the more the correlation differs from zero, the stronger the linear relationship between the two variables. The sign of the correlation coefficient indicates the direction of the relationship but does not aid in determining the strength.

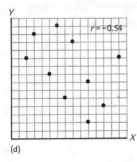

(a)

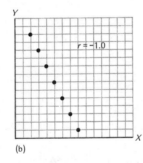

(b)

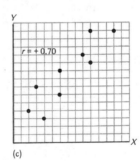

(c)

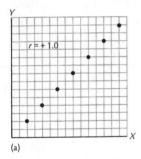

(d)

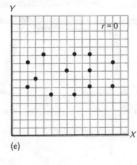

(e)

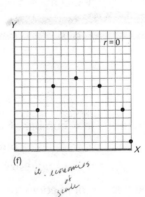

(f)

FIGURE 16–3 Correlations.

Figure 16–3 illustrates some possible correlations between two variables. Note that for the correlation coefficient to equal ± 1.0, all the (X, Y) points must be perfectly aligned. The more the points depart from a straight line, the weaker the correlation between the two variables.

The scatter plot of sales and advertising values for Bato Tool (Figure 16–2) indicates a *positive* linear relationship but *not* a *perfect* linear relationship. Table 16–3 lists the calculations necessary to determine the correlation coefficient for Bato Tool. The calculated correlation coefficient, r, is 0.734.

TABLE 16–3   Correlation coefficient calculations, Bato Tool Company

Sales Y	Advertising X	YX	Y²	X²
22	0.8	17.6	484	0.64
28	1.0	28.0	784	1.00
22	1.6	35.2	484	2.56
26	2.0	52.0	676	4.00
34	2.2	74.8	1,156	4.84
18	2.6	46.8	324	6.76
30	3.0	90.0	900	9.00
38	3.0	114.0	1,444	9.00
30	4.0	120.0	900	16.00
40	4.0	160.0	1,600	16.00
50	4.0	200.0	2,500	16.00
46	4.6	211.6	2,116	21.16
$\Sigma = 384$	$\Sigma = 32.8$	$\Sigma = 1,150.0$	$\Sigma = 13,368$	$\Sigma = 106.96$

$$r = \frac{n\Sigma XY - \Sigma X\Sigma Y}{\sqrt{[n(\Sigma X^2) - (\Sigma X)^2][n(\Sigma Y^2) - (\Sigma Y)^2]}} = \frac{12(1,150.0) - 384(32.8)}{\sqrt{[12(106.96) - (32.8)^2][12(13,368) - (384)^2]}}$$
$$= 0.734$$

Although a correlation coefficient of 0.734 seems quite high (relative to zero), remember that this value is based on a sample of 12 data points. To determine whether the linear relationship between sales and advertising is significant, we must test whether the sample data support or refute the hypothesis that the population correlation coefficient, ρ, is zero. The hypothesis that the population correlation coefficient is zero is tested by computing the following t statistic:

$$t = \frac{r}{\sqrt{\dfrac{1 - r^2}{n - 2}}} \qquad \textbf{(16–2)}$$

where: t = number of standard deviations r is from 0

r = sample correlation coefficient

n = sample size

The degrees of freedom for this test are $n - 2$.

Table 16–4 shows the test for an alpha level of 0.05. Based on these sample data, we should conclude there is a significant linear relationship between Bato Tool Company's advertising and sales.

TABLE 16–4 Correlation significance test, Bato Tool Company

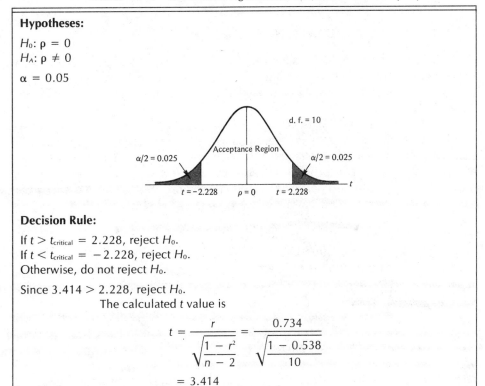

Hypotheses:

H_0: $\rho = 0$
H_A: $\rho \neq 0$

$\alpha = 0.05$

d. f. = 10

$\alpha/2 = 0.025$ Acceptance Region $\alpha/2 = 0.025$

$t = -2.228$ $\rho = 0$ $t = 2.228$

Decision Rule:

If $t > t_{critical} = 2.228$, reject H_0.
If $t < t_{critical} = -2.228$, reject H_0.
Otherwise, do not reject H_0.

Since $3.414 > 2.228$, reject H_0.
    The calculated t value is

$$t = \frac{r}{\sqrt{\dfrac{1 - r^2}{n - 2}}} = \frac{0.734}{\sqrt{\dfrac{1 - 0.538}{10}}}$$

$$= 3.414$$

Cause-and-Effect Interpretations

Care must be used when interpreting the correlation results. Even though we found a significant linear relationship between Bato Tool's sales and advertising, the correlation does not imply cause and effect. Although a change in advertising may, in fact, cause sales to change, **simply because the two variables are correlated does not guarantee a cause-and-effect situation.** Two seemingly unconnected variables will often be highly correlated. For example, over a period of time, teachers' salaries in North Dakota might be highly correlated with the price of grapes in Spain. Yet, we doubt that a change in grape prices will *cause* a corresponding change in salaries for teachers in North Dakota, or vice versa. When a correlation exists between two unrelated variables, the correlation is *spurious*. You should take great care to avoid basing conclusions on spurious correlations.

Harry Aims has a logical reason to believe that advertising and sales are related. In fact, marketing theory holds that a change in advertising expenditure might well cause a change in tool sales by enticing customers to purchase Bato tools rather than some other brand. However, the correlation alone does not prove that this cause-and-effect situation exists.

16-3
SIMPLE COEFFICIENT OF DETERMINATION

Harry Aims has determined that sales vary from quarter to quarter and that a positive linear relationship exists between sales and advertising. In Section 16–2 we showed that sales and advertising are linearly related with a sample correlation coefficient, r, of 0.734.

A major part of most managers' jobs is planning and budgeting. Therefore Harry would like to explain these changes in sales, because, if he could, he might be able to plan more effectively in such areas as inventory, sales expenses, and so forth. The extent to which advertising can explain the variation in sales is measured by the *simple coefficient of determination*. **The *simple coefficient of determination* is found by squaring the simple correlation coefficient.**

$$\text{Simple coefficient of determination} = r^2 \qquad (16\text{–}3)$$

Thus, if the simple correlation between advertising and sales is 0.734, the simple coefficient of determination is $(0.734)^2 = 0.538$. This means, for the sample data, that 53.8 percent of the variation in sales can be explained by knowing the level of advertising expenditure.

The simple coefficient of determination can vary from zero to 1.0. If two variables are perfectly correlated, the simple coefficient of determination will be 1.0, meaning that 100 percent of the variation in the dependent variable can be explained by knowing the value of the independent variable. Likewise, if the correlation is zero, none of the variation in Y can be accounted for by knowing X.

Testing the Significance of r^2

Because the correlation coefficient, r, is based on a sample of observations and is therefore subject to sampling error, r^2 is also subject to sampling error. Before Harry Aims attempts to use advertising to help explain the change in sales, he needs to test whether or not the population coefficient of determination could actually be zero. The F distribution is used to test the significance of r^2 as shown in Table 16–5. The test statistic is

$$F = \frac{r^2}{\dfrac{1 - r^2}{n - 2}} \qquad (16\text{–}4)$$

where: F = value from the F distribution with $D_1 = 1$ and $D_2 = n - 2$ degrees of freedom if the null hypothesis is true

r^2 = simple coefficient of determination

n = sample size

TABLE 16–5   Significance test, Bato Tool Company

Hypotheses:

H_0: The simple coefficient of determination, $\rho^2 = 0$.
H_A: The simple coefficient of determination, $\rho^2 \neq 0$.

$\alpha = 0.05$

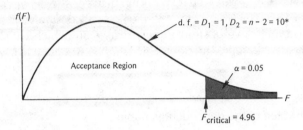

$$d.\,f. = D_1 = 1,\, D_2 = n - 2 = 10^*$$

Acceptance Region

$\alpha = 0.05$

$F_{\text{critical}} = 4.96$

Decision Rule:

If $F > F_{\text{critical}}$, reject H_0.
Otherwise, do not reject H_0.

Since $11.64 > 4.96$, reject H_0.

The F calculation is

$$F = \frac{r^2}{\dfrac{1 - r^2}{n - 2}} = \frac{0.538}{\dfrac{0.462}{10}}$$

$$= 11.64$$

*The degrees of freedom for hypothesis tests about the simple coefficient of determination are always $D_1 = 1$ and $D_2 = n - 2$.

As shown in Table 16–5, the decision is to reject the null hypothesis, and we conclude that advertising *does* explain a significant percentage of variation in sales.

If you have read this section and Section 16–2 carefully, you may have noticed that the tests of significance for r and r^2 are closely related. In fact, the calculated F value is the same as t^2. (Compare the values in Tables 16–4 and 16–5.) The two tests are equivalent.

If we conclude that the correlation is significantly different from zero, we will also conclude that the independent variable explains a significant proportion of the variation in the dependent variable. **Once we decide that two variables are linearly related, the independent variable can be used to explain changes in the dependent variable.**

16-4
SIMPLE LINEAR REGRESSION ANALYSIS

Harry Aims has decided that the relationship between advertising expenditures and sales is linear and, further, that for the sample data, the advertising level explains

53.8 percent ($r^2 = 0.538$) of the variation in sales. Harry would very much like to use this information to help predict his company's sales. Predicted sales levels would make his whole planning and budgeting process much easier.

The most common statistical method for prediction is *regression analysis.* When there are only two variables, a dependent variable such as sales and an independent variable such as advertising, the technique for prediction is called *simple regression analysis.* And, when the relationship between the dependent variable and the independent variable is linear, the technique is called *simple linear regression.*

The objective of simple linear regression (which we shall call simply regression analysis) is to represent the relationship between X and Y with a model of the form

$$Y_i = \beta_o + \beta_1 X_i + e_i \qquad\qquad (16\text{–}5)$$

where: Y_i = value of the dependent variable

X_i = value of the independent variable

β_o = Y-intercept

β_1 = slope of the regression line

e_i = error term (i.e., the difference between the actual Y value and the value of Y predicted by the model)

The simple linear regression model described in equation (16–5) has four assumptions:

1. Individual values of the dependent variable, Y, are statistically independent of one another.

2. For a given X, there can exist many values of Y. Further, the distribution of possible Y values for any X is normal.

3. The distributions of possible Y values have equal variances for all values of X.

4. The averages of the dependent variable, Y, for all values of the independent variable, ($\mu_Y|X$), can be connected by a straight line called the ***population regression model.***

Figure 16–4 illustrates assumptions 2, 3, and 4. The regression model (straight line) connects the averages of Y for each level of the independent variable,

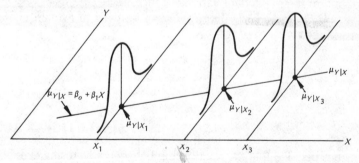

FIGURE 16–4   Graphical display of linear regression assumptions.

X. The regression line (like any other straight line) is determined by two values, β_o and β_1. These values are the **regression coefficients.** Value β_o identifies the Y-intercept, and β_1 the slope of the regression line. Under the regression assumptions, the coefficients define the true population model. For each observation, the actual value of the dependent variable, Y, for any X, is the sum of two components; that is,

$$Y_i = \underbrace{\beta_o + \beta_1 X_i}_{\text{Linear component}} + \underbrace{e_i}_{\text{Random component}}$$

The random component, e_i, may be positive or negative, depending upon whether a single value of Y for a given X falls above or below the regression line.

Meaning of the Regression Coefficients

Coefficient β_1, the slope of the regression line, gives the *average* change in the dependent variable, Y, for each unit change in X. The slope can be either positive or negative, depending on the relationship between X and Y. A positive slope of 12 ($\beta_1 = 12$) means that for a 1-unit increase in X, we can expect an average 12-unit increase in Y. Correspondingly, if the slope is a negative 12 ($\beta_1 = -12$), we can expect an average decrease of 12 units in Y for a 1-unit increase in X.

The Y-intercept, β_o, indicates the mean value of Y when X is zero. However, this interpretation holds only if the population could have X values of zero. When this cannot occur, β_o does not have a meaningful interpretation in the regresion model.

16-5

ESTIMATING THE SIMPLE REGRESSION MODEL—THE LEAST SQUARES APPROACH

Harry Aims of the Bato Tool Company has only a sample of quarterly data available, yet he has been able to establish a significant linear relationship between advertising and sales using correlation analysis. He has also determined that advertising explains a significant proportion of the variation in sales. Now he would like to estimate the *true* linear relationship between advertising and sales by determining the regression model for the 12 quarters of data he has available.

Figure 16–5 shows the scatter plot for advertising and sales. Harry needs to use these sample points to estimate β_o and β_1.

Estimation was fairly straightforward when discussed in Chapters 10 and 11. When we want a point estimate for μ_x, we simply calculate $\overline{X}$, the sample mean. Estimating the regression model is not quite so straightforward: although there is only one sample mean for a specific sample, there are an infinite number of possible regression lines for a set of points. For example, Figure 16–6 shows three different lines that pass through Bato Tool's advertising and sales points. These are but a few of the lines that could have been drawn. Which line should be used to estimate the true regression model?

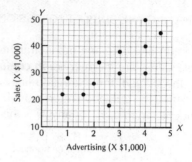

FIGURE 16–5 Plot of sales versus advertising expense, Bato Tool Company.

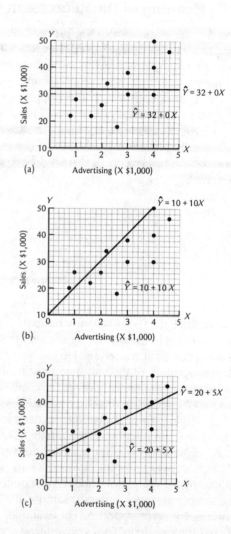

FIGURE 16–6 Potential regression lines, Bato Tool Company.

Since so many possible regression lines exist for a sample of data, we must establish a criterion for selecting the "best" line. The criterion used is the *least squares* criterion. **According to the least squares criterion, the "best" regression line is the one that minimizes the sum of squared distances between the observed (X, Y) points and the regression line.** Note that the distance between an (X, Y) point and the regression line is e_i, the random error.

Figure 16–7 shows how the random error is calculated when $X = 4.6$ and the regression line is $\hat{Y} = 20 + 5X$ (where $\hat{Y}$ is the estimated sales value). Notice that when $X = 4.6$, the difference between the regression line value $\hat{Y} = 43$ and the observed $Y = 46$ is $Y - \hat{Y} = 3$. Thus the random error for this line when X is 4.6 is 3.

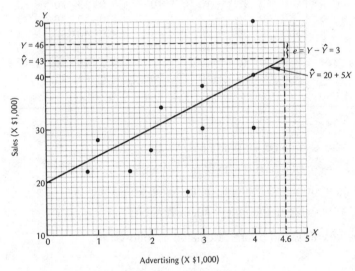

FIGURE 16–7 Computation of regression error, Bato Tool Company.

Table 16–6 shows the calculated errors and sum of squared errors for each of the three regression lines shown in Figure 16–6. Of these three potential regression models, the line with the equation $\hat{Y} = 20 + 5X$ has the smallest sum of squared errors. However, is this line the "best" of all possible lines? That is, would

$$\sum_{i=1}^{n}(Y_i - \hat{Y}_i)^2 = 539$$

be smaller than for any other line? One way to determine this is to calculate the sum of squared errors for other regression lines. However, since there are an infinite number of these lines, this approach is not feasible. Fortunately the calculus provides equations that can be used to directly determine the slope and intercept estimates such that $\Sigma(Y_i - \hat{Y})^2$ is minimized.

Let the estimated regression model be of the form

$$\hat{Y}_i = b_o + b_1X_i \qquad \textbf{(16–6)}$$

TABLE 16–6 Sum of squared errors, Bato Tool Company

Figure 16–6(a) $\hat{Y} = 32 + 0X$					Figure 16–6(b) $\hat{Y} = 10 + 10X$					Figure 16–6(c) $\hat{Y} = 20 + 5X$				
X	$\hat{Y}$	Y	$Y - \hat{Y}$	$(Y - \hat{Y})^2$	X	$\hat{Y}$	Y	$Y - \hat{Y}$	$(Y - \hat{Y})^2$	X	$\hat{Y}$	Y	$Y - \hat{Y}$	$(Y - \hat{Y})^2$
0.8	32	22	−10	100	0.8	18	22	4	16	0.8	24	22	−2	4
1.0	32	28	−4	16	1.0	20	28	8	64	1.0	25	28	3	6
1.6	32	22	−10	100	1.6	26	22	−4	16	1.6	28	22	−6	36
2.0	32	26	−6	36	2.0	30	26	−4	16	2.0	30	26	−4	16
2.2	32	34	2	4	2.2	32	34	2	4	2.2	31	34	3	9
2.6	32	18	−14	196	2.6	36	18	−8	64	2.6	33	18	−15	225
3.0	32	30	−2	4	3.0	40	30	−10	100	3.0	35	30	−5	25
3.0	32	38	6	36	3.0	40	38	−2	4	3.0	35	38	3	9
4.0	32	30	−2	4	4.0	50	30	−20	400	4.0	40	30	−10	100
4.0	32	40	8	64	4.0	50	40	−10	100	4.0	40	40	0	0
4.0	32	50	18	324	4.0	50	50	0	0	4.0	40	50	10	100
4.6	32	46	14	196	4.6	56	46	−10	100	4.6	43	46	3	9
				$\Sigma = 1{,}080$					$\Sigma = 884$					$\Sigma = 539$

where: $\hat{Y}_i$ = estimated, or predicted, Y value

b_o = unbiased estimate of the regression intercept

b_1 = unbiased estimate of the regression slope

X_i = value of the independent variable

Then the values of b_o and b_1 are calculated as follows:

$$b_1 = \frac{\Sigma XY - \dfrac{\Sigma X \Sigma Y}{n}}{\Sigma X^2 - \dfrac{(\Sigma X)^2}{n}} \tag{16–7}$$

and

$$b_o = \bar{Y} - b_1 \bar{X} \tag{16–8}$$

Table 16–7 shows the calculations necessary to estimate the population line for Bato Tool. Here, the "best" regression line, given the least squares criterion, is

$$\hat{Y}_i = 16.143 + 5.801X_i$$

Table 16–8 shows the predicted sales values and the errors and squared errors associated with this "best" regression line. **The random errors are commonly called residuals.** From Table 16–8, the sum of the squared residuals is 497.556. This is the smallest sum of squared residuals possible. Any other regression line through these 12 (X, Y) points will produce a larger sum of squared residuals.

TABLE 16–7   Least squares regression coefficients, Bato Tool Company

X	Y	XY	X^2
0.8	22	17.6	0.64
1.0	28	28.0	1.00
1.6	22	35.2	2.56
2.0	26	52.0	4.00
2.2	34	74.8	4.84
2.6	18	46.8	6.76
3.0	30	90.0	9.00
3.0	38	114.0	9.00
4.0	30	120.0	16.00
4.0	40	160.0	16.00
4.0	50	200.0	16.00
4.6	46	211.6	21.16
$\Sigma = 32.8$	$\Sigma = 384$	$\Sigma = 1{,}150.0$	$\Sigma = 106.96$

$$b_1 = \frac{\Sigma XY - \dfrac{\Sigma X \Sigma Y}{n}}{\Sigma X^2 - \dfrac{(\Sigma X)^2}{n}} = \frac{1{,}150.0 - \dfrac{32.8(384)}{12}}{106.96 - \dfrac{(32.8)^2}{12}}$$

$$= 5.801$$

Then

$$b_o = \overline{Y} - b_1\overline{X} = 32 - 5.801(2.733)$$
$$= 16.143$$

The least squares regression line is, therefore,

$$\hat{Y} = 16.143 + 5.801X_i$$

TABLE 16–8   Residuals and squared residuals, Bato Tool Company

X	Y	$\hat{Y}$	Residuals $Y - \hat{Y}$	Squared Residuals $(Y - \hat{Y})^2$
0.8	22	20.784	1.216	1.4786
1.0	28	21.944	6.056	36.6751
1.6	22	25.425	− 3.425	11.7306
2.0	26	27.746	− 1.746	3.0485
2.2	34	28.906	5.094	25.9488
2.6	18	31.226	−13.226	174.9271
3.0	30	33.547	− 3.548	12.5883
3.0	38	33.547	4.453	19.8292
4.0	30	39.348	− 9.348	87.3851
4.0	40	39.348	0.652	0.4251
4.0	50	39.348	10.652	113.4651
4.6	46	42.829	3.170	10.0489
			$\Sigma = 0.000$	$\Sigma = 497.5560$

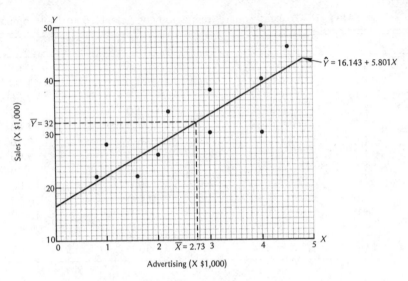

FIGURE 16–8   Least squares regression line, Bato Tool Company.

Figure 16–8 shows the scatter plot of sales and advertising and the least squares regression line for the Bato Tool Company.

Least Squares Regression Properties

Table 16–8 and Figure 16–8 illustrate several important properties of least squares regression:

1. The sum of the residuals from the least squares regression line is zero.

$$\sum_{i=1}^{n}(Y_i - \hat{Y}_i) = 0 \qquad \text{(16–9)}$$

2. The sum of the squared residuals is a minimum. That is,

$$\sum_{i=1}^{n}(Y_i - \hat{Y}_i)^2 \qquad \text{(16–10)}$$

   is minimized. This property provided the basis for developing the equations for b_o and b_1.

3. The simple regression line always passes through the mean of the Y variable, $\bar{Y}$, and the mean of the X variable, $\bar{X}$. This is illustrated in Figure 16–8. Thus, to draw any simple linear regression line, all you need to do is connect the least squares Y-intercept with the $(\bar{X}, \bar{Y})$ point.

4. The least squares coefficients are unbiased estimates of β_o and β_1. Thus the expected values of b_o and b_1 are β_o and β_1, respectively.

16-6
SIGNIFICANCE TESTS IN REGRESSION ANALYSIS

If you recall, Harry Aims found that his company's sales varied from quarter to quarter. The amount of total variation in the dependent variable, sales, is TSS (total sum of squares). From the analysis of variance chapter (Chapter 14), remember that

$$TSS = \sum_{i=1}^{n}(Y_i - \bar{Y})^2 \qquad (16\text{–}11)$$

where: TSS = total sum of squares

n = sample size

Y_i = ith value of the dependent variable

$\bar{Y}$ = average value of the dependent variable

The total sum of squares for Bato Tool's sales is calculated in Table 16–9. As you can see, the total variation in sales that needs to be explained is 1,080.

The least squares regression line *minimized* the sum of squared residuals. This value (see Table 16–8) is called the **sum of squares error** (**SSE**) and is calculated

$$SSE = \sum_{i=1}^{n}(Y_i - \hat{Y}_i)^2 \qquad (16\text{–}12)$$

TABLE 16–9 Calculation of total sum of squares, Bato Tool Company

Quarter	Y	Y²
1	22	484
2	28	784
3	22	484
4	26	676
5	34	1,156
6	18	324
7	30	900
8	38	1,444
9	30	900
10	40	1,600
11	50	2,500
12	46	2,116
	$\Sigma = 384$	$\Sigma = 13{,}368$

$$TSS = \Sigma(Y - \bar{Y})^2 = \Sigma Y^2 - \frac{(\Sigma Y)^2}{n} = 13{,}368 - \frac{(384)^2}{12}$$

$$= 1{,}080$$

where: n = sample size

Y_i = ith value of the dependent variable

$\hat{Y}_i$ = least squares estimated value for the average of Y for each given X value

The SSE represents the amount of variation in the dependent variable that is not explained by the least squares regression line. For Bato Tool, the amount of unexplained variation is 497.556. **The amount of variation in the dependent variable that is explained by the regression line is called the** *sum of squares regression (SSR)* and is calculated

$$SSR = TSS - SSE \tag{16-13}$$

where: SSR = sum of squares regression

TSS = total sum of squares

SSE = sum of squares error

For the Bato Tool Company, the least squares regression line calculated in Table 16-7 produces a sum of squares regression of

$$SSR = 1,080 - 497.556$$
$$= 582.444$$

In Section 16-3 we introduced the simple coefficient of determination. This value, the square of the simple correlation coefficient, is $r^2 = 0.538$ for Bato Tool. Harry Aims was able to interpret this r^2 to mean that advertising could explain 53.8 percent of the variation in sales using the linear model. We can calculate the r^2 a second way.

$$r^2 = \frac{SSR}{TSS} \tag{16-14}$$

where: r^2 = coefficient of determination

SSR = sum of squares regression

TSS = total sum of squares

For the Bato Tool Company,

$$r^2 = \frac{582.444}{1,080}$$

$$= 0.539 \quad \text{(Different from 0.538 due to rounding)}$$

Equation (16-14) shows that **the percentage of explained variation is the ratio of the sum of squares explained to the total sum of squares.**

Significance of the Regression Model

Before he uses the regression model to predict sales values, Harry should find out if the model itself is statistically significant. To test this, he uses the analysis of variance approach shown in Table 16-10. Based on the analysis of variance F test, he con-

TABLE 16–10 Significance test of the regression model, Bato Tool Company

Hypotheses:

H_0: The regression model does not explain any of the total variation in the dependent variable.

H_A: The regression model does explain a proportion of the total variation in the dependent variable greater than 0.0.

$\alpha = 0.05$

Source of Variation	SS	d.f.	MS	F Ratio
Regression	SSR = 582.444	K = 1	SSR/1 = 582.444	MSR/MSE* = 11.70
Unexplained (error)	SSE = 497.556	$n - 2$ = 10	SSE/$(n - 2)$ = 49.7556	
Total	TSS = 1,080.000	$n - 1$ = 11		

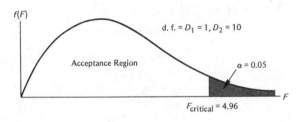

Decision Rule:

If $F > F_{\text{critical}} = 4.96$, reject H_0.

Otherwise, do not reject H_0.

Since $11.70 > 4.96$, we reject H_0 and conclude that the regression model explains a significant amount of variation in the dependent variable.

*MSR = mean square regression; MSE = mean square error.

cludes that the proportion of sales variation explained by the least squares regression line is greater than zero. The analysis of variance test has shown that the sample coefficient of determination is significantly greater than zero.

Note that the tests in Tables 16–5 and 16–10 are equivalent. In Table 16–5 we assumed that just the correlation between sales and advertising was known. In Table 16–10 the test is based on the calculated least squares regression line. The difference in the calculated F values is caused by rounding differences.

The reason for introducing two approaches for testing whether one independent variable can explain a significant proportion of the variation in the dependent variable is that the analysis of variance approach can be generalized for testing the significance of regression models with more than one independent variable. Regression models with more than one independent variable are called **multiple regression models** and are the subject of Chapter 17.

Significance of the Slope Coefficient

Another way of testing the significance of the simple linear regression model is to test whether the true regression slope is zero. Because the b_1 value is calculated from a sample, it is subject to sampling error. Thus, even though b_1 is not zero, we must determine whether its difference from zero is greater than would generally be attributed to sampling error.

If we selected several samples from the same population, and for each sample determined the least squares regression line, we would likely get lines with different slopes and different Y-intercepts. This is analogous to getting different sample means from different samples. And, just as the possible sample means have a standard deviation, the possible regression slopes have a standard deviation, which is

$$\sigma_{b_1} = \frac{\sigma_e}{\sqrt{\Sigma(X - \overline{X})^2}}$$

(16–15)

where: σ_{b_1} = **standard deviation of the regression slope** (called the **standard error of the slope**)

σ_e = **standard error of the estimate**

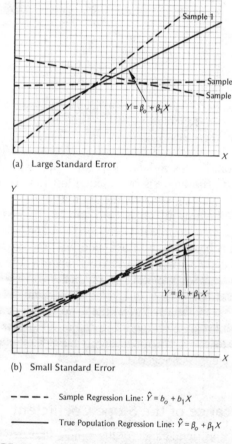

(a) Large Standard Error

(b) Small Standard Error

- - - - Sample Regression Line: $\hat{Y} = b_o + b_1 X$

——— True Population Regression Line: $\hat{Y} = \beta_o + \beta_1 X$

FIGURE 16–9 Standard error of the slope.

Because we are sampling from the population, we estimate σ_{b_1} by

$$S_{b_1} = \frac{S_e}{\sqrt{\Sigma X^2 - \dfrac{(\Sigma X)^2}{n}}} \qquad\qquad (16\text{--}16)$$

where: S_{b_1} = **estimate of the standard error of the least squares slope**

$$S_e = \sqrt{\frac{SSE}{n-2}} = \text{\textbf{sample standard error of the estimate}}$$

For Bato Tool Company, the estimate of the standard error of the slope is

$$S_{b_1} = \frac{\sqrt{\dfrac{497.556}{10}}}{\sqrt{106.96 - \dfrac{(32.8)^2}{12}}}$$

$$= 1.695$$

We have already determined the values needed to calculate S_{b_1}. The numerator (standard error of the estimate) is the square root of the mean square error

TABLE 16–11   Significance test for the regression slope, Bato Tool Company

Hypotheses:

H_0: $\beta_1 = 0$
H_A: $\beta_1 \neq 0$

$\alpha = 0.05$

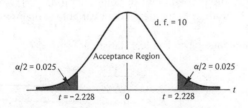

The calculated t is

$$t = \frac{b_1 - \beta_1}{S_{b_1}} = \frac{5.82 - 0}{1.695}$$

$$= 3.43$$

Decision Rule:

If $t > t_{\text{critical}} = 2.228$, reject H_0.
If $t < t_{\text{critical}} = -2.228$, reject H_0.
Otherwise, do not reject H_0.

Since $3.43 > 2.228$, we should reject the null hypothesis and conclude that the true slope is not zero.

in the analysis of variance (Table 16–10). The denominator is the denominator in the equation for b_1, the least squares slope.

If the standard error of the slope is large, the value of b_1 will be quite variable from sample to sample. On the other hand, if S_{b_1} is small, the slope will be less variable. However, regardless of the standard error, the average value of b_1 will equal β_1, the true regression slope, if the assumptions of the regression analysis are satisfied. Figure 16–9 illustrates what we mean. Notice that when the standard error is large, the sample slopes can take on values *much* different from the true population slope. As Figure 16–9(a) shows, a sample slope and the true population slope can even have different signs. However, when S_{b_1} is small, the sample regression lines will cluster closely around the true population line.

Because the sample regression slope will most likely not equal the true population slope, we must test to determine if the true slope could possibly be zero. **A slope of zero in the linear model means that the independent variable will not**

TABLE 16–12   Summary of simple regression steps

Step 1. Develop a scatter plot of Y and X. You are looking for a linear relationship between the two variables.
Step 2. Calculate the correlation coefficient, r. This measures the strength of the linear relationship between the two variables.
Step 3. Test to see whether r is significantly different from zero. The test statistic is $$t = \frac{r}{\sqrt{\dfrac{1 - r^2}{n - 2}}}$$
Step 4. Calculate the simple coefficient of determination, r^2. This value measures the proportion of variation in the dependent variable explained by the independent variable.
Step 5. Test the significance of r^2. The test statistic is $$F = \frac{r^2}{\dfrac{1 - r^2}{n - 2}}$$
Step 6. Calculate the least squares regression line for the sample data.
Step 7. Test to see whether the model is significant. The test statistic is $$F = \frac{\dfrac{SSR}{1}}{\dfrac{SSE}{n - 2}}$$
Step 8. Test to determine whether the true regression slope is zero. The test statistic is $$t = \frac{b_1 - \beta_1}{S_{b_1}}$$

Note: Steps 3, 5, 7, and 8 are equivalent tests for the simple regression model. Only one of these tests needs to be performed.

explain any variation in the dependent variable. To test the significance of a slope coefficient, we use the following t test:

$$t = \frac{b_1 - \beta_1}{S_{b_1}} \tag{16–17}$$

where: t = number of standard errors b_1 is from β_1

b_1 = sample regression slope coefficient

β_1 = hypothesized slope

S_{b_1} = estimate of the standard error of the slope

TABLE 16–13 Computer printout, Bato Tool Company

****

VARIABLE SELECTED IS X

SUM ØF SQUARES REDUCED IN THIS STEP 582.444

PRØPØRTIØN ØF VARIANCE ØF Y REDUCED 5393

CUMULATIVE SUM ØF SQUARES REDUCED 582.444

CUMULATIVE PRØPØRTIØN REDUCED 5393   (ØF 1080)

F FØR ANALYSIS ØF VARIANCE (D.F. = 1, 10) 11.7061
STANDARD ERRØR ØF ESTIMATE 7.05376

VARIABLE	REG. CØEFF.	STD. ERR.-CØEFF.	CØMPUTED T
2	5.80124	1.69556	3.42142

INTERCEPT   16.1433

ØBS. NØ.	Y ØBSERVED	Y ESTIMATED	RESIDUAL
1	28	21.9445	6.05547
2	22	25.4253	− 3.42527
3	26	27.7458	− 1.74576
4	34	28.906	5.09399
5	18	31.2265	− 13.2265
6	30	33.547	− 3.547
7	38	33.547	4.453
8	30	39.3482	− 9.34824
9	40	39.3482	.6511764
10	50	39.3482	10.6518
11	46	42.829	3.17102
12	22	20.7843	1.21572

****

This test has $n - 2$ degrees of freedom. Table 16–11 illustrates this test for the Bato Tool Company, which indicates we should reject the hypothesis that the true regression slope is zero. Thus advertising can be used to help explain the variation in Bato Tool Company sales.

Table 16–12 outlines the steps involved in developing a simple linear regression model and reviews the various tests of significance. You should recognize that the four tests used thus far to examine the linear relationship between X and Y are actually equivalent. Therefore the decision maker needs to perform only one of these tests, since they will all lead to the same conclusion. However you should be familiar with all four since they are all often used.

Computer Application

The calculations required to develop a simple linear regression model can be performed with only minor frustrations on a hand calculator. However many computer programs have been developed that will perform the calculations very quickly and with great accuracy. These computer programs will also calculate the values necessary to test the significance of the regression model. Table 16–13 is an example computer printout of the calculations made in this chapter.

16-7
REGRESSION ANALYSIS FOR PREDICTION

A principal use of regression analysis is **prediction.** In the Bato Tool Company example, Harry Aims wanted to predict sales by knowing advertising expenditures. Regression is used for predictive purposes in applications ranging from predicting demand to predicting production and output levels. For example, your state government may use regression to forecast annual tax revenues so that elected officials can establish state departmental budgets.

Using a regression model to predict the dependent variable is quite straightforward once the model has been found. The regression model for Bato Tool Company sales is

$$\hat{Y} = 16.143 + 5.801X$$

where: $\hat{Y}$ = point estimate for the average sales given X

X = level of advertising in thousands of dollars

To find $\hat{Y}$, the point estimate of expected sales, we substitute the specified advertising level into the regression model. For example, suppose Harry Aims learns that the company's marketing department has decided to spend \$2,500 ($X = 2.5$) on advertising during the next quarter. Then the point estimate of sales, $\hat{Y}$, is

$$\hat{Y} = 16.143 + 5.801(2.5) = 30.6455$$
$$= \$30,645.50$$

Confidence Interval for the Average Y Given X

In Chapter 10 we stated that decision makers cannot be confident of the accuracy of any point estimate. In fact, the best they can hope for is that the point estimate is close to the true value being estimated.

In some cases only a point estimate of the expected value of the dependent variable is required. However, in many other cases the decision maker will want a **confidence interval estimate.** For example, Harry Aims would like a 95 percent confidence interval estimate for average sales given that $2,500 is spent on advertising. The prediction interval for the expected value of a dependent variable, given a specific level of the independent variable, is determined by

$$\hat{Y} \pm t_{\alpha/2} \sqrt{\frac{SSE}{n-2}} \sqrt{\frac{1}{n} + \frac{(X_p - \overline{X})^2}{\sum X^2 - \frac{(\sum X)^2}{n}}} \qquad (16\text{–}18)$$

where: $\hat{Y}$ = point estimate of the dependent variable

t = interval coefficient with $n - 2$ d.f.

n = sample size

X_p = value of the independent variable used to arrive at $\hat{Y}$

$\overline{X}$ = mean of the independent variable observations in the sample

Thus, if $X_p = 2.5$, we get

$$30.6455 \pm 2.228 \sqrt{\frac{497.556}{10}} \sqrt{\frac{1}{12} + \frac{(2.5 - 2.733)^2}{106 - \frac{(32.8)^2}{12}}}$$

$$30.6455 \pm 4.626$$

$$26.019 \underline{\hspace{2cm}} 35.271$$

Therefore Harry Aims can be 95 percent confident that the average sales for a $2,500 advertising expenditure will be between $26,019 and $35,271. He might well use this information to help establish future budgeting policies and future inventory levels.

Confidence Interval for a Single Y Given X

The prediction interval just calculated is for the expected, or average, sales level given a $2,500 advertising expenditure. Harry Aims would likely be more interested in predicting the actual sales next quarter if his company spends $2,500 on advertising. Developing the interval within which he can be 95 percent confident next quarter's sales will fall requires only a slight modification of equation (16–18). This predictive interval is given by

$$\hat{Y} \pm t_{\alpha/2} \sqrt{\frac{SSE}{n-2}} \sqrt{1 + \frac{1}{n} + \frac{(X_p - \overline{X})^2}{\sum X^2 - \frac{(\sum X)^2}{n}}} \qquad (16\text{–}19)$$

For the Bato Tool Company, the 95 percent confidence interval estimate for next quarter's sales, given that $2,500 is spent on advertising, is

$$30.6455 \pm 2.228 \sqrt{\frac{497.556}{10}} \sqrt{1 + \frac{1}{12} + \frac{(2.5 - 2.733)^2}{106 - \frac{(32.8)^2}{12}}}$$

$$30.6455 \pm 16.3820$$

$$14.2635 \underline{\qquad\qquad} 47.0275$$

Thus Harry Aims can be 95 percent confident that sales next quarter will be between $14,263.50 and $47,027.50 if his company spends $2,500 on advertising. As you can see, this estimate has extremely poor precision. Although the regression model explains a significant proportion of variation in the dependent variable, it is relatively imprecise for predictive purposes. To improve the precision, Harry might decrease his confidence requirements or increase the sample size and redevelop the model.

Note that the prediction interval for a single value of the dependent variable is wider (less precise) than the interval for predicting the average value of the dependent variable. This will always be the case, as seen in equations (16–18) and (16–19). From an intuitive viewpoint, we should always come closer to predicting an average value than a single value (for example, although the average weight of the U.S. population will not be above 250 pounds, many people weigh more than that).

Note that the term $(X_p - \overline{X})^2$ has a particular effect on the confidence interval determined by both equations (16–18) and (16–19). The farther X_p (the value of the independent variable used to predict Y) is from $\overline{X}$, the greater $(X_p - \overline{X})^2$. Figure 16–10 shows two regression lines developed from two samples with the same set of X values. We have made both lines pass through the same $(\overline{X}, \overline{Y})$ point; however, they

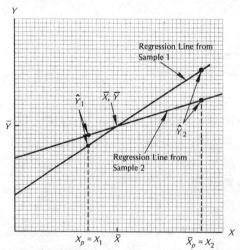

FIGURE 16–10   Regression lines, illustrating the increase in the potential variation in Y as X_p moves farther from $\overline{X}$.

have different slopes and intercepts. At $X_p = X_1$, the two regression lines give predictions of Y that are close to each other. However, for $X_p = X_2$, the predictions of Y are quite different. Thus, when X_p is close to $\overline{X}$, the problems caused by variations in regression slopes are not as great as when X_p is far from $\overline{X}$. Figure 16–11 shows the prediction intervals over the range of possible X_p values. The band around the estimated regression line bends away from the regression line as X_p moves in either direction from $\overline{X}$.

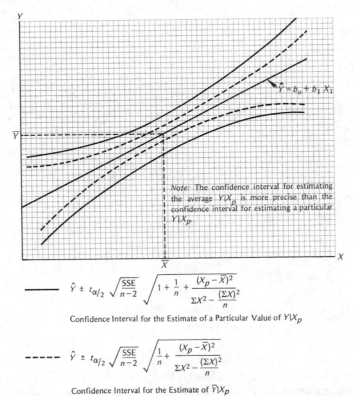

$$\hat{Y} \pm t_{\alpha/2} \sqrt{\frac{SSE}{n-2}} \sqrt{1 + \frac{1}{n} + \frac{(X_p - \overline{X})^2}{\Sigma X^2 - \frac{(\Sigma X)^2}{n}}}$$

Confidence Interval for the Estimate of a Particular Value of $Y|X_p$

$$\hat{Y} \pm t_{\alpha/2} \sqrt{\frac{SSE}{n-2}} \sqrt{\frac{1}{n} + \frac{(X_p - \overline{X})^2}{\Sigma X^2 - \frac{(\Sigma X)^2}{n}}}$$

Confidence Interval for the Estimate of $\overline{Y}|X_p$

FIGURE 16–11   Confidence intervals for $Y|X_p$, and $\overline{Y}|X_p$.

Things to Consider

Decision makers often use regression analysis as a predictive tool. When doing so, they should keep several important things in mind. One consideration is that **the conclusions and inferences made from a regression line apply only over the range of data contained in the sample used to develop the regression line**. For instance, in the Bato Tool Company example, advertising expenditures ranged from 0.8 to 4.6. Therefore predictions for sales based upon advertising expenditures between $800 and $4,600 are justified because the regression model was formed with data within that range. However, if Harry Aims attempts to predict sales with advertising levels outside the range $800–$4,600, he should understand that the relationship between advertising and sales may be different. For example, Figure 16–12 shows a case where the true relationship between advertising and sales is not linear but curvilinear. If a linear regression line were used to predict sales based upon advertising values beyond the relevant range of data, large overpredictions would result. Thus the range of data in the sample, if at all possible, should cover the range of data in the population. If this can be done, decision makers are more apt to recognize the true relationship between the two variables and be able to develop the appropriate regression model.

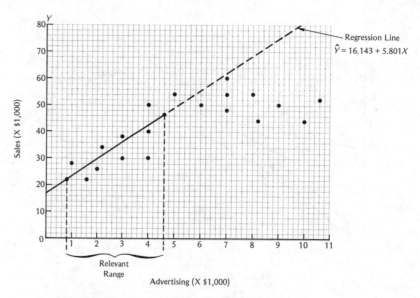

FIGURE 16–12   Problems with extrapolation in the least squares regression model, Bato Tool Company.

A second important consideration, one that was discussed earlier, involves correlation and causation. **The fact that a significant linear relationship exists between two variables does not imply that one variable causes the other. Although there may be a cause-and-effect relationship, decision makers should not infer such a relationship is present based only on regression and/or correlation analysis.**

Decision makers should also recognize that a cause-and-effect relationship between two variables is not necessary for regression analysis to be used for prediction. What matters is that the regression model accurately reflects the relationship between the two variables and that the relationship remains stable.

Finally, many users of regression analysis mistakenly believe that a high simple coefficient of determination (r^2) will guarantee that the regression model will be a good predictor. You should remember that r^2 measures the percentage of variation in the dependent variable explained by the independent variable. While the least squares criterion assures us that r^2 will be maximized, the r^2 applies only to the sample data used to develop the model. **Thus r^2 measures the "fit" of the regression line to the sample data. There is no guarantee that there will be an equally good fit with new data. The only true test of a regression model's predictive ability is how well the model actually predicts.**

16-8

REGRESSION ANALYSIS FOR DESCRIPTION

The purpose of regression and correlation analysis is often to *describe* rather than to predict. For example, the loan manager at a savings and loan might be interested in describing the relationship between a loan's term (number of months) and its dollar value. Although the loan manager has never studied this, she thinks a positive linear relationship exists between time and volume. Smaller loans would tend to be associated with shorter lending periods, whereas larger loans would be for longer periods.

A descriptive analysis using linear regression requires the same steps described in Table 16–10. **The key difference between using regression for description and using it for prediction is that for description, the decision maker concentrates on the significance of the regression slope coefficient, its sign and size, and its standard error.**

For example, suppose the loan manager sampled 60 accounts and found the following information:

$$Y = \text{length of loan period (months)}$$
$$X = \text{dollar amount of loan}$$
$$\Sigma(X - \overline{X})^2 = \$42,000$$
$$b_o = 7.5$$
$$b_1 = 0.120$$
$$SSE = 972$$
$$\text{d.f.} = n - 2 = 60 - 2 = 58$$

In performing the descriptive analysis, the loan officer might first test to see whether the slope coefficient is significantly different from zero.

$$t = \frac{b_1 - \beta_1}{\dfrac{\sqrt{SSE/(n-2)}}{\sqrt{\Sigma(X - \overline{X})^2}}} = \frac{0.120 - 0}{\dfrac{\sqrt{972/58}}{\sqrt{42,000}}}$$

$$= 6.007$$

For any reasonable level of alpha, $t = 6.007$ will exceed the critical value from the t table. Thus the loan officer can reject the hypothesis that the true regression slope is zero. Further, since the sign on the regression slope is positive, she can infer there is a positive linear relationship between the loan's term and its size.

Also, the loan officer would no doubt be interested in developing a confidence interval estimate for the regression slope and interpreting this interval. This would be done as follows:

$$b_1 \pm t_{\alpha/2} S_{b_1}$$

or, equivalently,

$$b_1 \pm t_{\alpha/2} \sqrt{\frac{\dfrac{SSE}{n-2}}{\Sigma(X - \overline{X})^2}} \qquad \text{(16–20)}$$

So, for a 95 percent confidence interval, she would arrive at

$$0.120 \pm 2.0 \sqrt{\frac{\frac{972}{58}}{42,000}}$$

$$0.120 \pm 2.0(0.0199)$$

$$0.120 \pm 0.0399$$

$$0.0801 \underline{\hspace{2cm}} 0.1599$$

The loan officer would be quite confident that, if the amount of a loan is increased by one dollar, the term will be increased by an average of from 0.0801 to 0.1599 month.

There are many other situations where the prime purpose of regression analysis is description. Economists use regression analysis for descriptive purposes as they search for a means of explaining our economy. Market researchers also use regression analysis, among other techniques, in an effort to describe the factor(s) that influence the demand for products.

16-9
REGRESSION ANALYSIS FOR CONTROL

Using regression analysis for **control** is quite common in industrial settings. For example, James Redmond, production manager for the Is-Sweet Sugar Company, is concerned that the Hilo, Hawaii, plant may be using an excessive amount of fresh water in producing packaged sugar. Fresh water is used for many purposes, including cleaning and transporting sugar cane, and cooling the production machinery. The sugar refinery at Hilo is designed to use an average of 10 gallons of water per pound of sugar in the range of 10,000–30,000 pounds per day and 8.75 gallons per pound for production over 30,000 pounds. The relationship between water used and sugar produced is depicted by the line in Figure 16–13.

In the past, the Hilo plant always produced between 10,000 and 30,000 pounds of sugar per day. Mr. Redmond believes that if the Hilo refinery is operating in control, the average increase in water used per pound increase in sugar production should be close to 10 gallons. To determine whether the production control is being maintained, Redmond has selected a random sample of 30 days' production during the past year and calculated the following regression values:

$$b_o = 70,400$$
$$b_1 = 24$$
$$S_{b_1} = 9$$
$$\text{d.f.} = 28$$

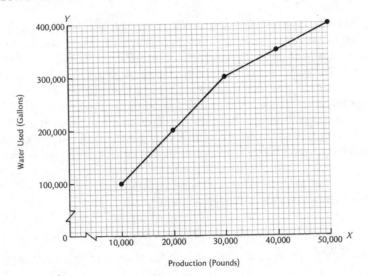

FIGURE 16–13 Plot of water used versus production, Is-Sweet Sugar Company.

To determine whether the production process is in control, a confidence interval estimate of the slope can be developed.

$$b_1 \pm t_{\alpha/2} S_{b_1}$$

For a 90 percent interval, we get

$$24 \pm 1.701(9)$$
$$24 \pm 15.30$$

8.7 gal ——————— 39.3 gal

Mr. Redmond is, therefore, quite certain that the true regression slope is between 8.7 and 39.3. This means that he feels an increase in sugar production of one pound will result in an average increase in water usage between 8.7 and 39.3 gallons. Since the standard of 10 gallons per pound is within this confidence interval, the regression analysis indicates that the production process should be considered in control. However Redmond would likely be concerned by the confidence interval width. While he feels the true slope could be 10, he also feels the true slope could be as high as 39.3. The lack of precision would be bothersome. Further, for given production levels, water usage tends to vary substantially. James Redmond will most likely seek to determine why this variation is occurring and whether or not such variation is expected.

16-10
CONCLUSIONS

In this chapter we have introduced the fundamental concepts of simple linear regression and correlation analysis. The techniques of regression and correlation analysis are widely used in business decision making and data analysis. Regression analysis can be applied as a tool for prediction, description, and control within an organization.

Correlation measures the strength of the linear relationship between two variables. The closer the correlation coefficient is to ± 1.0, the stronger the linear relationship between the two variables. When a dependent and an independent variable are highly correlated, the resulting simple linear regression model will tend to explain a substantial proportion of the variation in the dependent variable.

We have presented a wide variety of statistical tests in this chapter. As we have indicated, many of these tests are equivalent. In this chapter we have limited our discussion to the simple regression model. In the next chapter we introduce multiple regression analysis (more than one independent variable), and at that point the variety of statistical tests will become more useful.

chapter glossary

correlation   A quantitative measure of the linear relationship between two variables. The correlation ranges from $+1.0$ to -1.0. A correlation of ± 1.0 indicates a perfect linear relationship, whereas a correlation of zero indicates no linear relationship.

dependent variable   The variable to be predicted or explained in a regression model. This variable is assumed to be functionally related to the independent variable.

independent variable   A variable related to a response variable in a regression equation. The independent variable is used in a regression model to estimate the value of the dependent variable.

least squares criterion   The criterion for determining a regression line that minimizes the sum of squared residuals.

regression coefficients   In the simple regression model there are two coefficients: the intercept and the slope.

regression slope coefficient   The average change in the dependent variable for a unit change in the independent variable. The slope coefficient may be positive or negative depending upon the relationship between the two variables.

residual   The difference between the actual value of the dependent variable and the value predicted by the regression model.

scatter plot   A two-dimensional plot showing the (X, Y) values for each observation. The scatter plot is used as a picture of the relationship between two variables.

simple coefficient of determination   The square of the correlation coefficient. A measure of the percentage of variation in the dependent variable explained by the independent variable in the regression model.

simple regression analysis   A regression model that uses one independent variable to explain the variation in the dependent variable. The model takes the form

$$Y_i = \beta_o + \beta_1 X_i + e_i$$

spurious correlation   Correlation between two variables that have no known cause-and-effect connection.

Correlation coefficient

$$r = \frac{n\Sigma XY - \Sigma X\Sigma Y}{\sqrt{[n(\Sigma X^2) - (\Sigma X)^2][n(\Sigma Y^2) - (\Sigma Y)^2]}}$$

Test statistic for significance of the correlation coefficient

or #4

$$t = \frac{r}{\sqrt{\dfrac{1 - r^2}{n - 2}}}$$

Simple coefficient of determination

$$r^2 = \frac{SSR}{TSS}$$

Test statistic for significance of r^2

$$F = \frac{r^2}{\dfrac{1 - r^2}{n - 2}}$$

Least squares estimate of the regression slope coefficient

$$b_{o_1} = \frac{\Sigma XY - \dfrac{\Sigma X\Sigma Y}{n}}{\Sigma X^2 - \dfrac{(\Sigma X)^2}{n}}$$

Least squares estimate of the regression intercept coefficient

$$b_o = \overline{Y} - b_1\overline{X}$$

Standard error of the slope coefficient

$$S_{b_1} = \frac{S_e}{\sqrt{\Sigma(X - \overline{X})^2}}$$

Standard error of the estimate

$$S_e = \sqrt{\frac{SSE}{n - 2}}$$

Sum of squares error

$$SSE = \Sigma(Y_i - \hat{Y}_i)^2$$

Total sum of squares

$$TSS = \Sigma(Y_i - \overline{Y})^2$$

Test statistic for significance of the slope coefficient

$$t = \frac{b_1 - \beta_1}{S_{b_1}}$$

Test statistic for significance of the simple regression model

$$F = \frac{\dfrac{SSR}{1}}{\dfrac{SSE}{n-2}}$$

Prediction interval for average Y given X_p

$$\hat{Y} \pm t_{\alpha/2} \sqrt{\frac{SSE}{n-2}} \sqrt{\frac{1}{n} + \frac{(X_p - \bar{X})^2}{\Sigma X^2 - \dfrac{(\Sigma X)^2}{n}}}$$

Prediction interval for a particular Y given X_p

$$\hat{Y} \pm t_{\alpha/2} \sqrt{\frac{SSE}{n-2}} \sqrt{1 + \frac{1}{n} + \frac{(X_p - \bar{X})^2}{\Sigma X^2 - \dfrac{(\Sigma X)^2}{n}}}$$

solved problems

1. The Cal-Pit Water Company located in Pittsburg, California, has selected a random sample of 15 customers to see if it can develop a model for predicting water usage. Because of the short supply of water in the area, predicting water usage is an important part of the company's planning activities.

   The company wishes to see whether a simple linear regression model with family size as the independent variable, and water used as the dependent variable, can be a valid predictive tool. The data based upon last month's water usage are

Customer	Water Used Y	Family Size X
1	1,100 gal	3
2	1,425	5
3,	785	2
4	950	3
5	1,200	4
6	1,152	4
7	973	3
8	1,525	5
9	1,600	4
10	700	3
11	1,100	5
12	1,414	4
13	700	2
14	953	2
15	1,063	2

Solution:

The first step is to develop the scatter plot.

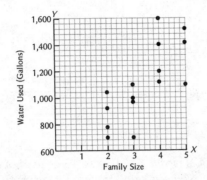

Although not perfect, there does appear to be a positive linear relationship between water used and family size. The next step is to calculate the correlation coefficient, r. Recall that r is a measure of the strength of the linear relationship between two variables and is calculated with the following equation:

$$r = \frac{n\Sigma XY - \Sigma X \Sigma Y}{\sqrt{[n(\Sigma X^2) - (\Sigma X)^2]\,[n(\Sigma Y^2) - (\Sigma Y)^2]}}$$

Instead of calculating this value by hand, a computer program was used, and the resulting printout is

```
*************************************************************************

STEP NUMBER 1

VARIABLE SELECTED IS . . . . . X2

SUM ØF SQUARES REDUCED IN THIS STEP . . . . . . . . . . . . . . . . . . . . . . . . . . . . 622130

PRØPØRTIØN ØF VARIANCE ØF Y REDUCED . . . . . . . . . . . . . . . . . . . . . . . . . . . . .551986


CUMULATIVE SUM ØF SQUARES REDUCED . . . . . . . . . . . . . . . . . . . . . . . . . . . . . . 622130

CUMULATIVE PRØPØRTIØN REDUCED . . . . . . . . . . . . . . . . . . . . . . . . . . . . . . . . .551986


MULTIPLE CØRRELATIØN CØEFFICIENT . . . . . . . . . . . . . . . . . . . . . . . . . . . . . . .742958

F FØR ANALYSIS ØF VARIANCE (D.F. = 1, 13) . . . . . . . . . . . . . . . . . . . . . . . . . 16.0169

STANDARD ERRØR ØF ESTIMATE . . . . . . . . . . . . . . . . . . . . . . . . . . . . . . . . . . 197.084


VARIABLE            REG. CØEFF.        STD. ERR.-CØEFF.         CØMPUTED T
2                   188.011            46.978                  4.00212


INTERCEPT   470.094

*************************************************************************
```

The correlation coefficient in this computer output is labeled as the multiple correlation coefficient. For a simple regression problem, the multiple correlation coefficient is the same as the simple correlation coefficient. In this case, $r = 0.743$. The test of significance is

$$H_0: \rho = 0$$
$$H_A: \rho \neq 0$$
$$\alpha = 0.05$$

$$t = \frac{r}{\sqrt{\dfrac{1 - r^2}{n - 2}}} = \frac{0.743}{\sqrt{\dfrac{1 - (0.743)^2}{13}}}$$

$$= 4.002$$

The critical t for $\alpha = 0.05$ and 13 degrees of freedom is 2.160, and, since $4.002 > 2.160$, we conclude that there is a significant linear relationship between the two variables.

From this output, we can test whether the model explains a significant proportion of the variation in the dependent variable. Note that calculated $r^2 = 0.552$.

Hypotheses:

$$H_0: \rho^2 = 0$$
$$H_A: \rho^2 \neq 0$$
$$\alpha = 0.05$$

$$F = \frac{r^2}{\dfrac{1 - r^2}{n - 2}} = \frac{0.552}{\dfrac{1 - 0.552}{13}}$$

$$= 16.0169 \qquad \text{(This value also appears on the computer output)}$$

Decision Rule:

If $F \leq F_{critical}$, accept H_0.
If $F > F_{critical}$, reject H_0.

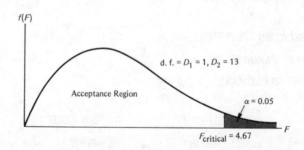

Since $16.0169 > 4.67$, we reject the null hypothesis and conclude that the model does explain a significant proportion of variation in water usage.

Given these two tests (which are equivalent), the next step is to calculate the regression coefficients. This was also done by the computer. The estimates are

$$Y = 470.094 + 188.011X$$

The standard error of the estimate is 197.084. The table of residuals printed by the computer program is

ØBS. NØ.	Y ØBSERVED	Y ESTIMATED	RESIDUAL
1	1100	1034.13	65.8713
2	1425	1410.15	14.8486
3	785	846.117	−61.1173
4	950	1034.13	−84.1287
5	1200	1222.14	−22.1401
6	1152	1222.14	−70.1401
7	973	1034.13	−61.1287
8	1525	1410.15	114.849
9	1600	1222.14	377.860
10	700	1034.13	−334.129
11	1100	1410.15	−310.151
12	1412	1222.14	191.860
13	700	846.117	−146.117
14	953	846.117	106.883
15	1063	846.117	216.883

Although our tests of significance have indicated that family size has a positive linear relationship to water usage, note that family size explained only 55 percent of the variation in water usage for the sample data. Also, the standard error of the estimate, 197.084, is quite large, indicating that any prediction interval developed will be imprecise.

2. Referring to solved problem 1, suppose a family of five is going to be added to the Cal-Pit Water Company's customer list. Using the regression model developed in problem 1, determine the 90 percent confidence interval for the number of gallons of water this family will use.

Solution:

The desired prediction interval format is

$$\hat{Y} \pm t_{\alpha/2} \sqrt{\frac{SSE}{n-2}} \sqrt{1 + \frac{1}{n} + \frac{(X_p - \bar{X})^2}{\Sigma X^2 - \frac{(\Sigma X)^2}{n}}}$$

To obtain the value of $\hat{Y}$, we substitute 5 for X in the regression model.

$$\hat{Y} = 470.097 + 188.011(5)$$
$$= 1,410.152$$

Thus the point estimate of water used by this family of five is roughly 1,410 gallons.

To determine the confidence interval, you should recall that $\sqrt{SSE/(n-2)}$ is the standard error of the estimate. This value is 197.084, and appeared on the computer printout shown earlier. Thus we have

$$1,410 \pm 1.771(197.084) \sqrt{1 + \frac{1}{15} + \frac{(X_p - \overline{X})^2}{\Sigma X^2 - \frac{(\Sigma X)^2}{n}}}$$

Our next step is to complete the calculations under the square root.

Family Size X	X^2
3	9
5	25
2	4
3	9
4	16
4	16
3	9
5	25
4	16
3	9
5	25
4	16
2	4
2	4
2	4
$\Sigma = 51$	$\Sigma = 191$

Then

$$\overline{X} = \frac{51}{15}$$

$$= 3.4$$

Thus the 90 percent prediction interval given $X_p = 5$ is

$$1,410 \pm 1.771(197.084) \sqrt{1 + \frac{1}{15} + \frac{(5 - 3.4)^2}{191 - \frac{2,601}{15}}}$$

$$1,410 \pm 384.28$$

$$1,025.72 \text{ gal} \underline{\hspace{2cm}} 1,794.28 \text{ gal}$$

The Cal-Pit Water Company can be confident that this family of five will use between 1,026 and 1,794 gallons of water per month. Although this might be helpful information, the precision of the estimate is very poor.

Problems marked with an asterisk (*) require extra effort.

1. Describe in your own words the following concepts:
   a. Coefficient of correlation
   b. Scatter diagram
   c. Least squares criterion
   d. Estimate of the regression coefficients
   e. Confidence interval for the regression estimate

2. Discuss how to determine which variable to call the dependent variable and which to define as the independent variable.

3. A national retail chain is experimenting with methods to increase sales of its house brand of cosmetics. A local store has been asked to help determine whether sales level is related to shelf space allocated to the house brand. The shelf space is randomly varied over the next ten weeks, and sales levels are recorded. The following data points are found:

Sales	Shelf Length
$ 868	4 ft
697	3
1,125	6
970	5
742	3
1,035	5
1,203	6
967	5
853	4
730	3

   a. Plot these values as a scatter plot. Analyze your graph.
   b. Determine the least squares line for these points. Plot the least squares line on your scatter plot.

Problems 4–8 relate to the data in problem 3.

4. For the least squares line, determine the correlation coefficient. Is this value statistically significant at the 0.05 level?

5. If the store manager were to decide to allow five linear feet of shelf space for cosmetics, determine the best estimate of the expected sales level. Determine a 95 percent confidence interval for both the average Y given X and the particular Y given X.

6. Determine both a 95 percent and a 98 percent interval for the slope of the least squares line. What does this slope tell the store manager?

7. The store manager has questions about the intercept of the least squares line. She asks, "Am I supposed to believe the results of this analysis when it also tells me I will sell cosmetics with no shelf space?" How would you answer her? Also discuss problems if the store manager asks about putting in 20 feet of shelf space.

8. How much of the variation in sales can be explained by your least squares regression model?

9. A regional farm equipment distributor has experienced great variability in yearly sales. Conventional industry wisdom states this variability is caused by the variability in farm family income. The distributor wants to know if this explanation applies to his sales. He has gathered data on his sales and farm family income since 1973. Note in the following that the income variables have also been expressed in 1973 constant dollars:

Year	Sales (× $1 million)	Income— Current $	Income— 1973 $
1973	412	$ 4,790	$4,202
1974	428	5,030	4,263
1975	531	6,504	5,288
1976	789	11,727	8,817
1977	674	9,232	6,114
1978	621	8,637	5,203
1979	581	7,203	4,093
1980	577	7,870	4,186

   a. Develop two simple regression models relating sales to both income in current dollars and income in constant dollars.
   b. Develop a scatter plot for both relations.
   c. Determine a least squares line for both relationships. Plot these lines on the scatter plots. Based on your least squares line and scatter plot, which relationship do you feel provides the best explanation of sales variability?

Problems 10–13 refer to the data presented in problem 9.

10. Determine the correlation coefficient for each of the least squares lines just calculated. Do these values indicate which line best fits the sales data? Is either correlation coefficient significant at the 0.05 level?

11. If 1981 per-family farm income is estimated to be $8,130 in current dollars, what is the best single point estimate of the equipment distributor's sales? Develop a 98 percent confidence interval sales estimate for the average sales level given current income of $8,130.

12. How much of the variability in sales can be explained by each of the least squares lines you have determined? How much practical significance would you attach to any prediction made using each of these models?

13. Determine a 95 percent confidence interval for the slope of both least squares regression lines.

*14. Henry Prince has served as a consultant to the Department of Education for several years. Recently he was asked to make a study of high school students to obtain information about how television viewing habits are related to academic performance. Possibly because Henry is always paid by the hour, he has decided to perform two studies. In the first he collected data from 50 randomly selected students on two variables: hours of television watched per week and grade-point average during a given period. In the second study he collected data from 100 students on two variables: number of hours per week working at a paying job and grade-point average during a given period.

Table P–1 shows a partial computer printout for the first study. Table P–2 shows a partial computer printout for the second study. Using the information in these printouts, and some insight of your own, answer problems 15–20. Your responsibility in *this* problem is to fill in the missing values in Tables P–1 and P–2.

TABLE P–1   Study 1 computer printout
****

CØRRELATIØN CØEFFICIENTS

	Y	X
Y	1.0000	− .4926
X	− .4926	1.0000

- -

DEPENDENT VARIABLE   Y   GRADE PØINT AVERAGE

INDEPENDENT VARIABLE   X   TV HØURS

- -

MEAN X   20.0

- -

R SQUARE   .2426

STANDARD ERRØR ØF ESTIMATE   .5200

- - - - - - - - - - - - - - -VARIABLES IN THE EQUATIØN- - - - - - - - - - - - - - - -

VARIABLE	B	STD. ERRØR B	T VALUE
X	− .0015	.000382	
(CØNSTANT)	2.5300		

****

TABLE P–2   Study 2 computer printout

```
****************************************************************************
DEPENDENT VARIABLE   Y   GRADE PØINT AVERAGE

INDEPENDENT VARIABLE  X   HØURS WØRKED

- - - - - - - - - - - - - - - - - - - - - - - - - - - - - - - - - - - - - - - - -

R SQUARE   .1600

- - - - - - - - - - - - - - - - - - - - - - - - - - - - - - - - - - - - - - - - -
```

ANALYSIS ØF VARIANCE	D.F.	SUM ØF SQUARES	MEAN SQUARE	F
REGRESSIØN				
RESIDUAL (UNEXPLAINED)	—	—		
	99	550.0		

```
- - - - - - - - - - - - - - -VARIABLES IN THE EQUATIØN- - - - - - - - - - - - - -
```

VARIABLE	B
X	.0100
(CØNSTANT)	2.2500

```
****************************************************************************
```

15. In his report to the Department of Education, Henry Prince has indicated that the number of hours a student watches television each week is not a significant variable for explaining the variation in student grade-point average. He states that he tested this at the $\alpha = 0.05$ level.

       Based upon the information provided, do you agree with Henry's conclusion? Discuss why or why not.

*16. In the random sample of 50 students in study 1, the number of hours of television ranged from a low of 8 to a high of 40 per week. In his report, Henry states that based upon the regression model, if no hours are spent watching television, students will have an average grade-point average of 2.530.

       Why do you suppose he came to this conclusion? Discuss whether you agree or disagree with him and indicate why.

17. In study 2, where hours worked ranged from zero hours to 20 hours, Henry Prince concluded that the independent variable, hours worked, and the dependent variable, grade-point average, are significantly correlated at the $\alpha = 0.05$ level. Therefore he has stated that a student can increase his or her grade-point average by working at a paying job and should be encouraged to do so. Further, the more hours the student works, the higher the grade-point average will be. Support or refute Henry's statement.

18. In his report on study 2, Henry has indicated that he tested the significance of the model using the analysis of variance approach. However he failed to include the results of the test in his report.

       Using the information in Table P–2, test the null hypothesis that the regression model is not significant using the analysis of variance approach. Use $\alpha = 0.05$. Discuss why a large F ratio should lead to the rejection of the null hypothesis. Also, using the information in the analysis of variance table, determine the standard error of the estimate and discuss briefly what it measures.

19. With respect to study 1, analyze the calculated regression slope coefficient. In your analysis, interpret this slope coefficient relative to this study. In addition, develop a 90 percent confidence interval estimate for the regression slope and discuss what this interval means.

*20. A senator from Missouri has read Henry's report and has expressed strong dissatisfaction with study 1. It seems the senator decided to apply the regression model to her own daughter. She observed that her daughter watched television an average of 15 hours each week and that her actual grade-point average during the observed period was 3.40. The senator claims the model is invalid because the predicted grade-point average for 15 hours of television is $\hat{Y} = 2.507$.

Comment on the senator's claim. In your comment introduce the senator to the concept of sampling error and the idea of confidence intervals. Does the senator's daughter have a grade-point average within the limits of a 95 percent confidence interval?

21. Cal Maxwell teaches ethical standards at a major university. Cal has grown to despise the grading procedure. In particular, he dislikes taking the time necessary to do what he considers an adequate job and always feels somewhat uncomfortable with the grades when he is finished.

Cal's school assigns grades on a 100-point system, and Cal decides to check the relationship between the grades he assigns and the students' averages up to the time they take his course. A sample of his most recent class yields the following results:

Cal's Grade	Student's Average
90	83
60	57
72	80
77	69
96	95
90	84
79	84
65	74
80	84
75	72

a. Develop a scatter plot for these data.
b. Develop the least squares regression line estimating the relationship between grades in Cal's class and previous academic performance.

*22. Cal Maxwell, in problem 21, is considering using a student's previous grades to help determine the grades he gives the student. How would you feel about your grade being determined in this manner? Discuss your answer in statistical, not emotional, terms. You may want to cover ideas such as explained variation, confidence intervals of the estimate, sampling problems, and so on.

23. Referring to problems 21 and 22, develop a 95 percent confidence interval estimate of the grade a particular student would receive in Cal Maxwell's class if her average in other classes is 95. Be sure to interpret your results.

24. With respect to problems 21 and 22, develop a 95 percent confidence interval for estimating the average grade students would be exected to get in Maxwell's class if their average in other classes is 95. Interpret your results.

*25. Referring to problems 23 and 24, comment on the difference in precision of the two interval estimates and discuss why this difference should be expected.

26. The Rio-River Railroad, headquartered in Santa Fe, New Mexico, is trying to devise a method for allocating fuel costs to individual railroad cars on a particular route between Denver and Santa Fe. The railroad feels that fuel consumption will increase as more cars are added to the train, but it is uncertain how much cost should be assigned to each additional car. In an effort to deal with this problem, the cost-accounting department has randomly sampled ten trips between the two cities and recorded the following data:

Rail Cars X	Fuel (units/mile) Y
18	55
18	50
35	76
35	80
45	117
40	90
37	80
50	125
40	100
27	75

Develop a scatter plot for these two variables and comment on the apparent relationship between fuel consumption and the number of rail cars on the train.

27. Referring to problem 26, compute the correlation coefficient between fuel consumption and train cars. Test statistically the hypothesis that the true correlation is zero using $\alpha = 0.05$. Comment on the results of this test. Do these results necessarily indicate that adding more cars will increase the fuel usage?

28. Referring to problem 26, develop the least squares regression model to help explain the variation in fuel consumption. Interpret the results and clearly show the least squares equation.

29. Referring to problems 26–28, how much of the variation in fuel consumption can be explained by knowing the number of rail cars on the train? What is the proper measure of variation explained?

*30. Suppose the fuel cost for the Rio-River Railroad in problems 26–29 is $10 per unit. What is the point estimate (based upon the regression model developed in problem 28) for the average cost of fuel per mile of each additional railroad car added to the train?

31. Referring to problems 26–30, develop a 95 percent confidence interval estimate for the average change in fuel cost per mile for each additional railroad car added to the train. Interpret your results. Assume the fuel cost is $10 per unit.

32. Referring to problems 26–31, what other variables might help explain the variation in train fuel consumption? Discuss.

==**cases**
==

16A STATE SOCIAL SERVICES

The State Social Services Department has been affected by a recent tax limitation bill passed by the state legislature. Although the director, Allan Bixby, made a logical defense of the department's increased needs, his budget was not increased this year. At the same time, the legislative committee overseeing the Social Services Department indicated it expected the level of services to remain the same. The exact words of one senator were: "You guys will have to learn to get more out of your people."

Unfortunately Allan and all other department heads are constrained by very stringent civil-service regulations. Once a person has been hired and has been on the job a year, he or she is extremely difficult to get rid of. Allan has concluded that his best chance to increase the productivity of the work force is to hire better people. Fortunately the turnover in his department is high enough that this could be a relatively effective method. The major problem is in devising a method to hire better people.

Claudette Chambers, chief researcher for the department, has devised an entrance examination she claims will separate good workers from poor workers. This test is designed to determine a person's attitude toward work and whether he or she is attracted by secure conditions or is result-oriented. Claudette claims the higher a person scores on the test, the more effective he or she will be on the job.

Claudette has given her test to a sample of workers presently with the department. As a measure of their productivity level, she used an average of their last two job evaluations. She found the following results:

Test Score (100 points possible)	Job Evaluation (50 points possible)
71	43
58	37
91	47
86	42
97	48
65	38
78	40
82	42
89	49
74	43
69	38

Claudette claims these figures show an obvious relationship between her test and job performance and recommends the test be used to screen all future applicants.

16B CONTINENTAL TRUCKING

Norm Painter is the newly hired cost analyst for Continental Trucking. Continental is a nationwide trucking firm, and until recently, most of its routes were driven under regulated rates. These rates were set to allow small trucking firms to earn an adequate profit, and there was little incentive to work to reduce costs by efficient management techniques. By far the greatest effort was trying to influence regulatory agencies to grant rate increases.

A recent rash of deregulation moves has made the long-distance trucking industry more competitive. Norm has been hired to analyze Continental's whole expense structure. As part of this study, Norm is looking at truck repair costs. Since the trucks are involved in long hauls, they inevitably break down. Up to now, little preventive maintenance has been done, and if a truck broke down in the middle of a haul, either a replacement tractor was sent or the haul was finished by an independent contractor. The truck was then repaired at the nearest local shop. Norm is sure this procedure has been much more expensive than if major repairs had been made before they caused trucks to fail.

Norm feels some method needs to be found for determining when preventive maintenance is needed. He feels that fuel consumption is a good indicator of possible breakdowns. The idea is that as the trucks begin running badly, they will consume more fuel. Unfortunately the major determinants of fuel consumption are the weight of the truck and head winds. Norm picks a sample of a single truck model and gathers data relating fuel consumption to truck weight. All trucks in the sample were in good condition. He separates the data by direction of the haul, feeling winds tend to blow predominately out of the west.

East-West Haul		West-East Haul	
Miles/Gallon	Haul Weight	Miles/Gallon	Haul Weight
4.1	41,000 lb	4.3	40,000 lb
4.7	36,000	4.5	37,000
3.9	37,000	4.8	36,000
4.3	38,000	5.2	38,000
4.8	32,000	5.0	35,000
5.1	37,000	4.7	42,000
4.3	46,000	4.9	37,000
4.6	35,000	4.5	36,000
5.0	37,000	5.2	42,000
		4.8	41,000

Although he can gather future data on fuel consumption and haul weight rapidly, now that Norm has these data, he is not quite sure what to do with them.

references

AAKER, DAVID. *Multivariate Analysis in Marketing: Theory and Application.* Belmont, Calif.: Wadsworth, 1971.

DRAPER, N. R. and H. SMITH. *Applied Regression Analysis.* New York: Wiley, 1966.

KLEINBAUM, DAVID G. and LAWRENCE L. KUPPER. *Applied Regression Analysis and Other Multivariate Methods.* North Scituate, Mass.: Duxbury, 1978.

NETER, JOHN and WILLIAM WASSERMAN. *Applied Linear Statistical Models.* Homewood, Ill.: Irwin, 1974.

RICHARDS, LARRY and JERRY LACAVA. *Business Statistics: Why and When.* New York: McGraw-Hill, 1978.

17

Introduction to Multiple Regression Analysis

We indicated in Chapter 16 that decision makers often need to consider the relationship between two variables when analyzing a problem. We introduced simple linear regression and correlation analysis, which together provide a basis for analyzing two variables and their relationship to each other.

As you might expect, decision makers' problems are not limited to only two variables. Most practical situations involve analyzing the relationship between three or more variables. For example, a vice-president of planning for an automobile manufacturer would be interested in the relationship between her company's automobile sales and the variables that influence those sales. Included in her analysis might be such independent or explanatory variables as automobile price, competitors' sales, and advertising; and such economic variables as disposable personal income, the inflation rate, and the unemployment rate. The simple regression and correlation analysis techniques discussed in Chapter 16 do not allow analysis of more than two variables. This chapter introduces the extension of the simple regression methods—***multiple regression analysis.***

471

chapter objectives

In this chapter we will introduce multiple regression analysis. We will discuss how a multiple regression model is developed and how it should be applied and interpreted in a business setting.

We will also discuss how to incorporate both qualitative and quantitative variables in a multiple regression model. Finally, we will consider some potential problems in using multiple regression analysis and the possible consequences if a regression model is improperly developed.

student objectives

After studying the material in this chapter, you should be able to:

1. Recognize the need for multiple regression analysis in a decision-making situation.
2. Analyze the computer output for a multiple regression model and interpret the regression statistics.
3. Test hypotheses about the significance of a multiple regression model and test the significance of the independent variables in the model.
4. Recognize potential problems when using multiple regression analysis, including the problems of multicollinearity.
5. Incorporate qualitative variables into the regression model by using dummy variables.

17-1

A NONQUANTITATIVE ANALOGY FOR MULTIPLE REGRESSION ANALYSIS

Suppose for a minute you are a manager solely responsible for completing a labor-intensive job. Figure 17–1 illustrates in symbolic form that the job is to fill the box as full as possible.

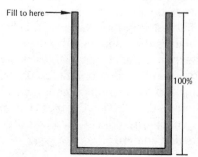

FIGURE 17–1 The symbolic job.

Since managers supervise other people, your first step would be to hire the people necessary to do the job. You might start by searching for and screening applicants. You would be interested in each applicant's ability to do the job and work with other employees. Since all the prospective applicants are not equal, you would likely hire the best available person given your wage limitations.

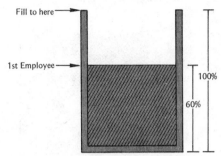

FIGURE 17–2   The symbolic job, one employee.

Figure 17–2 illustrates the first employee's contribution to completing the task. Alone, he or she can perform 60 percent of the total work—the box is 60 percent full (or 40 percent empty depending upon your outlook). Although you might be satisfied with this performance, the box is not yet full.

Your next step is to search for a second person to help fill the box. Ideally this person would *complement* the first employee. He or she could bring new skills and talents to the job such that his or her total effort would be allotted to the undone 40 percent. Unfortunately, perfect complements are hard to locate. Quite likely, you would have to settle for someone who would add some new skills, but who would also overlap your first employee. You would hire the person who would add the most while overlapping the least. You would expect some overlap; however, too much would result in the two employees "stumbling" over each other, with productivity actually going down. Your job as a manager is to decide whether adding a second employee is cost-justified.

Suppose a second person is hired and together the two employees complete 75 percent of the task (see Figure 17–3). Figure 17–3 shows both an overlap and a net addition to task completion. As you can see, because of the overlap, you would not get full productivity from the latest employee, and quite likely you would lose some productivity from the first.

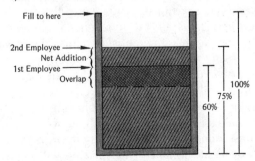

FIGURE 17–3   The symbolic job, two employees.

You continue trying to fill the box until you reach the point indicated in Figure 17–4. Now the box has been 83 percent filled by five employees. Based on the available applicants and a diligent recruiting effort, you conclude that this is as far as you can go. The box will remain 17 percent empty. To the extent this is satisfactory, you have succeeded as a manager.

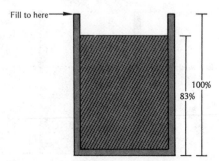

FIGURE 17–4   The symbolic job, five employees.

The concept discussed thus far was originally presented by the economist David Ricardo as **_diminishing marginal productivity._** However management writers, starting with those involved in the Hawthorne Studies, have recognized an additional factor. That is, in this case, that not all groups of five workers will be able to fill the box 83 percent full. As a manager, you are not only interested in hiring five people, but in hiring the _right_ group of five. Therefore what is important in adding new workers is not only their individual abilities but also how they work together as a group. You would search for the best combination of workers to perform the task.

If you have followed this example, you already understand the basic logic of multiple regression analysis. Our analogy, although not perfect, closely parallels the steps in developing a multiple regression model.

17-2
THE ANALOGY EXPLAINED IN REGRESSION TERMS

Think of the empty box (the job) as the _total variation_ in a dependent variable, like the variation in automobile sales or the variation in the appraised values of pieces of property. The job applicants are the _independent variables_ that could explain this variation. There may be many potential variables. However, because of data-collection problems, many potentially good variables may have to be eliminated, just as the manager is often unable to hire a good applicant because of cost or other factors. Other variables will simply not be correlated with the dependent variable and thus have nothing to offer by way of explaining variation in the dependent variable. These are analogous to employees with the wrong job skills.

The best independent variable, by itself, is the one most highly correlated with the dependent variable. This variable will explain the greatest per-

centage of the total variation in the dependent variable. After the first independent variable has been selected, the net benefit of each of the other variables will depend on its relationship with the first independent variable as well as with the dependent variable. Ideally the independent variables are just that—*independent*. However, to be statistically independent, the variables must be uncorrelated with one another. Just as we rarely find individuals whose talents are totally nonoverlapping, we rarely find a group of explanatory variables that are uncorrelated.

In selecting the regression model's independent variables, we try to pick those that are highly correlated with the dependent variable and minimally correlated with each other. We recognize the potential problems if the "overlap" among independent variables is too great. Later in this chapter we will discuss some indications of whether correlation among independent variables is causing problems.

Finally, we must measure the cost-effectiveness of adding each additional independent variable to the model. As we will show, the cost of including a variable is a loss in degrees of freedom. The *adjusted coefficient of determination* will be used to determine the cost-effectiveness of including each additional variable.

Developing a Multiple Regression Model

A simple regression model has the form

$$Y_i = \beta_o + \beta_1 X_i + e_i \tag{17-1}$$

where: β_o = regression intercept

β_1 = regression slope

e_i = random error

If we assume that the expected value of $e_i = 0$, the regression model is

$$E[Y] = \beta_o + \beta_1 X_1 \tag{17-2}$$

The simple regression model is characterized by two variables: Y, the *dependent variable*, and X_1, the *independent* or *explanatory variable*. The single independent variable explains some variation in the dependent variable, but unless X and Y are perfectly correlated, the proportion explained will be less than 100 percent. This means the error term, e_i, will be present. Recall that e_i is the *residual* and is the difference between the true regression line and the actual Y value. In Chapter 16 we assumed that these e_i values have a mean of zero and a standard deviation called the *standard error of the estimate*. If this standard error is too large, the regression model may not be very useful for prediction.

As we have implied, in multiple regression analysis additional independent variables are added to the regression model to explain some of the yet-unexplained variation in the dependent variable. Adding appropriate additional variables should thereby reduce the standard error of the estimate.

The general format of a multiple regression model is

$$Y_i = \beta_o + \beta_1 X_{1i} + \beta_2 X_{2i} + \ldots + \beta_K X_{Ki} + e_i \tag{17-3}$$

where: β_o = regression constant

β_1 = regression coefficient for variable X_1

.

.

.

β_K = regression coefficient for variable X_K

K = number of independent variables

e_i = random error

There are three general assumptions of the linear multiple regression model:

1. The random errors are normally distributed.
2. The mean of the random error terms is zero.
3. The error terms have a constant variance, σ^2, for all combined values of the independent variables.

Since the random error terms have a mean of zero, the expected value of Y for given values of $X_1, X_2, X_3, \ldots, X_K$ is

$$E[Y] = \beta_o + \beta_1 X_1 + \beta_2 X_2 + \beta_3 X_3 + \ldots + \beta_K X_K \qquad (17\text{--}4)$$

As you can see, this model is an extension of the simple regression model of Chapter 16 and equation (17–2). The principal difference is that, whereas equation (17–2) for the simple model is the equation for a straight line in a two-dimensional space (see Figure 17–5), the multiple regression model forms a **hyperplane** (or **response surface**) through multidimensional space. Each regression coefficient represents a slope. When there are only two independent variables, the regression plane can be drawn as shown in Figure 17–6. If there are more than two independent variables, the regression model cannot be visually represented, but the concept remains intact. In the next section we show how a multiple regression model is developed.

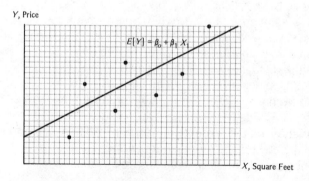

FIGURE 17–5 Simple regression model.
(dependent variable—price; independent variable—square feet)

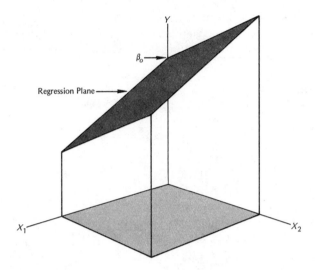

FIGURE 17–6   Illustration of a regression plane.

17-3
MOON REALTY'S APPRAISAL MODEL

Allen Rogers is a managing partner in Moon Realty, a fairly large real estate firm in Idaho. Over the past few years the real estate market in Idaho has been booming. Prices on residential and commercial properties have been soaring. However, in the residential area particularly, Moon Realty has seen a growing problem. Homeowners are having increasing difficulty knowing the value of their property when it comes time to sell. This is a problem for Moon Realty as well as for the homeowner. An overpriced house is costly for the firm to handle due to its low sale potential. Conversely, a home priced below market loses money for the homeowner and commissions for the real estate firm.

To protect his firm's interests and to provide a needed service for the company's clients, Rogers would like to develop a residential real estate appraisal model. He has read about cities and counties that have developed successful models using multiple regression analysis. He thinks his firm might also use this approach.

Through the cooperation of the local multiple listing service, Rogers has acquired a random sample of 531 houses that have sold within the last three months. He has data on the following variables for each house:

$$Y = \text{sales price}$$
$$X_1 = \text{square feet}$$
$$X_2 = \text{age of house}$$
$$X_3 = \text{number of bedrooms}$$
$$X_4 = \text{number of bathrooms}$$
$$X_5 = \text{number of fireplaces}$$

location (dummy variable)

Table 17–1 illustrates how the data would be arranged. Table 17–2 presents the computer output of the mean and standard deviation for each variable. The standard deviation for price is very high, indicating a large variation in selling prices for houses.

TABLE 17–1 Moon Realty data

Observation	Y	X_1	X_2	X_3	X_4	X_5
1	21,400	1,410	3.0	2.0	1.0	0
2	37,275	1,725	5.0	3.0	2.5	1
.	.	.	.	.	.	.
.	.	.	.	.	.	.
.	.	.	.	.	.	.
531	47,175	2,250	1.0	4.0	2.75	2

TABLE 17–2 Means and standard deviations, Moon Realty

```
********************************************************************************
```

VARIABLE	MEAN	STD. DEV.	CASES
Y	45009.6139	14290.2132	531
X1	1715.5273	588.1123	531
X2	4.3936	7.7781	531
X3	3.4038	.0555	531
X4	1.9143	.6094	531
X5	.9209	.5608	531

```
********************************************************************************
```

Correlation Matrix

The first step in developing the appraisal model is to examine the relationship between each independent variable and the dependent variable, sales price. **As with simple regression analysis, we can measure the strength of the linear relationship between any two variables by calculating the correlation coefficient for each pair of X, Y variables** using

$$r = \frac{n\Sigma XY - \Sigma X\Sigma Y}{\sqrt{[n(\Sigma X^2) - (\Sigma X)^2][n(\Sigma Y^2) - (\Sigma Y)^2]}} \qquad (17\text{–}5)$$

The correlation matrix found by the computer program is shown in Table 17–3. Note that the correlation between each X and the Y variable is given, plus the correlation between each pair of independent variables. The correlation matrix is very useful for determining which independent variables are likely to help explain variation in the dependent variable. We look for correlations close to ± 1.0. Also, we can use the correlation matrix to determine the extent of overlap between the independent variables.

TABLE 17-3 Correlation matrix computer output, Moon Realty

	Y	X1	X2	X3	X4	X5
Y	1.000	.841	-.068	.494	.720	.599
X1	.841	1.000	.054	.644	.680	.589
X2	-.068	.054	1.000	.007	-.149	.086
X3	.494	.644	.007	1.000	.551	.338
X4	.720	.680	-.149	.551	1.000	.518
X5	.599	.589	.086	.338	.518	1.000

In Chapter 16 we discussed the significance test for the correlation coefficient, which is

$$t = \frac{r}{\sqrt{\dfrac{1 - r^2}{n - 2}}} \qquad (17\text{--}6)$$

If the calculated t value falls in the rejection region located in either tail of the t distribution, we conclude that the correlation is significantly different from zero.

For example, the correlation between house price and X_1, square feet, is $r = 0.841$. The t test will lead us to conclude that there is a significant linear relationship between house size and house price. The bigger the house, the higher the price tends to be. However the correlation between price and X_3, home age, is insignificant. Thus we suspect age will *not* be a good independent variable to use in the regression model. Further t tests will show that other independent variables are significantly correlated with the dependent variable and represent possible explanatory variables.

The Regression Model

Allen Rogers's goal is to develop a regression model to predict the appropriate selling price for a home using certain measurable characteristics. The first attempt at developing the model will be to run a multiple regression computer program using all available independent variables except X_2, age. The resulting output is shown in Table 17-4. Thus the multiple regression model is

$$Y = 11{,}549.14 + 16.44X_1 - 2{,}845.89X_3 + 6{,}599.24X_4 + 2{,}507.93X_5$$

To obtain a sales price point estimate for any house, we could substitute values for X_1, X_3, X_4, and X_5 into this regression model. For example, suppose a person arrives at Moon Realty with a house with the following characteristics:

$$X_1 = \text{square feet} = 2{,}100$$
$$X_3 = \text{number of bedrooms} = 4$$
$$X_4 = \text{number of baths} = 1.75$$
$$X_5 = \text{number of fireplaces} = 2$$

TABLE 17–4 Regression model computer output, Moon Realty
****

R SQUARE   .76686

ADJUSTED R SQUARE   .76509

STANDARD ERROR   6926.16662

- -

ANALYSIS ØF VARIANCE	D.F.	SUM ØF SQUARES	MEAN SQUARE	F
REGRESSIØN	4	82998243602.67404	20749560900.66851	432.53678
RESIDUAL (UNEXPLAINED)	526	25233158443.18288	47971784.11251	

- - - - - - - - - - - - - - - - VARIABLES IN THE EQUATIØN- - - - - - - - - - - - - - -

VARIABLE	B	STD. ERRØR B	T
X1 SQ. FT.	16.43956	.84287	19.504
X3 BEDRØØMS	− 2845.89318	616.91611	− 4.613
X4 BATHS	6599.24009	604.46151	9.367
X5 FIREPLACES	2507.93424	681.93968	3.677
(CØNSTANT)	11549.14261		

****

The point estimate for the sales price is

$$\text{Price} = 11{,}549.14 + 16.44(2{,}100) - 2{,}845.89(4) + 6{,}599.24(1.75)$$
$$+ 2{,}507.93(2.0)$$
$$= \$51{,}233.32$$

Inferences about the Regression Model

Before Allen Rogers and his real estate firm actually use this regression model to estimate the value of a house, there are several questions that need answers:

1. Is the overall model significant?
2. Are the individual variables significant?
3. Is the standard error of the estimate too large to provide meaningful results?
4. Do the signs on the regression coefficients make sense?

We shall answer each of these questions in order.

Is the Overall Model Significant?   In Chapter 16 we introduced the coefficient of determination, which measures the percentage of variation in the dependent variable that can be accounted for by the independent variables. With multiple regression, we use a similar value, **R square,** which is calculated as follows:

$$\text{R square} = R^2 = \frac{\text{sum of squares regression}}{\text{total sum of squares}} = \frac{\text{SSR}}{\text{TSS}} \qquad \textbf{(17–7)}$$

The R square (***multiple coefficient of determination***) for Moon Realty is given in the computer printout in Table 17–4 as 0.76686. Therefore almost 77 percent of the variation in sales price can be explained by the four independent variables in the regression model.

We showed in Table 16–10 of Chapter 16 that an analysis of variance F test can be used to test the overall significance of the regression model. The computer printout in Table 17–4 provides the analysis of variance test for the Moon Realty example. To test the model's significance, we compare the calculated $F = 432.54$ with a table F value for a given alpha level and 4 and 526 degrees of freedom. If we specify alpha at 0.01, the test is as shown in Figure 17–7.

Hypotheses:

H_0: The regression model does not explain a significant proportion of the total variation in the
    dependent variable (model is not significant). ($\beta_1 = \beta_3 = \beta_4 = \beta_5 = 0$)
H_A: The regression model does explain a significant proportion of the total variation in the
    dependent variable (model is significant). (At least one $\beta_j \neq 0; j = 1, 3, 4, 5$)

$\alpha = 0.01$

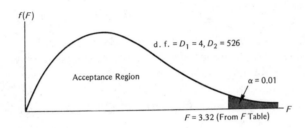

Decision Rule:

If $F > F_{critical} = 3.32$, reject H_0.
Otherwise, do not reject H_0.

The calculated F (see Table 17–4) is 432.54. Since $432.54 > 3.32$, the null hypothesis should be rejected.

FIGURE 17–7   Significance test, Moon Realty.

Clearly, based upon the analysis of variance F test, Allen Rogers should conclude that the regression model *does* explain a significant proportion of the variation in sales price.

The computer printout in Table 17–4 also provides a measure called the ***adjusted R square.*** This is calculated

$$\text{Adjusted } R \text{ square} = R_A^2 = 1 - (1 - R^2)\left(\frac{n - 1}{n - K - 1}\right) \tag{17–8}$$

where: n = sample size

        K = number of independent variables in the model

In the real estate example,

$$R_A^2 = 1 - (1 - 0.76686)\left(\frac{531 - 1}{531 - 4 - 1}\right)$$

$$= 0.76509$$

Adding more independent variables to the regression model can only increase R^2. However the cost in terms of losing degrees of freedom may not justify adding an additional variable. The R_A^2 value takes into account this cost and adjusts the R^2 value accordingly. Therefore, if a variable is added that does not contribute its "fair share," the R_A^2 will actually decline.* Decision makers should pay attention to both R^2 and R_A^2 when analyzing a regression model.

Are the Individual Variables Significant? We have concluded that the overall model is significant. This means that *at least one* independent variable explains a significant proportion of the variation in sales price. This does not mean that *all* the variables are significant.

We can test the significance of each independent variable using a *t* test as discussed in Chapter 16. The calculated *t* value for each variable is provided on the computer printout in Table 17–4. The test for each variable is performed in Figure 17–8. These *t* tests are *conditional* tests. This means the hypothesis that the value of each slope coefficient is zero is made recognizing that the other independent variables are already in the model. Based on the *t* tests in Figure 17–8, we conclude that all four independent variables in the model are significant.

Is the Standard Error of the Estimate Too Large? The standard error of the estimate (S_e) measures the dispersion of observed sales values, Y, around values predicted by the regression model. Sometimes, even though the model has a high R^2, the standard error of the estimate will be too large to provide adequate precision for the prediction interval. A rule of thumb we have found useful is to examine the range†

$$\pm 2S_e$$

If this range is acceptable from a practical viewpoint, the standard error of the estimate should be considered acceptable.

In the Moon Realty example, as shown in Table 17–4, the standard error is 6,926.17. Thus the rough prediction range is

$$\pm\ 2(6,926.17)$$
$$\pm\ \$13,852.34$$

From a practical viewpoint, this range is *not* acceptable. The error is over $13,000 in either direction. Not many homeowners would be willing to have their selling

---

*You might think of R_A^2 as similar to the net income on an income statement and R^2 as the total revenue. Thus R_A^2 is total revenue less expenses.

†The actual confidence interval for prediction of a new observation requires the use of matrix algebra as follows:

$$\hat{Y}_p \pm S_e \sqrt{1 + [(\mathbf{X_p'})(\mathbf{X'X})^{-1}(\mathbf{X_p})]}$$

where: $\hat{Y}_p$ = point estimate given values for the independent variables

$\mathbf{X_p}$ = vector of independent variable values used to calculate $\hat{Y}_p$

$\mathbf{X}$ = matrix of independent variable values used to develop the regression model

Refer to *Applied Linear Statistical Models* by Neter and Wasserman (1974) for further discussion.

Hypotheses:

H_0: $\beta_i = 0$, given all other variables are already in the model
H_A: $\beta_i \neq 0$, given all other variables are already in the model

$\alpha = 0.01$

Decision Rule:

If $-2.576 \leq t \leq 2.576$, accept H_0.
Otherwise, reject H_0.

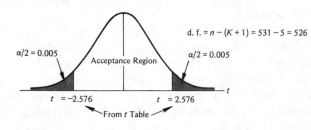

d. f. $= n - (K + 1) = 531 - 5 = 526$

$\alpha/2 = 0.005$ Acceptance Region $\alpha/2 = 0.005$

$t = -2.576$ $t = 2.576$
From t Table

The test is:

For β_1: Calculated t (from printout) $= 19.504$.
        Since $19.504 > 2.576$, reject H_0.

For β_3: Calculated $t = -4.613$.
        Since $-4.613 < -2.576$, reject H_0.

For β_4: Calculated $t = 9.367$.
        Since $9.367 > 2.576$, reject H_0.

For β_5: Calculated $t = 3.677$.
        Since $3.677 > 2.576$, reject H_0.

FIGURE 17–8 Significance test for a single independent variable, Moon Realty.

price set by a model with this possible error. Allen Rogers needs to take steps to reduce the standard error of the estimate. In subsequent sections of this chapter, we discuss some ways to reduce this value.

Do the Signs on the Regression Coefficients Make Sense? Even if a regression model is significant, and if each independent variable is significant, decision makers should still examine the regression coefficients for reasonableness. That is, does any coefficient have an unexpected sign?

Before answering this question for the Moon Realty example, we should review what the regression coefficients mean. First, the constant term, β_o, is the model's Y-intercept. If the data used to develop the regression model contain values of X_1, X_3, X_4, and X_5 that are simultaneously zero (such as would be the case for vacant land), the constant is the mean value of Y given $X_1 = 0$, $X_3 = 0$, $X_4 = 0$, and $X_5 = 0$. Under these conditions β_o would equal the average value of a vacant lot. However, in the Moon Realty example no vacant land was in the sample, so the constant has no particular meaning.

The coefficient for square feet, β_1, indicates the average change in sales price corresponding to a change in house size of one square foot, holding the other independent variables constant. The value shown in Table 17–4 for β_1 is 16.44.

The coefficient is positive, indicating that an increase in house size is associated with an increase in sales price. This relationship is expected.

Likewise, the coefficients for X_4, number of bathrooms, and X_5, number of fireplaces, are positive, indicating that an increase in their numbers is also associated with an increased price. This is also expected. However the coefficient on variable X_3, number of bedrooms, is $-2,845.89$, meaning that if we hold the other variables constant but increase the number of bedrooms by one, the average price will *drop* by $2,845.89. This relationship may not be reasonable given the real estate market and buyers' attitudes.

Referring to the correlation matrix in Table 17–3, the correlation between variable X_3 and Y, the sales price, is $+0.494$. This indicates that without considering the other independent variables, the linear relationship between number of bedrooms and sales price is positive. But why does the regression coefficient turn out negative? The answer lies in what is called *multicollinearity*. *Multicollinearity* **occurs when the independent variables are themselves correlated.** For example, X_3 and the other independent variables have the following correlations:

$$r_{X_3, X_1} = 0.644$$

$$r_{X_3, X_4} = 0.551$$

$$r_{X_3, X_5} = 0.338$$

In each case, the correlation is significant. Therefore the other variables in the model are overlapping X_3. This overlapping is multicollinearity. Multicollinearity can cause problems like we have seen in this example, where the regression coefficient sign is clearly opposite of what we would expect.

The problems caused by multicollinearity, and how to deal with them, continue to be of prime concern to theoretical statisticians. From a decision maker's viewpoint, you should be aware that multicollinearity can (and usually does) exist and recognize the basic problems it can cause. Some of the most obvious problems, and indications of severe multicollinearity, are:

1. Incorrect signs on the coefficients

2. A change in the values of the previous coefficients when a new variable is added to the model

3. The change to insignificant of a previously significant variable when a new variable is added to the model

4. An increase in the standard error of the estimate when a variable is added to the model

Mathematical approaches exist for dealing with multicollinearity and reducing its impact. Although these procedures are beyond the scope of this text, one suggestion is to eliminate the variable(s) that are the chief cause of the multicollinearity problems. In the Moon Realty example, we would drop X_3 from the analysis. This variable is highly correlated with X_1, X_4, and X_5 and has low correlation with Y.

Dealing with multicollinearity problems requires a great deal of experience. In this text we simply want to make you aware that such problems may exist, and should be considered before the regression model is used.

17-4

DUMMY VARIABLES IN REGRESSION ANALYSIS

In many cases decision makers may want to use a *qualitative* variable as an independent variable in a regression model. For example, in a model for predicting individual income, a potential variable might be sex—male or female. In another example with GNP (gross national product) as the dependent variable, an interesting independent variable might be the U.S. President's political party—Republican or Democrat. In still another example, where the dependent variable is the number of dollars spent by women on cosmetics, such independent variables as marital status—single, married, widowed, divorced, or separated—and employment status—full-time, part-time, retired, unemployed, or other—are variables that may help to explain the variation in the dependent variable.

The problem with qualitative variables in a regression analysis is in assigning values to the outcomes. What value should we assign to a male as opposed to a female? What about assigning values for Republican versus Democrat? These classifications do not have unique numerical values, and different decision makers might well assign different values, which could affect the regression analysis.

To overcome this problem, **qualitative variables are incorporated into a regression analysis by using *dummy variables*.** For example, for the qualitative variable sex, which has two categories—male or female—a single dummy variable is created as follows:

If sex = male, $X_2 = 0$.
If sex = female, $X_2 = 1$.

For the political party variable, the dummy variable is

If party = Republican, $X_2 = 1$.
If party = Democrat, $X_2 = 0$.

Although the qualitative variable is coded (0, 1), it makes no difference which attribute is assigned zero and which is assigned 1.

When the qualitative variable has more than two possible categories, a series of dummy variables must be used. For example, since marital status could have five categories: single, married, separated, divorced, and widowed, we would develop $5 - 1 = 4$ dummy variables as follows:

If single, $X_2 = 1$.
Otherwise, $X_2 = 0$.

If married, $X_3 = 1$.
Otherwise, $X_3 = 0$.

If separated, $X_4 = 1$.
Otherwise, $X_4 = 0$.

If divorced, $X_5 = 1$.
Otherwise, $X_5 = 0$.

For a widowed person, X_2, X_3, X_4, and X_5 would all be zero. By default, the person must belong to the remaining category, "widowed."‡

We can show the effect of including a dummy variable in a regression model by starting with the following simple regression equation:

Sales price = 9,915.78 + 20.45(square feet)

Figure 17–9 illustrates this regression line for sales price as a function of square feet.

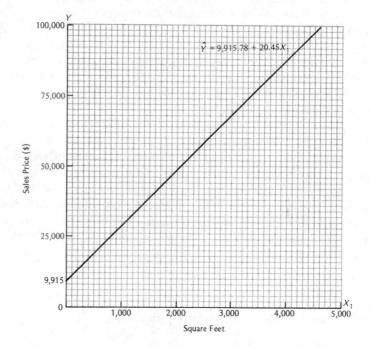

FIGURE 17–9 Simple regression line, Moon Realty.

Now, suppose a qualitative variable defining whether or not the house has central air conditioning is added. The variable would be

If air conditioning, $X_6 = 1$.
If no air conditioning, $X_6 = 0$.

Then the regression model would be

$$\hat{Y} = b_o + b_1X_1 + b_6X_6$$

---

‡ The mathematical reason that the number of dummy variables must be one less than the number of possible responses is called the **dummy variable trap.** Perfect multicollinearity is introduced, and the least squares regression estimates cannot be obtained if the number of dummy variables equals the number of possible categories.

Table 17–5 shows the computer output for the Moon Realty regression model with square feet and air conditioning as the independent variables. The model is

$$\hat{Y} = 8,739.21 + 18.11X_1 + 2,455.16X_6$$

TABLE 17–5 Regression model with dummy variable, Moon Realty
****

R SQUARE .7583

ADJUSTED R SQUARE .7573

STANDARD ERROR 7038.7899

- -

ANALYSIS ØF VARIANCE	D.F.	SUM ØF SQUARES	MEAN SQUARE	F
REGRESSIØN	2	82071871520.3	41035935760.1	828.44
RESIDUAL (UNEXPLAINED)	528	26159530525.8	49544563.5	

- - - - - - - - - - - - - - - VARIABLES IN THE EQUATIØN - - - - - - - - - - - - - - -

VARIABLE	B	STD. ERRØR B	T
X1 SQ. FT.	18.11	.70240	25.78
X6 AIR	2455.16	651.32152	3.76
(CØNSTANT)	8739.21		

****

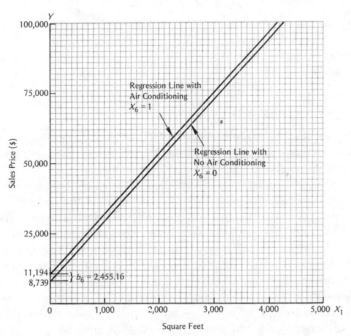

FIGURE 17–10 Impact of a dummy variable on the regression model, Moon Realty.

Since variable X_6 can be only zero or 1, incorporating a dummy variable is equivalent to finding two regression equations with *equal* slopes but *different* intercepts.

$$\hat{Y} = 8,739.21 + 18.11X_1 \quad \text{if } X_6 = 0$$

and

$$\hat{Y} = 11,194.37 + 18.11X_1 \quad \text{if } X_6 = 1$$

The impact of the dummy variable is illustrated in Figure 17–10. The difference in intercepts indicates that, holding square feet constant, the price of a house is on the average $2,455.16 higher if it has central air conditioning.

Improving the Real Estate Model

As we showed in the previous section, the real estate appraisal model developed thus far by Allen Rogers is not satisfactory. It has a large standard error of the estimate and problems with multicollinearity. Before abandoning the appraisal model concept, Rogers decides to attempt to improve the model by finding data for other variables that might help explain the variation in sales price.

Two variables that Rogers feels affect sales price are the location and whether the house has central air conditioning. If we can identify whether a house has air conditioning, we can use the dummy-variable concept just discussed and code this variable

If air conditioning, $X_6 = 1$.
If no air conditioning, $X_6 = 0$.

Data for the location variable is more difficult to come by. There are two high schools in the city. Rogers has determined in which high school district each house is located. He hopes this breakdown of location will improve the regression model. The location variable is coded

If Douglas High, $X_7 = 1$.
If Biltmore High (not Douglas High), $X_7 = 0$.

Table 17–6 presents the computer output for the latest regression run with variables X_6 and X_7 included and variable X_3, number of bedrooms, excluded. Recall that X_3 was apparently giving rise to multicollinearity problems.

This latest regression model has a smaller standard error and higher R^2 than the original model shown in Table 17–4. However the dummy variable for location, X_7, has a t value of -0.53. We must conclude that the regression coefficient on this variable is not different from zero since -0.53 does not exceed the critical t value for $\alpha = 0.01$ and 525 degrees of freedom of -2.576 (see Figure 17–8). Therefore the location variable (high school district) is not helpful in determining the sales price of houses. In fact, the model as it now stands is still of little use to Rogers since the standard error is again quite large. However the standard error of the estimate should be judged on a *relative* rather than an *absolute* basis. A standard error of the estimate that is too large in one situation may be quite acceptable in another. For instance, while $S_e = \$6,514$ may be too large in an appraisal model,

TABLE 17–6 Improved regression model, Moon Realty

```
**************************************************************************
```

R SQUARE	.7940
ADJUSTED R SQUARE	.7920
STANDARD ERRØR	6514.1353

- -

ANALYSIS ØF VARIANCE	D.F.	SUM ØF SQUARES	MEAN SQUARE	F
REGRESSIØN	5	859535732448.3	17190714649.6	405.5
RESIDUAL (UNEXPLAINED)	525	22277828797.0	42433959.6	

- - - - - - - - - - - - - -VARIABLES IN THE EQUATIØN- - - - - - - - - - - - - -

VARIABLE		B	STD. ERRØR B	T
X1	SQ. FT.	16.93	.82041	20.63
X4	BATHS	4154.21	612.30521	6.78
X5	FIREPLACES	2010.45	692.57183	2.90
X6	AIR	1972.11	604.21073	3.26
X7	SCHØØLS	− 692.20	1304.52751	− .53

```
**************************************************************************
```

this same value would be viewed as "wonderful" in a model where U.S. GNP was the dependent variable.

This example illustrates that although a regression model may pass the statistical tests of significance, it may not be functional. Good appraisal models can be developed using multiple regression analysis provided more detail is available about such characteristics as finish quality, landscaping, location, neighborhood characteristics, and so forth. The cost and effort required to obtain these data can be very high.

Developing a multiple regression model is more an art than a science. The real decisions revolve around how to select the best set of independent variables for the model.

17-5
STEPWISE REGRESSION ANALYSIS

In the Moon Realty example, we began with five independent variables plus the dependent variable, sales price. After examining the correlation matrix, we eliminated the age variable, X_2. Finally, we used a computer routine to develop the multiple regression model shown in Table 17–4. This process is similar to hiring four people at one time to perform a job and then observing how well they do. Once the model had been developed, we were left to analyze its statistical and functional validity. We use the term ***one-shot regression*** for this approach of bringing all independent variables into the model at one step.

Another method for developing a regression model is called ***stepwise regression***. ***Stepwise regression,*** **as the name implies, develops the least squares regression equation in steps, either through** *backward elimination* **or through** *forward selection.*

Backward Elimination

The ***backward elimination*** stepwise method begins by developing a one-shot regression model using all independent variables. Then a *t* test for significance is performed on each regression coefficient at a specified alpha level. Provided at least one *t* value is in the acceptance region (H_0: $\beta_i = 0$ is accepted), the variable with the *t* value closest to zero is removed, and another one-shot regression model is developed with the remaining independent variables. **The backward elimination continues until all independent variables remaining in the model have coefficients that are significantly different from zero.** The models at each step are printed by the computer routine.

(a)   No independent variables in the model.

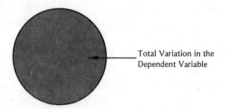

(b)   Enter the variable with the highest correlation.

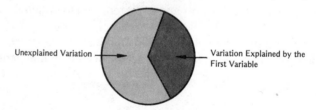

(c)   Enter the variable that explains the most of the unexplained variation (has the highest coefficient of partial determination).

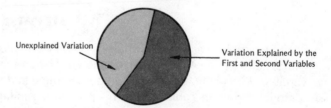

(d)   Continue adding variables based on how much of the unexplained variation they account for until the next variable explains an insignificant proportion.

FIGURE 17–11   Graphical   representation   of   the   forward   stepwise procedure.

The advantage of backward elimination is that the decision maker has the opportunity to look at all the independent variables in the model before removing the "insignificant" variables.

Forward Selection

Whereas the backward elimination procedure begins with a regression model containing all variables and eliminates insignificant variables in a stepwise fashion, the *forward selection* procedure works in the opposite direction. The forward selection procedure begins by selecting a single independent variable from all available. The independent variable selected at step 1 is the variable that is most highly correlated with the dependent variable. At step 2, a second independent variable is selected based upon its ability to explain the remaining unexplained variation in the dependent variable. This is analogous to the step-by-step hiring of people to perform a job.

TABLE 17–7 Data, B. T. Longmont Company

	Temperature X_1	No. Sales Made X_2	No. Days X_3	No. Employees X_4	Dollars Lost Y
1	58.8	7,107	21	129	3,067
2	65.2	6,373	22	141	2,828
3	70.9	6,796	22	153	2,891
4	77.4	9,208	20	166	2,994
5	79.3	14,792	25	193	3,082
6	81.0	14,564	23	189	3,898
7	71.9	11,964	20	175	3,502
8	63.9	13,526	23	186	3,060
9	54.5	12,656	20	190	3,211
10	39.5	14,119	20	187	3,286
11	44.5	16,691	22	195	3,542
12	43.6	14,571	19	206	3,125
13	56.0	13,619	22	198	3,022
14	64.7	14,575	22	192	2,922
15	73.0	14,556	21	191	3,950
16	78.9	18,573	21	200	4,488
17	79.4	15,618	22	200	3,295

TABLE 17–8 Correlation matrix, B. T. Longmont Company

	X1	X2	X3	X4	Y
X1	1.000	−.024	.438	−.082	.286
X2		1.000	.096	.920	.628
X3			1.000	.032	−.089
X4				1.000	.413
Y					1.000

The independent variable selected at each step is the variable with the highest *coefficient of partial determination.* The coefficient of determination measures the proportion of variation explained by all the independent variables in the model. By contrast, **the *coefficient of partial determination* measures the marginal contribution of a single independent variable, given that other independent variables are in the model.**

Thus, after the first variable (say, X_1) is selected, R^2 will indicate the percentage of variation explained by this variable. The forward selection routine will then compute all possible two-variable regression models with X_1 included, and determine the R^2 for each model. The coefficient of partial determination at step 2 is the proportion of unexplained variation (after X_1 is in the model) that is explained by the additional variable. The independent variable that adds the most to R^2, given the variables already in the model, is the one selected. This process continues until either all independent variables have been entered or the remaining independent variables do not add appreciably to R^2. This procedure is outlined in Figure 17–11 (p. 490).

The forward selection stepwise method serves one more important function. If two or more variables overlap, a variable selected in an early step may become insignificant when other variables are added at later steps. The forward selection procedure will drop this insignificant variable from the model.

TABLE 17–9   Step 1 of the regression model, B. T. Longmont Company

```
********************************************************************************
```

STEP NUMBER 1

VARIABLE ENTERED   X2

R SQUARE   .3942

ADJUSTED R SQUARE   .3538

STANDARD ERROR   359.0574

```
- - - - - - - - - - - - - - - - - - - - - - - - - - - - - - - - - - - - - - - -
```

ANALYSIS OF VARIANCE	D.F.	SUM OF SQUARES	MEAN SQUARE	F
REGRESSION	1	1258790.00	1258790.0	9.764
RESIDUAL (UNEXPLAINED)	15	1933835.00	128922.3	

```
- - - - - - - - - - - - - - - - - VARIABLES IN THE EQUATION - - - - - - - - - - - - - - - -
```

VARIABLE	B	STD. ERROR B	T
X2 SALES	.07687	.02460	3.12
(CONSTANT)	2316.51855		

```
- - - - - - - - - - - - - - - VARIABLES NOT IN THE EQUATION - - - - - - - - - - - - - - -
```

VARIABLE	PARTIAL R SQUARE
X1 TEMP.	.14932
X3 DAYS	.03704
X4 EMPLOYEES	.28828

```
********************************************************************************
```

The forward selection stepwise procedure is widely used in decision-making applications and is generally recognized as a useful regression method. However, care should be exercised when using this procedure since it is easy to rely too heavily on the automatic selection process. Decision makers still must use common sense in applying regression analysis to make sure they have usable regression models.

A Stepwise Regression Example

The B. T. Longmont Company operates a large retail department store in San Francisco. Like other department stores, Longmont has incurred heavy losses due to shoplifting and pilferage. The store's security manager wants to develop a regression model to explain the monthly dollar loss from these factors. The variables the security manager is interested in are

X_1 = average monthly temperature (°F)

X_2 = number of sales transactions

X_3 = number of days per month the store is open

X_4 = number of persons on the store's monthly payroll

Y = monthly dollar loss due to shoplifting and pilferage

TABLE 17–10 Step 2 of the regression model, B. T. Longmont Company

****

STEP NUMBER 2

VARIABLE ENTERED   X4

R SQUARE   .5689

ADJUSTED R SQUARE   .5073

STANDARD ERROR   313.5452

- -

ANALYSIS OF VARIANCE	D.F.	SUM OF SQUARES	MEAN SQUARE	F
REGRESSION	2	1816276.0	908138.0	9.23
RESIDUAL (UNEXPLAINED)	14	1376349.0	98310.6	

- - - - - - - - - - - - - - - VARIABLES IN THE EQUATION - - - - - - - - - - - - - -

VARIABLE	B	STD. ERROR B	T
X2 SALES	.19661	.05468	3.59
X4 EMPLOYEES	− 21.60054	9.07085	− 2.38
(CONSTANT)	4706.38203		

- - - - - - - - - - - - - - - VARIABLES NOT IN THE EQUATION - - - - - - - - - - - - - -

VARIABLE	PARTIAL R SQUARE
X1 TEMP.	.13306
X3 DAYS	.10452

****

Table 17–7 (p. 491) lists the data for these variables for a random sample of 17 months. The correlation matrix of the data is presented in Table 17–8 (p. 491).

Note that variable X_2, number of sales transactions, is most highly correlated with dollars lost. Using the forward stepwise selection procedure, X_2 will be the first variable selected. The computer output for the model at step 1 is shown in Table 17–9 (p. 492).

At step 1, variable X_2, number of monthly sales transactions, explains 0.3942 of the variation in the dependent variable. The overall model is significant at the 0.05 alpha level because the calculated $F = 9.764$ exceeds the table F value with 1 and 15 degrees of freedom.

Now, look at the bottom of Table 17–9 to the section "variables not in the equation." The next variable added is the one that can contribute most to R^2, given that variable X_2 is already in the model. This variable will be the one with the highest coefficient of partial determination. As you can see, the variable selected in step 2 should be X_4, number of employees, since it can explain almost 29 percent of the remaining unexplained variation in the dependent variable. Tables 17–10, 17–11, and 17–12 show steps 2, 3, and 4 of the stepwise regression.

The model at step 4 explains 74.8 percent of the variation and is significant at the 0.05 level ($F = 8.90$). In analyzing the independent variables using the t test, we find all four significant at the alpha = 0.05 level. However we find

TABLE 17–11   Step 3 of the regression model, B. T. Longmont Company

****

STEP NUMBER 3

VARIABLE ENTERED	X1	
R SQUARE	.6263	
ADJUSTED R SQUARE	.5400	
STANDARD ERROR	302.9600	

- -

ANALYSIS OF VARIANCE	D.F.	SUM OF SQUARES	MEAN SQUARE	F
REGRESSION	3	1999422.0	6664740.0	7.26
RESIDUAL (UNEXPLAINED)	13	1193203.0	91784.8	

- - - - - - - - - - - - - - - VARIABLES IN THE EQUATION - - - - - - - - - - - - - - -

VARIABLE	B	STD. ERROR B	T
X2 SALES	.18667	.05330	3.50
X4 EMPLOYEES	− 19.67947	8.86950	−2.21
X1 TEMP.	8.01620	5.67487	1.41
(CONSTANT)	3964.87085		

- - - - - - - - - - - - - - - VARIABLES NOT IN THE EQUATION - - - - - - - - - - - - - - -

VARIABLE	PARTIAL R SQUARE
X3 DAYS	.32490

****

TABLE 17–12 Step 4 of the regression model, B. T. Longmont Company

```
*******************************************************************************
STEP NUMBER 4

VARIABLE ENTERED        X3

R SQUARE                .7480

ADJUSTED R SQUARE       .6640

STANDARD ERROR      258.9106

- - - - - - - - - - - - - - - - - - - - - - - - - - - - - - - - - - - - - - - -

ANALYSIS OF VARIANCE     D.F.     SUM OF SQUARES     MEAN SQUARE        F

REGRESSION                4         2388207.0          597051.7        8.90
RESIDUAL (UNEXPLAINED)   12          804417.0           67034.7

- - - - - - - - - - - - - - - VARIABLES IN THE EQUATION - - - - - - - - - - - - - - -

VARIABLE                 B          STD. ERROR B           T

X2   SALES             .20053         .04591             4.36
X4   EMPLOYEES      - 21.25937        7.60824           - 2.79
X1   TEMP.           13.57066         5.37026             2.52
X3   DAYS          - 119.80313       49.74646           - 2.41

(CONSTANT)           6286.20703

- - - - - - - - - - - - - - - VARIABLES NOT IN THE EQUATION - - - - - - - - - - - - - - -

VARIABLE                                  PARTIAL R SQUARE
*******************************************************************************
```

some evidence of multicollinearity in the model. For instance, at step 4 the coefficient on X_4, number of employees, is negative, yet the correlation matrix shows a positive correlation between dollar loss and number of employees. Also, as we move from step to step in the output, the regression coefficients change. In addition, the significance of variable X_1 improves when variable X_3 is added in the final step. These are all signs of correlation between the independent variables. We suggest the security manager be cautious about any conclusions he might reach based on this model.

This example has illustrated stepwise regression analysis and has shown what a computer output for such an analysis looks like. The model was developed using the forward selection procedure; however the model at step 4 is exactly the same as it would be using the one-shot approach with all four independent variables. The order in which variables enter makes no difference as long as all variables are in the model.

17-6
CONCLUSIONS

Multiple regression is an extension of simple regression analysis. In multiple regression, two or more independent variables are used to explain the variation in the dependent variable. Just as a manager searches for the "best" combination of employees to perform a job, the decision maker using multiple regression analysis searches for the "best" combination of independent variables to explain variation in the dependent variable.

Our presentation of multiple regression analysis has largely been an analysis of computer printouts. As a decision maker, you will almost assuredly not be required to manually develop the regression model, but you will have to judge its applicability based on a computer printout. The programs we have used in Chapter 16 and 17 are representative of the many available. You no doubt will encounter printouts that look somewhat different from those shown in this text, and some of the terms used may differ slightly. However the basic information will be the same, as will be the inferences you can make from the model.

In this chapter we have discussed the difference between R^2 and adjusted R^2, and also the difference between statistical significance and practical significance. As a decision maker, you must recognize that a regression model can be statistically significant yet have no practical use because the standard error of the estimate is too large or multicollinearity impacts too heavily.

chapter glossary

adjusted R^2 (R_A^2)   A measure of the percentage of explained variation in the dependent variable that takes into account the relationship between the number of cases and the number of independent variables in the regression model. Whereas R^2 will always increase when an independent variable is added, adjusted R^2 (R_A^2) will decrease if the added variable does not reduce the unexplained variation enough to offset the loss of degrees of freedom.

correlation matrix   A table showing the pairwise correlations between all variables (dependent and independent).

dummy variables   Variables in a regression model that have two categories, valued zero and 1. If a qualitative variable has r multiple categories, $r - 1$ dummy variables are formed to represent the qualitative variable in the analysis.

multicollinearity   Correlation among the independent variables. Usually the term is used when the intercorrelation is high.

multiple coefficient of determination (R^2)   The percentage of variation in the dependent variable explained by the independent variables in the regression model.

multiple regression model   A regression model having two or more independent variables with a regression equation of the form

$$Y = \beta_o + \beta_1 X_1 + \beta_2 X_2 + \beta_3 X_3 + \ldots + \beta_K X_K + e$$

regression plane   The multiple regression equivalent of the simple regression line. The plane has a different slope for each independent variable.

standard error of the estimate   The square root of the mean square residual in the analysis of variance table for a regression model. The standard error measures the dispersion of the actual values of the dependent variable around the fitted regression plane.

===chapter formulas

Standard error of the estimate

$$S_e = \sqrt{MSE} = \sqrt{\frac{SSE}{n - K - 1}}$$

Correlation coefficient

$$r = \frac{n\Sigma XY - \Sigma X\Sigma Y}{\sqrt{[n(\Sigma X^2) - (\Sigma X)^2][n(\Sigma Y^2) - (\Sigma Y)^2]}}$$

Coefficient of multiple determination

$$R^2 = \frac{SSR}{TSS}$$

Adjusted R^2

$$R_A^2 = 1 - (1 - R^2)\left(\frac{n - 1}{n - K - 1}\right)$$

===problems

Problems marked with an asterisk (*) require extra effort.

1. Discuss in your own words each of the following:
   a. The difference between simple and multiple regression analysis. Which type do you think will be more useful and why?
   b. The meaning of the estimates of the regression slope coefficients.
   c. The coefficient of multiple determination.

2. Discuss the method by which stepwise regression analysis enters explanatory variables. What determines the first independent variable to be entered?

3. The managerial development director of a major corporation is trying to determine what personal abilities are necessary for a manager to move from middle- to upper-level management. Although she has been relatively successful predicting who will move rapidly from lower- to middle-management levels, she has had difficulty determining the characteristics necessary to move to the next major level. For a long time the director has heard that the most glaring deficiency in college graduates entering the company is in communication skills, so she decides to measure whether these skills may be a determining factor.

   The director decides to try to develop a multiple regression relationship between job ratings and communication ability. She picks a random sample of middle-level managers who have been in their present positions less than five years but more than one year. These managers are given a series of cases to analyze and asked to present both written and verbal recommendations. They are rated by a group of top-level managers on their analyses and on their written and verbal presentations. These ratings are then compared with the latest employee rating. The data are as follows:

	Y	X_1	X_2	X_3
N Employee	Job Rating	Case Analysis Score	Written Presentation Score	Verbal Presentation Score
1	87	8.4	8.7	9.2
2	93	8.2	9.4	9.4
3	91	9.3	9.7	9.5
4	85	7.9	8.1	8.7
5	86	8.1	8.3	8.8
6	97	9.4	9.3	9.6
7	90	9.1	9.0	9.2
8	93	8.9	9.2	9.5
9	88	8.6	8.4	8.5
10	96	9.7	9.5	9.6
11	86	8.3	7.9	8.4
12	89	8.7	8.5	8.7
13	94	9.2	9.1	9.6
14	91	8.1	9.5	9.2
15	95	9.3	9.1	9.7

Use a computer routine available at your school to determine the multiple regression equation for these data.

Problems 4–11 refer to the data given in problem 3.

4. One of the assumptions of multiple regression is that the independent variables are not correlated with each other. Is this assumption satisfied for these data? What do you check to see if multicollinearity is a problem? Use $\alpha = 0.05$.

5. Does the multiple regression model you have estimated show a significant relationship between job ratings and the three independent variables measured? How did you measure this significance? Test with $\alpha = 0.05$.

6. If you were a middle-level manager, would you be willing to have your job rating determined just on the basis of your performance on these three independent variables? Explain in statistical terms why or why not.

7. Discuss how much of the variation in job rating is explained by the three independent variables. How do you measure this factor?

8. Are all the independent variables significant in your multiple regression relationship? How can you tell?

9. As a test, the development director gives the same cases to a group of middle-level managers without knowing their job ratings. One of the managers received the following scores:

Case analysis	9.1
Written presentations	9.4
Verbal presentations	9.3

Based on these data, what is the best estimate of the job rating this manager received?

10. The personnel director comments that perhaps the regression model just developed would be a good tool to use before hiring new employees. What do you think of this idea?

11. One manager who participated in this study is concerned with his job rating and would like to know how much his job rating should change if his written presentation score increased by a full point. You are to develop a 95 percent confidence interval for the regression coefficient for the independent variable written presentation. Be sure you interpret this interval.

The following information applies to problems 12–21.

The J. J. McCracken Company has authorized its marketing research department to make a study of customers who have been issued a McCracken charge card. The marketing research department hopes to be able to identify the significant variables that explain the variation in purchases. Once these variables are determined, the department intends to try to attract new customers who would be predicted to have a high volume of purchases. REGR, CORR, DESC

TABLE P-1 Data, McCracken Company

No. Purchases Y	Age X_1	Family Income X_2	Family Size X_3
75	42	$29,000	4
129	36	25,000	2
105	38	25,000	2
42	54	17,000	3
17	49	15,000	5
26	55	19,500	3
144	25	24,000	2
100	24	14,000	1
92	30	11,000	1
58	35	12,000	2
111	27	29,000	3
146	29	38,000	2
93	38	19,500	4
68	40	24,000	3
11	36	22,500	2
50	22	10,200	1
55	25	14,000	3
88	69	19,200	4
100	54	52,000	4
86	48	21,400	3
105	30	26,000	2
121	27	18,250	3
14	62	10,250	3
37	50	18,100	2
43	26	24,500	4

Twenty-five customers were selected at random and values for the following variables were recorded:

Y = average monthly purchases at McCracken

X_1 = customer age

X_2 = customer family income

X_3 = family size

Table P–1 illustrates the data.

A computer program was used to perform the multiple regression analysis. Tables P–2 through P–5 show the results of the computer run.

TABLE P–2   Correlation matrix, McCracken Company

	Y	X1	X2	X3
Y	1.000	− .4057	.4591	− .2444
X1		1.000	.0512	.5037
X2			1.000	.2718
X3				1.000

TABLE P–3   Step 1 of stepwise regression, McCracken Company

```
*************************************************************************************

STEP NUMBER 1

VARIABLE ENTERED        X2

R SQUARE            .2107

ADJUSTED R SQUARE     .1763

STANDARD ERROR    36.3553

- - - - - - - - - - - - - - - - - - - - - - - - - - - - - - - - - - - - - - - - - -

ANALYSIS OF VARIANCE        D.F.     SUM OF SQUARES      MEAN SQUARE        F

REGRESSION                   1          8118.48            8118.48        6.14
RESIDUAL (UNEXPLAINED)      23         30399.32            1321.70

- - - - - - - - - - - - - - VARIABLES IN THE EQUATION - - - - - - - - - - - - - - -

VARIABLE                     B              STD. ERROR B              T

X2  INCOME                .00199             .000803               2.478

(CONSTANT)              33.7544

*************************************************************************************
```

TABLE P–4   Step 2 of stepwise regression, McCracken Company

****

STEP NUMBER 2

VARIABLE ENTERED   X1

R SQUARE   .3955

ADJUSTED R SQUARE   .3905

STANDARD ERRØR   32.5313

ANALYSIS ØF VARIANCE	D.F.	SUM ØF SQUARES	MEAN SQUARE	F
REGRESSIØN	2	15235.4	7617.7	7.19
RESIDUAL (UNEXPLAINED)	22	23282.4	1058.29	

---------------VARIABLES IN THE EQUATIØN ---------------

VARIABLE	B	STD. ERRØR B	T
X2 INCØME	.00208	.000719	2.899
X1 AGE	− 1.31807	.508267	− 2.593
(CØNSTANT)	82.8875		

****

TABLE P–5   Step 3 of stepwise regression, McCracken Company

****

STEP NUMBER 3

VARIABLE ENTERED   X3

R SQUARE   .4322

ADJUSTED R SQUARE   .3510

STANDARD ERRØR   32.2724

ANALYSIS ØF VARIANCE	D.F.	SUM ØF SQUARES	MEAN SQUARE	F
REGRESSIØN	3	16646.1	5548.6	5.33
RESIDUAL (UNEXPLAINED)	21	21871.7	1040.5	

---------------VARIABLES IN THE EQUATIØN ---------------

VARIABLE	B	STD. ERRØR B	T
X2 INCØME	.00233	.000745	3.132
X1 AGE	− .97047	.586042	− 1.655
X3 FAMILY	− 8.7233	7.49549	− 1.163
(CØNSTANT)	87.7897		

****

12. A first step in regression analysis often involves developing a scatter plot of the data. Develop the scatter plots of all the possible pairs of variables and with a brief statement, indicate what each plot says about the relationship between the two variables.

13. Table P–2 illustrates the correlation matrix. Develop the decision rule for testing the significance of each coefficient. Which, if any, correlations are not significant? Use $\alpha = 0.05$.

14. At step 1 of the output (see Table P–3), the variable X_2, family income, was brought into the model. Discuss why this happened.

15. Test the significance of the regression model at step 1 of the computer printout. Justify the alpha level you have selected.

16. Develop a 95 percent confidence level for the slope coefficient for the family income variable at step 1 of the model. Be sure to interpret this confidence interval.

17. Describe the regression model at step 2 (see Table P–4) of the analysis. In your discussion, be sure to discuss the effect of adding a new variable on the standard error of the estimate and on R^2.

*18. Suppose the manager of McCracken's marketing department questions the appropriateness of adding a second variable. How would you respond to her question? Use the information in Table P–4 in your response.

19. Table P–5 presents the third, and final, step in the regression analysis. Test statistically the significance of each independent variable in the model at an $\alpha = 0.05$ level. Also test the hypothesis that all slope coefficients are zero at the $\alpha = 0.05$ level. Why can the overall model be significant while some individual variables are not significant?

*20. If you look carefully at the results shown in Tables P–3 through P–5, you can see that the value of the slope coefficient for variable X_2, family income, changes each time a new variable is added to the regression model. Discuss why this change takes place.

*21. Analyze the regression model at step 3 and the intermediate results at steps 1 and 2. Write a report to the marketing manager pointing out the strengths and weaknesses of the model. Be sure to comment on the department's goal of being able to use the model to predict customers who will purchase high volumes from McCracken.

22. A publishing company in New York is attempting to develop a model that it can use to help predict textbook sales for books it is considering for future publication. The marketing department has collected data on several variables from a random sample of 15 books. These data are

Volumes Sold Y	Pages X_1	Competing Books X_2	Advertising Budget X_3	Age of Author X_4
15,000	176	5	$25,000	49
140,000	296	10	83,000	57
75,000	483	7	40,000	29
100,000	811	14	29,000	37
26,000	302	9	52,000	35
33,000	411	15	33,000	43
59,000	333	7	19,000	51
103,000	602	4	37,000	62
88,000	504	12	51,000	33
10,000	204	3	30,000	50
9,000	376	4	19,000	26
77,000	600	7	41,000	40
59,000	400	3	26,000	44
183,000	597	8	51,000	59
16,000	126	1	27,000	38

Use an available computer routine to develop the correlation matrix showing the correlation between all possible pairs of variables. Test statistically to determine which independent variables are significantly correlated with the dependent variable, book sales. Use $\alpha = 0.05$.

23. Referring to problem 22, develop a multiple regression model containing all four independent variables. Show clearly the regression coefficients.

24. Referring to problems 22 and 23, how much of the total variation in book sales can be explained by these four independent variables? Would you conclude that the model is significant at the 0.05 level?

25. Referring to problems 22–24, develop a 95 percent confidence interval for each regression coefficient and interpret these confidence intervals.

26. Referring to problem 25, which of the independent variables can be concluded to be significant in explaining the variation in book sales? Test using $\alpha = 0.05$.

27. The publishing company in problems 22–26 recently came up with some additional data for the 15 books in the original sample. Two new variables, production expenditures (X_5) and number of prepublication reviewers (X_6) have been added. These additional data are

Book	X_5	X_6
1	$38,000	5
2	86,000	8
3	59,000	3
4	80,000	9
5	29,500	3
6	31,000	3

7	40,000	5
8	69,000	4
9	51,000	4
10	34,000	6
11	20,000	2
12	80,000	5
13	60,000	5
14	87,000	8
15	29,000	3

Calculate the correlation between each of these additional variables and the dependent variable, book sales. You will have to use the data from problem 22.

28. Referring to problem 27, test the significance of the correlation coefficients using $\alpha = 0.05$. Comment on your results.

29. Referring to problems 22–28, develop a multiple regression model that includes all six independent variables. Which, if any, variables would you recommend be retained if this model is going to be used to predict book sales for the publishing company? For any statistical tests you might perform, use $\alpha = 0.05$. Discuss your results.

30. Referring to problem 29, use the analysis of variance approach to test the null hypothesis that all slope coefficients are zero. Test with $\alpha = 0.05$. What do these results mean? Discuss.

31. Referring to problems 22–30, does it appear that multicollinearity problems are present in the model? Discuss the potential consequences of multicollinearity with respect to the regression model.

cases

17A   DYNAMIC SCALES, INC.

In 1975 Stanley Ahlon and three financial partners formed Dynamic Scales, Inc. The company was based on an idea Stanley had for developing a scale to weigh trucks in motion and thus eliminate the need for every truck to stop at weigh stations along highways. This dynamic scale would be placed in the highway approximately one-quarter mile from the regular weigh station. The scale would have a minicomputer which would automatically record truck speed, axle weights, and climate variables including temperature, wind, and moisture. Stanley Ahlon and his partners felt that state transportation departments in the United States would be the primary market for such a scale.

Like many technological advances, developing the dynamic scale has been difficult. When the scale finally proved "accurate" for trucks traveling 40 miles per hour, it would not perform for trucks traveling at higher speeds. However, eight months ago Stanley announced that the dynamic scale was ready to be field-tested by the Nebraska State Department of Transportation under a grant from the federal government.

Stanley explained to his financial partners, and to Nebraska transportation officials, that the dynamic weight would not exactly equal the static weight (truck weight on a static scale), but that he was sure a statistical relationship between dynamic weight and static weight could be determined, which would make the dynamic scale useful.

Nebraska officials, along with people from Dynamic Scales, Inc., installed a dynamic scale on a major highway in Nebraska. Each month for six months, data were collected for a random sample of trucks weighed on both the dynamic scale and a static scale. Table C–1 presents these data.

TABLE C–1 Test data, Dynamic Scales, Inc.

Month	Front-Axle Static Weight	Front-Axle Dynamic Weight	Truck Speed	Temperature	Moisture
Jan	1,800 lb	1,625 lb	52 mi/h	21°F	0.00%
	1,311	1,904	71	17	0.15
	1,504	1,390	48	13	0.40
	1,388	1,402	50	19	0.10
	1,250	1,100	61	24	0.00
Feb	2,102	1,950	55	26	0.10
	1,410	1,475	58	32	0.20
	1,000	1,103	59	38	0.15
	1,430	1,387	43	24	0.00
	1,073	948	59	18	0.40
Mar	1,502	1,493	62	34	0.00
	1,721	1,902	67	36	0.00
	1,113	1,415	48	42	0.21
	978	983	59	29	0.32
	1,254	1,149	60	48	0.00
Apr	994	1,052	58	37	0.00
	1,127	999	52	34	0.21
	1,406	1,404	59	40	0.40
	875	900	47	48	0.00
	1,350	1,275	68	51	0.00
May	1,102	1,120	55	52	0.00
	1,240	1,253	57	57	0.00
	1,087	1,040	62	63	0.00
	993	1,102	59	62	0.10
	1,408	1,400	67	68	0.00
June	1,420	1,404	58	70	0.00
	1,808	1,790	54	71	0.00
	1,401	1,396	49	83	0.00
	933	1,004	62	88	0.40
	1,150	1,127	64	81	0.00

Once the data were collected, the next step was to determine if, based on this test, the dynamic scale measurements could be used to predict static weights. A complete report will be submitted to the U.S. government and to Dynamic Scales, Inc.

references

DEMMERT, HENRY and MARSHALL MEDOFF. "Game Specific Factors and Major League Baseball Attendance: An Econometric Study." *Santa Clara Business Review* (1977): 49–56.

DRAPER, N. R. and H. SMITH. *Applied Regression Analysis.* New York: Wiley, 1966.

GLOUDEMANS, ROBERT J. and DENNIS MILLER. "Multiple Regression Analysis Applied to Residential Properties." *Decision Sciences* 7 (April 1976): 294–304.

JOHNSON, J. *Econometric Methods.* New York: McGraw-Hill, 1972.

NETER, JOHN and WILLIAM WASSERMAN. *Applied Linear Statistical Models.* Homewood, Ill.: Irwin, 1974.

SEARLE, S. R. and W. H. HAUSMAN. *Matrix Algebra for Business and Economics.* New York: Wiley, 1970.

18

Time Series Analysis

Managers have at least one thing in common with people walking in a forest: they can't spend much time looking behind themselves for fear of walking into a "tree." However decision makers generally spend some time reviewing past decisions, and the results of those decisions, in order to gain a perspective for making decisions that will affect the futures of their organizations.

Forecasting the future of an organization is somewhat like forecasting the weather. What the weather will be like tomorrow is, of course, related to what it is like today. The corporate "weather" is generally measured as profit, earnings per share, price/earnings ratio, or some other quantitative characteristic of the industry within which the business functions. The historical record of a firm's performance can be charted to provide an indication of the past "weather" conditions as measured over time. Whether or not this historical record will provide insight about what the future holds depends on many factors. However decision makers should understand how to analyze the past if they expect to incorporate past information into future decisions.

chapter objectives

In this chapter we shall introduce the fundamentals of time series analysis. Time series analysis is the process by which a set of data measured over time is analyzed. The goals of this chapter are to describe the various components of a time series and indicate how to recognize these components. In addition, we will point out what the components mean to decision makers who intend to use information from the time series in future decision making. We will also present the four separate components of time series analysis and provide business examples of each component. And we will introduce the concepts involved in developing index numbers. Finally, we will discuss some common statistical techniques used to help decision makers use past data to forecast future events.

student objectives

After studying the material in this chapter, you should be able to:

1. Define a trend component and recognize if a trend component is present in a time series.
2. Define a seasonal component and determine if a seasonal component is present in a time series.
3. Define a cyclical component and determine whether the time series contains a cyclical component.
4. Define a random or irregular component.
5. Produce appropriate index numbers for applications requiring comparisons involving changes over time.
6. Discuss the differences between judgmental and statistical forecasts.
7. Use regression analysis to make a statistical forecast.

18-1
TIME SERIES COMPONENTS

The world is filled with forecasters. They range from stock market investors to small businesspeople to government officials to you trying to forecast the weather. If we spray weed killer on our lawn just before an unexpected rain, the consequences are not devastating; we are out a few dollars for the weed killer. Unfortunately the results of a poor managerial decision are often more severe and cannot be rectified by an additional application of "weed spray." When faced with trying to predict the future, a manager, like all of us, would like to connect what is going to happen with something that can be seen happening now.

A large number of factors can affect the results of a managerial decision. Identifying these factors, and then measuring them, is in many cases impossible since their importance changes and new factors are continually added. However, although the factors that affect the future are uncertain, often the past offers a good indication of what the future will be. Before decision makers utilize past information to make a decision about the future, they must recognize how to extract the meaningful information from all available past data. In this chapter we discuss the basic components of a *time series* and illustrate, through example, what each means. We also discuss briefly some of the forecasting techniques that are available to decision makers.

All time series contain at least one of four time series components:

1. **Long-term trend**
2. **Seasonal**
3. **Cyclical**
4. **Irregular or random**

Time series analysis involves breaking down data measured over time into one or more of these components.

Long-Term Trend Component

The *trend* component is the long-term increase or decrease in a variable being measured over time. Increases in such diverse factors as annual tissue paper production in the United States (Figure 18–1) and percent participation of females over 20 in the labor force (Figure 18–2) show very definite long-term trends. In these two cases, not much variation exists in addition to the trend. In other cases, a strong trend component is evident, but variation exists around the long-term increase or decrease. Employment in the United States, shown in Figure 18–3, is such a case.

In today's world, organizations are facing increased planning problems caused mainly by changing technology, complicated government regulations, and uncertain foreign competition. A combination of these and other factors has forced most organizations into increasing the time spans of their planning cycles. Most organizations, when considering capital expenditures, look at a time frame of from three to seven years into the future. When you are planning to start spending money now to meet a demand existing in three to seven years, you better have a reason to

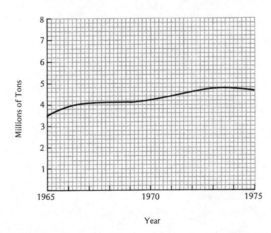

FIGURE 18–1   U.S. tissue paper production (1965–1975).

SOURCE: Based on a chart originally appearing in the June 1976 issue of FORTUNE Magazine.

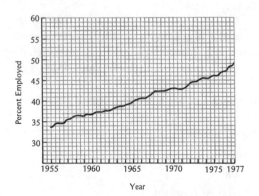

FIGURE 18–2   Percentage of females 20 years and over in the labor force (1955–1977).

SOURCE: *Business Conditions Digest,* U.S. Department of Commerce, January 1978, p. 52.

believe the demand will be there. Because long-term forecasting is becoming increasingly important, the trend component in time series analysis is important to all organizations.

Seasonal Component

Some organizations and industries are affected not only by long-term trends but also by *seasonal* variation. **The *seasonal* component represents those changes in a time**

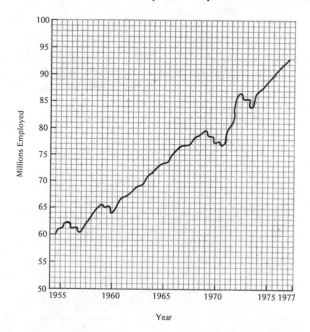

FIGURE 18–3   Total employed in the United States (1955–1977). (millions)

SOURCE: *Business Conditions Digest,* U.S. Department of Commerce, January 1978, p. 52.

series that occur at the same time every year. Figure 18–4 shows wheel-type tractor production in the United States. Tractor production seems to have two yearly peaks, a major one in the spring and a minor one in the fall. Also, valleys occur every summer and at the end of the year.

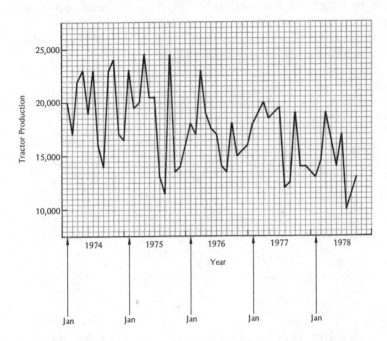

FIGURE 18–4 Number of U.S. wheel-type tractors shipped (1974–1978).

SOURCE: *Current Industrial Reports,* U.S. Department of Commerce, September 1978.

Organizations facing seasonal variations, like the motor vehicle industry, are often interested in knowing how well, or poorly, they are doing relative to the normal seasonal variation. For instance, while most retailers expect sales volume to be higher in December than November, they are interested in whether December sales are higher or lower than normal relative to November sales. The Department of Employment expects unemployment to increase in June because recent graduates are just arriving on the job market and because schools have let out for the summer. The question is whether the increase is more or less than expected.

Organizations that are affected by seasonal variation need to identify and measure this seasonality to help with planning for temporary increases or decreases in labor requirements, inventory, training, periodic maintenance, and so forth. In addition, these organizations need to know if the seasonal variation they experience occurs at more or less than the average rate.

Cyclical Component

Data collected annually obviously cannot have a seasonal component; however they can contain certain **cyclical** effects. **Cyclical effects in a time series are represented**

by wavelike fluctuations around a long-term trend. These fluctuations are generally thought to be caused by pulsations in factors such as interest rates, money supply, consumer demand, inventory levels, national and international market conditions, and other government policies. Cyclical fluctuations repeat themselves in a general pattern in the long term, but occur with differing frequencies and intensities. Thus they can be isolated, but not *totally* predicted. Figure 18–5 shows Standard and Poor's 500 Stock Index for the years 1966–1976. Note the strong cyclic effect with periods of high values and periods of low values.

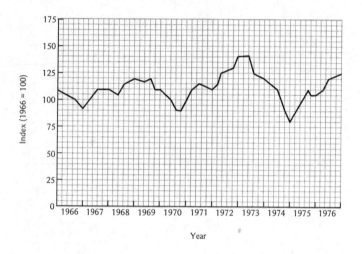

FIGURE 18–5   Standard and Poor's 500 Stock Index (1966–1976).

Firms affected by cyclical fluctuations are those particularly vulnerable to unexpected changes in the economy. Not only are cyclical effects generally unpredictable, but the overall effect of each cycle on individual organizations is different. Thus, even though you know what happened to your firm during the last cycle, you have no guarantee the effect will be the same the next time.

Irregular or Random Component

The *irregular* or *random* component in a time series is that part of the series that cannot be attributed to any of the three previously discussed components. Random fluctuations can be caused by many factors, such as weather, political events, and other human and nonhuman actions. For example, severe winters affect vast segments of the economy, and political statements or actions of governments or regulatory agencies often introduce an unpredicted element into an organization's environment.

Two types of irregular fluctuations may exist in a time series. *Minor* irregularities show up as sawtoothlike patterns around the long-term trend. These minor irregular fluctuations are caused by many factors and individually are not significant in an organization's long-term operations. *Major* irregularities are significant one-time, unpredictable changes in the time series due to such external and uncontrollable factors as an oil embargo, war, droughts, and so forth. Figure 18–6 illustrates both minor and major irregular fluctuations in U.S. crude-oil reserves. The discovery in 1970 of the Alaskan oil fields caused a major increase in confirmed U.S. oil reserves.

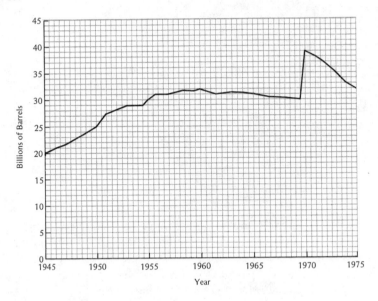

FIGURE 18–6 U.S. crude-oil reserves (1945–1975).

SOURCE: American Petroleum Institute, *Reserves of Crude Oil, Natural Gas Liquids, and Natural Gas in the United States and Canada,* May 1976, p. 85.

Almost all industries and organizations are affected by irregular components. Obviously organizations dealing with products that can be influenced by variations in the weather are interested in this component. This includes most agricultural and mining concerns. Some organizations, like insurance companies and certain governmental agencies, are directly concerned with eliminating the risk to other organizations caused by unforeseen irregular factors.

18-2
ANALYZING THE VARIABILITY OF HISTORICAL DATA

Decision makers can think about time series analysis in much the same manner they can think about analysis of variance and regression analysis. The time series they are analyzing has some inherent variability, and they would like to explain as much of that variability as possible. Four separate factors, or components (long-term trend, seasonal, cyclical, and irregular), help explain that variability. The purpose of time

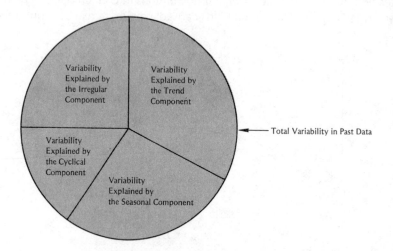

The problem is how to best separate each component from the others so that each can be analyzed. Some general approaches to this problem will be addressed in the remainder of this chapter. Only the most basic methods are presented. Refer to *Applied Time Series Analysis* by Nelson (1973) for discussions of more sophisticated approaches to analyzing a time series.

18-3
ANALYZING THE TREND COMPONENT

The trend component identifies the long-term growth or decline in the time series. Long-term growth patterns have a wide variety of shapes. The **first-degree, exponential, modified exponential,** and **Gompertz curves** are some of the most common trend patterns. Figure 18–7 illustrates these curves and indicates some areas where these curves have been used to describe long-term trends.

Many methods are available to fit trend lines to a series of data. The easiest is simply to graph the data points and draw the trend line freehand. This approach often provides adequate information about the long-term trend.

Another way of fitting a trend line to a set of data is to use least squares regression as discussed in Chapters 16 and 17. For example, suppose a financial analyst is comparing the performance of two large banking chains during the period 1975–1980. He is interested in comparing trends in the two banks. Annualized quarterly net incomes per share for the two banking groups are shown in Table 18–1.

The analyst can use simple regression analysis to measure the long-term trend of the annualized earnings per share. The fitted model is

$$\hat{Y} = b_o + b_1t$$

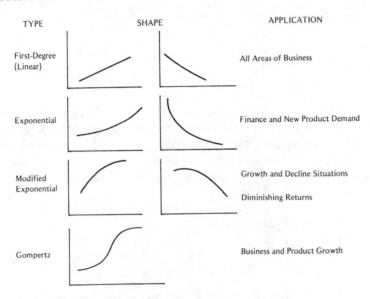

FIGURE 18–7 Long-term trend patterns.

TABLE 18–1 Annualized quarterly net incomes per share for bank A and bank B

Bank A		Bank B	
1975	1978	1975	1978
1st—105	1st—163	1st—115	1st—183
2nd—108	2nd—166	2nd—121	2nd—187
3rd—113	3rd—173	3rd—125	3rd—194
4th—116	4th—177	4th—132	4th—200
1976	1979	1976	1979
1st—120	1st—181	1st—136	1st—207
2nd—125	2nd—187	2nd—142	2nd—212
3rd—131	3rd—192	3rd—150	3rd—220
4th—136	4th—196	4th—158	4th—226
1977	1980	1977	1980
1st—141	1st—203	1st—164	1st—231
2nd—147	2nd—210	2nd—168	2nd—239
3rd—151	3rd—217	3rd—174	3rd—245
4th—156	4th—224	4th—178	4th—250

Note: The net income was calculated quarterly. The table shows this quarterly income represented on an annual basis (annualized).

where: $\hat{Y}$ = estimated annualized earnings per share

t = time periods in quarters ($t = 1, 2, 3, \ldots, 24$)

Note that the independent variable in the regression analysis is time, t, where for the first quarter of 1975, $t = 1$, and for the fourth quarter of 1980, $t = 24$. The regression results are as follows:

Bank A	Bank B
$b_o = 95.71$	$b_o = 108.33$
$b_1 = 5.137$	$b_1 = 5.857$
$r^2 = 0.997$	$r^2 = 0.998$

The high r^2 values for both banks indicate that the simple linear model, with time as the independent variable, fits the annualized earnings per share very well. Figure 18–8 shows the time series plots for the two banks and the least squares trend lines. As can be seen, the earnings-per-share time series for both banks during the time period studied exhibit little seasonal or cyclical variation.

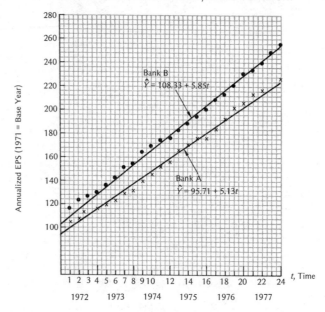

FIGURE 18–8 Annualized earnings per share for bank A and bank B.

The analyst would conclude from the least squares trends not only that bank B begins with a higher annualized earnings per share than bank A, but also that growth has been faster for bank B than for bank A.

18-4
ANALYZING THE SEASONAL COMPONENT

Decision makers would have a much easier time of it if their operating environment contained no variation, but most often variation does exist. This is particularly true of industries and firms that are affected by a seasonal factor. For instance, although food processors would like to have a continual supply of fresh products, the supply is tied to the growing season. Lumber mill operators would like to have a continual supply of trees. However tree harvesting is affected by snow depths and muddy ground during the winter and early spring. Toy store owners have a very seasonal business, with the big majority of sales occurring in the two or three months immediately before Christmas.

For many organizations, either the demand for their product or service, or their source of supply, is highly dependent on the time of year. Since seasonal variation is considered normal in many businesses, the real questions are how this variation affects the planning process and whether the observed variation is more or less than expected. A seasonal index known as the **ratio to moving average** can be calculated to measure seasonal variation in a time series.

For example, the Folks Toy Company is faced with a highly seasonal demand for its toys. Although some toys are sold throughout the year, sales peak sharply during October, November, and December, and fall off sharply in the spring and summer. The number of toy units sold during each month for the three-year period 1976–1979 is shown in Table 18–2. Figure 18–9 shows the time series sales data for the Folks Toy Company.

TABLE 18–2 Monthly toy sales, Folks Toy Company
(toy units)

Month	1977	1978	1979
Jan	3,200	3,500	3,500
Feb	3,300	3,600	3,700
Mar	3,400	3,300	3,400
Apr	3,200	3,400	3,200
May	3,500	3,900	4,000
June	3,400	4,000	4,400
July	3,700	4,000	5,000
Aug	4,000	3,900	4,800
Sept	4,800	4,900	5,900
Oct	5,600	5,800	6,400
Nov	5,900	6,200	7,200
Dec	6,600	7,000	7,800

The ratio to moving average is only one of several methods of determining a seasonal index, but is widely used because of its statistical and practical advantages. The ratio to moving average method begins with what is called a **multiplicative model.**

$$Y = T \times S \times C \times I$$

where: Y = value of the time series

T = trend value

S = seasonal value

C = cyclical value

I = irregular value

The multiplicative model assumes that each time series value is determined by the relationships between the four components. The ratio to moving average first attempts to estimate $T \times C$ by calculating a **12-month moving average.**

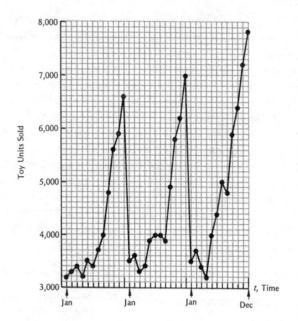

FIGURE 18–9   Monthly toy sales, Folks Toy Company.

A 12-month moving average is computed from 12 successive monthly time series values. The moving average is then located in the middle of the 12 values. The next moving average value is determined by dropping the first time series value and adding the next, yet-unused value. Table 18–3 contains the moving averages for the Folks Toy Company.

      Once the 12-month moving average is calculated, the next step is to find a **centered 12-month moving average.** We do this by finding the mean of each successive pair of moving averages. For example, the first centered 12-month moving average, corresponding to the seventh period (that is, July 1977), is the average of 4,217 and 4,242.

      The next step is to divide the original data by the corresponding centered 12-month moving average. This ratio is called the **ratio to moving average.** These ratios approximate the $S \times I$ factor in the multiplicative model and are listed in Table 18–3. For example, the ratio corresponding to July 1977 is

$$\frac{3,700}{4,229.5} = 0.857 = 87.5\%$$

      Before attempting to separate the irregular component, we must determine whether the ratio to moving averages are stable from year to year. If they do not appear to be stable, there is little value in trying to arrive at an index value for a particular month. In fact, such a value might be misleading.

      In the Folks Toy Company example, we have only two years of ratios to examine. Figure 18–10 shows the lines connecting the ratios for each year. As can be seen in Figure 18–10, the seasonal ratios for the two years have the same basic turning points. December is the seasonal high, and March and April represent the seasonal low. Because of the consistency in the ratios between years, we will extract the irregular influences.

TABLE 18–3 Computation of ratio to moving average, Folks Toy Company

Year and Month	Units Sold	12-Month Moving Total	12-Month Moving Average	Centered 12-Month Moving Average	Ratio to Moving Average (%)
1977					
Jan	3,200				
Feb	3,300				
Mar	3,400				
Apr	3,200				
May	3,500				
June	3,400	50,600	4,217		
July	3,700	50,900	4,242	4,229.5	87.5
Aug	4,000	51,200	4,267	4,254.5	94.0
Sept	4,800	51,100	4,258	4,262.5	112.6
Oct	5,600	51,300	4,275	4,266.5	131.3
Nov	5,900	51,700	4,308	4,291.5	137.5
Dec	6,600	52,300	4,358	4,333.0	152.3
1978					
Jan	3,500	52,600	4,383	4,370.5	80.0
Feb	3,600	52,500	4,375	4,379.0	82.2
Mar	3,300	52,600	4,383	4,379.0	75.4
Apr	3,400	52,800	4,400	4,391.5	77.4
May	3,900	53,100	4,425	4,412.5	88.4
June	4,000	53,500	4,458	4,441.5	90.0
July	4,000	53,500	4,458	4,458.0	89.7
Aug	3,900	53,600	4,467	4,462.5	87.4
Sept	4,900	53,700	4,475	4,471.0	109.6
Oct	5,800	53,500	4,458	4,466.5	129.9
Nov	6,200	53,600	4,467	4,462.5	138.9
Dec	7,000	54,000	4,500	4,483.5	145.1
1979					
Jan	3,500	55,000	4,583	4,541.5	77.1
Feb	3,700	55,900	4,658	4,620.5	80.1
Mar	3,400	56,900	4,742	4,700.0	72.3
Apr	3,200	57,500	4,792	4,767.0	67.1
May	4,000	58,500	4,875	4,833.5	82.7
June	4,400	59,300	4,942	4,908.5	89.6
July	5,000				
Aug	4,800				
Sept	5,900				
Oct	6,400				
Nov	7,200				
Dec	7,800				

One method to eliminate the irregular component is to take the ***normalized*** average of the ratio to moving averages as shown in Table 18–4. **The *normalization* is performed by multiplying each average seasonal index by the ratio of 1,200 over the sum of these seasonal indexes.** This forces the normalized indexes to sum to 1,200 (12 months × an average of 100).

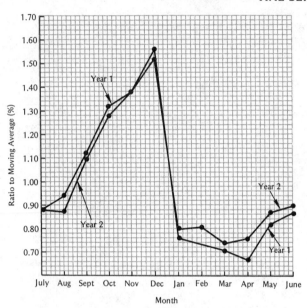

FIGURE 18–10 Seasonal patterns, Folks Toy Company (July 1977–June 1979).

Figure 18–11 graphs the seasonal index values for each month for the Folks Toy Company. Looking at Figure 18–11 and Table 18–4, we can see that December's normalized seasonal pattern is about 55 percent greater than the yearly average, whereas April's pattern is about 27 percent below the yearly average.

This method of eliminating irregular fluctuations should be applied only when we are willing to assume that the irregular fluctuations are caused by purely random circumstances. In cases where this assumption cannot be made, more sophisticated methods must be used. Chou (1975) presents a more detailed treatment of the methods for separating irregular influences from the seasonal component in a time series.

TABLE 18–4 Seasonal index, Folks Toy Company

Year	July	Aug	Sept	Oct	Nov	Dec	Jan	Feb	Mar	Apr	May	June	
July 1977– June 1978	87.5	94.0	112.6	131.3	137.5	152.3	80.0	82.2	75.4	77.4	88.4	90.0	
July 1978– June 1979	89.7	87.4	109.6	129.9	138.9	156.1	77.1	80.1	72.3	67.1	82.7	89.6	
Total	177.2	181.4	222.2	261.2	276.4	308.4	157.1	162.3	147.7	144.5	171.1	179.6	
Average seasonal index (%) =	88.6	90.7	111.1	130.6	138.2	154.2	78.5	81.1	73.8	72.2	85.5	89.8	$\Sigma = 1,194.3$
Normalization factor = 1,200/1,194.3 = 1.0047													
Normalized average seasonal index (%) =	89.0	91.1	111.6	131.1	138.8	154.9	78.9	81.6	74.1	72.6	85.9	90.2	$\Sigma = 1,200.0$

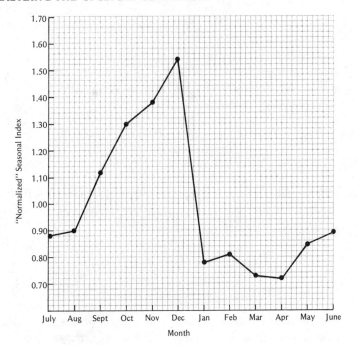

FIGURE 18–11 Seasonal index, Folks Toy Company.

18-5
ANALYZING THE CYCLICAL COMPONENT

Cyclical variations in time series data do not repeat themselves in a regular pattern as do seasonal factors, but they cannot be considered random variations in the data either. Although cyclical variations generally show some recognizable pattern and are repetitious, they always differ in both *intensity* and *timing*. **By *intensity* of a cycle we mean the height from its crest to its trough. *Timing* is the frequency with which the crests and troughs occur.** Therefore cyclical components can be isolated and analyzed, but, unfortunately, cannot be accurately predicated.

Many industries are influenced by cyclical patterns in their environment. The most important cyclical factor for most organizations is the cycle of economic factors generally referred to as the business cycle. While some economists recently thought governmental action could essentially eliminate cyclical swings in the economy, this component is still apparently with us and is very important for many organizations. The organizations hardest hit by the cyclical component are those connected with items purchased with discretionary income (appliances, cars, travel, and so forth). These are items people can often postpone purchasing, and consequently are those most affected by a downturn in the economy.

The cyclical component is isolated by first removing the trend and seasonal factors from the time series data. While many complicated methods exist to isolate the cyclical component, the general procedure can be demonstrated graphically as shown in Figure 18–12.

      We will utilize the sales data for Folks Toy Company to illustrate one method for isolating the cyclical component in a time series. Table 18–5 shows the values involved in the calculations. The monthly sales data for the three years 1977–1979 are contained in column (1).

      Column (2) contains the trend values determined by the least squares regression trend line,

$$\hat{Y} = 3,353.17 + 64.09t$$

Note that this line is calculated using units sold as the dependent variable and $t = 1, 2, 3, \ldots, 36$ as the independent variable. Thus, for example, June 1977, which is the sixth time period ($t = 6$), has a trend value of

$$\hat{Y} = 3,353.17 + 64.09(6)$$
$$= 3,737.7$$
$$\text{Trend value} = 3,738$$

      Column (3) contains the seasonal index for the month as determined in Table 18–4. Note that the seasonal index for July 1977 is the same as for July 1978 and July 1979.

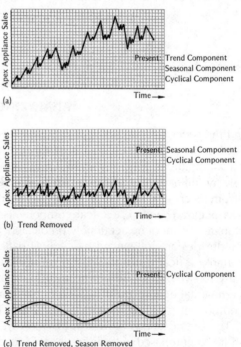

FIGURE 18–12   Isolating the cyclical component, Apex Appliances.

TABLE 18–5 Cyclical component index, Folks Toy Company

Year and Month	Units Sold Y (1)	Trend Value T (2)	Seasonal Index Ratio S (3)	Statistical Normal $T \times S$ (4)	Cyclical-Irregular Component (%) $\dfrac{Y}{T \times S}100$ (5)
1977					
Jan	3,200	3,417	0.789	2,696.9	119.6
Feb	3,300	3,481	0.815	2,837.4	116.3
Mar	3,400	3,545	0.741	2,629.9	129.2
Apr	3,200	3,610	0.725	2,620.5	122.1
May	3,500	3,674	0.859	3,157.6	110.8
June	3,400	3,738	0.902	3,373.3	100.7
July	3,700	3,802	0.890	3,383.6	109.3
Aug	4,000	3,866	0.911	3,522.4	113.5
Sept	4,800	3,930	1.116	4,385.9	109.4
Oct	5,600	3,994	1.311	5,237.6	106.9
Nov	5,900	4,058	1.388	5,633.7	104.7
Dec	6,600	4,122	1.549	6,385.8	103.3
1978					
Jan	3,500	4,186	0.789	3,303.9	105.9
Feb	3,600	4,250	0.815	3,464.2	103.9
Mar	3,300	4,314	0.741	3,200.5	103.1
Apr	3,400	4,379	0.726	3,178.8	106.9
May	3,900	4,443	0.859	3,818.7	102.1
June	4,000	4,507	0.902	4,062.5	98.3
July	4,000	4,571	0.890	4,068.1	98.3
Aug	3,900	4,635	0.911	4,223.2	92.3
Sept	4,900	4,699	1.116	5,244.3	93.4
Oct	5,800	4,763	1.311	6,246.2	92.8
Nov	6,200	4,827	1.388	6,701.4	92.5
Dec	7,000	4,891	1.549	7,577.3	92.3
1979					
Jan	3,500	4,955	0.789	3,910.9	89.5
Feb	3,700	5,019	0.815	4,091.1	90.4
Mar	3,400	5,084	0.741	3,777.0	90.1
Apr	3,200	5,148	0.726	3,737.2	85.6
May	4,000	5,212	0.859	4,479.8	89.2
June	4,400	5,276	0.902	4,761.6	92.4
July	5,000	5,340	0.890	4,752.6	105.2
Aug	4,800	5,404	0.911	4,924.0	97.4
Sept	5,900	5,468	1.116	6,102.6	96.6
Oct	6,400	5,532	1.311	7,254.7	88.2
Nov	7,200	5,596	1.388	7,769.1	92.6
Dec	7,800	5,661	1.549	8,768.7	88.9

The next step in separating the cyclical component from the trend and seasonal components is to calculate the *statistical normal values* by multiplying the trend values by the seasonal index values.

The cyclical component, which also contains the irregular fluctuations, is determined for each time period by dividing the statistical normal values into the original units sold [column (1)]. To transfer this ratio to a percent form, we multiply by 100. These percentages are shown in column (5) of Table 18–5.

Although some decision makers attempt to smooth out the irregular fluctuations by taking a three- or five-month moving average of the percentages in column (5), this is generally considered unacceptable. Thus we suggest that the cyclical and irregular components be analyzed as a single component.

18-6
ANALYZING THE IRREGULAR COMPONENT

Irregular components are those fluctuations in a time series that cannot be attributed to any of the three previously discussed components. Irregular influences may be caused by any number of one-time factors. For instance, a severe winter, a summer drought, a civil war in a country supplying raw materials, and many other factors can cause irregular changes in the time series. While these events would likely cause a large irregular fluctuation, many small, unrelated events can combine to cause random variation around the time series.

Extreme irregular variations can cause organizations a great deal of trouble. In fact, if the irregularities are severe enough, they can cause the organization to go out of business. When possible, most organizations buy insurance against the monetary effects of severe irregular effects.

Small irregular variations cause lesser problems. One of these problems crops up when the time series is being analyzed. Decision makers will generally attempt to smooth out the minor irregularities using a moving average approach as discussed earlier. The goal is to eliminate as much as possible the irregular influences so that the true trend, seasonal, and cyclical components can be recognized.

18-7
INDEX NUMBERS

When evaluating data collected over time, decision makers must often compare one figure or data point with others measured at different times. A common procedure for making relative comparisons is to construct a series of *index numbers* using a base to which all other data values can be compared.

The Armstrong Corporation is considering purchasing a small textile mill in Georgia. The present mill owners stress as a positive attribute the mill's rapid sales growth over the past ten years.

Table 18–6 shows the mill sales data for the years 1971–1980. To construct a *sales index* over the ten-year period, we first select a base year. Suppose we use 1971 as the base year. We set 1971 sales equal to 100. This means that 1971 sales were 100 percent of 1971 sales. Next, we compute each subsequent year's sales as a percentage of 1971 sales. To do this, we divide each year's sales by sales in the base period and multiply the quotient by 100. This is shown in Table 18–7.

TABLE 18–6 Textile sales data, Armstrong Corporation example

Year	Sales
1971	$14.0 million
1972	15.2
1973	17.8
1974	21.4
1975	24.6
1976	30.5
1977	29.8
1978	32.4
1979	37.2
1980	39.1

TABLE 18–7 Textile sales index, Armstrong Corporation example

Year	Sales	Index Number
1971	$14.0 million	100.0
1972	15.2	108.5
1973	17.8	127.1
1974	21.4	152.8
1975	24.6	175.7
1976	30.5	217.8
1977	29.8	212.8
1978	32.4	231.4
1979	37.2	265.7
1980	39.1	279.3

By examining the index numbers in Table 18–7, the Armstrong Corporation can see that 1980 sales were up 179.3 percent over 1979 sales. However Armstrong cannot compare two index numbers, such as those for 1979 and 1980, to each other. That is, sales were not up 13.6 percent between 1979 and 1980, simply because this is the difference in the index numbers for the two years. The true percent increase between 1979 and 1980 is found by dividing the difference in index numbers by the earlier index number and then multiplying by 100.

$$\frac{279.3 - 265.7}{265.7}100 = 5.1\%$$

The Problem of Different Bases

Suppose the Armstrong Corporation is also interested in a second textile company. This company's sales for 1977–1980 are listed in Table 18–8. With 1977 as the base year, Table 18–8 also provides the index numbers for the years 1977–1980. We see that sales in 1980 were up 67 percent over the base period.

TABLE 18–8 Second textile company sales, Armstrong Corporation example

Year	Sales	Index Number
1977	$19.4 million	100.0
1978	26.2	135.0
1979	21.4	110.3
1980	32.4	167.0

Because base periods for the two textile companies differ, the indexes for the two companies cannot be compared. To make a logical comparison of the two textile companies, we must shift the base year of one firm so that both share a common base period. As long as the sales data are unweighted, we can accomplish the shift by dividing the existing index number by the index number of the common base year and multiplying by 100. Table 18–9 provides the revised index numbers for the first textile company using 1977 as the base year.

TABLE 18–9 Revised index numbers for the first textile company, Armstrong Corporation example

Year	Sales	Original Index Base 1971	Revised Index Base 1977
1971	$14.0 million	100.0	$\frac{100.0}{212.8}100 = 47.0$
1972	15.2	108.5	$\frac{108.5}{212.8}100 = 51.0$
1973	17.8	127.1	$\frac{127.1}{212.8}100 = 59.7$
1974	21.4	152.8	$\frac{152.8}{212.8}100 = 71.8$
1975	24.6	175.7	$\frac{175.7}{212.8}100 = 82.6$
1976	30.5	217.8	$\frac{217.8}{212.8}100 = 102.3$
1977	29.8	212.8	$\frac{212.8}{212.8}100 = 100.0$
1978	32.4	231.4	$\frac{231.4}{212.8}100 = 108.7$
1979	37.2	265.7	$\frac{265.7}{212.8}100 = 124.9$
1980	39.1	279.3	$\frac{279.3}{212.8}100 = 131.3$

Comparison of the index numbers in Tables 18–8 and 18–9 shows us that, since 1977, the second company clearly has had a greater percent increase in sales than the first. In order for us to be able to make this kind of comparison, the two indexes must have a common base.

Commonly Used Index Numbers

Decision makers may encounter a variety of index numbers in their normal activities. Federal and state governmental agencies produce index numbers that help to measure such economic variables as retail prices, wholesale prices, stock market activity, production output, inventory levels, and the value of the dollar. The next few paragraphs briefly describe these common index numbers.

Consumer Price Index. To most of us, inflation has come to mean increased prices and less purchasing power for our dollar. The *Consumer Price Index (CPI)* constructed by the U.S. Department of Labor, Bureau of Labor Statistics, attempts to measure the overall change in retail prices for goods and services.

The CPI provides perhaps the most recognized example of the problem of changing index bases as discussed in the last section. The CPI measures the relative change in some standard group of goods and services. To make the series consistent over time, this group of goods and services should remain constant. However, if it is to reflect typical buying habits of consumers, it must also reflect these changing habits. Therefore the standard group is periodically updated. In the most recent (1977) change, two indexes were constructed. The first, the *index for wage earners and clerical workers,* is the same as has been constructed in previous years. The second, the *index for all urban households,* is a new index that was created to better represent this major segment of the economy. The purpose of having two indexes is to provide more useful economic information; however, changing consumption patterns will undoubtedly cause future revisions in this important series.

Most states and even some cities compute their own consumer price index or market basket index to provide information about relative price changes. In fact, many companies and governmental agencies base employee pay increases on these retail price indexes.

Wholesale Price Index. The U.S. Department of Labor, Bureau of Labor Statistics, also constructs an index of prices of goods sold in U.S. primary markets. Primary markets are those where the first significant large-volume purchase for each item is made. Their measures determine the *Wholesale Price Index.*

Many groups—governmental, industrial, and academic—study the relationship between the Wholesale Price Index and the CPI. Generally, increases in the Wholesale Price Index are followed by a delayed increase in the CPI.

Industrial Production Index. The Federal Reserve Board constructs the *Industrial Production Index.* This index measures change in the volume of industrial production, including manufacturing, mining, and utilities. The purpose of this index is to keep us apprised of productivity levels through various groupings of products and materials.

Stock Market Indexes.   Probably the best known stock market index is the ***Dow-Jones Industrial Average.*** This index is computed each day the New York Stock Exchange is open and indicates the condition of the market in general. The Dow-Jones Industrial Average is determined by the daily closing stock prices for 30 selected industrial companies, adjusted for any dividends or mergers that take place for these companies. Of course, stocks not included in the "DOW" may perform much differently than would be indicated by the index.

Another stock market index is the ***Standard and Poor's 425.*** This index is determined by stock prices of 425 industrials, not all of which are blue-chip companies. Many market analysts prefer the Standard and Poor's 425 to the "DOW" because they feel it better represents the market as a whole.

18-8
FORECASTING

Forecasting is the practice of making predictions about future events. Forecasting is a part of every functional business area and thus is a major responsibility of every business decision maker. The production manager must forecast when a new product will be ready for marketing. The financial planner must develop a forecast of company revenues and expenses in preparing pro forma financial statements. The marketing manager must forecast product demand in a market area to determine how many salespeople to assign to the area. The inventory control manager must use a demand forecast in determining the optimal order point and order quantity.

While these are only a few instances in which business forecasting is required, they should give you an idea of how important forecasting is in the decision-making process. This section briefly introduces some elementary forecasting methods. For a much more complete discussion of forecasting techniques, consult Anderson (1975) and Armstrong (1978).

Forecasting methods can be roughly grouped into two categories: ***statistical*** and ***judgmental. Statistical* procedures attempt to determine a formal model for using data gathered on past and current events to predict what will happen in the future. *Judgmental* forecasts are made when either past data are not available, or the event being forecast is totally new, and hence what happened in the past is really not connected with the future.**

In practice, the primary factor that determines the extent to which statistical and judgmental forecasts are used in a given situation seems to be how much uncertainty the decision maker sees in the future. When the decision maker sees the future as mostly an extension of the past, statistical models should be used. The more the future is expected to differ from the past, the more judgment should be incorporated into the decision. Many practical situations involve a combination of both techniques.

Judgmental Forecasts—Advantages and Disadvantages

The advantage of judgmental forecasting is that it employs the experience base of the decision maker along with qualitative and quantitative data. **A judgmental forecast has no specific data requirements and is not tied directly to the past. The major disadvantage of judgmental forecasting is that there are no statistical means to test the methodology used to arrive at the forecast.**

Once a decision maker has decided a judgment forecast is needed, the question is whose judgment to use. Although many people have knowledge that will aid in making a forecast, most managers would like to base their decisions on the forecast of *experts* in a field. From an organizational point of view, these experts seem to fall into particular groups:

1. **Top Management.**   In many organizations, top-level managers are a good source of forecast information. This is particularly true if the manager has a lot of experience in the field or the organization.
2. **The Sales Force.**   All organizations ultimately exist because they supply a product or service to meet a demand. Since demand keeps changing, the sales force often gets a feel for this demand faster than any other group.
3. **Consumers.**   The people who ultimately buy the product or service of an organization are often the "experts" on their perceived future needs.
4. **Consultants.**   Often what is a new situation for one organization has been experienced closely by another organization. Consultants are often familiar with what was tried, and often worked, someplace else and can provide the company with the necessary experience base.

This list is not all-inclusive. For many organizations, the expert chosen depends on the product or service being sold. Pharmaceutical manufacturers seek the judgment of physicians; electronics manufacturers seek the judgment of electrical engineers; textbook publishers seek the judgment of instructors; and on and on. The point is, judgment forecasts may be the best, or even the only, forecasts that can be made.

Statistical Forecasting Techniques

Statistical forecasting techniques assist decision makers in obtaining objective forecasts based upon historical data. In Chapters 16 and 17 we showed how regression analysis can be used for prediction, description, and control. If passage of time is accounted for in the model, regression analysis also provides a tool for statistical forecasting.

Regression forecasting models range from very simple to very complex. Possibly the most basic forecasting model is the simple linear time series model of the form

$$\hat{Y} = b_o + b_1 t \qquad \qquad \textbf{(18–1)}$$

where: $\hat{Y}$ = estimated value of the variable of interest

b_1 = long-term trend coefficient

t = time

This model is based exclusively on past observations of the Y variable. We used this model in Section 18–3 to determine the time series trend component. The model can also be used to forecast future time periods. For example, the data in Table 18–10 show the number of patients treated at the Montpier Hospital emergency room for the years 1960–1979. Figure 18–13 shows the time series plot of these data.

TABLE 18–10   Emergency room data, Montpier Hospital (1960–1979)

Time Period t	Year	No. Patients
1	1960	1,459
2	1961	1,620
3	1962	1,783
4	1963	1,794
5	1964	2,191
6	1965	2,423
7	1966	2,677
8	1967	2,815
9	1968	3,093
10	1969	3,190
11	1970	4,550
12	1971	5,166
13	1972	5,820
14	1973	5,809
15	1974	6,622
16	1975	8,092
17	1976	9,426
18	1977	10,554
19	1978	10,998
20	1979	11,433

Notice in Table 18–10 and Figure 18–13 the sharp upward trend in the number of patients treated in the emergency room. A number of factors may be causing this growth. For instance, the county population has been growing steadily, although accurate numbers are not available. Also, more doctors are living and practicing in the area. While such variables as population and number of doctors might be good indicators of future emergency room demand, past data for these variables are difficult or impossible to obtain. Also, since the administrator is concerned with future emergency room demand, future values for these variables would be required. Thus time (which itself does not cause anything to happen) can be used in place of these other factors, which also change with time. The least squares linear trend model developed from 20 years of data is

Number of emergency room patients = $-663.274 + 546.574t$

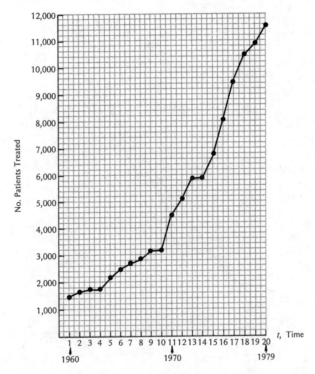

FIGURE 18–13 Emergency room data, Montpier Hospital.

This model has an R^2 of 0.913, a standard error of 1,026.61, and an F ratio of 188.498. Thus we can infer that the model is significant in explaining the variation in emergency room patients over time.

Future emergency room demand is forecasted by substituting values for t and solving for $\hat{Y}$. Thus the forecast is made by simply *extrapolating* the linear trend into the future. This is shown in Figure 18–14, and forecasts for 1980–1984 are contained in Table 18–11. You should recognize the dangers involved with extrapolation. Should the trend change, the forecasts will obviously be either too high or too low.

Regression Models with Indicator Variables. Even though the simple linear time series model just described may provide acceptable forecasts, it may not be adequate. An alternative regression-based forecasting technique measures the dependent variable one or more periods ahead of the independent variables. For example, the R. B. Sawyer Lumber Company needs to forecast lumber sales a year from now in order to contract for logs this year. The sales manager has established the following regression model:

$$\text{Sales}_t = b_o + b_1(\text{price}_{t-1}) + b_2(\text{prime interest rate}_{t-1}) + b_3(\text{housing starts}_{t-1})$$

Note that the independent variables, lumber price, prime rate, and housing starts, are measured in time $t - 1$, whereas the dependent variable, lumber sales, is measured in time t. These independent variables are called *leading indicator variables*

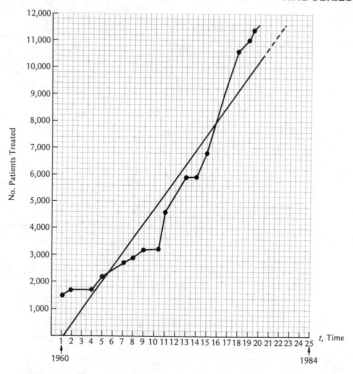

FIGURE 18–14   Emergency room forecast model, Montpier Hospital.

since their values in one time period help explain the dependent variable's value in subsequent time periods.

Forecasting models using leading indicator variables have high intuitive appeal. The relationships are logical, and decision makers may place more confidence in the resulting forecasts than in those from another forecasting method with less intuitive appeal. However, as with any other regression-based forecasting model, just because the model works well for past data is no guarantee it will work well in the future. If the relationships between the independent variables and the dependent variable change, the indicator variables may provide poor forecasts.

Autoregressive Models

Autoregressive models assume that the best forecast of what will happen one period from now is made by considering a combination of what is happening now plus what

TABLE 18–11   Emergency room forecasts for 1980–1984, linear model, Montpier Hospital

Year	t	Forecast
1980	21	10,815
1981	22	11,361
1982	23	11,908
1983	24	12,455
1984	25	13,001

has happened in the past. The models assume that the forecast can be made by some combination of past observations. For a linear autoregressive model, the forecast equation is

$$\hat{Y}_{t+1} = b_o + b_1 Y_t + b_2 Y_{t-1} + \ldots + b_n Y_{t-n+1}$$

where: $\hat{Y}_{t+1}$ = forecast made for period $t + 1$

Y_t = observed value in period t

. .

. .

. .

Y_{t-n+1} = observed value in period $t - n + 1$

Other Statistical Forecasting Techniques.   In addition to the regression-based models, many other statistical methods exist for providing forecasts based upon past data. For instance, the ***exponential smoothing model*** [see Nelson (1973)] often provides good short-term forecasts and is used by inventory control managers for sales forecasting. Whereas regression models weight all past observations equally in determining the forecast model, the exponential smoothing technique weights each past observation differently. This allows more-current values in the time series to have more influence in determining the forecast for future periods.

Another forecasting model, useful when the time series is seasonal, is the ***exponential weighted moving average model*** [see Winters (1960)]. This model uses past data to develop separate estimates for both the long-term trend and the seasonal component and then combines these estimates to forecast future periods. As with the exponential smoothing model, the exponential weighted moving average model weights current data more heavily than older data in determining the forecast value.

Testing a Model by Holding Out Data

Forecasters often run into trouble by using a set of data to construct a model and then testing the model on the same data. Mathematically, we can construct a model to fit perfectly a set of data by including a number of independent variables equal to the number of data points. However the model that is perfect on past data may not work well on future data. Although the ultimate test of a model is with future data, unfortunately these data are not available when needed. The next best alternative is to randomly separate the past data into two groups and use one group to develop the model and the second group to test the model. A model that accurately fits this second group of past data is not guaranteed to be a good forecasting model, but this approach is preferred over testing the model with the data used to develop the model.

Testing alternative models depends on having past data to forecast and compare. If no past data exist, or if the past data do not seem to apply to the present situation or forecast period, decision makers have no real alternative but to use some form of judgment forecast.

18-9
CONCLUSIONS

In this chapter we have introduced the basics of time series analysis. We defined the four components of a time series: long-term trend, seasonal component, cyclical component, and irregular component.

Decision makers need to examine variables measured over time in an effort to learn about the past. We study the past to make better decisions about the future. However, the future may or may not reflect the past, and care should be used in making conclusions about the future based solely on historical data.

We have illustrated some basic techniques for extracting the long-term trend, seasonal, and cyclical components from a time series. The techniques discussed here represent only a few of the many time series analytic procedures that have been developed and are applied in decision-making situations.

Both the strengths and weaknesses of time series analysis can be pointed out by the following analogy. Imagine yourself walking west from Kansas City two hundred years ago—backwards. There are no maps and no roads. You don't know what to expect as you go. You may have a pretty good idea of what lies ahead by looking at what type of country you have just passed through. Because you're walking backwards, you may stumble slightly as the terrain changes in small amounts. However, after getting wet a few times, you will even recognize when you're coming to a river. Unfortunately nothing you have seen in your entire trip will prepare you for when you hit Colorado and back into the Rocky Mountains. Your past experiences simply will not indicate this new and drastically different experience.

Thus it is with time series analysis. We can look backwards by analyzing the time series data. This analysis will provide "good" information about the future as long as the future looks pretty much like the past. However time series analysis is of little help if the future departs drastically from the past. No matter how sophisticated our analysis, we cannot use the past to foresee new and unusual events of the future. Decision makers should recognize both the strengths and weaknesses of time series analysis. If they do, they will find it a valuable tool in decision making.

In this chapter we have also discussed the concept of index numbers. An index number is generally constructed to help make comparisons between a process at different points in times, or to compare two different processes at the same point in time. Finally, we introduced some elementary forecasting concepts.

chapter glossary

cyclical component   The periodic movements in a time series usually caused by economic factors. The frequency and intensity of the cyclical components are not totally predictable.

irregular component   The changes in the time series that are unpredictable and cannot be attributed to a trend, seasonal, or cyclical factor.

seasonal component   The increases and decreases in the time series that occur at predetermined times of the year with predictable intensities.

seasonal index   The ratio found by dividing the observed value of a period by the value of an average period.

time series   A series of measurements taken of a variable at different times. In most applications, the time periods are uniform.

trend component   The long-run average increase or decrease in the time series.

=problems

Problems marked with an asterisk (*) require extra effort.

1. Which of the four main time series components do you think would be of most interest to each of the following industries? Justify your answers.
   a. Office equipment
   b. Recreational vehicles
   c. Tourism
   d. Heavy construction equipment
   e. Banking
   f. Car insurance

2. Your company feels its sales might be influenced by the normal cyclical trends in the economy. How would you go about determining whether this is true or not? If you did find a cyclical component, how would you use this information to aid the company in its decision-making process?

3. The monthly sales (rounded to the nearest thousand) for the Ripley Department Store are as follows:

Month	1979	1980
Jan	$ 60	$ 65
Feb	70	75
Mar	80	85
Apr	90	95
May	100	110
June	110	120
July	120	130
Aug	130	140
Sept	140	145
Oct	120	130
Nov	100	115
Dec	80	90

Calculate the monthly ratio to moving average for Ripley. Then discuss why a store might want to use these ratios.

4. The following are the hourly and weekly earnings of manufacturing workers in the United States for selected years:

Year	Hourly Earnings	Weekly Earnings
1909	$0.19	$ 10
1914	0.22	11
1920	0.55	26
1925	0.54	24
1930	0.55	23
1935	0.54	20
1940	0.66	25
1945	1.02	44
1950	1.44	58
1955	1.86	76
1960	2.26	90
1965	2.61	108
1970	3.36	134
1975	4.81	190

Plot this set of data on a graph. What type of long-term trend seems most apparent? How would you go about removing the trend component from this set of data?

5. A student in class comments that the data in problem 4 prove workers are much better off now than they were in the beginning of the century. Comment on this statement. Could these figures be used to measure how well off the workers are? How?

6. Fit simple time series regression lines to the data given in problem 4. Comment on how well these lines explain the variation in both hourly and weekly earnings.

7. The following are birthrates (per 1,000) in the United States between 1930 and 1976:

Year	Birthrate	Year	Birthrate
1930	21.3	1954	25.2
1931	20.2	1955	24.9
1932	19.5	1956	25.1
1933	18.4	1957	25.2
1934	19.0	1958	24.5
1935	18.7	1959	24.3
1936	18.4	1960	23.8
1937	18.7	1961	23.5
1938	19.2	1962	22.6
1939	18.8	1963	21.9
1940	19.4	1964	21.2
1941	20.3	1965	19.6
1942	22.2	1966	18.5
1943	22.7	1967	17.9
1944	21.3	1968	17.6
1945	20.4	1969	17.9
1946	24.1	1970	18.2

1947	26.5	1971	17.2
1948	24.8	1972	15.6
1949	24.5	1973	14.9
1950	23.9	1974	14.9
1951	24.8	1975	14.7
1952	25.0	1976	14.7
1953	24.9		

a. Do you see a long-term trend in these data? If so, what?

b. Is there a cyclical component in these data?

c. Based on these data, do you see any evidence that the long-term birthrate in the United States is permanently reduced?

d. What problems do you see in making forecasts of future birthrates from just these data?

8. Using sources available in your school's library, select a variable of interest and collect data for as many periods as possible. Identify the components present in these data and comment on how you might go about forecasting future values of this variable.

9. Sunrise Sports has experienced rapidly expanding retail sales. Its sales levels for the past 12 years are

Year	Sales	Year	Sales
1969	$1.9 million	1975	$ 8.6 million
1970	3.1	1976	9.3
1971	2.8	1977	11.0
1972	4.5	1978	13.9
1973	5.7	1979	16.6
1974	5.8	1980	19.4

a. Plot this time series on ordinary graph paper.

b. Explain the trend in these data. What quantitative tool do you think would assist you in making this description?

10. Construct a least squares regression line to fit the data in problem 9. How do you feel this line explains the variation in past sales data? Comment on any patterns you see in the relationship between the actual sales values and those values predicted by the regression analysis.

11. The Green Valley Farmer's Cooperative has gathered the following data on quarterly ice cream sales by the coop's creamery. All values are in millions of dollars.

Quarter	1975	1976	1977	1978	1979	1980
1st	1.1	1.4	1.3	1.6	1.9	2.4
2nd	2.4	2.8	2.9	3.3	3.8	4.4
3rd	2.9	3.6	3.9	4.2	4.8	5.1
4th	1.7	1.9	2.0	2.6	3.1	3.2

a. Graph these time series data.
b. Determine the quarterly moving average and centered quarterly moving average for these sales levels.
c. Determine the ratio to moving average figures for these data.
d. Normalize the quarterly ratios.

12. Ellial's Quality Discount Store has applied for a line of credit with the First National Bank. This line of credit is to be used primarily for financing inventory purchases. As part of the financial application, Ellial's has been asked to provide monthly inventory levels for the past five years. These levels (in millions of dollars) are:

Month	1976	1977	1978	1979	1980
Jan	5.2	4.7	6.6	7.1	7.0
Feb	3.3	2.9	4.0	4.0	6.2
Mar	2.8	3.0	3.6	2.6	4.3
Apr	5.3	6.3	7.2	8.0	9.5
May	9.4	10.0	11.4	7.8	12.5
June	2.6	4.3	4.0	5.4	6.4
July	6.2	7.7	8.0	9.3	8.6
Aug	7.2	7.5	6.8	8.2	8.4
Sept	6.8	5.8	6.8	7.9	6.9
Oct	9.7	9.6	8.9	9.3	9.8
Nov	13.6	13.9	14.2	16.1	16.5
Dec	11.8	11.9	12.7	13.8	14.6

a. Determine the seasonal index number for each month using the ratio to moving average method.
b. Is there enough consistency between years to make you comfortable using seasonal index numbers?
***c.** Ellial's will finance 90 percent of its monthly inventory through bank borrowing. The company has been able to get money at the prime rate plus 2 percent. However the interest must be paid monthly, and the value of the loan can change monthly. Estimate the value of interest payments for the next year.

*13. You have been hired recently by a farm-feed manufacturer. This organization is a relatively small regional operation that is beginning to feel some competition from major national manufacturers. The company has decided to institute some modern manufacturing and inventory policies. The first step is to decide just what influenced past sales. No one presently in the organization has experience with this analysis, and the personnel director discovers you have a statistical background. You are to outline the steps you would take to analyze the past sales.

14. The Wayne Construction Company specializes in middle-income housing. The following data show quarterly new housing starts for Wayne Construction:

Quarter	1976	1977	1978	1979	1980
1st	110	97	113	101	130
2nd	112	98	123	110	122
3rd	107	98	117	128	132
4th	95	105	103	136	145

a. Plot these data on a graph.

b. Which of the time series components seems to be most prevalent in these data? Why?

15. Tom Wayne, the general manager of Wayne Constuction (see problem 14), would like to have an indication of how his company has grown over the last five years. Construct a series of quarterly indexes to provide this information.

16. Four years ago Central State College instituted a computer preregistration system. Although the vast majority of students take advantage of this system, each quarter a significant number of students go through the manual registration procedure. Roberta Ashford, the financial vice-president, has kept track of the number of students manually registering each quarter. These data are:

Quarter	1976	1977	1978	1979	1980
Fall	1,325	1,267	1,245	1,130	1,039
Winter	1,128	1,138	1,074	1,003	967
Spring	814	820	753	692	622

a. Plot these data on a graph.

b. Separate the trend and seasonal components from the data.

c. Roberta Ashford is trying to determine how many workers will be needed at the manual registration next year. Which of the time series components would be most useful in helping her make this decision?

17. The president of Central State College is interested in whether an increasing or decreasing number of students are still going through manual registration. Roberta needs to construct a set of index numbers for the registration data in problem 16. Construct the series she needs.

18. The cyclical component of the national economy receives a great deal of attention. Private and governmental forecasters continually try to determine just why the cyclical component behaves as it does. Go to the library, and by studying the "Business Roundup" section of *Fortune*, Federal Reserve publications, the *Wall Street Journal*, or an economic forecasting text, determine what factors are considered when discussing the time series cyclical component and which are the most important.

*19. You have been hired by a local manufacturer primarily because of your extensive knowledge of statistics. Both the production and sales manager are interested in forecasting the probable future demand and sales for their products. Although they have extensive past records of operations and sales, they have no idea how to proceed. Discuss how you would go about making the forecasts.

20. As discussed in this chapter, many firms find their forecasting job made easier by finding leading indicators for their product or service. What leading indicators would you look at for the following industries?
    a. Children's clothes
    b. Automotive parts
    c. Color television sets
    d. Pet food
    e. Meat packing
    f. Industrial equipment exporting
    g. Weapons

21. Referring to problem 3, use the data provided to develop a simple linear time series forecasting model. Reserve the November and December sales values to test the model's forecasting ability. Comment on the model's appropriateness as a forecasting tool for the Ripley Department Store.

22. Referring to problem 3, develop an autoregressive forecasting model of the form

$$\hat{Y}_t = b_o + b_1 Y_{t-1}$$

Hold out the December sales data to test the forecasting ability of this autoregressive model. Comment on the results of this forecast and make suggestions for improving the model as a forecasting tool for the Ripley Department Store.

*23. Referring to the forecasting models developed in problems 21 and 22, if you were faced with a choice between the two, which would you select for predicting January 1981 sales? Justify your response.

*24. Consider the data presented in problem 7. Develop a linear time series model of the form

$$\hat{Y}_t = b_o + b_1 t$$

Comment on the validity of using this model for forecasting birthrates in the United States. What potential problems do you see in using a linear model?

*25. Colleges and universities, like private businesses, need accurate projections of their demand. A college or university needs to have a good idea of total enrollment at least a year in advance to plan for faculty to cover the required sections and to have an idea of the revenue that will be generated from student tuition.

You are to first determine a list of potential "indicator" variables to use for projecting total enrollment on a one-year basis. Next, collect as many years' worth of data as possible and develop a regression model using the "indicator" variables to predict total enrollment at your school.

=cases

18A MEDICAL CENTER HOSPITAL

R. T. Trusty, the administrator of Medical Center Hospital, recently hired U. R. Regis Consulting to study the medical center's admissions and develop a model for predicting future admissions. Because of the emphasis on long-range planning, Mr. Trusty needs a good admissions forecasting model. U. R. Regis collected data for the past ten years on the following variables:

VAR01 Admissions in year t
VAR02 Area population in year t
VAR03 Number of doctors in the area in year t
VAR04 Crime rate per 1,000 people in year t
VAR05 Average annual temperature in year t
VAR06 Number of vehicles registered in the county in year t
VAR07 Hospital capacity in year t
VAR08 Average daily patient cost in year t

U. R. Regis Consulting provided a printout of its work but no written report or evaluation of the model's usefulness.

Since Mr. Trusty is a little shaky on regression analysis, he would like you to write a short, but accurate synopsis of the printouts in Tables C–1 through C–4. Be sure to indicate any attributes the model has relative to Medical Center Hospital's needs, as well as any potential problems with the model.

TABLE C–1 Means and standard deviations

****

VARIABLE	MEAN	STD. DEV.	CASES
VAR01	5545.2000	2392.0319	10
VAR02	35860.0000	12655.9779	10
VAR03	22.8000	6.5625	10
VAR04	2.9700	1.0338	10
VAR05	65.6000	3.6878	10
VAR06	18563.5000	7398.9617	10
VAR07	6940.0000	2531.6661	10
VAR08	109.4000	23.3248	10

****

TABLE C–2   Correlation matrix

	VAR01	VAR02	VAR03	VAR04	VAR05	VAR06	VAR07	VAR08
VAR01	1.000	.948	.932	.280	.308	.945	.944	.962
VAR02	.948	1.000	.955	.175	.060	.922	.892	.913
VAR03	.932	.955	1.000	.191	.239	.836	.869	.905
VAR04	.280	.175	.191	1.000	.460	.207	.378	.298
VAR05	.308	.060	.239	.460	1.000	.167	.294	.387
VAR06	.945	.922	.836	.207	.167	1.000	.856	.893
VAR07	.944	.892	.869	.378	.294	.856	1.000	.940
VAR08	.962	.913	.905	.298	.387	.893	.940	1.000

TABLE C–3   Step 1 of the regression output

```
***************************** MULTIPLE REGRESSIØN  *****************************

DEPENDENT VARIABLE   VAR01   ADMISSIØNS

VARIABLE(S) ENTERED ØN STEP NUMBER 1   VAR08   CØST

- - - - - - - - - - - - - - - - - - - - - - - - - - - - - - - - - - - - - - - -

MULTIPLE R          .96258

R SQUARE            .92657

ADJUSTED R SQUARE   .91739

STANDARD ERRØR    687.51747

- - - - - - - - - - - - - - - - - - - - - - - - - - - - - - - - - - - - - - - -

ANALYSIS ØF VARIANCE      D.F.      SUM ØF SQUARES       MEAN SQUARE         F

REGRESSIØN                 1       47714909.45544      47714909.45544    100.94542
RESIDUAL (UNEXPLAINED)     8        3781442.14456        472680.26807

- - - - - - - - - - - - - - VARIABLES IN THE EQUATIØN  - - - - - - - - - - - - -

VARIABLE                   B              STD. ERRØR B               F

VAR08                   98.71624            9.82529              100.945

(CØNSTANT)            -5254.35671

- - - - - - - - - - - - - - VARIABLES NØT IN THE EQUATIØN  - - - - - - - - - - -

VARIABLE                            PARTIAL                        F

VAR02                               .62262                      4.431
VAR03                               .53046                      2.741
VAR04                              -.02570                       .005
VAR05                              -.25797                       .499
VAR06                               .70780                      7.028
VAR07                               .43235                      1.609

*******************************************************************************
```

TABLE C–4 Step 2 of the regression output

```
****************************** MULTIPLE REGRESSIØN ******************************
VARIABLE(S) ENTERED ØN STEP NUMBER 2   VAR06   VEHICLES
```

- -

MULTIPLE R	.98151
R SQUARE	.96336
ADJUSTED R SQUARE	.95289
STANDARD ERRØR	519.20279

- -

ANALYSIS ØF VARIANCE	D.F.	SUM ØF SQUARES	MEAN SQUARE	F
REGRESSIØN	2	49609350.84590	24804675.42295	92.01519
RESIDUAL (UNEXPLAINED)	7	1887000.75410	269571.53630	

- - - - - - - - - - - - VARIABLES IN THE EQUATIØN - - - - - - - - - - - - - - - -

VARIABLE	B	STD. ERRØR B	F
VAR08	59.67294	16.49146	13.093
VAR06	.13782	.05199	7.028
(CØNSTANT)	− 3541.42269		

- - - - - - - - - - - - - -VARIABLES NØT IN THE EQUATIØN - - - - - - - - - - - -

VARIABLE	PARTIAL	F
VAR02	.36412	.917
VAR03	.61276	3.607
VAR04	.10130	.062
VAR05	.07230	.032
VAR07	.50339	2.036

```
*******************************************************************************
```

=references

ANDERSON, O. D. *Time Series Analysis and Forecasting*. Boston: Butterworths, 1975.

ANDERSON, T. W. *The Statistical Analysis of Time Series*. New York: Wiley, 1971.

ARMSTRONG, J. S. *Long Range Forecasting: From Crystal Ball to Computer*. New York: Wiley, 1978.

BOX, G. E. P. and G. M. JENKINS. *Time Series Analysis, Forecasting, and Control*. San Francisco: Holden-Day, 1969.

CHOU, YA-LUN. *Statistical Analysis*. New York: Holt, Rinehart and Winston, 1975.

GROSS, CHARLES W. and ROBIN T. PETERSON. *Business Forecasting*. Boston: Houghton Mifflin, 1976.

HAMBURG, MORRIS. *Statistical Analysis for Decision Making*. 2nd ed. New York: Harcourt Brace Jovanovich, 1977.

JOHNSON, J. *Econometric Methods*. New York: McGraw-Hill, 1972.

LAPIN, LAWRENCE. *Statistics for Modern Business Decisions*. New York: Harcourt Brace Jovanovich, 1978.

MENDENHALL, WILLIAM and JAMES REINMUTH. *Statistics for Management and Economics*. North Scituate, Mass.: Duxbury, 1978.

MONTGOMERY, D. C. and L. A. JOHNSON. *Forecasting and Time Series Analysis*. New York: McGraw-Hill, 1976.

NELSON, CHARLES R. *Applied Time Series Analysis*. San Francisco: Holden-Day, 1973.

WINTERS, P. R. "Forecasting Sales by Exponentially Weighted Moving Averages." *Management Science* 6 (1960): 324–42.

19

Introduction to Decision Analysis

A concept repeated many times in this book is that business decision makers must make decisions in an environment packed with uncertainty. Production managers face uncertainty in areas such as quality control, production scheduling, and inventory control. Marketing managers cross paths with uncertainty any time they make a decision to enter a new market area with an established product, or when introducing a new product. Accounting auditors deal with uncertainty when they base their audit on a sample of their client's financial transactions.

In Chapters 1–18 we introduced the fundamentals of classical statistics. The techniques we have presented, if applied properly, should help decision makers deal more effectively with uncertainty. The classical statistical methods are useful tools for managers making "objective" decisions about a population after measuring a sample from the population.

The classical approach assumes that the only pertinent information about the population of interest is contained in the sample. However many decision makers feel that the sample is only one source of information, and that such "subjective" forms of information as expert opinion and personal experience can, and should, be included in the decision process. The formal means by which subjectivity is included in the decision process is known as *decision analysis,* or *Bayesian decision analysis.* This process is named after the mathematician the Reverend Thomas Bayes, who formulated Bayes' rule for conditional probability (see Chapter 5). Pioneering work in the area of decision analysis was performed by Professors Ronald Howard, Howard Raiffa, and Robert Schlaifer. These men are primarily responsible for making the theory of decision analysis available to business decision makers.

In Chapters 19 and 20 we introduce some concepts of decision analysis and illustrate how these concepts can be used to help managers make "better" decisions under conditions of uncertainty.

chapter objectives

The objective of this chapter is to introduce three different decision-making environments: decision making under certainty, decision making under risk, and decision making under uncertainty. Managers who can classify their decision environments correctly will have taken a large step in choosing among the available set of decision tools.

In this chapter we will discuss how to include subjective probability assessments in the decision process. We will also introduce the expected value criterion for deciding between alternatives in a business decision framework and show how decision tree analysis can be used in making decisions under risk and uncertainty.

student objectives

After studying the material presented in this chapter, you should be able to:

1. Recognize the limitations of making decisions under certainty.
2. Understand the difference between making decisions under certainty, risk, and uncertainty.
3. Understand the basic differences, and similarities, between the risk environment and the uncertainty environment.
4. Understand the principles behind the expected value criterion and be able to apply it in a decision-making problem.
5. Set up a problem in a decision tree format.

19-1
DECISION MAKING UNDER CERTAINTY

Sometimes managers can make decisions when they are certain of the results of selecting each alternative. When this is the case, the decision makers are operating in a certain environment. The following example demonstrates a certainty decision environment.

The Spudnick Corporation produces frozen french fries for many large fast-food chains. Jack Dale, the production manager at the Spudnick eastern region plant, has just received a mailgram from Teresa Powers, a company salesperson. Powers has negotiated an order from a new customer for one million pounds of french fries at $0.50 per pound. However this customer wants crinkle-cut fries rather than the regular fries that Spudnick now produces. Powers points out that Spudnick already has the equipment necessary to modify the cutting process to produce crinkle-cuts. Her mailgram indicates that she needs a decision very soon or the customer will take his business elsewhere. She reminds Dale that Spudnick has spent a great deal of money trying to sell to this customer and all will be lost if Dale decides not to take the order.

To make a decision, Dale checks with his industrial engineering and accounting staffs and learns that production costs, including raw materials, will be $0.44 per pound after the process modification is made.

Dale's performance evaluation is based to a large extent on his plant's profit-and-loss statement. Thus Dale might use the certainty model presented in Table 19–1 for making his decision. Note that each item in Table 19–1 is assumed to be known with certainty. Thus, if the actual expenditures and revenues occur as expected, Dale's decision is straightforward: he should accept the order. Some managers may not be logical in their decisions, but decision analysis assumes that the decision makers are logical and will act accordingly. Therefore, once the certainty model has been correctly specified, the "best" decision is evident.

TABLE 19–1 Profit-or-loss decision table

	Alternative	
	---	---
Revenues and Expenses	Accept Order	Do Not Accept Order
Revenue (1 million lb at $0.50/lb)	$500,000	0
Production cost (1 million lb at $0.44/lb)	(440,000)	0
Selling cost (5% of revenue)*	(25,000)	0
Administrative overhead (1% of revenue)*	(5,000)	0
Profit or loss	$ 30,000	0

*These are variable costs that are allocated to any order based upon corporate policy and therefore impact the plant's income statement.

19-2
DECISION MAKING UNDER RISK

A second decision environment involves decision making under risk. In a risk environment, decision makers must decide among alternative actions while faced with several states of nature, or possible outcomes. Decision makers have no control over which state of nature will occur, but the chances of each state occurring are assumed known.

For example, the Barton Construction Company is considering whether to bid on a project for the state of Maryland. State officials have specified the bid price at $300,000. To avoid vendor pressure, the state will make the selection by a random drawing from the submitted proposals. Only four companies are authorized bidders on this type of project.

After several phone calls, Dan Barton, owner-manager of Barton Construction, learns that only two other companies will submit a bid. If Barton bids, the chance of winning the contract is one-third. Dan Barton has calculated the following costs:

Cost of submitting a bid	$ 10,000
Cost of performing the work	$250,000

Thus Barton Construction faces two possible outcomes if it submits a bid:

1. *If Barton wins the bid:*

$$\text{Profit} = \$300,000 - \$250,000 - \$10,000$$
$$= \$40,000$$

2. *If Barton loses the bid:*

$$\text{Loss} = -\$10,000$$

The probability of making $40,000 is one-third, and the probability of losing $10,000 is two-thirds. Although Dan Barton can determine the probability associated with each possible outcome, he has no way of knowing which payoff will actually occur. **Thus a risk environment is an environment in which the decision maker knows the possible outcomes and their probabilities, but does not know which outcome will occur.**

Decisions made using classical statistics are often considered examples of decision making under risk. For example, suppose the credit manager in a bank wishes to know the average account balance for her bank's 10,000 Master Charge accounts. As we discussed in Chapter 10, the manager would likely select a random sample of accounts and develop a statistical confidence interval. The credit manager would know (based upon the central limit theorem) that the distribution of possible sample means will be approximately normal with mean μ_x and standard deviation $\sigma_x/\sqrt{n}$. Thus she would know the probability distribution associated with her estimation problem. However, before she selects the sample, she does not know which sample mean will actually occur. Figure 19–1 illustrates the distribution of possible

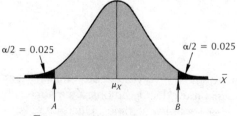

An $\overline{X}$ between A and B will produce a 95% confidence interval that actually includes the true population mean.

FIGURE 19–1 Estimation under risk.

sample means. Figure 19–1 also shows the range within which a sample mean will produce a 95 percent confidence interval that includes the population mean, μ_x. The chance of the confidence interval not including the true value of μ_x is 0.05. Thus the credit manager knows the probabilities of getting both a "good" confidence interval and a "bad" confidence interval but does not know which of the two she actually has. Therefore the credit manager is making an estimate under risk.

19-3
DECISION MAKING UNDER UNCERTAINTY

Uncertainty exists for decisions where the possible outcomes are known but the probabilities associated with the outcomes are not known. For example, the manager of Big Sky Lumber Company in Ronan, Montana, was recently forced to make a decision under uncertainty. Big Sky Lumber sells virtually all types of building supplies to contractors and do-it-yourself builders. In the spring, Bill Dickson, the owner-manager, received a special order of 1,500 board feet of 2″ × 6″ knotty pine, tongue-and-groove decking. The special-order customer paid $675 when placing the order. This represented complete payment. However the buyer said he would not pick up the material until October. Bill Dickson guaranteed the material would be available by October.

In July another customer, who was building a cabin at a nearby lake, arrived at Big Sky Lumber. He was looking for 1,200 board feet of 2″ × 6″ knotty pine, tongue-and-groove decking material. Because decking requires special setups, lumber mills make it only on special order. No other local lumber store had any decking material in stock. However Big Sky Lumber had just received the special order of 1,500 board feet. The new customer was desperate for the decking because his cabin construction was stopped until the decking was installed. He was willing to pay cash for 1,200 board feet.

Bill Dickson hates to lose sales about as much as any struggling small businessperson. He is faced with two alternatives:

1. Sell 1,200 board feet to the new customer and hope he can get the reorder in by October.
2. Don't sell to the new customer and lose out on this sale and maybe future sales from that customer.

Many factors affect Bill's decision:

1. He is uncertain whether he can replace the decking by October. He really has no past experience to use to accurately measure his uncertainty.
2. If he can replace the decking, he is uncertain about what its price will be relative to what he paid for the special-order material. There-

fore, if he *does* sell to the new customer, he doesn't know for sure what price to charge.

3. He is uncertain how the original customer will react if the material (already paid for) is not available in October.

As you can see, what might appear to be a simple problem actually involves a decision in a very uncertain environment. Instances involving uncertainty abound in the business world. Most decisions made in a rapidly changing environment involve uncertainty. In this chapter, and Chapter 20, we discuss decision situations under uncertainty. Our objective is to present some basic techniques for making decisions in such situations.

19-4
SUBJECTIVE PROBABILITY ASSESSMENT

We have discussed three decision environments: certainty, risk, and uncertainty. For each environment, decision makers must know the outcomes that can result from a decision. In a certainty environment, there is only one outcome associated with each decision (with a probability of occurrence of 1.0). However, in both the risk and uncertainty environments, more than one possible outcome exists.

The decision-making tools employed when dealing with risk and uncertainty environments are basically the same. These tools require that probabilities be assigned to possible decision outcomes. In a risk environment, the outcome probabilities are known. In an uncertain environment, the probabilities must be subjectively assessed.

There are three requirements for subjective probability assessments:

1. The probabilities must be nonnegative.
2. The probabilities must sum to 1.0.
3. The probabilities must accurately reflect the decision maker's state of mind about the chance of each outcome occurring.

The first two requirements are restatements of probability rules 1 and 2 from Chapter 5. The third requirement states that decision makers must quantify all available information about the possible outcomes in the form of a probability. If the information is based primarily on past history (or a sample), the assessed probabilities should be formed as relative frequency of occurrence statements. For example, suppose the golf pro at Toke-A-Tee Golf Club has determined that the number of pairs of golf shoes he will sell on any day can vary from zero to 4. A history of the last 200 days' sales is given in Table 19–2. If all days are the same, the golf pro might make the sales assessments using only the relative frequencies shown in the table.

In other cases decision makers have little or no sample information and must make purely subjective probability assessments. These subjective assessments must still include all available relevant information.

TABLE 19–2 Relative frequency approach to probability assessments, Toke-A-Tee Golf Club

Demand	Days	Probability
0	100	0.50
1	30	0.15
2	30	0.15
3	20	0.10
4	20	0.10
	$\Sigma = 200$	$\Sigma = 1.00$

If decision makers understand probability, they might assign probabilities directly to the possible outcomes. For example, in deciding whether to drill a new gas well, the exploration manager for Intermountain Gas Company might be able to assign a 0.70 chance to finding gas and a 0.30 chance to not finding gas. Some managers can think more easily in terms of betting odds. Such statements as "I think the odds are 4 to 1 that we will find natural gas if we drill" can be translated to probabilities by the relationship

$$P(X) = \frac{a}{a + b} \tag{19–1}$$

where: X = event of interest

odds of X are a to b

Then, for the natural gas example, the probability of finding gas is

$$P(gas) = \frac{4}{4 + 1} = \frac{4}{5}$$
$$= 0.80$$

Sometimes decision makers have trouble pinpointing an exact probability. Suppose the exploration manager feels the probability of finding gas if Intermountain drills is somewhere between 0.60 and 0.90 but cannot identify a single probability value. Nothing seems to remove vagueness faster than placing a bet. So when a single probability value cannot be identified, the decision maker could be asked whether he or she would place a bet on finding gas with 4 to 1 odds. If the decision maker would bet, the probability is 0.80 or higher. If not, the probability is below 0.80. The decision maker would continue in this manner until finding the betting odds for which he or she is totally indifferent. These odds would then reflect the true subjective probability.

This presentation of probability assessment is only a basic discussion. More advanced texts on Bayesian inference and analysis by Winkler (1972), Schlaifer (1969), and Raiffa (1968) present more in-depth treatments of subjective probability assessment methods.

As we have said, the subjective probability assessment that a specific outcome will occur must reflect the decision maker's state of mind. Thus there can be no right or wrong subjective probability assessment.

19-5
DECISION-MAKING CRITERIA

The prime managerial concern is the future outcome of today's decisions. Since no one can predict the future, rarely can managers predict with certainty the outcome of their decisions. This is what makes decisions risky or uncertain. A basic premise in decision analysis is that **a "good" decision does not insure a "good" outcome.** Decision makers who cannot accept this premise will end up with ulcers from second-guessing their own decisions.

A "good" decision is one that is made using all available information to satisfy criteria established by the decision maker. The final decision depends upon the decision criteria. Here, we introduce two classes of decision-making criteria: *nonprobabilistic* and *probabilistic*.

Nonprobabilistic Criteria for Decisions under Uncertainty

The Ajax Electronics Company must decide how many television sets to produce this year. The firm's cost accountants have determined that the annual fixed costs are $350,000. The variable cost of producing each television set is $300, and the sale price is $700. Company market analysts have determined that demand will be one of the following levels:

$$700$$
$$800$$
$$900$$
$$1,000$$
$$1,100$$

Due to yearly model changes, there is no market for unsold television sets.

The production manager has to establish a production level and therefore must determine the profits for each possible production level. She uses the following format to arrive at these profits, or **payoffs:**

$$\text{Payoff} = \text{TR} - \text{FC} - \text{VC} \qquad (19\text{--}2)$$

where: TR = total revenue = (sales price)(number of units sold)

FC = fixed cost

VC = variable cost = (cost per unit)(number of units produced)

For example, if demand is 700 television sets and 700 are produced, the payoff is

$$\text{Payoff} = \$700(700) - \$350,000 - \$300(700)$$
$$= -\$70,000$$

If 800 sets are produced and 1,000 demanded, the payoff is

$$\text{Payoff} = \$700(800) - \$350,000 - \$300(800)$$
$$= \$560,000 - \$350,000 - \$240,000$$
$$= -\$30,000$$

The production manager can find a payoff for each combination of output and demand and can construct the **payoff table** shown in Table 19–3.

TABLE 19–3 Payoff table, Ajax Electronics Company

		Demand Level (State of Nature)			
	700	800	900	1,000	1,100
700	−$ 70,000	−$ 70,000	−$70,000	−$70,000	−$70,000
800	−$100,000	−$ 30,000	−$30,000	−$30,000	−$30,000
Production Level 900	−$130,000	−$ 60,000	$10,000	$10,000	$10,000
1,000	−$160,000	−$ 90,000	−$20,000	$50,000	$50,000
1,100	−$190,000	−$120,000	−$50,000	$20,000	$90,000

Nonprobabilistic decision rules are used when the manager has the payoff table but cannot assess probabilities for the possible demand levels. Several rules exist to help the manager make a "good" decision about how many television sets to produce. The **maximin** rule is one.

MAXIMIN RULE

For each option, find the minimum possible payoff and then select the option that has the greatest minimum payoff.

Table 19–3 shows that the minimum payoffs for the Ajax Electronics production levels are

Production Level	Minimum Payoff
700	−$ 70,000
800	− 100,000
900	− 130,000
1,000	− 160,000
1,100	− 190,000

According to the maximin criterion, the best decision is to produce 700 sets because the worst possible outcome (−$70,000) is greater than for any other production level. Note that the maximin rule is very pessimistic. It assumes the worst will happen, and the decision maker's goal is to cut the losses. A manager might choose to use the maximin criterion if his firm is in financial trouble and he must make certain that even if the worst possible outcome occurs, further operations are possible.

The **maximax** decision criterion is the opposite of the maximin criterion. The maximax criterion assumes the best is going to happen and that the decision maker seeks to maximize the good fortune.

MAXIMAX RULE

For each option, find the maximum possible payoff and then select the option that has the greatest maximum payoff.

The maximum profits for the Ajax Electronics production levels are

Production Level	Maximum Payoff
700	− $70,000
800	− 30,000
900	10,000
1,000	50,000
1,100	90,000

Thus the "best" decision under the maximax criterion is to produce 1,100 television sets.

The maximin and the maximax criteria are straightforward and easy to use. However they are often criticized because they fail to include any information about the probabilities of the possible outcomes. To use this information, the decision maker should employ probabilistic criteria to make the "best" decision.

Probabilistic Criteria for Decisions under Uncertainty

Ajax Electronic's production manager is informed by the marketing department that the following probabilities have been assessed for the demand levels:

Demand Level	Probability
700	0.05
800	0.10
900	0.20
1,000	0.40
1,100	0.25
	$\Sigma = 1.00$

These probabilities reflect the marketing department manager's state of mind and were subjectively determined.

The fundamental probabilistic criterion is the **expected value criterion.** We discussed the concept of expected value in Chapter 6. The expected value of a probability distribution is the *long-run* average value of that distribution. Thus, the expected value criterion is based on the long-run average of a probability distribution.

EXPECTED VALUE CRITERION

Given a probability distribution of payoffs, select the option that yields the greatest expected payoff or the minimum expected loss.

To incorporate the demand-level probabilities into his decision-making process, the manager would calculate the expected payoff for each production level. For example, the expected payoff for a production level of 900 television sets is found by

$$E[X] = \sum_{i=1}^{S} X_i P(X_i) \qquad\qquad (19\text{--}3)$$

where: $E[X]$ = expected value of X (payoff or loss)
   S = number of different payoff levels
   X_i = *i*th possible payoff
   $P(X_i)$ = probability of the *i*th payoff

If the production level is 900,

$$E[\text{payoff}] = -\$130,000(0.05) + (-\$60,000)(0.10) + \$10,000(0.20)$$
$$+ \$10,000(0.40) + \$10,000(0.25)$$
$$= -\$4,000$$

Therefore, on the average, Ajax Electronics can expect to lose $4,000 if it decides to produce 900 television sets. Table 19–4 illustrates the expected payoffs associated with each possible production level. Using the expected value criterion, the "best" decision is to produce 1,000 television sets.

TABLE 19–4   Expected values, Ajax Electronics

			Demand Level			
		700	800	900	1,000	1,100	
				Probability			Expected Value
		0.05	0.10	0.20	0.40	0.25	↓
	700	−$ 70,000	−$ 70,000	−$70,000	−$70,000	−$70,000	−$70,000
	800	−$100,000	−$ 30,000	−$30,000	−$30,000	−$30,000	−$33,500
Production Level	900	−$130,000	−$ 60,000	$10,000	$10,000	$10,000	−$ 4,000
	1,000	−$160,000	−$ 90,000	−$20,000	$50,000	$50,000	$11,500*
	1,100	−$190,000	−$120,000	−$50,000	$20,000	$90,000	−$ 1,000

*Highest expected value—produce 1,000.

The expected value criterion allows decision makers to incorporate additional information, in the form of probabilities, into a decision. Because of this feature, expected value analysis plays a key role in decisions under risk or uncertainty.

19-6
DECISION TREE ANALYSIS

When decision analysis is applied to an actual problem, the process can become quite complex. The decision maker must identify the outcomes for each decision alternative, and must also assess probabilities associated with each outcome, assign cash flows in the form of payoffs and costs, and somehow keep the sequence of outcomes and decisions in the proper chronological order. *Decision tree analysis* is a technique to aid the decision maker in this process. **The decision tree provides a "road map" of the decision problem.**

Very few managerial situations allow single isolated decisions. Most decisions are made in a dynamic, evolving environment. For instance, Squelch Electronics Corporation designs and manufactures a variety of electronic components. Recently a salesperson in the new-products division approached his vice-president with a proposition from ACR Television, Inc. If Squelch will foot the expense of building a prototype color television tuner, ACR will test the tuner and may purchase 5,000 of them. ACR buys its current tuner for $40 each. Based on Squelch's reputation, and the salesperson's persuasiveness, ACR has agreed to test the prototype tuner if Squelch decides to produce it. If the prototype outperforms the current tuner, ACR will purchase 5,000 tuners at $40 each.

Squelch engineers have informed the vice-president there is a 60 percent chance they can design a tuner that will outperform ACR's current tuner. The engineers also estimate that the prototype design cost will be about $25,000, and the construction cost about $10,000. The engineers further estimate that if the 5,000 tuners are produced using primarily manual labor, the fixed setup cost will be $20,000, and the per-unit variable production cost will be $24. The engineers are sure that tuners assembled by hand will perform as well as the prototype. However the variable production cost can be reduced to $22 if a flow-solder technique is used. Squelch will have to invest $10,000 in equipment for this flow-solder technique.

If flow-solder is used, there is a 30 percent chance the production tuner will not perform as well as the prototype. Fortunately the engineers are confident that any problems in the flow process would be discovered in time to switch to hand assembly and still meet the ACR deadline.

The first decision facing Squelch's vice-president is whether to produce the prototype. However this decision will likely depend on the following decision about whether to use flow-solder assembly. This simplified example has the *act-event-act-event* sequence typical of many managerial decisions. Whereas the vice-president has control over the acts, or decisions, she has no control over the subsequent events. But these events determine the appropriate following act. For instance, if the vice-president decides to build the prototype, she may or may not get a model that will outperform the competition, but only if the prototype is good will she have to consider the flow-solder process.

Constructing the Decision Tree

Decision trees help the decision maker logically structure the decision problem. Structuring involves properly sequencing the acts and events in the decision problem. An *act* **is a decision or alternative within the decision maker's control. An** *event* **is a state of nature beyond the decision maker's control.**

To construct the decision tree, the decision maker must analyze the possible acts and events chronologically. The tree will be formed by a series of branches that represent the chronological order of the acts and events. The first act in the decision problem facing Squelch's vice-president is to build, or not build, the prototype. This act is shown in Figure 19–2. The small square at the fork represents an act, or decision, for the decision maker.

The decision tree is completed for each initial branch by adding all other acts and events in their chronological order. For example, if the lower branch ("don't develop prototype") is selected, ACR will not purchase tuners from Squelch. Figure 19–3 shows the tree with the lower branch completed. A small circle indicates an event.

Figure 19–4 shows the completed upper act branch ("develop proto-type") with the correct sequence of acts and events. As the tree shows, if a prototype is developed, it will either be better than the existing tuner, in which case ACR will order 5,000, or the prototype will not be better, and ACR will not order. If ACR does

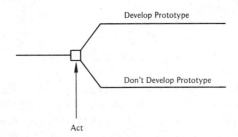

FIGURE 19–2 Initial act in the decision tree, Squelch Corporation.

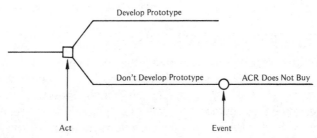

FIGURE 19–3 Complete lower branch in the decision tree, Squelch Corporation.

not order the tuners, Squelch has no further acts or events. However, if ACR orders the tuners, Squelch must decide (an act) whether to use manual assembly or the flow-solder process. If the decision is to use manual assembly, all tuners will be satisfactory, and no further acts or events exist. However, if the flow-solder alternative is selected, it will either work satisfactorily, or it won't (an event). If the flow process doesn't work, Squelch will have to switch to the manual alternative. Eventually ACR will receive 5,000 tuners, and the Squelch Corporation will receive payment.

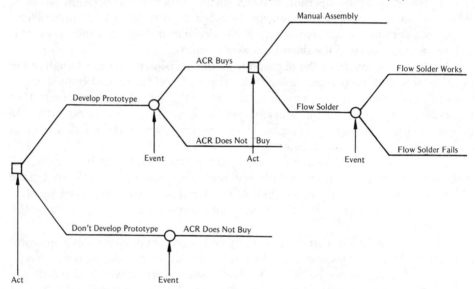

FIGURE 19–4 Complete upper act branch in the decision tree, Squelch Corporation.

As you can see, constructing a decision tree is not a complex process. If you begin on the left and add the appropriate act and event branches toward the right, when the tree is completed it should include all possible decision paths.

In an actual application, once you have formulated the initial decision tree, have someone else familiar with the problem review it to make sure no branches have been left out. This is especially important if the tree is complex. An accurate decision tree is essential. However, constructing the tree is only the first step in decision tree analysis.

Assigning Cash Flows

A decision tree is useful for visualizing the problem at hand, but by itself is not a complete decision-making tool. The next step is to assign cash flows to each tree segment and arrive at a cash value for each branch. These final cash values are called **end values** since they represent the values at the ends of the decision branches.

Figure 19–5 shows the cash flows for each act and event branch in the Squelch decision tree. The lower branch shows that if the Squelch Corporation does not develop a prototype, there is no cost or revenue. Consequently the net cash flow, or end value, is $0. However the upper branch ("develop prototype") involves several cash inflows and outflows. The first outflow is the development (prototype) cost.

$$
\begin{aligned}
\text{Design cost} &= -\$25{,}000 \\
\text{Construction cost} &= \underline{-\ \ 10{,}000} \\
\text{Prototype cost} &= -\$35{,}000
\end{aligned}
$$

The development cost ($-\$35{,}000$) is placed on the development branch.

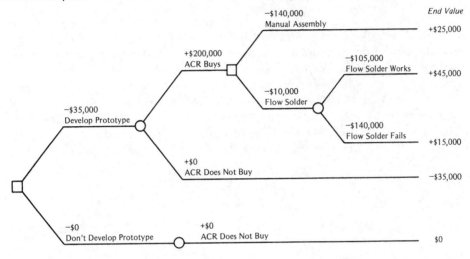

FIGURE 19–5 Cash-flow assignments, Squelch Corporation.

Next comes the event of ACR buying or not buying. If the prototype fails, ACR will not buy, and the end value is $-\$35{,}000$. However, if ACR does buy, the cash inflow is

$$5{,}000 \text{ tuners} \times \$40/\text{tuner} = \$200{,}000$$

The $200,000 cash inflow is placed on the appropriate branch in Figure 19–5.

Following the "ACR buys" branch comes the act associated with whether to use manual or flow-solder assembly. If manual assembly is used, the production cash outflow is

$$
\begin{aligned}
\text{Fixed cost} &= -\$\ 20{,}000 \\
\text{Variable cost (5,000 units} \times \$24/\text{unit)} &= \underline{-\ 120{,}000} \\
\text{Total production cost} &= -\$140{,}000
\end{aligned}
$$

The $-\$140{,}000$ cash outflow is shown on the "manual assembly" branch. If flow-solder assembly is used, there is a $-\$10{,}000$ cash outflow for equipment, and the process will either prove successful or it won't. If flow solder works, the production cost becomes

$$5{,}000 \text{ units} \times \$22/\text{unit} = \$110{,}000$$

This cash outflow is also shown in Figure 19–5. However, if the flow-solder parts prove inadequate, Squelch will be forced to manually assemble all the tuners as we showed earlier. In this case, the cash outflow, or production cost, is $-\$140{,}000$.

Figure 19–5 also shows the end value for each branch. The end value is the sum of the cash inflows and outflows for the branch. The end values range from a high of +$45,000 to a low of −$35,000 and represent the possible cash results from the decisions facing the Squelch vice-president.

Assigning Probabilities and Determining the Expected Value

The next step in the process is to assign probabilities to the decision's events, or states of nature. As we discussed in Section 19–4, subjectively assessing probabilities is an important part of decision analysis under uncertainty. In this example, the Squelch engineers have used their experience and abilities to arrive at the event probabilities. However these probabilities reflect the decision makers' state of mind, and, of course, other decision makers could assign other probabilities. Figure 19–6 shows the decision tree with the event probabilities included. Remember, if the prototype is produced, there is a 0.60 chance it will successfully pass ACR's test and ACR will place an order. If flow-solder assembly is used, there is a 0.70 chance it will work.

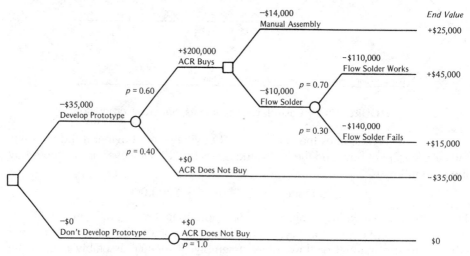

FIGURE 19–6 Subjective probabilities, Squelch Corporation.

The best decision using the expected value criterion is to select the alternative with the highest expected profit or the lowest expected cost. However, finding this alternative requires the decision maker to *fold back* the decision tree.

The folding back process involves starting at the right-hand side of the tree and moving to the left until the first act fork is reached. There are two rules for folding back a decision tree:

1. For all event forks, find the expected value.
2. For all act forks, pick the alternative with the larger expected value.

To see how these rules apply, consider Figure 19–7, which is part of Squelch's overall decision tree.

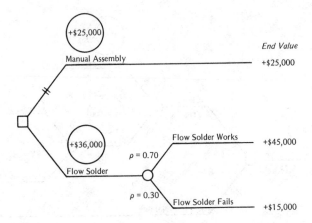

FIGURE 19–7 Step 1 in folding back the decision tree, Squelch Corporation.

The end value associated with the manual-assembly alternative is +$25,000. Since there are no events after this decision, its expected value is +$25,000. However, to find the expected value for the flow-solder alternative, we must weight the two possible payoffs, +$45,000 and +$15,000, by their probabilities.

$$E[\text{payoff flow}] = 0.70(+\$45,000) + 0.30(+\$15,000)$$
$$= +\$36,000$$

Since +$36,000 exceeds +$25,000, if the Squelch vice-president ever has to decide between using manual or flow-solder construction, her "best" decision is to use flow-solder. (The two hashmarks through the manual alternative indicate that this branch should be ignored.)

The next step is to move farther left and determine the expected value of each alternative at the next act fork. From Figure 19–8, this fork involves the "develop prototype"/"don't develop prototype" alternative. If the "don't build" alternative is selected, the end value is $0. Since the probability associated with this outcome is 1.0, the expected value of this decision branch is $0. However, for the "develop prototype" alternative, we must find the weighted average of the "ACR buys"/"ACR does not buy" event.

$$E[\text{payoff}|\text{prototype}] = P(\text{ACR buys})(\text{payoff}) + P(\text{ACR does not buy})(\text{payoff})$$
$$= 0.60(+\$36,000) + 0.40(-\$35,000)$$
$$= +\$7,600$$

Note that we used +$36,000 as the payoff for the event "ACR buys." This was done because +$36,000 is the expected value of the best decision ("flow-solder") given this branch. This is the essence of folding back the tree and shows clearly that the initial decision about building the prototype is dependent on the later production decision.

We have placed the +$7,600 and $0 in circles to show that they are the expected values of the alternatives. Since +$7,600 is greater than $0, the "best"

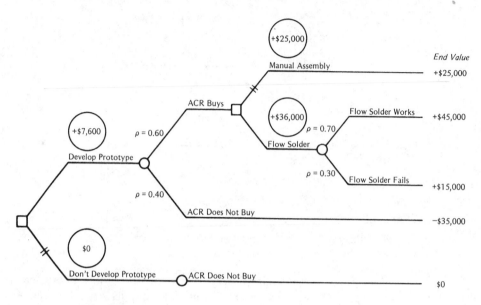

FIGURE 19–8 Final step in folding back the decision tree, Squelch Corporation.

decision for the vice-president is to develop the prototype. Later, if the prototype is acceptable, flow-solder production should be used.

19-7
CONCLUSIONS

In this chapter we have discussed the three environments in which business decisions are made: certainty, risk, and uncertainty. When uncertainty exists, not only do decision makers have no control over which state of nature will occur, they do not even know for sure the probability that any state of nature will occur. To incorporate probability into the uncertainty environment, decision makers must be willing to introduce subjectivity into the process. However the subjective probabilities managers assess must reflect their states of mind about the chances any particular state of nature will occur.

In this chapter we have also introduced some nonprobabilistic decision criteria and illustrated the expected value criterion for decision making when probabilities are employed.

We also indicated that the ''best'' decision does not always result in the ''best'' outcome if the decision is made under conditions of uncertainty or risk. However, by using the appropriate tool, decision makers can increase the chances the decision will have good results. One such tool is the decision tree, which provides a framework for more complex decision analysis.

A listing of statistical terms introduced in this chapter is presented in the chapter glossary. You should understand these terms before you go on to Chapter 20.

chapter glossary

certainty A decision environment in which the results of selecting each alternative are known before the decision is made.

decision tree A diagram that illustrates the chronological ordering of actions and events in a decision analysis problem. Each act and event is depicted by a branch on the decision tree.

expected value criterion A decision criterion that employs probability: select the alternative that will produce the greatest long-run average payoff or minimum long-run average loss.

maximax criterion An optimistic decision criterion for dealing with uncertainty when probability is not employed: for each option, find the maximum possible payoff and select the option that produces the greatest maximum payoff.

maximin criterion A conservative decision criterion for dealing with uncertainty without using probabilities: for each option, find the minimum possible payoff and select the option with the greatest minimum payoff.

payoff table A two-way table that shows the payoff (profit or loss) for each combination of alternative and state of nature.

risk A decision environment in which the possible outcomes are known for each alternative. Although decision makers have no control over which outcome will occur, they do know the probability of each outcome occurring.

uncertainty A decision environment in which the possible outcomes are known but the decision makers do not know the probability of each outcome actually occurring.

solved problems

1. Korman Industries produces record albums that are sold to the public through the mail. Because of economies of scale and scheduling problems, Korman's policy is to produce all copies of a given record in one production run. If the demand exceeds the amount produced, the customer's money is returned along with a coupon good for $0.40 on any future Korman album. If Korman makes too many albums, the extras are sold to a department store chain for $0.50 each. This price is half the variable production cost of a record.

Korman has recently agreed to pay $200,000 for the rights to a particular record. The company will sell this record for $4.98. Its market-research department predicts that one of the following demand levels will occur:

20,000
40,000
60,000
80,000

a. Set up a payoff table showing the payoff for each combination of production and demand.

b. Use the maximin rule to arrive at the appropriate decision. Contrast this decision with the one made if the maximax rule is used.

c. The following probabilities have been subjectively assessed for the various demand levels:

Demand Level	Probability
20,000	0.10
40,000	0.30
60,000	0.40
80,000	0.20
	$\Sigma = 1.00$

Use the expected value criterion to arrive at the best decision for Korman.

Solutions:

a. To set up the payoff table, we determine the payoff for each combination of production and demand. The following formula can be used:

Payoff $=$ $-$ fixed cost $-$ variable cost $+$ revenue $+$ salvage $-$ penalty

where

Fixed cost $=$ \$200,000

Variable cost $=$ \$1.00/record

Revenue $=$ \$4.98/record sold

Salvage $=$ \$0.50/unsold record

Penalty $=$ \$0.40/unfilled demand

The payoff table is

	Demand Level (State of Nature)			
	20,000	40,000	60,000	80,000
Production Level 20,000	$-$\$120,400	$-$\$128,400	$-$\$136,400	$-$\$144,400
40,000	$-$\$130,400	$-$\$ 40,800	$-$\$ 48,800	$-$\$ 56,800
60,000	$-$\$140,400	$-$\$ 50,800	\$ 38,800	\$ 30,800
80,000	$-$\$150,400	$-$\$ 60,800	\$ 28,800	\$118,400

b. The maximin rule states that for each alternative, find the minimum possible payoff and then select the option that has the greatest minimum payoff. If we do this for the Korman payoff table, we find the following values:

Production Level	Minimum Payoff
20,000	$-$\$144,400
40,000	$-$ 130,400
60,000	$-$ 140,400
80,000	$-$ 150,400

Using the maximin rule, we should produce 40,000 records since this level minimizes the maximum losses. However, if we use the maximax rule, which says to find the maximum possible return for each alternative and select the alternative with the greatest maximum payoff, we find the following values:

Production Level	Maximum Payoff
20,000	− $120,400
40,000	− 40,800
60,000	38,800
80,000	118,400

The decision using the maximax rule is to produce 80,000 records. This level will produce the maximum possible payoff.

The primary difference between the maximin and the maximax decision criterion is that the maximin is very conservative and the maximax is very optimistic. They are both nonprobabilistic criteria.

c. The expected value criterion uses probability assessments to arrive at the decision that maximizes average payoffs. The formula for finding the expected payoff for each alternative is

$$E[\text{payoff}] = \Sigma XP(X)$$

where: X = payoff for each demand level

$P(X)$ = probability of each demand level occurring

Consequently, we find the following payoff for each alternative:

Production Level	Expected Payoff
20,000	− $134,000
40,000	− 56,160
60,000	− 7,600
80,000	1,920

For example, for the production level of 60,000,

$$E[\text{payoff}] = (-\$140,400)(0.10) + (-\$50,800)(0.30) + \$38,800(0.40)$$
$$+ \$30,800(0.20)$$
$$= -\$7,600$$

The best decision is to pick the production level with the highest expected payoff. Therefore Korman should produce 80,000 records because that production level yields the greatest expected payoff.

2. The S. Claus Tree Company has acquired the lease rights to a large timber area in western Oregon from which it can cut up to one million Christmas trees. The

problem is how many trees to cut and ship to retail outlets during the coming December. If the company doesn't cut enough, it loses potential business. If it cuts more than are demanded, it is out the costs of cutting and shipping these unused trees. As you might guess, there is not a big market for unsold Christmas trees after Christmas.

S. Claus's lease costs $50,000 per year and $2 per tree cut. Cost accountants have determined that the cost of cutting and shipping a tree averages another $1. Marketing experts have estimated this year's demand and its probability distribution as follows:

Demand Level	Probability
50,000	0.10
100,000	0.40
125,000	0.20
150,000	0.20
200,000	0.10
	$\Sigma = 1.00$

Assuming the S. Claus trees sell for an average of $8 each, determine the payoff table and use the expected value criterion to find the number of trees the company should cut this year.

Solution:

To construct a payoff table for the S. Claus Tree Company, we use the following equation:

Payoff = revenue − fixed lease cost − variable lease cost − variable cut cost

where

$$\text{Revenue} = \$8/\text{tree sold}$$
$$\text{Lease fixed cost} = \$50,000$$
$$\text{Variable lease cost} = \$2/\text{tree cut}$$
$$\text{Variable cut cost} = \$1/\text{tree cut}$$

		Demand Level				
		50,000	100,000	125,000	150,000	200,000
	50,000	$200,000	$200,000	$200,000	$200,000	$200,000
	100,000	$ 50,000	$450,000	$450,000	$450,000	$450,000
Cut Level	125,000	−$ 25,000	$375,000	$575,000	$575,000	$575,000
	150,000	−$100,000	$300,000	$500,000	$700,000	$700,000
	200,000	−$250,000	$150,000	$350,000	$550,000	$950,000

The expected payoff associated with each cut alternative is:

Cut Level	Expected Payoff
50,000	$200,000
100,000	$410,000
125,000	$435,000
150,000	$420,000
200,000	$310,000

For example, if the cutting level is 125,000 trees, the expected payoff is

$$E[\text{payoff}] = (-\$25,000)(0.10) + \$375,000(0.40) + \$575,000(0.50)$$
$$= \$435,000$$

Then, according to the expected value criterion, the S. Claus Tree Company should cut 125,000 trees because this level produces the highest expected profit.

=problems

Problems marked with an asterisk (*) require extra effort.

1. In this chapter we have stated that the "best" decision does not always produce the "best" outcome. Provide two examples that illustrate the accuracy of this statement.

2. Discuss the concept of nonprobabilistic decision criteria and indicate why they might be used in decision making under uncertainty.

3. In your own words, discuss the three decision-making environments and demonstrate your understanding by providing an example of each which points up their basic differences.

4. How are decision making under risk and decision making under uncertainty similar?

5. The Pick-It-Up Garbage Company has an option to take over the garbage collection in a new neighborhood. There are 450 customers who will pay $4 per month for service. Pick-It-Up will have to hire a new person at $900 per month and lease a new truck at $300 per month. Gas and other expenses will run $200 per month.
   Given the information presented here, what decision environment is indicated? Why? What are Pick-It-Up's alternatives?

6. Considering the information supplied in problem 5, decide whether or not Pick-It-Up should take on the new neighborhood. Make sure you show why you made the decision.

7. The Home-Sweet-Home Corporation builds condominiums in Dallas, Texas. Presently the firm is considering a new condominium complex. The question is how many units to build.

   Keeping the problem as simple as possible, assume that the land cost will be $100,000 and that the cost of building will be $120,000, plus $48,000 for each unit. Because of design problems, Home-Sweet-Home must build the units in blocks of ten. Given that the company has decided to build, the production levels are

   10
   20
   30
   40
   50

   Each unit sells for $80,000 if sold in the regular market. However, if demand does not meet supply, Home-Sweet-Home will be forced to auction the units, and the average price will be $21,000.

   Set up a payoff table assuming the possible demand levels are

   0
   10
   20
   30
   40
   50

8. Referring to problem 7, what is the "best" decision if the maximin criterion is used? Show how you arrived at your answer.

9. Referring to problem 7, what is the "best" decision if the maximax criterion is used? Show how you arrived at your answer.

10. Referring to problem 7, assume that the real estate marketing consultants Home-Sweet-Home has hired have assessed the following probability distribution for condominium demand:

Demand	Probability
0	0.05
10	0.10
20	0.25
30	0.25
40	0.20
50	0.15
	$\Sigma = 1.00$

   Using the expected value criterion, how many units should Home-Sweet-Home build?

11. The production manager for a beer company has been asked to help upper management in deciding whether or not to market a new dark beer. If the new beer is successful, the company will make $1 million, but if the beer is a failure, the company stands to lose $625,000. The production manager feels that there is a 60 percent chance that the new beer will be successful.

Construct the payoff table for the decision facing the beer company.

12. Referring to problem 11, what decision should the company make if it wishes to maximize its expected payoff?

13. Referring to problems 11 and 12, suppose marketing personnel have determined that there are several levels of success and failure possible with the new beer. These possibilities (states of nature) and the assessed probabilities are

Success Level	Probability
Excellent ($1,200,000)	0.30
Good ($1,000,000)	0.30
Fair ($100,000)	0.20
Poor (− $625,000)	0.20
	$\Sigma = 1.00$

Develop the payoff table that reflects these states of nature.

14. Given the payoff table in problem 13, what decision should the company make if it wishes to maximize its expected payoff?

15. The owner of a large service station must decide before winter arrives how many gallons of antifreeze to order and have in inventory. Demand for antifreeze will depend upon the winter temperatures. The service station is located in an area with usually mild temperatures, but if cold weather hits, the car owners in the area panic and immediately demand antifreeze.

The antifreeze costs the station owner $2.50 per gallon, and is sold for $6.25 a gallon. Any antifreeze left over after the winter season will be sold to a wholesaler for $1.00 per gallon. Based upon past history, the owner feels that the demand will be one of the following with the associated probability:

Demand Level	Probability
500	0.10
1,000	0.30
1,500	0.40
2,000	0.20
	$\Sigma = 1.00$

Develop a payoff table that indicates the possible alternatives and associated payoffs for each state of nature.

16. Referring to problem 15, what level of antifreeze should the station owner stock in inventory if he wishes to maximize his expected profit?

17. The bakery manager at a large supermarket must decide how many large, expensive, fancy cakes to have the bakers make each morning. The cakes cost $2.25 each to make and sell for $8.00. Any leftover cakes can be sold for $1.25 each. Past sales records indicate that demand should have the following probability distribution:

Demand Level	Probability
0	0.10
1	0.30
2	0.25
3	0.20
4	0.10
5	0.05
6 or more	0
	$\Sigma = 1.00$

Develop a payoff table for this decision problem.

18. Referring to problem 17, how many expensive cakes should the manager tell the bakers to prepare each day?

*19. The Continental Automobile Agency of Levelport, Washington, is considering two alternative advertising plans. The first plan will take advantage of the radio and television media and is projected to cost $40,000. The marketing department has analyzed this advertising approach and, based upon a joint meeting with Continental's sales managers, believes that the following probability distribution on revenue increases is accurate:

Increased Revenue	Probability
$ 20,000	0.10
30,000	0.10
40,000	0.20
60,000	0.20
80,000	0.20
90,000	0.10
100,000	0.10
	$\Sigma = 1.00$

The second plan is to use billboard and poster advertising at a cost of $12,000. However the marketing department feels that this approach would not be as effective in generating added revenue as the first approach. This is reflected in the following probability distribution:

Increased Revenue	Probability
$ 4,000	0.10
5,000	0.10
10,000	0.25
20,000	0.25
30,000	0.10
40,000	0.10
50,000	0.10
	$\Sigma = 1.00$

Because of the fear of "overkill," the company has decided not to try both approaches simultaneously. However, the managers did agree that if the billboard approach was chosen and it resulted in $10,000 or less in added revenues, they could switch to the television-radio approach at the original cost but with all potential revenue levels reduced by $5,000. They assume the probabilities would not be affected.

a. Set up the appropriate decision tree for the decision problem facing the Continental Automobile Agency.

b. Using the criterion of maximizing expected increase in net revenues, what decision strategy should Continental take?

20. The Balbado Corporation owns and operates a large orchard on the West Coast. Before the apple season, Balbado was approached by a grocery chain which proposed that Balbado sell 300,000 pounds of apples at a fixed price of $0.10 per pound. The grocery chain would pay for all shipping costs to get the apples to its central warehouse. The problem facing Balbado is whether to accept this offer, or to attempt to market its entire crop on the open market.

An uncertainty exists regarding the size of the upcoming crop. If such factors as weather are good, the crop will be large (600,000 pounds). On the other hand, the crop may be small (400,000 pounds). Based upon available information, Balbado executives have assessed a 0.60 probability that the season will be good.

The price of apples on the open market depends on the size of the crop on both the West Coast and the East Coast. The following matrix indicates the possible open-market prices along with the assessed probabilities. Assume that the crop size on the East Coast is independent of the crop size on the West Coast.

		East Coast Crop Size	
		Large	Small
West Coast Crop Size	Large	$0.04 p = 0.18	$0.08 p = 0.42
	Small	$0.07 p = 0.12	$0.15 p = 0.28

Assuming that Balbado will be able to sell its entire crop in any case, should it accept the grocery chain's offer if it wishes to maximize its expected revenue?

*21. Referring to problem 20, suppose Balbado feels it will be able to negotiate the following new deal with the grocery chain. The contract price paid to Balbado would be reduced to $0.09 per pound. Balbado commits to deliver 150,000 pounds at this price with the option of committing to supply the remaining 150,000 pounds at $0.09 per pound after it determines what the West Coast crop size will be. This decision must be made before the size of the East Coast crop is known, however.

To negotiate this deal, Balbado would have to retain the services of legal council for a fee estimated to be $2,000.

Using any applicable information from problem 20, determine whether Balbado should propose this new deal to the grocery chain. Base your decision on expected revenues.

*22. Referring to problems 20 and 21, if you were the purchasing manager for the grocery chain and Balbado did make the offer outlined in problem 21, would you recommend that it be accepted if your only other alternative were to buy 300,000 pounds of apples on the open market? Assume that you have assessed the same probabilities as Balbdo regarding crop size. Base your conclusion on expected costs.

cases

19A ROCKSTONE INTERNATIONAL, PART 1

Rockstone International is one of the world's largest diamond brokers. The firm purchases "stones" from South Africa which must be cut and polished for sale in the United States and Europe. The diamond business has been very profitable, and from all indications it will continue that way. However, R. B. Penticost, president and chief executive officer for Rockstone International, has stressed the need for effective management decisions throughout the organization if Rockstone is to continue to be profitable and competitive.

Normally R. B. does not involve himself in personnel decisions, but today's decision is not typical. Beth Harkness, Rockstone's personnel manager, is considering whether or not to hire Hans Marquis, "world-famous diamond cutter," to fill the opening left when Omar Barboa, "former world-famous diamond cutter," broke both his hands in a freak skateboard accident almost one month ago. Hans Marquis, if hired, will be paid on a commission basis at the rate of $5,000 per stone he successfully cuts. Because of professional pride, Hans will accept nothing if he unsuccessfully cuts a stone (smashes it to bits).

In the past, the decision to hire Hans would have been simple. If he was available, hire him. However, within the last six months, the Lictenstien Corporation located in Pretoria, South Africa, developed the world's first diamond-cutting machine. This machine, which can be leased for $1 million per year, is guaranteed to successfully cut 90 percent of the stones.

Although Hans Marquis has an excellent reputation, Rockstone International cannot be sure about his percentage of successful cuts due to the extreme secrecy in the diamond business. Hans claims that his success rate is 95 percent, but he has been known to exaggerate. Rockstone executives, including Penticost, have made the following assessments based upon all the information they could obtain:

Success Rate	Probability
0.97	0.10
0.95	0.40
0.90	0.30
0.85	0.10
0.80	0.10
	$\Sigma = 1.00$

Rockstone purchases each stone at a cost of $15,000. If a stone is successfully cut, four diamonds can be salvaged at an average sales price of $35,000 each. Harry Winkler, sales and purchasing manager, has indicated that 100 stones will need to be cut this year.

R. B. Penticost figures there must be some way to decide whether to hire Hans Marquis or to lease the new machine.

references

BAIRD, BRUCE F. *Introduction to Decision Analysis.* North Scituate, Mass.: Duxbury, 1978.

BROWN, R. V., A. S. KAHR, and C. PETERSON. *Decision Analysis for the Manager.* New York: Holt, Rinehart and Winston, 1974.

BROWN, REX V. "Do Managers Find Decision Theory Useful?" *Harvard Business Review* 48 (May–June 1970): 78–89.

KRISTY, JAMES E. "Managing Risk and Uncertainty." *Management Review,* September 1978, pp. 15–22.

MAGEE, J. F. "Decision Trees for Decision Making." *Harvard Business Review* 42 (July–August 1974): 126–38.

PARSONS, ROBERT. *Statistical Analysis.* New York: Harper and Row, 1974.

RAIFFA, HOWARD. *Decision Analysis: Introductory Lectures on Choices under Uncertainty.* Reading, Mass.: Addison-Wesley, 1968.

SCHLAIFER, ROBERT. *Analysis of Decisions under Uncertainty.* New York: McGraw-Hill, 1969.

WINKLER, ROBERT L. *Introduction to Bayesian Inference and Decision.* New York: Holt, Rinehart and Winston, 1972.

20

Bayesian Posterior Analysis

In Chapter 19 we introduced the three basic decision-making environments, and the expected value criterion as a tool for making decisions in an environment of risk or uncertainty. Peter F. Drucker, who has probably written more about business management than any other person, emphasizes that the normal management operating environments are risk and uncertainty. He also states that managers need quantitative tools to help use the continual flow of information they receive about their risky or uncertain environment.

Bayesian decision makers accept the premise that their states of mind (and therefore subjective probabilities) can and should change as new information affecting the decision is obtained. Decision makers need a formal basis for combining prior information and new information to make the "best" decision. *Bayesian posterior analysis* provides such a basis and is the subject of this chapter.

chapter objectives

The objective of this chapter is to introduce the Bayesian approach to revising the probabilities assigned to states of nature in a decision environment. We shall use

575

examples to show how added information is used with the Bayesian technique to revise prior information.

student objectives

After studying the material presented in this chapter, you should be able to:

1. Recognize applications of Bayesian posterior analysis in the business world.
2. Apply the revision process to find new probabilities based upon new information.
3. Understand how the Bayesian revision procedure incorporates new information into a decision process.

20-1

BAYES' RULE REVISITED

In Chapter 5 we presented Bayes' rule as a method for finding conditional probabilities. Bayes' rule is particularly important when certain conditional probabilities cannot be calculated directly. Consider the following example.

Merrit Electronics manufactures electronic calculators at three locations. Further, 50 percent of the calculators are produced at location A, 30 percent at location B, and 20 percent at location C. Thus

$$P(A) = 0.50$$
$$P(B) = 0.30$$
$$P(C) = \underline{0.20}$$
$$\Sigma = \overline{1.00}$$

The quality control systems differ at the three locations, and the proportions of defective calculators produced also differ: location A produces 5 percent defectives, and location B and location C produce 7 and 12 percent defectives, respectively. Thus

$$P(\text{defective}|A) = 0.05$$
$$P(\text{defective}|B) = 0.07$$
$$P(\text{defective}|C) = 0.12$$

Merrit Electronics requires each manufacturing location to be a profit center. Therefore the cost of replacing or repairing defective calculators must be allocated to the three locations. Once a defective calculator has been returned for replacement or repair, Merrit must determine the probability it was made at location A, B, or C. Although Merrit knows $P(\text{defective}|A)$, what the firm really wants is

$$P(A|\text{defective}) = ?$$
$$P(B|\text{defective}) = ?$$
$$P(C|\text{defective}) = ?$$

Once these probabilities are known, Merrit could allocate costs on a percentage basis as determined by these probabilities. Using Bayes' rule, we can find the following allocation percentages:

$$P(A|\text{defective}) = \frac{P(A)P(\text{defective}|A)}{P(\text{defective})}$$

$$P(B|\text{defective}) = \frac{P(B)P(\text{defective}|B)}{P(\text{defective})}$$

$$P(C|\text{defective}) = \frac{P(C)P(\text{defective}|C)}{P(\text{defective})}$$

First, we find $P(\text{defective})$ by summing the joint probabilities of defective calculators and plants.

$$P(\text{defective}) = P(A \text{ and defective}) + P(B \text{ and defective}) + P(C \text{ and defective})$$
$$= P(A)P(\text{defective}|A) + P(B)P(\text{defective}|B) + P(C)P(\text{defective}|C)$$

Therefore

$$P(\text{defective}) = 0.50(0.05) + 0.30(0.07) + 0.20(0.12) = 0.025 + 0.021 + 0.024$$
$$= 0.07$$

so

$$P(A|\text{defective}) = \frac{P(A)P(\text{defective}|A)}{P(\text{defective})} = \frac{0.50(0.05)}{0.07}$$
$$= 0.3571$$

$$P(B|\text{defective}) = \frac{P(B)P(\text{defective}|B)}{P(\text{defective})} = \frac{0.30(0.07)}{0.07}$$
$$= 0.3000$$

$$P(C|\text{defective}) = \frac{P(C)P(\text{defective}|C)}{P(\text{defective})} = \frac{0.20(0.12)}{0.07}$$
$$= 0.3429$$

Thus the appropriate allocation of replacement or repair costs should be

Location A—35.71%

Location B—30.00%

Location C—34.29%

Bayesian Statistics and the Human Thought Process

Bayes' rule allows decision makers to incorporate new information into the decision-making process. Most people, when making nonrepetitive decisions in a changing environment, follow a relatively constant decision process. They make a tentative decision, gather new information, use this new information to modify the tentative decision, gather more new information, use this new information to update the decision, and so on until the time comes to act. Then they follow the course of action dictated by their decision criteria.

Suppose, before going to bed at night, a contractor hears the weather forecaster predict that a cold front will move in tomorrow morning and the chance of rain will increase to 50 percent through the day. The contractor may tentatively decide to cancel the concrete he had ordered for tomorrow. However he will modify this decision based on the weather conditions in the morning. If the sky is dark and overcast the next day, he most likely will decide to cancel the concrete. However, if the sky is clear and blue, he will likely decide not to cancel. Thus he will either affirm or change his tentative decision based on new information. Up until the point at which he must make the final decision, you can bet he will be watching the sky to collect current information.

As another example, the transportation manager for a large corporation has tentatively decided to change the brand of tires she specifies on the company's fleet of sales cars. She has based this decision on an article in a trade journal. However, before making the decision final, she decides to test a few new tires. If the test indicates that the new brand is more economical, the manager may decide, based on the journal article and the test, to change the brand of tires. However, if the test results are negative, she may revise her original decision and decide not to change brands.

A major strength of Bayesian decision analysis is that it provides managers with a systematic method for using new information to revise prior opinions. In the next section we discuss how to revise prior opinions (or probabilities) based on new information.

20-2
BAYESIAN POSTERIOR ANALYSIS

The Bartlett Corporation supplies fluorescent lights to office buildings in California. Because of its efforts over the past 20 years to stress a quality product at a fair price, Bartlett has developed a large clientele. Like most other organizations, Bartlett continually seeks to improve its profit situation. Thus Bartlett's purchasing agent continually seeks fluorescent light sources that will produce higher profits.

Bartlett currently buys its lights from Brightday, Inc., a U.S. firm. Brightday has produced a good product that, based upon historical records, has the following probability distribution for defectives:

Proportion Defective	Probability
0.01	0.50
0.02	0.40
0.03	0.10
	$\Sigma = 1.00$

Bartlett guarantees the fluorescent lights it supplies and replaces any defectives at no cost to the customer. Bartlett accountants estimate the overall cost of each replaced defective, including damaged goodwill, at $5.

Bartlett's marketing department has used multiple regression analysis to forecast an expected demand of 100,000 lights for the coming year. If Bartlett purchases all 100,000 lights from Brightday, the probability distribution for defective lights is

No. Defective	Cost	Probability
1,000	$ 5,000	0.50
2,000	10,000	0.40
3,000	15,000	0.10
		$\Sigma = 1.00$

Since Bartlett ultimately will be interested in costs, we have also determined the probability distribution for the cost of replacing defective lights.

$$\text{Cost} = \$5 \times \text{number of defectives}$$

Using the cost distribution, we can find the expected replacement cost if Bartlett buys 100,000 lights from Brightday.

Cost	$P(X)$	$XP(X)$
$ 5,000	0.50	$2,500
10,000	0.40	4,000
15,000	0.10	1,500
	$\Sigma = 1.00$	$\Sigma = \$8,000 =$ expected cost

While attending an international convention for fluorescent light dealers, Bartlett representatives were approached by a West German light manufacturer. The West German offered to supply Bartlett with 100,000 lights at a cost, including shipping, that is $0.25 per light less than the current price charged by Brightday.

Through other inquiries, and their personal experience with European fluorescent light manufacturers, the Bartlett representatives developed the following subjective probability distribution for defectives from this manufacturer:

Proportion Defective	Subjective Probability
0.01	0.10
0.02	0.10
0.04	0.30
0.08	0.30
0.10	0.10
0.15	0.10
	$\Sigma = 1.00$

This distribution indicates that the Bartlett representatives feel the West German lights will likely contain more defectives than the Brightday lights.

Assuming these two suppliers are Bartlett's only alternatives, which supplier should Bartlett select if it uses the expected value criterion? To make this decision, Bartlett must find the expected cost of the West German alternative. The expected cost of replacing defective lights is found by the following. (The $5 replacement cost still holds.)

Proportion Defective	No. Defective	Cost X	Probability P(X)	XP(X)
0.01	1,000	$ 5,000	0.10	$ 500
0.02	2,000	10,000	0.10	1,000
0.04	4,000	20,000	0.30	6,000
0.08	8,000	40,000	0.30	12,000
0.10	10,000	50,000	0.10	5,000
0.15	15,000	75,000	0.10	7,500
			$\Sigma = 1.00$	$\Sigma = \$32,000$ = expected cost

If we compare only expected replacement costs, we find

$$E[\text{cost}|\text{Brightday}] = \$8,000$$
$$E[\text{cost}|\text{West German}] = \$32,000$$

Since $8,000 < $32,000, the choice clearly is to stay with Brightday. This decision assumes no difference in the price Bartlett pays per light. However the West German lights are $0.25 less than the Brightday lights. One hundred thousand lights purchased from Brightday would cost $25,000 (100,000 × $0.25) more than the same number of lights from the West German manufacturer. Therefore we must add the purchase price difference to Brightday's $8,000 expected replacement cost. Now

$$E[\text{cost}|\text{Brightday}] = \$8,000 + \$25,000$$
$$= \$33,000$$
$$E[\text{cost}|\text{West German}] = \$32,000$$

Since $33,000 > $32,000, the best decision, using expected values, is to purchase from the West German manufacturer. This decision leads to an expected savings of $1,000.

Note that this decision process is consistent with the expected value criterion discussed in Chapter 19. However, suppose the long-term business relationship with Brightday has been important to the Bartlett Corporation, and, before this relationship is severed, Bartlett's management would like a little more information. Specifically, Bartlett would like to sample some of the West German supplier's lights and combine the sample information with the subjective prior assessments to make a "better" decision. **The method of incorporating sample information into the decision process is called *Bayesian posterior analysis.***

Constructing a Posterior Table

Bartlett representatives decide to test 50 lights before making their final decision, and they find 6 defectives. Assuming the binomial probability distribution represents the process that produces fluorescent lights, we can use Bayesian posterior analysis through the following steps:

1. The possible outcomes, in terms of proportion defective, are

Proportion Defective
0.01
0.02
0.04
0.08
0.10
0.15

2. The prior probabilities for these potential outcomes, as subjectively assessed by Bartlett representatives, were

Proportion Defective	Prior Probability
0.01	0.10
0.02	0.10
0.04	0.30
0.08	0.30
0.10	0.10
0.15	0.10
	$\Sigma = 1.00$

3. The sample of $n = 50$ lights produced $X_1 = 6$ defectives.

4. The probability of observing 6 defectives in a sample of 50 depends, of course, on the fraction of defective lights in the lot of 100,000. But according to step 1, this fraction is estimated to range from 0.01 to 0.15. Therefore we must determine the conditional probability of finding 6 defectives in a sample of 50 for each possible defective rate. Since the production process can be represented by a binomial distribution, the table in Appendix A can be used to determine these conditional probabilities.

Proportion Defective p	Prior Probability	Conditional Probability $P(n = 50, X_1 = 6\|proportion\ defective)$
0.01	0.10	0.0000
0.02	0.10	0.0004
0.04	0.30	0.0108
0.08	0.30	0.1063
0.10	0.10	0.1541
0.15	0.10	0.1419
	$\Sigma = 1.00$	

The conditional probabilities in this table are all found in the usual way:

1. Go to the binomial table for $n = 50$.
2. Find the row for $X_1 = 6$.
3. In the column headed $p = 0.04$, find $P(n = 50, X_1 = 6 | p = 0.04) = 0.0108$. In a like manner, find $P(n = 50, X_1 = 6 | p = 0.08) = 0.1063$; $P(n = 50, X_1 = 6 | p = 0.10) = 0.1541$; and so forth.

5. The next step is to find the joint probabilities of the defective levels based on the prior probabilities and the observed sample results. To do this, we apply the multiplication rule of probability.

$$\text{Joint probability} = P(\text{proportion defective and sample result})$$
$$= P(\text{proportion defective})P(\text{sample results} | \text{proportion defective})$$
$$= (\text{prior probability})(\text{conditional probability})$$

This is the numerator in Bayes' rule. We find the joint probabilities as follows:

Proportion Defective	Prior Probability	Conditional Probability	Joint Probability
0.01	0.10	0	0(0.10) = 0
0.02	0.10	0.0004	0.0004(0.10) = 0.00004
0.04	0.30	0.0108	0.0108(0.30) = 0.00324
0.08	0.30	0.1063	0.1063(0.30) = 0.03189
0.10	0.10	0.1541	0.1541(0.10) = 0.01541
0.15	0.10	0.1419	0.1419(0.10) = 0.01419
	$\Sigma = 1.00$		$\Sigma = 0.06477$

By summing the joint probabilities, we find the overall probability of 6 defective lights in a sample of 50, given the possible defective levels listed in the left column. Thus 0.06477 is the overall probability of finding 6 defectives in a sample of 50 given the prior states of nature. This probability is the divisor in Bayes' rule.

6. The last step in combining the prior probabilities and the sample information is to find the posterior probability for each possible outcome. These posterior probabilities are the revised probabilities of the possible defective levels given the new sample information. They are found by dividing each joint probability by the sum of the joint probabilities.

Proportion Defective	Prior Probability	Conditional Probability	Joint Probability	Posterior Probability
0.01	0.10	0	0	0
0.02	0.10	0.0004	0.00004	$\dfrac{0.00004}{0.06477} = 0.000618$
0.04	0.30	0.0108	0.00324	$\dfrac{0.00324}{0.06477} = 0.050023$
0.08	0.30	0.1063	0.03189	$\dfrac{0.03189}{0.06477} = 0.492358$
0.10	0.10	0.1541	0.01541	$\dfrac{0.01541}{0.06477} = 0.237918$
0.15	0.10	0.1419	0.01419	$\dfrac{0.01419}{0.06477} = 0.219083$
	$\Sigma = 1.00$		$\Sigma = 0.06477$	$\Sigma = 1.000000$

These posterior probabilities reflect Bartlett's revised state of mind about the proportion of defectives produced by the West German manufacturer. These probabilities incorporate both new and prior information. From the Bayesian viewpoint, it makes sense to use these to reevaluate the purchase decision. We therefore find a new expected value.

Proportion Defective	No. Defective	Cost X	Posterior Probability $P(X)$	$XP(X)$
0.01	1,000	$ 5,000	0	$ 0
0.02	2,000	10,000	0.000618	6.18
0.04	4,000	20,000	0.050023	1,000.46
0.08	8,000	40,000	0.492358	19,694.32
0.10	10,000	50,000	0.237918	11,895.90
0.15	15,000	75,000	0.219083	16,431.22
				$\Sigma = \$49,028.08$

Thus, the expected cost of selecting the West German manufacturer is $49,028.08. Since $33,000 < $49,028.08, this posterior analysis indicates that the "best" decision, using expected values, is to buy from Brightday.

Note that the original decision has changed based on this Bayesian posterior analysis. The sample information (6 defectives in a sample of 50) indicated that the probability of finding either 8, 10, or 15 percent defectives is greater than the Bartlett representatives originally thought.

The Manufacturer Asks for a Recount

As we have indicated throughout this text, whenever sampling occurs, we can expect sampling error. Suppose the West German manufacturer believes he has been victimized by an unlucky sample. He is willing to pay for another sample of 50 lights in the hope that Bartlett will change its decision. Of course, since there is no cost to Bartlett, Bartlett is happy to obtain more information. However, Bartlett managers assure the manufacturer that they will not ignore the first sample. In fact, Bartlett will view the posterior probabilities just calculated as the new prior probabilities in the subsequent analysis. Suppose the second sample of 50 lights contain 4 defectives. The new revised probabilities are determined by using the same six steps just discussed. The results are

Proportion Defective	Prior Probability	Conditional Probability $P(n=50, X_2=4 \vert proportion\ defective)$	Joint Probability	Posterior Probability
0.01	0	0.0015	0	0
0.02	0.000618	0.0145	0.000009	0.000055
0.04	0.050023	0.0902	0.004512	0.027795
0.08	0.492356	0.2037	0.100293	0.617819
0.10	0.237918	0.1809	0.043039	0.265126
0.15	0.219083	0.0661	0.014481	0.089205
	$\Sigma = 1.000000$		$\Sigma = 0.162334$	

These posterior probabilities can now be used to determine the expected cost of purchasing 100,000 bulbs from the West German manufacturer.

Proportion Defective	No. Defective	Cost X	Posterior Probability $P(X)$	$XP(X)$
0.01	1,000	$ 5,000	0	$ 0
0.02	2,000	10,000	0.000055	0.55
0.04	4,000	20,000	0.027795	555.90
0.08	8,000	40,000	0.617819	24,712.76
0.10	10,000	50,000	0.265125	13,256.30
0.15	15,000	75,000	0.089205	6,690.38
				$\Sigma = \$45,215.89$ = expected cost

Since $\$33,000 < \$45,215.89$, using expected values, Bartlett should still continue to purchase from Brightday.

Two Small Samples or One Large Sample?

In the example, Bartlett selected two successive samples of $n = 50$. Bartlett used the results from the first sample to revise the probabilities for the possible defective rates. These revised or posterior probabilities became the prior probabilities for the second sample. Using results from the second sample, Bartlett representatives again revised the probabilities. Note that selecting two successive samples of 50 and finding $X_1 = 6$ and $X_2 = 4$ defectives provides no more (or no less) information than if one sample of $n = 100$ had been selected and $X_1 = 10$ defectives had been found. This is shown as follows:

Proportion Defective	Prior Probability	Conditional Probability $P(n = 100, X_1 = 10\|$proportion defective)	Joint Probability	Posterior Probability
0.01	0.10	0	0	0
0.02	0.10	0	0	0
0.04	0.30	0.0046	0.00138	0.02775
0.08	0.30	0.1024	0.03072	0.61774
0.10	0.10	0.1319	0.01219	0.26523
0.15	0.10	0.0444	0.00444	0.08928
	$\Sigma = 1.00$		$\Sigma = 0.04973$	

These posterior probabilities (with small differences due to rounding) are the same as the posterior probabilities found after the second sample of 50 lights was selected. Thus the expected cost of buying from the West German manufacturer is the same regardless of whether Bartlett selected one sample of $n = 100$ and found $X_1 = 10$ defectives or two samples of $n = 50$ and found $X_1 = 6$ and $X_2 = 4$ defectives.

A Nonbinomial Example

In the previous example we assumed that the binomial distribution could be used to determine the conditional probabilities of observing the sample information. However not all decision-making situations can be described by the binomial distribution. Suppose we use another example to demonstrate how Bayesian posterior analysis can be used when the underlying process is not binomial.

The Hamilton-Rock Company sells tax-sheltered annuities (TSAs). A tax-sheltered annuity allows employees who qualify to have part of their monthly paychecks deposited directly with an authorized company. These deposits are not subject to income taxes until the money is withdrawn.

The marketing manager for Hamilton-Rock is concerned each time she hires a new salesperson. She would like to hire only successful salespeople. Based on her past experience, the marketing manager feels TSA salespeople can be divided into four categories: "excellent," "good," "fair," and "poor." By her definition, an "excellent" salesperson will average one TSA sale per day, a "good" salesperson will average one sale every two days, a "fair" salesperson will average one TSA sale every five days, and a "poor" salesperson will average only one sale every ten days. If we assume that the number of tax-sheltered annuities sold can be reasonably approximated by a Poisson probability distribution, we can describe each salesperson category by λ, the average sales per day. Thus we get

Salesperson Classification	λ
Excellent	1.0
Good	0.5
Fair	0.2
Poor	0.1

Suppose the marketing manager has hired a new salesperson. Based on a complete review of this person's credentials, and a lengthy personal interview, the manager feels the new salesperson has a 20 percent chance of being "excellent"; a 40 percent chance of being "good"; a 30 percent chance of being "fair"; and a 10 percent chance of being "poor." Thus the manager's prior distribution for λ is

λ	Prior Probability
1.0	0.20
0.5	0.40
0.2	0.30
0.1	0.10
	$\Sigma = 1.00$

The standard agreement Hamilton-Rock has with each new employee is that the position is temporary for the first ten working days. After the ten-day trial the employee either will be given a permanent contract or released, as determined by the marketing manager. Of course, during the ten-day trial, sample information is collected on the new person's sales ability.

Suppose the new person sells seven TSA accounts during the first ten days. Bayesian posterior analysis can be used to incorporate this sample information into the decision-making process by using it to revise the prior probabilities. The procedure is as follows:

1. After the prior probability distribution has been assessed, the Poisson distribution table in Appendix B can be used to arrive at the following conditional probabilities:

Class	λ	Prior Probability	Conditional Probability $P(X_1 = 7\|\lambda t)$
Excellent	1.0	0.20	0.0901
Good	0.5	0.40	0.1044
Fair	0.2	0.30	0.0034
Poor	0.1	0.10	0.0001
		$\Sigma = 1.00$	

The conditional probabilities in this table are all found in the same way:

1. Go to the Poisson table.

2. Find the row for $X_1 = 7$.

3. In the column headed $\lambda t = 10$ (one per day $\times$ ten days), find $P(X_1 = 7|\lambda t = 10) = 0.0901$. In a like manner, find $P(X_1 = 7|\lambda t = 0.5(10) = 5) = 0.1044$; $P(X_1 = 7|\lambda t = 0.2(10) = 2) = 0.0034$; and so forth.

2. Find the joint probability of the salesperson making seven sales given each possible state of nature. This joint probability is found by multiplying the prior probability by the associated conditional probability.

Class	Prior Probability	Conditional Probability	Joint Probability
Excellent	0.20	0.0901	0.01802
Good	0.40	0.1044	0.04176
Fair	0.30	0.0034	0.00102
Poor	0.10	0.0001	0.00001
	$\Sigma = 1.00$		$\Sigma = 0.06081$

3. Develop each revised or posterior probability by dividing each joint probability by the sum of the joint probabilities.

Class	λ	Prior Probability	Conditional Probability	Joint Probability	Posterior Probability
Excellent	1.0	0.20	0.0901	0.01802	$\dfrac{0.01802}{0.06081} = 0.29633$
Good	0.5	0.40	0.1044	0.04176	$\dfrac{0.04176}{0.06081} = 0.68673$
Fair	0.2	0.30	0.0034	0.00102	$\dfrac{0.00102}{0.06081} = 0.01677$
Poor	0.1	0.10	0.0001	0.00001	$\dfrac{0.00001}{0.06081} = 0.00017$
		$\Sigma = 1.00$		$\Sigma = 0.06081$	$\Sigma = 1.00000$

The sample information has almost totally convinced the marketing manager that the new salesperson is neither "fair" nor "poor." Before the ten-day trial period, the expected daily sales for the new person were

$$E[\lambda] = \Sigma \lambda P(\lambda) = 1.0(0.20) + 0.5(0.40) + 0.2(0.30) + 0.1(0.10)$$
$$= 0.4700$$

After the trial period, the expected sales per day are

$$E[\lambda] = \Sigma \lambda P(\lambda) = 1.0(0.29633) + 0.5(0.68673) + 0.2(0.01677)$$
$$+ 0.1(0.00017)$$
$$= 0.6431$$

Hamilton-Rock pays all new salespeople $40 per day regardless of their sales level. Hamilton-Rock's profit on each TSA sale is $100. The expected daily profit or loss for a new salesperson is

$$E[\text{profit}] = E[\lambda](\$100) - \$40$$

Thus, before the ten-day trial, the expected daily profit from the new salesperson was

$$E[\text{profit}] = 0.47(\$100) - \$40 = \$47 - \$40$$
$$= \$7$$

After ten days, the expected profit is

$$E[\text{profit}] = 0.6431(\$100) - \$40 = \$64.31 - \$40$$
$$= \$24.31$$

Thus, if the marketing manager was willing to hire the new salesperson based on the prior probabilities (expected profit = $7.00), she should be even more willing after the ten-day trial (expected profit = $24.31).

To test your intuitive understanding of Bayesian posterior analysis, what do you think would happen to the expected profit value if the new person sold only two TSA accounts during the ten-day trial? The expected profit under this condition can be determined precisely.

Class	λ	Prior Probability	Conditional Probability $P(X_1 = 2 \mid \lambda t)$	Joint Probability	Posterior Probability
Excellent	1.0	0.20	0.0023	0.00046	0.00344
Good	0.5	0.40	0.0842	0.03368	0.25183
Fair	0.2	0.30	0.2707	0.08121	0.60722
Poor	0.1	0.10	0.1839	0.01839	0.13751
		$\Sigma = 1.00$		$\Sigma = 0.13374$	$\Sigma = 1.00000$

The expected sales and profit, given these posterior probabilities, are

$$E[\text{sales}] = 1.0(0.00344) + 0.5(0.25183) + 0.2(0.60722) + 0.1(0.13751)$$
$$= 0.2645$$
$$E[\text{profit}] = 0.2645(\$100) - \$40 = \$26.45 - \$40$$
$$= -\$13.55 \quad (\text{Loss})$$

Thus, if only two sales were made during the trial period, the marketing manager likely would not want to retain the new salesperson.

On your own, determine the minimum number of sales that must be made during the trial period before a person will be retained. The only requirement for retention is that the expected profit must exceed $0.

20-3
BAYESIAN POSTERIOR ANALYSIS WITH SUBJECTIVE CONDITIONALS

In the previous section we discussed Bayesian posterior analysis. We showed how prior probabilities can be revised using sample information. Whereas the prior probabilities were subjectively assessed, the conditional probabilities were based on known probability distributions: the binomial in one case and the Poisson in another. In this section we show how Bayesian posterior analysis can be used when the conditional probabilities are also subjectively assessed.

Several project managers at the Actal Corporation are considering writing a proposal for an atomic breeder-reactor. This proposal, which will be submitted to the Department of Energy, will be very expensive to develop. These managers are concerned that the federal government may decide to place a moratorium on new atomic power production. If this happens, the money spent to develop the proposal will be wasted. Based on all information available on March 1, the managers assess the following probabilities:

Event	Prior Probability
Moratorium	0.20
No moratorium	0.80
	$\Sigma = 1.00$

Based upon these prior probabilities, the managers decide to develop the proposal. However, two weeks later the managers learn that the Department of Energy has transferred two atomic energy experts. These experts had worked in the proposal review department. Although this could mean nothing, there is a great deal of speculation within Actal that it means the moratorium is going to be ordered.

At the weekly progress meeting, the decision to develop the breeder-reactor proposal is reevaluated. After careful consideration, the managers agree that if the moratorium is going to occur, there is a 90 percent chance these employees would have been transferred. In addition they feel that if the moratorium does not occur, the chances the employees would have been transferred is 40 percent. The managers have subjectively assessed the conditional probabilities of this new information, given each possible event. Bayesian posterior analysis can be used as follows:

Event	Prior Probability	Conditional Probability	Joint Probability	Posterior Probability
Moratorium	0.20	0.90	0.18	0.36
No moratorium	0.80	0.40	0.32	0.64
	$\Sigma = 1.00$		$\Sigma = 0.50$	$\Sigma = 1.00$

The new information has changed the managers' thinking about the chances of a moratorium; however, the decision is still to continue preparing the proposal.

Subjective conditional probabilities can be assessed at any time and are dependent upon all previous information. For example, suppose these managers later learn that the Department of Energy has hired five geothermal experts. The managers then assess the following conditional probabilities:

P(hiring 5 geothermal|transferred 2 atomic, moratorium) = 0.70
P(hiring 5 geothermal|transferred 2 atomic, no moratorium) = 0.10

These assessments reflect the fact that two atomic energy specialists were transferred before the five geothermal experts were hired.

To include this latest information in the analysis, the posterior probabilities from the previous step become the prior probabilities at this step, leading to the following table:

Event	Prior Probability	Conditional Probability	Joint Probability	Posterior Probability
Moratorium	0.36	0.70	0.252	0.797
No moratorium	0.64	0.10	0.064	0.203
	$\Sigma = 1.00$		$\Sigma = 0.316$	$\Sigma = 1.000$

With this information, the assessed probability is almost 0.80 that a moratorium on atomic power production will take place. Based on this assessment, the managers might elect to stop the proposal and channel their resources in another direction. (Keep in mind that future information might once again cause Actal to revive the proposal.)

20-4
THE VALUE OF INFORMATION

In the previous sections we indicated how new information, either objective (that is, sample data) or subjective, can be combined with prior information in the decision situation. This new information can cause a decision to change or can support the decision that would have been made with the prior information only. However, obtaining new information costs money, and to determine whether the new information is cost-justified, we must have some idea of its value. This section provides an overview of how the value of new information can be determined.

Certainty is the ideal decision-making environment. Under certainty managers can make the "best" decisions and know that the "best" outcomes will result. However decision makers forced to act under risk or uncertainty must expect that "good" decisions will not always be associated with "good" outcomes. In the managerial world, the difference between a "good" and a "poor" outcome is often measured in terms of revenues or costs. Since "poor" outcomes can occur from "good" decisions, there is a cost associated with being uncertain. This cost is the *cost of uncertainty* (we will formally define this term later).

The "Best" Decision with No Sample Information

The Tree-Light Corporation supplies Christmas tree ornaments to retail stores across the United States. This year the company has two potential sources of ornaments. One supplier guarantees that in lot sizes of 50,000 ornaments, no more than 3 percent will be defective. The second supplier makes no guarantee, but will sell the ornaments for $0.02 less per ornament than the first supplier. The Tree-Light managers have assessed the following probabilities for defectives produced by the second supplier:

Proportion Defective	Probability
0.02	0.40
0.03	0.20
0.10	0.30
0.20	0.10
	$\Sigma = 1.00$

Tree-Light offers free replacement for any defective ornament. The replacement cost is $0.50 per defective.

Using this information, we can determine which supplier Tree-Light should select if the criterion is to minimize expected cost. For the first supplier, the expected cost has two terms: the expected cost of replacing defective ornaments and a cost differential because the ornaments are more expensive. If we take the 3 percent maximum defective level as the true level, the expected cost for a lot of 50,000 is calculated as follows:

$$\text{Expected replacement cost} = (0.03)(50,000)(\$0.50)$$
$$= \$750$$

and

$$\text{Cost of being more expensive} = 50,000(\$0.02)$$
$$= \$1,000$$

so

$$E[\text{cost first supplier}] = \$750 + \$1,000$$
$$= \$1,750$$

The expected cost for the second supplier is only the expected replacement cost. Since the percentage of defectives for the second supplier is not known, we must use the assessed probabilities.

Proportion Defective	No. Defective in 50,000	Cost	Probability
0.02	1,000	$ 500	0.40
0.03	1,500	750	0.20
0.10	5,000	2,500	0.30
0.20	10,000	5,000	0.10
			$\Sigma = 1.00$

Therefore the expected cost for the second supplier is

$$E[\text{cost second supplier}] = \$500(0.40) + \$750(0.20) + \$2,500(0.30)$$
$$+ \$5,000(0.10)$$
$$= \$1,600$$

Since the criterion is the lowest expected cost, the "best" decision is to select the second supplier.

The "Best" Decision with Perfect Information

While the "best" decision with no sample information for Tree-Light has an expected cost of $1,600, for any lot of 50,000 the actual cost will be $500, $750, $2,500, or $5,000 (see the previous cost table). Therefore, if Tree-Light had some perfect source of information, the ideal decision would be:

If the second supplier will ship a lot with 2% or 3% defectives, buy from that supplier.
However, if the second supplier will ship a lot with 10% or 20% defectives, buy from the first supplier.

Even the ideal decision, made with perfect information, has a cost. This is

$$E[cost] = \$500(0.40) + \$750(0.20) + \$1,750(0.30) + \$1,750(0.10)$$
$$= \$1,050$$

Perfect information will not eliminate defects; it will simply tell us what level of defectives we can expect. Thus 40 percent of the time, the second supplier will produce 2 percent defectives; 10 percent of the time, it will produce 20 percent defectives; and so forth.

The Expected Value of Perfect Information

The expected cost with perfect information is less than the expected cost of the "best" decision under uncertainty ($1,600 in the Tree-Light example). **The *cost of uncertainty* is the difference between the expected cost under uncertainty and the expected cost given perfect information.** That is,

$$\text{Cost of uncertainty} = E[cost|\text{best decision under uncertainty}] - E[cost|\text{best decision under certainty}]$$

For the Tree-Light example,

$$\text{Cost of uncertainty} = \$1,600 - \$1,050$$
$$= \$550$$

If someone offered to provide perfect information to Tree-Light (maybe the second supplier will count the defectives before shipping), Tree-Light would be willing to spend a maximum of $550 for the perfect information. Any more than $550 would increase the total perfect information cost above the cost of the best decision under uncertainty. This $550 is also called the ***expected value of perfect information.***

Cost of Uncertainty for Profit Maximization

The previous example was a cost-minimization problem. The cost of uncertainty also applies when we are dealing with payoff maximization.

In Chapter 19 we discussed a decision facing Ajax Electronics. The firm was trying to decide how many television sets to produce. The payoff table constucted in Chapter 19 is reproduced as Table 20–1. This table shows the payoffs for levels of production and demand. The expected payoff for each production level is listed in the right-hand column.

TABLE 20–1   Expected payoffs, Ajax Electronics

		Demand Level				
	700	800	900	1,000	1,100	
			Probability			Expected Payoff ↓
	0.05	0.10	0.20	0.40	0.25	
700	−$ 70,000	−$ 70,000	−$70,000	−$70,000	−$70,000	−$70,000
800	−$100,000	−$ 30,000	−$30,000	−$30,000	−$30,000	−$33,500
Production Level 900	−$130,000	−$ 60,000	$10,000	$10,000	$10,000	−$ 4,000
1,000	−$160,000	−$ 40,000	−$20,000	$50,000	$50,000	$11,500
1,100	−$190,000	−$120,000	−$50,000	−$20,000	$90,000	$ 5,000

If Ajax Electronics can receive perfect information about which demand level will occur, it can always select the optimal production level. For example, if the production manager knew that 700 televisions were going to be demanded, she would produce 700. (*Note:* Ajax would lose $70,000 on each production run and obviously could not stay in business very long.) However, if demand were known to be 1,100, Ajax would produce 1,100 sets, for a profit of $90,000. Table 20–2 lists the profit associated with the optimal action for each demand level.

TABLE 20–2   Optimal payoffs, Ajax Electronics

Demand	Production	Profit	Probability
700	700	−$70,000	0.05
800	800	− 30,000	0.10
900	900	10,000	0.20
1,000	1,000	50,000	0.40
1,100	1,100	90,000	0.25
			$\Sigma = 1.00$

Then, given that Ajax Electronics had perfect information about demand, the expected profit would be

$$E[\text{profit}|\text{perfect information}] = -\$70,000(0.05) - \$30,000(0.10) + \$10,000(0.20)$$
$$+ \$50,000(0.40) + \$90,000(0.25)$$
$$= \$38,000$$

Once again, the actual demand level is out of Ajax's control. However, since the expected payoff for the "best" decision under uncertainty (produce 1,000 television sets) is $11,500, the cost of being uncertain is $38,000 − $11,500 = $26,500. This is also the expected value of perfect information.

Value of Sample Information

The value of perfect information is the maximum amount decision makers should pay to be told which state of nature will occur. In practice, perfect information is rare. Consequently the only real use for the value of perfect information is as an upper limit on what we would be willing to pay for less-than-perfect information, such as that which would be derived from a sample.

A prime source of information is a random sample. As we discussed in Chapters 9 and 10, random sampling is always subject to error. Therefore less-than-perfect information will be generated from a sample. Gathering sample information takes both time and money; thus, before decision makers agree to pay for the sample, they should know its actual value.

Although we will not discuss the calculations involved in determining the value of a sample, we will outline the basic steps. **The value of a sample is the difference between the cost of uncertainty before the sample is observed (*prior cost of uncertainty*) and the cost of uncertainty after the sample (*posterior cost of uncertainty*).** The prior cost of uncertainty is computed using prior probabilities as we have discussed in this section. The posterior cost of uncertainty is computed in the same manner except that the Bayesian revision process is employed and the posterior probabilities are used in computing the posterior cost of uncertainty.

The method described here deals with finding the value of a sample after that sample has been taken. The problem of computing the value of a sample before the results are known is much more involved. More involved yet is determining the sample size that will maximize the net gain from sampling. Texts by Parsons (1974) and Winkler (1972) contain excellent discussions of these situations.

20-5
CONCLUSIONS

In this chapter we have introduced Bayesian posterior analysis. We have illustrated how sample information can be used to revise subjectively assessed prior distributions. The Bayesian decision maker feels that this revision process is the core of the managerial decision-making process.

We limited our discussion to decisions involving discrete events and discrete probability distributions. More advanced texts, such as *Introduction to Bayesian Inference and Decisions* by Robert Winkler (1972), present the techniques for dealing with continuous events.

In this chapter we have introduced several new terms that are frequently used during Bayesian posterior analysis. These are listed in the chapter glossary. You should become familiar with each one.

chapter glossary

Bayes' rule   If A is an event that occurs if and only if either B or $\bar{B}$ occurs, where $\bar{B}$ includes all events that are not in B, then

$$P(B|A) = \frac{P(A \text{ and } B)}{P(A)} = \frac{P(A|B)P(B)}{P(A|B)P(B) + P(A|\bar{B})P(\bar{B})}$$

conditional probability   The probability of event A occurring given that another event, B, has occurred. The conditional probability represents the chances of observing the sample information given each possible state of nature.

joint probability   The probability of the simultaneous occurrence to two or more events. For example, joint probability represents the likelihood of observing the sample information and each state of nature.

prior probability   The probability of an event occurring before sample evidence relevant to the event has been observed.

posterior probability   The probability of an event modified in light of experimental evidence. Posterior probabilities are usually determined by employing Bayes' rule.

posterior table   A table showing the states of nature, the prior probability of each state, the conditional probability of each state given the sample information, the joint probability of each state given the prior and conditional probabilities, and the posterior probability for each state.

subjective conditional probability   A probability that represents the chances of observing A given that B has occurred. This probability is assessed subjectively.

solved problems

1. The Batlow Newspaper Corporation is considering setting up a division to publish a weekly magazine that summarizes the major news stories of the previous week. Batlow estimates that the weekly fixed cost of establishing the division will be $300,000 and that the variable cost per magazine printed will be $0.50. The magazine will sell for $1.50. Further, the marketing department estimates that of Batlow's one million customers, the following possibilities exist with respect to the percentage who will buy this new magazine:

Proportion Buying	Prior Probability
0.20	0.10
0.30	0.30
0.40	0.40
0.50	0.20
	$\Sigma = 1.00$

a. Based only upon this prior information, should Batlow go ahead with the magazine if its goal is to maximize expected profits?

b. Suppose a random sample of 20 newspaper customers is selected and 5 indicate they will buy the new magazine if it is published. Assuming the binomial distribution applies, determine the revised or posterior probability for each possible proportion of customers who will buy the magazine.

c. Given the sample information, what should the Batlow Corporation do?

Solutions:

a. The first step is to determine the payoff for each possible level of demand as follows:

Proportion Buying	Payoff	Probability
0.20	− $100,000	0.10
0.30	0	0.30
0.40	100,000	0.40
0.50	200,000	0.20
		$\Sigma = 1.00$

The expected payoff, or profit, associated with publishing the new magazine is

$$E[\text{profit}] = -\$100,000(0.10) + \$0(0.30) + \$100,000(0.40)$$
$$+ \$200,000(0.20)$$
$$= \$70,000$$

Since $\$70,000 > \$0$, Batlow should publish the magazine if its objective is to maximize expected profits.

b. To revise the prior probabilities, we set up the following table and calculate the conditional probabilities using the binomial distribution:

Proportion Buying	Prior Probability	Conditional Probability $P(n = 20, X_1 = 5\|\text{proportion buying})$	Joint Probability	Posterior Probability
0.20	0.10	0.1746	0.01746	0.1680
0.30	0.30	0.1789	0.05367	0.5164
0.40	0.40	0.0746	0.02984	0.2871
0.50	0.20	0.0148	0.00296	0.0285
	$\Sigma = 1.00$		$\Sigma = 0.10393$	$\Sigma = 1.0000$

c. To determine the "best" decision for Batlow using posterior analysis, we find the expected profit of the magazine alternative using the posterior probabilities.

$$E[\text{profit}] = -\$100,000(0.1680) + \$0(0.5164) + \$100,000(0.2871)$$
$$+ \$200,000(0.0285)$$
$$= \$17,610$$

Although the Bayesian posterior analysis has produced a smaller expected profit, $17,610 > $0, so the "best" decision is to publish the news summary magazine.

---

2. The Telephone Company is considering whether or not to install a new computerized telephone system in Central City. The new system will offer several services such as call forwarding and conference call capability. The new computer can be leased for $1 million per year, and the annual incremental cost per user is estimated at $72. The Telephone Company will charge each customer served $10 per month, or $120 per year.

There are currently 400,000 telephone customers in Central City, and for The Telephone Company to break even, 20,834 customers, or 5.2 percent of all customers, will have to purchase the extra services. The company's marketing department has assessed the following probability distribution for the proportion who will buy the service:

Proportion Buying	Probability
0.02	0.20
0.04	0.30
0.06	0.40
0.08	0.10
	$\Sigma = 1.00$

a. Determine the payoffs for each possible proportion buying.
b. Analyze the problem facing The Telephone Company using only prior information. Determine the "best" decision using the expected value criterion.

Solutions:

a.

$$\text{Payoff} = (P)(400,000)(\$120 - \$72) - \$1,000,000$$

where: P = proportion buying the new service

Proportion Buying	Alternative 1 Payoff if New System Is Implemented	Alternative 2 Payoff if New System Is Not Implemented
0.02	−$616,000	$0
0.04	− 232,000	0
0.06	152,000	0
0.08	536,000	0

b. Using the expected value criterion, we weight the payoff for each possible proportion by the associated probability.

$$E[\text{payoff alternative 1}] = -\$616,000(0.20) + (-\$232,000)(0.30)$$
$$+ \$152,000(0.40) + \$536,000(0.10)$$
$$= -\$78,400$$
$$E[\text{payoff alternative 2}] = \$0$$

Thus the "best" decision using only prior information is not to install the new system since the expected annual payoff for the new system is $-\$78,400$, compared to $\$0$ for not installing.

problems

Problems marked with an asterisk (*) require extra effort.

1. Discuss the statement that a subjectively assessed prior probability is really a subjective posterior probability because it is based upon information available to the decision maker.

2. Discuss in your own words how Bayes' rule for conditional probability is used in posterior decision analysis.

*3. When a prior probability distribution is revised based upon a sample, the results of a large sample will tend to dominate the prior probabilities more than results of a smaller sample. Discuss this statement.

4. We have stated that Bayesian posterior analysis is one way of formally modeling the human decision process. Comment on this statement and give some examples of decisions you have made using posterior analysis on an informal basis.

5. The operations manager of a toy-manufacturing plant is faced with the decision of whether to buy a new assembly machine or to have the current machine repaired at a cost of $\$2,800$. The new machine will cost $\$11,000$ but can decrease the cost of assembling each toy to $\$1.00$ per unit. The current machine, after repair, will assemble the toys at a cost of $\$1.50$ each. The toy for which the machine will be used sells for $\$3.25$ each. Because of unpredictable consumer buying habits, the demand for this toy is not known with certainty. The demand is expected to occur according to the following probability distribution:

Demand Level	Probability
10,000	0.05
15,000	0.15
20,000	0.20
30,000	0.40
50,000	0.20
	$\Sigma = 1.00$

Construct a payofff table listing the decision alternatives and the payoff associated with each.

6. Refer to problem 5. What decision should the operations manager make if the objective is to maximize the expected payoff? Provide a practical justification for your answer.

7. Referring to problems 5 and 6, suppose an independent market-research firm has been hired by the toy manufacturer. The results of the marketing study are

examined by the toy company managers. Basically the report is optimistic, but the managers don't want to ignore their prior assessments. Rather, they wish to include both the prior assessments and the market survey. Consequently the toy manufacturer assesses the conditional probability of observing the results of the market survey given each possible demand level. These conditional probabilities are

Demand Level	Conditional Probability P(optimistic survey|demand)
10,000	0.05
15,000	0.10
20,000	0.30
30,000	0.50
50,000	0.80

Using this information, determine the revised or posterior probability associated with each demand level.

8. Using the posterior probabilities calculated in problem 7, what decision should the operations manager make about fixing the current machine or buying the new one?

9. When the Arkansas Milling Company has correctly set up its manufacturing process, 5 percent of the products produced are defective. When the process is set up incorrectly, 20 percent of the products are defective. Past experience indicates that the chance the process will be set up properly is 0.85.

Suppose a sample of ten items has been randomly selected and four have been found defective. Use this sample information to revise the probability that the machine is set up correctly.

10. Referring to problem 9, suppose a second sample of 15 was selected and 6 were found defective. Combine this sample information with the revised probabilities found in problem 9 to arrive at new posterior probabilities.

11. How would the posterior probabilities calculated in problem 10 differ from the posterior probabilities that would be determined if one sample of 25 was selected and 10 defectives observed? Discuss.

12. The Lubtree Corporation has just developed a drug called Bedsorine. If the drug is a "good" product, improvement will be shown by 80 percent of the users. If the drug is a "fair" product, 60 percent of the users will show improvement, and if it is a "poor" product, 30 percent will show improvement. Based upon laboratory study, the Lubtree Corporation feels that the following probabilities apply:

Quality Level	Probability
Good	0.30
Fair	0.30
Poor	0.40
	$\Sigma = 1.00$

Suppose a random sample of 25 patients are given the drug and 14 show improvement. Determine the posterior probability for each state of nature.

13. Referring to problem 12, if two more samples of 25 were selected with the results that 12 improved and 15 improved, respectively, determine the posterior probability for each state of nature.

14. With respect to problems 12 and 13, the Lubtree Corporation has a policy of not putting a drug on the market unless it has in excess of a 90 percent chance of being at least "fair." Using the posterior probabilities determined in problem 13, should Lubtree market the drug Bedsorine?

15. Discuss in your own words the concept of perfect information. If possible, provide some business examples that illustrate perfect information.

16. Discuss why the value of sample information cannot exceed the value of perfect information.

17. The King Construction Company has the opportunity to undertake a major development project in the inner city of Washington, D.C. If the project is very successful, King Construction will net $2.3 million. If the project is a failure, King Construction will lose $1.2 million. The engineering and planning departments have worked out the following probability assessments for the possible outcomes:

State of Nature	Prior Probability
Very successful ($2,300,000)	0.30
Marginally successful ($250,000)	0.30
Failure ($-$1,200,000)	0.40
	$\Sigma = 1.00$

If the King Construction Company does not take this development opportunity, it will select a second project with a sure profit of $125,000.

Based only on the prior probabilities, what decision should the King Construction Company make if it wishes to maximize expected profit?

18. Referring to problem 17, suppose it were possible to obtain perfect information about this inner-city development project. What would its value be to the King Construction Company? What is the cost to King Construction of being uncertain?

19. Referring to problems 17 and 18, if King Construction could hire consultants to provide some information about the outcomes of the proposed project, what is the absolute maximum King should be willing to pay for this information? Under what conditions would the company pay this much?

20. One of the major automobile manufacturers is considering purchasing windshields for one car model from the Acme Glass Company. Currently this auto manufacturer is buying the windshields from Windpro, Inc.

Because of the way they are installed, windshields have to be able to withstand a specific level of impact without breaking. Windpro charges $201 for each windshield. Over the past few years 5 percent of Windpro windshields have broken during installation. Glass companies do not refund for broken windshields because the price is set to reflect the possibility of breakage.

Acme proposes to charge $200 for each windshield and claims its product is more reliable than Windpro's. Based upon initial testing, the automobile manufacturer has assessed the following probability distribution for the fraction of defectives produced by Acme:

Proportion Defective	Prior Probability
0.03	0.20
0.04	0.30
0.05	0.20
0.06	0.20
0.07	0.10
	$\Sigma = 1.00$

If the auto maker wants to buy 100,000 windshields, what decision should it make using only prior information?

21. With reference to problem 20, suppose it is possible to obtain perfect information about the proportion of defective windshields produced by Acme. What would be the value of the perfect information?

22. Referring to problems 20 and 21, suppose Acme has offered a free sample of 100 windshields to the auto maker for testing. The test results in 6 windshields failing the test. Given this sample information, what is the "best" decision for the auto maker?

*23. Considering the information in problem 22, what is the value of this sample information after the sample has been selected?

24. Bogie Basin Ski Resort managers are considering whether or not to have a year-end "bash" at the main lodge for their season lift ticket customers. They have 4,000 season lift ticket holders and figure the following distribution represents those who would attend the party:

Proportion Attending	Probability
0.20	0.30
0.25	0.40
0.30	0.20
0.35	0.10
	$\Sigma = 1.00$

Cost estimates indicate a fixed setup and advertising cost of $10,000 plus a variable cost of $10 per person attending. Tickets will be sold for $20.

Using only this information, should the "bash" be held? Also determine the expected value of perfect information.

20A   QUALITY BAKERY

Quality Bakery, managed by R.D. Poteet, makes specialty breads for distribution in retail outlets in several western states. Poteet has recently read an article in a trade magazine that described a sequential quality control sampling system. The sequential sampling system involves a process whereby a specified number of loaves are tested and the numbers of good and defective loaves are recorded. Based upon the results of this sample, the bakery can decide to test more loaves in the batch or stop sampling and make an accept/reject decision regarding the batch. Depending upon the cost of sampling and the time available, the sequential sampling can take on any number of stages with any number of loaves tested at each stage until the sample information indicates that a terminal decision should be made.

Poteet has been somewhat concerned that his bakery's quality control is not what it once was. A suggestion in the journal article was to introduce the system by sampling one loaf and, if required, one more. Thus, initially the sampling system would be limited to two stages and a total of two loaves tested.

At Quality Bakery, a batch contains 400 loaves. Cost accountants have determined that each loaf costs Quality $0.10 to make, including labor and raw materials. Packaging and shipping account for an additional $0.01 per loaf. (If a batch is rejected, the loaves are not packaged or shipped.) The selling price of the loaves is $0.25 each.

If a defective loaf is sold, the cost to Quality in lost goodwill is estimated to be $1 per loaf. Based upon past experience, the following percent defectives are possible and should occur with a frequency represented by the associated probabilities:

Proportion Defective	Prior Probability
0.00	0.40
0.01	0.30
0.02	0.20
1.00	0.10
	$\Sigma = 1.00$

This probability distribution indicates that a batch will be either good, pretty good, or completely defective. This results because of the mixing process. A totally bad batch will occur if an ingredient is left out or if bad yeast is used in the mixing process.

R.D. Poteet is attempting to establish the decision rules under the two-stage sampling plan with a sample size of one loaf at each stage. He has decided to ignore sampling costs.

20B   ARTISTIC AMERICANS, INC.

Artistic Americans, Inc. is a company that has been formed to produce television specials. Artistic has just been approached with an idea for a new special. The cost of producing this special is estimated at $825,000. To finance this type of special,

Artistic must secure television advertising contracts to underwrite part or all of the cost. Because of past performance, the advertisers have formed the following financing policy:

Show Rating	Percent Financing
A	100
B	80
C	40
D	0

Thus, if the show is given an "A" rating by Troody and Good's Rating Service, Artistic will be able to contract with the advertisers for the full $825,000. (Note that the advertisers pay for the right to advertise and receive no other money if the show is a successs.) If the rating is a "B" or "C," Artistic will be forced to finance the remaining part of the production cost itself. Unfortunately the rating cannot be determined until after the show has been taped.

The television networks will accept the show for a one-time presentation if it has an "A," "B," or "C" rating by Troody and Good's, with the following payment schedule:

Show Rating	Payment
A	$500,000
B	400,000
C	150,000

Thus, if the special receives an "A" rating, it will be 100 percent financed by the advertisers, and the network will pay Artistic $500,000 for rights to a one-time showing.

Suppose Artistic's management has assessed the following probabilities with respect to the potential ratings for the latest proposed special:

Show Rating	Prior Probability
A	0.20
B	0.30
C	0.40
D	0.10
	$\Sigma = 1.00$

A. L. MacMillan is responsible for making a decision for Artistic and wishes to present the analysis in a logical way to the other members of the management team.

20C   AMERICAN NATIONAL, PART 1

American National provides consumer credit to people all over the world in the form of an American National credit card. Eligible persons receive an American National card for a small annual fee.

Each month American National encloses a "special purchase" offer in its cardholder statement. Currently American National is considering offering its customers the opportunity to purchase a set of fancy carving knives at a "reduced" price of $15. (Customers can have the amount added to their next statement.)

American National must pay $200,000 for the right to sell the knives plus $10 per set sold. Currently American National has 100,000 customers. The marketing department has assessed the following probabilities for potential demand:

Proportion Buying	Probability
0.10	0.10
0.15	0.30
0.20	0.30
0.25	0.20
0.30	0.10
	$\Sigma = 1.00$

Fred Turner, American National's executive vice-president, is currently trying to determine whether to offer the knives and also the absolute maximum he would be willing to pay for a market-research study.

*20D   AMERICAN NATIONAL, PART 2

The marketing department at American National has stated that the cost of sampling the customers to estimate the proportion who will buy the carving knives (see case 20C) is $3,000 fixed and $25 per person sampled if the work is done by American National employees.

Fred Turner wants his eventual decision to be based on as much information as possible. He has asked his executive assistant to determine whether a sample of $n = 50$ would produce information worth its cost.

*20E   AMERICAN NATIONAL, PART 3

After hearing his executive assistant's report (see case 20D), Fred Turner received a phone call from an old Army pal. This friend is now a market-research consultant with Elmers, Smith, and O'Brady of Dallas, Texas. During the call, Fred mentioned his problem with the knives (cases 20C and 20D). The consultant friend offered to randomly sample 20 of American National's customers and report back with the percentage who will buy.

Fred has agreed to call back later that day with a price he is willing to pay for the information that could be obtained from the sample of 20 customers.

*20F   ROCKSTONE INTERNATIONAL, PART 2

Charley O'Finley, Rockstone's corporate controller, happened to walk into the conference room just as R. B. Penticost had made the decision whether or not to hire

Hans Marquis (see case 19A). R. B. filled O'Finley in on the analysis he and Beth Harkness had used in arriving at the decision.

O'Finley was impressed by the methods used. However he offered the following suggestion: "Why not hire Marquis on a temporary basis and have him cut five stones? Then, after seeing these results, you could make the decision." He added, "It seems to me that this approach could reduce our cost of uncertainty.

R. B. liked the idea but wondered how this could be done. He also wondered whether the costs involved would be worth the added information.

references

BAIRD, BRUCE F. *Introduction to Decision Analysis*. North Scitaute, Mass.: Duxbury, 1978.

GREEN, PAUL. "Bayesian Decision Theory in Advertising." *Journal of Advertising Research* 2 (1962): 33–41.

HARTLEY, H. O. "In Dr. Bayes' Consulting Room." *American Statistician* 17 (1963): 22–24.

JONES, J. M. *Introduction to Decision Theory*. Homewood, Ill.: Irwin, 1977.

PARSONS, ROBERT. *Statistical Analysis*. New York: Harper and Row, 1974.

RAIFFA, HOWARD. *Decision Analysis: Introductory Lectures on Choices under Uncertainty*. Reading, Mass.: Addison-Wesley, 1968.

SCHLAIFER, ROBERT. *Analysis of Decisions under Uncertainty*. New York: McGraw-Hill, 1969.

_____. *Computer Programs for Elementary Decision Analysis*. Cambridge, Mass.: Harvard University Press, 1971.

SORENSON, J. E. "Bayesian Analysis in Auditing." *Accounting Review* 44 (1969): 555–61.

THOMPSON, G. E. *Statistics for Decisions*. Boston: Little Brown, 1972.

WINKLER, R. L. *Introduction to Bayesian Inference and Decision*. New York: Holt, Rinehart and Winston, 1972.

appendixes

607

Binomial Distribution Table

$$P(X_1) = \frac{n!}{X_1!(n - X_1)!} p^{X_1}(1 - p)^{n - X_1}$$

N = 1

X1	P=.01	P=.02	P=.03	P=.04	P=.05	P=.06	P=.07	P=.08	P=.09	P=.10	N-X1
0	.9900	.9800	.9700	.9600	.9500	.9400	.9300	.9200	.9100	.9000	1
1	.0100	.0200	.0300	.0400	.0500	.0600	.0700	.0800	.0900	.1000	0
	Q=.99	Q=.98	Q=.97	Q=.96	Q=.95	Q=.94	Q=.93	Q=.92	Q=.91	Q=.90	

X1	P=.11	P=.12	P=.13	P=.14	P=.15	P=.16	P=.17	P=.18	P=.19	P=.20	N-X1
0	.8900	.8800	.8700	.8600	.8500	.8400	.8300	.8200	.8100	.8000	1
1	.1100	.1200	.1300	.1400	.1500	.1600	.1700	.1800	.1900	.2000	0
	Q=.89	Q=.88	Q=.87	Q=.86	Q=.85	Q=.84	Q=.83	Q=.82	Q=.81	Q=.80	

X1	P=.21	P=.22	P=.23	P=.24	P=.25	P=.26	P=.27	P=.28	P=.29	P=.30	N-X1
0	.7900	.7800	.7700	.7600	.7500	.7400	.7300	.7200	.7100	.7000	1
1	.2100	.2200	.2300	.2400	.2500	.2600	.2700	.2800	.2900	.3000	0
	Q=.79	Q=.78	Q=.77	Q=.76	Q=.75	Q=.74	Q=.73	Q=.72	Q=.71	Q=.70	

X1	P=.31	P=.32	P=.33	P=.34	P=.35	P=.36	P=.37	P=.38	P=.39	P=.40	N-X1
0	.6900	.6800	.6700	.6600	.6500	.6400	.6300	.6200	.6100	.6000	1
1	.3100	.3200	.3300	.3400	.3500	.3600	.3700	.3800	.3900	.4000	0
	Q=.69	Q=.68	Q=.67	Q=.66	Q=.65	Q=.64	Q=.63	Q=.62	Q=.61	Q=.60	

X1	P=.41	P=.42	P=.43	P=.44	P=.45	P=.46	P=.47	P=.48	P=.49	P=.50	N-X1
0	.5900	.5800	.5700	.5600	.5500	.5400	.5300	.5200	.5100	.5000	1
1	.4100	.4200	.4300	.4400	.4500	.4600	.4700	.4800	.4900	.5000	0
	Q=.59	Q=.58	Q=.57	Q=.56	Q=.55	Q=.54	Q=.53	Q=.52	Q=.51	Q=.50	

N = 2

X1	P=.01	P=.02	P=.03	P=.04	P=.05	P=.06	P=.07	P=.08	P=.09	P=.10	N-X1
0	.9801	.9604	.9409	.9216	.9025	.8836	.8649	.8464	.8281	.8100	2
1	.0198	.0392	.0582	.0768	.0950	.1128	.1302	.1472	.1638	.1800	1
2	.0001	.0004	.0009	.0016	.0025	.0036	.0049	.0064	.0081	.0100	0
	Q=.99	Q=.98	Q=.97	Q=.96	Q=.95	Q=.94	Q=.93	Q=.92	Q=.91	Q=.90	

X1	P=.11	P=.12	P=.13	P=.14	P=.15	P=.16	P=.17	P=.18	P=.19	P=.20	N-X1
0	.7921	.7744	.7569	.7396	.7225	.7056	.6889	.6724	.6561	.6400	2
1	.1958	.2112	.2262	.2408	.2550	.2688	.2822	.2952	.3078	.3200	1
2	.0121	.0144	.0169	.0196	.0225	.0256	.0289	.0324	.0361	.0400	0
	Q=.89	Q=.88	Q=.87	Q=.86	Q=.85	Q=.84	Q=.83	Q=.82	Q=.81	Q=.80	

X1	P=.21	P=.22	P=.23	P=.24	P=.25	P=.26	P=.27	P=.28	P=.29	P=.30	N-X1
0	.6241	.6084	.5929	.5776	.5625	.5476	.5329	.5184	.5041	.4900	2
1	.3318	.3432	.3542	.3648	.3750	.3848	.3942	.4032	.4118	.4200	1
2	.0441	.0484	.0529	.0576	.0625	.0676	.0729	.0784	.0841	.0900	0
	Q=.79	Q=.78	Q=.77	Q=.76	Q=.75	Q=.74	Q=.73	Q=.72	Q=.71	Q=.70	

X1	P=.31	P=.32	P=.33	P=.34	P=.35	P=.36	P=.37	P=.38	P=.39	P=.40	N-X1
0	.4761	.4624	.4489	.4356	.4225	.4096	.3969	.3844	.3721	.3600	2
1	.4278	.4352	.4422	.4488	.4550	.4608	.4662	.4712	.4758	.4800	1
2	.0961	.1024	.1089	.1156	.1225	.1296	.1369	.1444	.1521	.1600	0
	Q=.69	Q=.68	Q=.67	Q=.66	Q=.65	Q=.64	Q=.63	Q=.62	Q=.61	Q=.60	

X1	P=.41	P=.42	P=.43	P=.44	P=.45	P=.46	P=.47	P=.48	P=.49	P=.50	N-X1
0	.3481	.3364	.3249	.3136	.3025	.2916	.2809	.2704	.2601	.2500	2
1	.4838	.4872	.4902	.4928	.4950	.4968	.4982	.4992	.4998	.5000	1
2	.1681	.1764	.1849	.1936	.2025	.2116	.2209	.2304	.2401	.2500	0
	Q=.59	Q=.58	Q=.57	Q=.56	Q=.55	Q=.54	Q=.53	Q=.52	Q=.51	Q=.50	

N = 3

X1	P=.01	P=.02	P=.03	P=.04	P=.05	P=.06	P=.07	P=.08	P=.09	P=.10	N-X1
0	.9703	.9412	.9127	.8847	.8574	.8306	.8044	.7787	.7536	.7290	3
1	.0294	.0576	.0847	.1106	.1354	.1590	.1816	.2031	.2236	.2430	2
2	.0003	.0012	.0026	.0046	.0071	.0102	.0137	.0177	.0221	.0270	1
3	.0000	.0000	.0000	.0001	.0001	.0002	.0003	.0005	.0007	.0010	0
	Q=.99	Q=.98	Q=.97	Q=.96	Q=.95	Q=.94	Q=.93	Q=.92	Q=.91	Q=.90	

X1	P=.11	P=.12	P=.13	P=.14	P=.15	P=.16	P=.17	P=.18	P=.19	P=.20	N-X1
0	.7050	.6815	.6585	.6361	.6141	.5927	.5718	.5514	.5314	.5120	3
1	.2614	.2788	.2952	.3106	.3251	.3387	.3513	.3631	.3740	.3840	2
2	.0323	.0380	.0441	.0506	.0574	.0645	.0720	.0797	.0877	.0960	1
3	.0013	.0017	.0022	.0027	.0034	.0041	.0049	.0058	.0069	.0080	0
	Q=.89	Q=.88	Q=.87	Q=.86	Q=.85	Q=.84	Q=.83	Q=.82	Q=.81	Q=.80	

X1	P=.21	P=.22	P=.23	P=.24	P=.25	P=.26	P=.27	P=.28	P=.29	P=.30	N-X1
0	.4930	.4746	.4565	.4390	.4219	.4052	.3890	.3732	.3579	.3430	3
1	.3932	.4015	.4091	.4159	.4219	.4271	.4316	.4355	.4386	.4410	2
2	.1045	.1133	.1222	.1313	.1406	.1501	.1597	.1693	.1791	.1890	1
3	.0093	.0106	.0122	.0138	.0156	.0176	.0197	.0220	.0244	.0270	0
	Q=.79	Q=.78	Q=.77	Q=.76	Q=.75	Q=.74	Q=.73	Q=.72	Q=.71	Q=.70	

X1	P=.31	P=.32	P=.33	P=.34	P=.35	P=.36	P=.37	P=.38	P=.39	P=.40	N-X1
0	.3285	.3144	.3008	.2875	.2746	.2621	.2500	.2383	.2270	.2160	3
1	.4428	.4439	.4444	.4443	.4436	.4424	.4406	.4382	.4354	.4320	2
2	.1989	.2089	.2189	.2289	.2389	.2488	.2587	.2686	.2783	.2880	1
3	.0298	.0328	.0359	.0393	.0429	.0467	.0507	.0549	.0593	.0640	0
	Q=.69	Q=.68	Q=.67	Q=.66	Q=.65	Q=.64	Q=.63	Q=.62	Q=.61	Q=.60	

X1	P=.41	P=.42	P=.43	P=.44	P=.45	P=.46	P=.47	P=.48	P=.49	P=.50	N-X1
0	.2054	.1951	.1852	.1756	.1664	.1575	.1489	.1406	.1327	.1250	3
1	.4282	.4239	.4191	.4140	.4084	.4024	.3961	.3894	.3823	.3750	2
2	.2975	.3069	.3162	.3252	.3341	.3428	.3512	.3594	.3674	.3750	1
3	.0689	.0741	.0795	.0852	.0911	.0973	.1038	.1106	.1176	.1250	0
	Q=.59	Q=.58	Q=.57	Q=.56	Q=.55	Q=.54	Q=.53	Q=.52	Q=.51	Q=.50	

TABLES

N = 4

X1	P=.01	P=.02	P=.03	P=.04	P=.05	P=.06	P=.07	P=.08	P=.09	P=.10	N−X1
0	.9606	.9224	.8853	.8493	.8145	.7807	.7481	.7164	.6857	.6561	4
1	.0388	.0753	.1095	.1416	.1715	.1993	.2252	.2492	.2713	.2916	3
2	.0006	.0023	.0051	.0088	.0135	.0191	.0254	.0325	.0402	.0486	2
3	.0000	.0000	.0001	.0002	.0005	.0008	.0013	.0019	.0027	.0036	1
4	.0000	.0000	.0000	.0000	.0000	.0000	.0000	.0000	.0001	.0001	0
	Q=.99	Q=.98	Q=.97	Q=.96	Q=.95	Q=.94	Q=.93	Q=.92	Q=.91	Q=.90	

X1	P=.11	P=.12	P=.13	P=.14	P=.15	P=.16	P=.17	P=.18	P=.19	P=.20	N−X1
0	.6274	.5997	.5729	.5470	.5220	.4979	.4746	.4521	.4305	.4096	4
1	.3102	.3271	.3424	.3562	.3685	.3793	.3888	.3970	.4039	.4096	3
2	.0575	.0669	.0767	.0870	.0975	.1084	.1195	.1307	.1421	.1536	2
3	.0047	.0061	.0076	.0094	.0115	.0138	.0163	.0191	.0222	.0256	1
4	.0001	.0002	.0003	.0004	.0005	.0007	.0008	.0010	.0013	.0016	0
	Q=.89	Q=.88	Q=.87	Q=.86	Q=.85	Q=.84	Q=.83	Q=.82	Q=.81	Q=.80	

X1	P=.21	P=.22	P=.23	P=.24	P=.25	P=.26	P=.27	P=.28	P=.29	P=.30	N−X1
0	.3895	.3702	.3515	.3336	.3164	.2999	.2840	.2687	.2541	.2401	4
1	.4142	.4176	.4200	.4214	.4219	.4214	.4201	.4180	.4152	.4116	3
2	.1651	.1767	.1882	.1996	.2109	.2221	.2331	.2439	.2544	.2646	2
3	.0293	.0332	.0375	.0420	.0469	.0520	.0575	.0632	.0693	.0756	1
4	.0019	.0023	.0028	.0033	.0039	.0046	.0053	.0061	.0071	.0081	0
	Q=.79	Q=.78	Q=.77	Q=.76	Q=.75	Q=.74	Q=.73	Q=.72	Q=.71	Q=.70	

X1	P=.31	P=.32	P=.33	P=.34	P=.35	P=.36	P=.37	P=.38	P=.39	P=.40	N−X1
0	.2267	.2138	.2015	.1897	.1785	.1678	.1575	.1478	.1385	.1296	4
1	.4074	.4025	.3970	.3910	.3845	.3775	.3701	.3623	.3541	.3456	3
2	.2745	.2841	.2933	.3021	.3105	.3185	.3260	.3330	.3396	.3456	2
3	.0822	.0891	.0963	.1038	.1115	.1194	.1276	.1361	.1447	.1536	1
4	.0092	.0105	.0119	.0134	.0150	.0168	.0187	.0209	.0231	.0256	0
	Q=.69	Q=.68	Q=.67	Q=.66	Q=.65	Q=.64	Q=.63	Q=.62	Q=.61	Q=.60	

X1	P=.41	P=.42	P=.43	P=.44	P=.45	P=.46	P=.47	P=.48	P=.49	P=.50	N−X1
0	.1212	.1132	.1056	.0983	.0915	.0850	.0789	.0731	.0677	.0625	4
1	.3368	.3278	.3185	.3091	.2995	.2897	.2799	.2700	.2600	.2500	3
2	.3511	.3560	.3604	.3643	.3675	.3702	.3723	.3738	.3747	.3750	2
3	.1627	.1719	.1813	.1908	.2005	.2102	.2201	.2300	.2400	.2500	1
4	.0283	.0311	.0342	.0375	.0410	.0448	.0488	.0531	.0576	.0625	0
	Q=.59	Q=.58	Q=.57	Q=.56	Q=.55	Q=.54	Q=.53	Q=.52	Q=.51	Q=.50	

N = 5

X1	P=.01	P=.02	P=.03	P=.04	P=.05	P=.06	P=.07	P=.08	P=.09	P=.10	N−X1
0	.9510	.9039	.8587	.8154	.7738	.7339	.6957	.6591	.6240	.5905	5
1	.0480	.0922	.1328	.1699	.2036	.2342	.2618	.2866	.3086	.3280	4
2	.0010	.0038	.0082	.0142	.0214	.0299	.0394	.0498	.0610	.0729	3
3	.0000	.0001	.0003	.0006	.0011	.0019	.0030	.0043	.0060	.0081	2
4	.0000	.0000	.0000	.0000	.0000	.0001	.0001	.0002	.0003	.0004	1
5	.0	.0000	.0000	.0000	.0000	.0000	.0000	.0000	.0000	.0000	0
	Q=.99	Q=.98	Q=.97	Q=.96	Q=.95	Q=.94	Q=.93	Q=.92	Q=.91	Q=.90	

X1	P=.11	P=.12	P=.13	P=.14	P=.15	P=.16	P=.17	P=.18	P=.19	P=.20	N−X1
0	.5584	.5277	.4984	.4704	.4437	.4182	.3939	.3707	.3487	.3277	5
1	.3451	.3598	.3724	.3829	.3915	.3983	.4034	.4069	.4089	.4096	4
2	.0853	.0981	.1113	.1247	.1382	.1517	.1652	.1786	.1918	.2048	3
3	.0105	.0134	.0166	.0203	.0244	.0289	.0338	.0392	.0450	.0512	2
4	.0007	.0009	.0012	.0017	.0022	.0028	.0035	.0043	.0053	.0064	1
5	.0000	.0000	.0000	.0001	.0001	.0001	.0001	.0002	.0002	.0003	0
	Q=.89	Q=.88	Q=.87	Q=.86	Q=.85	Q=.84	Q=.83	Q=.82	Q=.81	Q=.80	

X1	P=.21	P=.22	P=.23	P=.24	P=.25	P=.26	P=.27	P=.28	P=.29	P=.30	N−X1
0	.3077	.2887	.2707	.2536	.2373	.2219	.2073	.1935	.1804	.1581	5
1	.4090	.4072	.4043	.4003	.3955	.3898	.3834	.3762	.3685	.3601	4
2	.2174	.2297	.2415	.2528	.2637	.2739	.2836	.2926	.3010	.3087	3
3	.0578	.0648	.0721	.0798	.0879	.0962	.1049	.1138	.1229	.1323	2
4	.0077	.0091	.0108	.0126	.0146	.0169	.0194	.0221	.0251	.0283	1
5	.0004	.0005	.0006	.0008	.0010	.0012	.0014	.0017	.0021	.0024	0
	Q=.79	Q=.78	Q=.77	Q=.76	Q=.75	Q=.74	Q=.73	Q=.72	Q=.71	Q=.70	

X1	P=.31	P=.32	P=.33	P=.34	P=.35	P=.36	P=.37	P=.38	P=.39	P=.40	N−X1
0	.1564	.1454	.1350	.1252	.1160	.1074	.0992	.0916	.0845	.0778	5
1	.3513	.3421	.3325	.3226	.3124	.3020	.2914	.2808	.2700	.2592	4
2	.3157	.3220	.3275	.3323	.3364	.3397	.3423	.3441	.3452	.3456	3
3	.1418	.1515	.1613	.1712	.1811	.1911	.2010	.2109	.2207	.2304	2
4	.0319	.0357	.0397	.0441	.0488	.0537	.0590	.0646	.0706	.0768	1
5	.0029	.0034	.0039	.0045	.0053	.0060	.0069	.0079	.0090	.0102	0
	Q=.69	Q=.68	Q=.67	Q=.66	Q=.65	Q=.64	Q=.63	Q=.62	Q=.61	Q=.60	

X1	P=.41	P=.42	P=.43	P=.44	P=.45	P=.46	P=.47	P=.48	P=.49	P=.50	N−X1
0	.0715	.0656	.0602	.0551	.0503	.0459	.0418	.0380	.0345	.0313	5
1	.2484	.2376	.2270	.2164	.2059	.1956	.1854	.1755	.1657	.1563	4
2	.3452	.3442	.3424	.3400	.3369	.3332	.3289	.3240	.3185	.3125	3
3	.2399	.2492	.2583	.2671	.2757	.2838	.2916	.2990	.3060	.3125	2
4	.0834	.0902	.0974	.1049	.1128	.1209	.1293	.1380	.1470	.1562	1
5	.0116	.0131	.0147	.0165	.0185	.0206	.0229	.0255	.0282	.0312	0
	Q=.59	Q=.58	Q=.57	Q=.56	Q=.55	Q=.54	Q=.53	Q=.52	Q=.51	Q=.50	

N = 6

X1	P=.01	P=.02	P=.03	P=.04	P=.05	P=.06	P=.07	P=.08	P=.09	P=.10	N−X1
0	.9415	.8858	.8330	.7828	.7351	.6899	.6470	.6064	.5679	.5314	6
1	.0571	.1085	.1546	.1957	.2321	.2642	.2922	.3164	.3370	.3543	5
2	.0014	.0055	.0120	.0204	.0305	.0422	.0550	.0688	.0833	.0984	4
3	.0000	.0002	.0005	.0011	.0021	.0036	.0055	.0080	.0110	.0146	3
4	.0000	.0000	.0000	.0000	.0001	.0002	.0003	.0005	.0008	.0012	2
5	.0000	.0000	.0000	.0000	.0000	.0000	.0000	.0000	.0000	.0001	1
6	.0	.0	.0000	.0000	.0000	.0000	.0000	.0000	.0000	.0000	0
	Q=.99	Q=.98	Q=.97	Q=.96	Q=.95	Q=.94	Q=.93	Q=.92	Q=.91	Q=.90	

X1	P=.11	P=.12	P=.13	P=.14	P=.15	P=.16	P=.17	P=.18	P=.19	P=.20	N−X1
0	.4970	.4644	.4336	.4046	.3771	.3513	.3269	.3040	.2824	.2621	6
1	.3685	.3800	.3888	.3952	.3993	.4015	.4018	.4004	.3975	.3932	5
2	.1139	.1295	.1452	.1608	.1762	.1912	.2057	.2197	.2331	.2458	4
3	.0188	.0236	.0289	.0349	.0415	.0486	.0562	.0643	.0729	.0819	3
4	.0017	.0024	.0032	.0043	.0055	.0069	.0086	.0106	.0128	.0154	2
5	.0001	.0001	.0002	.0003	.0004	.0005	.0007	.0009	.0012	.0015	1
6	.0000	.0000	.0000	.0000	.0000	.0000	.0000	.0000	.0000	.0001	0
	Q=.89	Q=.88	Q=.87	Q=.86	Q=.85	Q=.84	Q=.83	Q=.82	Q=.81	Q=.80	

X1	P=.21	P=.22	P=.23	P=.24	P=.25	P=.26	P=.27	P=.28	P=.29	P=.30	N−X1
0	.2431	.2252	.2084	.1927	.1780	.1642	.1513	.1393	.1281	.1176	6
1	.3877	.3811	.3735	.3651	.3560	.3462	.3358	.3251	.3139	.3025	5
2	.2577	.2687	.2789	.2882	.2966	.3041	.3105	.3160	.3206	.3241	4
3	.0913	.1011	.1111	.1214	.1318	.1424	.1531	.1639	.1746	.1852	3
4	.0182	.0214	.0249	.0287	.0330	.0375	.0425	.0478	.0535	.0595	2
5	.0019	.0024	.0030	.0036	.0044	.0053	.0063	.0074	.0087	.0102	1
6	.0001	.0001	.0001	.0002	.0002	.0003	.0004	.0005	.0006	.0007	0
	Q=.79	Q=.78	Q=.77	Q=.76	Q=.75	Q=.74	Q=.73	Q=.72	Q=.71	Q=.70	

	P=.31	P=.32	P=.33	P=.34	P=.35	P=.36	P=.37	P=.38	P=.39	P=.40	N−X1
X1											
0	.1079	.0989	.0905	.0827	.0754	.0687	.0625	.0568	.0515	.0467	6
1	.2909	.2792	.2673	.2555	.2437	.2319	.2203	.2089	.1976	.1866	5
2	.3267	.3284	.3292	.3290	.3280	.3261	.3235	.3201	.3159	.3110	4
3	.1957	.2061	.2162	.2260	.2355	.2446	.2533	.2616	.2693	.2765	3
4	.0660	.0727	.0799	.0873	.0951	.1032	.1116	.1202	.1291	.1382	2
5	.0119	.0137	.0157	.0180	.0205	.0232	.0262	.0295	.0330	.0369	1
6	.0009	.0011	.0013	.0015	.0018	.0022	.0026	.0030	.0035	.0041	0
	Q=.69	Q=.68	Q=.67	Q=.66	Q=.65	Q=.64	Q=.63	Q=.62	Q=.61	Q=.60	

	P=.41	P=.42	P=.43	P=.44	P=.45	P=.46	P=.47	P=.48	P=.49	P=.50	N−X1
X1											
0	.0422	.0381	.0343	.0308	.0277	.0248	.0222	.0198	.0176	.0156	6
1	.1759	.1654	.1552	.1454	.1359	.1267	.1179	.1095	.1014	.0938	5
2	.3055	.2994	.2928	.2856	.2780	.2699	.2615	.2527	.2436	.2344	4
3	.2831	.2891	.2945	.2992	.3032	.3065	.3091	.3110	.3121	.3125	3
4	.1475	.1570	.1666	.1763	.1861	.1958	.2056	.2153	.2249	.2344	2
5	.0410	.0455	.0503	.0554	.0609	.0667	.0729	.0795	.0864	.0937	1
6	.0048	.0055	.0063	.0073	.0083	.0095	.0108	.0122	.0138	.0156	0
	Q=.59	Q=.58	Q=.57	Q=.56	Q=.55	Q=.54	Q=.53	Q=.52	Q=.51	Q=.50	

N =   7

	P=.01	P=.02	P=.03	P=.04	P=.05	P=.06	P=.07	P=.08	P=.09	P=.10	N−X1
X1											
0	.9321	.8681	.8080	.7514	.6983	.6485	.6017	.5578	.5168	.4783	7
1	.0659	.1240	.1749	.2192	.2573	.2897	.3170	.3396	.3578	.3720	6
2	.0020	.0076	.0162	.0274	.0406	.0555	.0716	.0836	.1061	.1240	5
3	.0000	.0003	.0008	.0019	.0036	.0059	.0090	.0128	.0175	.0230	4
4	.0000	.0000	.0000	.0001	.0002	.0004	.0007	.0011	.0017	.0026	3
5	.0000	.0000	.0000	.0000	.0000	.0000	.0000	.0001	.0001	.0002	2
6	.0	.0000	.0000	.0000	.0000	.0000	.0000	.0000	.0000	.0000	1
	Q=.99	Q=.98	Q=.97	Q=.96	Q=.95	Q=.94	Q=.93	Q=.92	Q=.91	Q=.90	

	P=.11	P=.12	P=.13	P=.14	P=.15	P=.16	P=.17	P=.18	P=.19	P=.20	N−X1
X1											
0	.4423	.4087	.3773	.3479	.3206	.2951	.2714	.2493	.2288	.2097	7
1	.3827	.3901	.3946	.3965	.3960	.3935	.3891	.3830	.3756	.3670	6
2	.1419	.1596	.1769	.1936	.2097	.2248	.2391	.2523	.2643	.2753	5
3	.0292	.0363	.0441	.0525	.0617	.0714	.0816	.0923	.1033	.1147	4
4	.0036	.0049	.0066	.0086	.0109	.0136	.0167	.0203	.0242	.0287	3
5	.0003	.0004	.0006	.0008	.0012	.0016	.0021	.0027	.0034	.0043	2
6	.0000	.0000	.0000	.0000	.0001	.0001	.0001	.0002	.0003	.0004	1
7	.0000	.0000	.0000	.0000	.0000	.0000	.0000	.0000	.0000	.0000	0
	Q=.89	Q=.88	Q=.87	Q=.86	Q=.85	Q=.84	Q=.83	Q=.82	Q=.81	Q=.80	

	P=.21	P=.22	P=.23	P=.24	P=.25	P=.26	P=.27	P=.28	P=.29	P=.30	N−X1
X1											
0	.1920	.1757	.1605	.1465	.1335	.1215	.1105	.1003	.0910	.0824	7
1	.3573	.3468	.3356	.3237	.3115	.2989	.2860	.2731	.2600	.2471	6
2	.2850	.2935	.3007	.3067	.3115	.3150	.3174	.3186	.3186	.3177	5
3	.1263	.1379	.1497	.1614	.1730	.1845	.1956	.2065	.2169	.2269	4
4	.0336	.0389	.0447	.0510	.0577	.0648	.0724	.0803	.0886	.0972	3
5	.0054	.0066	.0080	.0097	.0115	.0137	.0161	.0187	.0217	.0250	2
6	.0005	.0006	.0008	.0010	.0013	.0016	.0020	.0024	.0030	.0036	1
7	.0000	.0000	.0000	.0000	.0001	.0001	.0001	.0001	.0002	.0002	0
	Q=.79	Q=.78	Q=.77	Q=.76	Q=.75	Q=.74	Q=.73	Q=.72	Q=.71	Q=.70	

	P=.31	P=.32	P=.33	P=.34	P=.35	P=.36	P=.37	P=.38	P=.39	P=.40	N−X1
X1											
0	.0745	.0672	.0606	.0546	.0490	.0440	.0394	.0352	.0314	.0280	7
1	.2342	.2215	.2090	.1967	.1848	.1732	.1619	.1511	.1407	.1306	6
2	.3156	.3127	.3088	.3040	.2985	.2922	.2853	.2778	.2699	.2613	5
3	.2363	.2452	.2535	.2610	.2679	.2740	.2793	.2838	.2875	.2903	4
4	.1062	.1154	.1248	.1345	.1442	.1541	.1640	.1739	.1838	.1935	3
5	.0286	.0326	.0369	.0416	.0466	.0520	.0578	.0640	.0705	.0774	2
6	.0043	.0051	.0061	.0071	.0084	.0098	.0113	.0131	.0150	.0172	1
7	.0003	.0003	.0004	.0005	.0006	.0008	.0009	.0011	.0014	.0016	0
	Q=.69	Q=.68	Q=.67	Q=.66	Q=.65	Q=.64	Q=.63	Q=.62	Q=.61	Q=.60	

TABLES

X1	P=.41	P=.42	P=.43	P=.44	P=.45	P=.46	P=.47	P=.48	P=.49	P=.50	N−X1
0	.0249	.0221	.0195	.0173	.0152	.0134	.0117	.0103	.0090	.0078	7
1	.1211	.1119	.1032	.0950	.0872	.0798	.0729	.0664	.0604	.0547	6
2	.2524	.2431	.2336	.2239	.2140	.2040	.1940	.1840	.1740	.1641	5
3	.2923	.2934	.2937	.2932	.2918	.2897	.2867	.2830	.2786	.2734	4
4	.2031	.2125	.2216	.2304	.2388	.2468	.2543	.2612	.2676	.2734	3
5	.0847	.0923	.1003	.1086	.1172	.1261	.1353	.1447	.1543	.1641	2
6	.0196	.0223	.0252	.0284	.0320	.0358	.0400	.0445	.0494	.0547	1
7	.0019	.0023	.0027	.0032	.0037	.0044	.0051	.0059	.0068	.0078	0
	Q=.59	Q=.58	Q=.57	Q=.56	Q=.55	Q=.54	Q=.53	Q=.52	Q=.51	Q=.50	

N = 8

X1	P=.01	P=.02	P=.03	P=.04	P=.05	P=.06	P=.07	P=.08	P=.09	P=.10	N−X1
0	.9227	.8508	.7837	.7214	.6634	.6096	.5596	.5132	.4703	.4305	8
1	.0746	.1389	.1939	.2405	.2793	.3113	.3370	.3570	.3721	.3826	7
2	.0026	.0099	.0210	.0351	.0515	.0655	.0888	.1087	.1288	.1488	6
3	.0001	.0004	.0013	.0029	.0054	.0089	.0134	.0189	.0255	.0331	5
4	.0000	.0000	.0001	.0002	.0004	.0007	.0013	.0021	.0031	.0046	4
5	.0000	.0000	.0000	.0000	.0000	.0000	.0001	.0001	.0002	.0004	3
6	.0	.0000	.0000	.0000	.0000	.0000	.0000	.0000	.0000	.0000	2
	Q=.99	Q=.98	Q=.97	Q=.96	Q=.95	Q=.94	Q=.93	Q=.92	Q=.91	Q=.90	

X1	P=.11	P=.12	P=.13	P=.14	P=.15	P=.16	P=.17	P=.18	P=.19	P=.20	N−X1
0	.3937	.3596	.3282	.2992	.2725	.2479	.2252	.2044	.1853	.1678	8
1	.3892	.3923	.3923	.3897	.3847	.3777	.3691	.3590	.3477	.3355	7
2	.1684	.1872	.2052	.2220	.2376	.2518	.2646	.2758	.2855	.2936	6
3	.0416	.0511	.0613	.0723	.0839	.0959	.1084	.1211	.1339	.1468	5
4	.0064	.0087	.0115	.0147	.0185	.0228	.0277	.0332	.0393	.0459	4
5	.0006	.0009	.0014	.0019	.0026	.0035	.0045	.0058	.0074	.0092	3
6	.0000	.0001	.0001	.0002	.0002	.0003	.0005	.0006	.0009	.0011	2
7	.0000	.0000	.0000	.0000	.0000	.0000	.0000	.0000	.0001	.0001	1
8	.0000	.0000	.0000	.0000	.0000	.0000	.0000	.0000	.0000	.0000	0
	Q=.89	Q=.88	Q=.87	Q=.86	Q=.85	Q=.84	Q=.83	Q=.82	Q=.81	Q=.80	

X1	P=.21	P=.22	P=.23	P=.24	P=.25	P=.26	P=.27	P=.28	P=.29	P=.30	N−X1
0	.1517	.1370	.1236	.1113	.1001	.0899	.0806	.0722	.0646	.0576	8
1	.3226	.3092	.2953	.2812	.2670	.2527	.2386	.2247	.2110	.1977	7
2	.3002	.3052	.3087	.3108	.3115	.3108	.3089	.3058	.3017	.2965	6
3	.1596	.1722	.1844	.1963	.2076	.2184	.2285	.2379	.2464	.2541	5
4	.0530	.0607	.0689	.0775	.0865	.0959	.1056	.1156	.1258	.1361	4
5	.0113	.0137	.0165	.0196	.0231	.0270	.0313	.0360	.0411	.0467	3
6	.0015	.0019	.0025	.0031	.0038	.0047	.0058	.0070	.0084	.0100	2
7	.0001	.0002	.0002	.0003	.0004	.0005	.0006	.0008	.0010	.0012	1
8	.0000	.0000	.0000	.0000	.0000	.0000	.0000	.0000	.0001	.0001	0
	Q=.79	Q=.78	Q=.77	Q=.76	Q=.75	Q=.74	Q=.73	Q=.72	Q=.71	Q=.70	

X1	P=.31	P=.32	P=.33	P=.34	P=.35	P=.36	P=.37	P=.38	P=.39	P=.40	N−X1
0	.0514	.0457	.0406	.0360	.0319	.0281	.0248	.0218	.0192	.0168	8
1	.1847	.1721	.1600	.1484	.1373	.1267	.1166	.1071	.0981	.0896	7
2	.2904	.2835	.2758	.2675	.2587	.2494	.2397	.2297	.2194	.2090	6
3	.2609	.2668	.2717	.2756	.2786	.2805	.2815	.2815	.2806	.2787	5
4	.1465	.1569	.1673	.1775	.1875	.1973	.2067	.2157	.2242	.2322	4
5	.0527	.0591	.0659	.0731	.0808	.0888	.0971	.1057	.1147	.1239	3
6	.0118	.0139	.0162	.0188	.0217	.0250	.0285	.0324	.0367	.0413	2
7	.0015	.0019	.0023	.0028	.0033	.0040	.0048	.0057	.0067	.0079	1
8	.0001	.0001	.0001	.0002	.0002	.0003	.0004	.0004	.0005	.0007	0
	Q=.69	Q=.68	Q=.67	Q=.66	Q=.65	Q=.64	Q=.63	Q=.62	Q=.61	Q=.60	

X1	P=.41	P=.42	P=.43	P=.44	P=.45	P=.46	P=.47	P=.48	P=.49	P=.50	N−X1
0	.0147	.0128	.0111	.0097	.0084	.0072	.0062	.0053	.0046	.0039	8
1	.0816	.0742	.0672	.0608	.0548	.0493	.0442	.0395	.0352	.0313	7
2	.1985	.1880	.1776	.1672	.1569	.1469	.1371	.1275	.1183	.1094	6
3	.2759	.2723	.2679	.2627	.2568	.2503	.2431	.2355	.2273	.2188	5
4	.2397	.2465	.2526	.2580	.2627	.2665	.2695	.2717	.2730	.2734	4
5	.1332	.1428	.1525	.1622	.1719	.1816	.1912	.2006	.2098	.2187	3
6	.0463	.0517	.0575	.0637	.0703	.0774	.0848	.0926	.1008	.1094	2
7	.0092	.0107	.0124	.0143	.0164	.0188	.0215	.0244	.0277	.0312	1
8	.0008	.0010	.0012	.0014	.0017	.0020	.0024	.0028	.0033	.0039	0
	Q=.59	Q=.58	Q=.57	Q=.56	Q=.55	Q=.54	Q=.53	Q=.52	Q=.51	Q=.50	

N = 9

X1	P=.01	P=.02	P=.03	P=.04	P=.05	P=.06	P=.07	P=.08	P=.09	P=.10	N-X1
0	.9135	.8337	.7602	.6925	.6302	.5730	.5204	.4722	.4279	.3874	9
1	.0830	.1531	.2116	.2597	.2985	.3292	.3525	.3695	.3809	.3874	8
2	.0034	.0125	.0262	.0433	.0629	.0840	.1061	.1285	.1507	.1722	7
3	.0001	.0006	.0019	.0042	.0077	.0125	.0186	.0261	.0348	.0446	6
4	.0000	.0000	.0001	.0003	.0006	.0012	.0021	.0034	.0052	.0074	5
5	.0000	.0000	.0000	.0000	.0000	.0001	.0002	.0003	.0005	.0008	4
6	.0	.0000	.0000	.0000	.0000	.0000	.0000	.0000	.0000	.0001	3
7	.0	.0	.0000	.0000	.0000	.0000	.0000	.0000	.0000	.0000	2
	Q=.99	Q=.98	Q=.97	Q=.96	Q=.95	Q=.94	Q=.93	Q=.92	Q=.91	Q=.90	

X1	P=.11	P=.12	P=.13	P=.14	P=.15	P=.16	P=.17	P=.18	P=.19	P=.20	N-X1
0	.3504	.3165	.2855	.2573	.2316	.2082	.1869	.1676	.1501	.1342	9
1	.3897	.3884	.3840	.3770	.3679	.3569	.3446	.3312	.3169	.3020	8
2	.1927	.2119	.2295	.2455	.2597	.2720	.2823	.2908	.2973	.3020	7
3	.0556	.0674	.0800	.0933	.1069	.1209	.1349	.1489	.1627	.1762	6
4	.0103	.0138	.0179	.0228	.0283	.0345	.0415	.0490	.0573	.0661	5
5	.0013	.0019	.0027	.0037	.0050	.0066	.0085	.0108	.0134	.0165	4
6	.0001	.0002	.0003	.0004	.0006	.0008	.0012	.0016	.0021	.0028	3
7	.0000	.0000	.0000	.0000	.0000	.0001	.0001	.0001	.0002	.0003	2
8	.0000	.0000	.0000	.0000	.0000	.0000	.0000	.0000	.0000	.0000	1
	Q=.89	Q=.88	Q=.87	Q=.86	Q=.85	Q=.84	Q=.83	Q=.82	Q=.81	Q=.80	

X1	P=.21	P=.22	P=.23	P=.24	P=.25	P=.26	P=.27	P=.28	P=.29	P=.30	N-X1
0	.1199	.1069	.0952	.0846	.0751	.0665	.0589	.0520	.0458	.0404	9
1	.2867	.2713	.2558	.2404	.2253	.2104	.1960	.1820	.1685	.1556	8
2	.3049	.3061	.3056	.3037	.3003	.2957	.2899	.2831	.2754	.2668	7
3	.1891	.2014	.2130	.2238	.2336	.2424	.2502	.2569	.2624	.2668	6
4	.0754	.0852	.0954	.1060	.1168	.1278	.1388	.1499	.1608	.1715	5
5	.0200	.0240	.0285	.0335	.0389	.0449	.0513	.0583	.0657	.0735	4
6	.0036	.0045	.0057	.0070	.0087	.0105	.0127	.0151	.0179	.0210	3
7	.0004	.0005	.0007	.0010	.0012	.0015	.0020	.0025	.0031	.0039	2
8	.0000	.0000	.0001	.0001	.0001	.0001	.0002	.0002	.0003	.0004	1
9	.0000	.0000	.0000	.0000	.0000	.0000	.0000	.0000	.0000	.0000	0
	Q=.79	Q=.78	Q=.77	Q=.76	Q=.75	Q=.74	Q=.73	Q=.72	Q=.71	Q=.70	

X1	P=.31	P=.32	P=.33	P=.34	P=.35	P=.36	P=.37	P=.38	P=.39	P=.40	N-X1
0	.0355	.0311	.0272	.0238	.0207	.0180	.0156	.0135	.0117	.0101	9
1	.1433	.1317	.1206	.1102	.1004	.0912	.0826	.0747	.0673	.0605	8
2	.2576	.2478	.2376	.2270	.2162	.2052	.1941	.1831	.1721	.1612	7
3	.2701	.2721	.2731	.2729	.2716	.2693	.2660	.2618	.2567	.2508	6
4	.1820	.1921	.2017	.2109	.2194	.2272	.2344	.2407	.2462	.2508	5
5	.0818	.0904	.0994	.1086	.1181	.1278	.1376	.1475	.1574	.1672	4
6	.0245	.0284	.0326	.0373	.0424	.0479	.0539	.0603	.0671	.0743	3
7	.0047	.0057	.0069	.0082	.0098	.0116	.0136	.0158	.0184	.0212	2
8	.0005	.0007	.0008	.0011	.0013	.0016	.0020	.0024	.0029	.0035	1
9	.0000	.0000	.0000	.0001	.0001	.0001	.0001	.0002	.0002	.0003	0
	Q=.69	Q=.68	Q=.67	Q=.66	Q=.65	Q=.64	Q=.63	Q=.62	Q=.61	Q=.60	

X1	P=.41	P=.42	P=.43	P=.44	P=.45	P=.46	P=.47	P=.48	P=.49	P=.50	N-X1
0	.0087	.0074	.0064	.0054	.0046	.0039	.0033	.0028	.0023	.0020	9
1	.0542	.0484	.0431	.0383	.0339	.0299	.0263	.0231	.0202	.0176	8
2	.1506	.1402	.1301	.1204	.1110	.1020	.0934	.0853	.0776	.0703	7
3	.2442	.2369	.2291	.2207	.2119	.2027	.1933	.1837	.1739	.1641	6
4	.2545	.2573	.2592	.2601	.2600	.2590	.2571	.2543	.2506	.2461	5
5	.1769	.1863	.1955	.2044	.2128	.2207	.2280	.2347	.2408	.2461	4
6	.0819	.0900	.0983	.1070	.1160	.1253	.1348	.1445	.1542	.1641	3
7	.0244	.0279	.0318	.0360	.0407	.0458	.0512	.0571	.0635	.0703	2
8	.0042	.0051	.0060	.0071	.0083	.0097	.0114	.0132	.0153	.0176	1
9	.0003	.0004	.0005	.0006	.0008	.0009	.0011	.0014	.0016	.0020	0
	Q=.59	Q=.58	Q=.57	Q=.56	Q=.55	Q=.54	Q=.53	Q=.52	Q=.51	Q=.50	

N = 10

X1	P=.01	P=.02	P=.03	P=.04	P=.05	P=.06	P=.07	P=.08	P=.09	P=.10	N-X1
0	.9044	.8171	.7374	.6648	.5987	.5386	.4840	.4344	.3894	.3487	10
1	.0914	.1667	.2281	.2770	.3151	.3438	.3643	.3777	.3851	.3874	9
2	.0042	.0153	.0317	.0519	.0746	.0988	.1234	.1478	.1714	.1937	8
3	.0001	.0008	.0026	.0058	.0105	.0168	.0248	.0343	.0452	.0574	7
4	.0000	.0000	.0001	.0004	.0010	.0019	.0033	.0052	.0078	.0112	6
5	.0000	.0000	.0000	.0000	.0001	.0001	.0003	.0005	.0009	.0015	5
6	.0000	.0000	.0000	.0000	.0000	.0000	.0000	.0000	.0001	.0001	4
7	.0	.0000	.0000	.0000	.0000	.0000	.0000	.0000	.0000	.0000	3

| | Q=.99 | Q=.98 | Q=.97 | Q=.96 | Q=.95 | Q=.94 | Q=.93 | Q=.92 | Q=.91 | Q=.90 | |

X1	P=.11	P=.12	P=.13	P=.14	P=.15	P=.16	P=.17	P=.18	P=.19	P=.20	N-X1
0	.3118	.2785	.2484	.2213	.1969	.1749	.1552	.1374	.1216	.1074	10
1	.3854	.3798	.3712	.3603	.3474	.3331	.3178	.3017	.2852	.2684	9
2	.2143	.2330	.2496	.2639	.2759	.2856	.2929	.2980	.3010	.3020	8
3	.0706	.0847	.0995	.1146	.1298	.1450	.1600	.1745	.1883	.2013	7
4	.0153	.0202	.0260	.0326	.0401	.0483	.0573	.0670	.0773	.0881	6
5	.0023	.0033	.0047	.0064	.0085	.0111	.0141	.0177	.0218	.0264	5
6	.0002	.0004	.0006	.0009	.0012	.0018	.0024	.0032	.0043	.0055	4
7	.0000	.0000	.0000	.0001	.0001	.0002	.0003	.0004	.0006	.0008	3
8	.0000	.0000	.0000	.0000	.0000	.0000	.0000	.0000	.0000	.0001	2
9	.0000	.0000	.0000	.0000	.0000	.0000	.0000	.0000	.0000	.0000	1

| | Q=.89 | Q=.88 | Q=.87 | Q=.86 | Q=.85 | Q=.84 | Q=.83 | Q=.82 | Q=.81 | Q=.80 | |

X1	P=.21	P=.22	P=.23	P=.24	P=.25	P=.26	P=.27	P=.28	P=.29	P=.30	N-X1
0	.0947	.0834	.0733	.0643	.0563	.0492	.0430	.0374	.0326	.0282	10
1	.2517	.2351	.2188	.2030	.1877	.1730	.1590	.1456	.1330	.1211	9
2	.3011	.2984	.2942	.2885	.2816	.2735	.2646	.2548	.2444	.2335	8
3	.2134	.2244	.2343	.2429	.2503	.2563	.2609	.2642	.2662	.2668	7
4	.0993	.1108	.1225	.1343	.1460	.1576	.1689	.1798	.1903	.2001	6
5	.0317	.0375	.0439	.0509	.0584	.0664	.0750	.0839	.0933	.1029	5
6	.0070	.0088	.0109	.0134	.0162	.0195	.0231	.0272	.0317	.0368	4
7	.0011	.0014	.0019	.0024	.0031	.0039	.0049	.0060	.0074	.0090	3
8	.0001	.0002	.0002	.0003	.0004	.0005	.0007	.0009	.0011	.0014	2
9	.0000	.0000	.0000	.0000	.0000	.0000	.0001	.0001	.0001	.0001	1
10	.0000	.0000	.0000	.0000	.0000	.0000	.0000	.0000	.0000	.0000	0

| | Q=.79 | Q=.78 | Q=.77 | Q=.76 | Q=.75 | Q=.74 | Q=.73 | Q=.72 | Q=.71 | Q=.70 | |

X1	P=.31	P=.32	P=.33	P=.34	P=.35	P=.36	P=.37	P=.38	P=.39	P=.40	N-X1
0	.0245	.0211	.0182	.0157	.0135	.0115	.0098	.0084	.0071	.0060	10
1	.1099	.0995	.0898	.0808	.0725	.0649	.0578	.0514	.0456	.0403	9
2	.2222	.2107	.1990	.1873	.1757	.1642	.1529	.1419	.1312	.1209	8
3	.2662	.2644	.2614	.2573	.2522	.2462	.2394	.2319	.2237	.2150	7
4	.2093	.2177	.2253	.2320	.2377	.2424	.2461	.2487	.2503	.2508	6
5	.1128	.1229	.1332	.1434	.1536	.1636	.1734	.1829	.1920	.2007	5
6	.0422	.0482	.0547	.0616	.0689	.0767	.0849	.0934	.1023	.1115	4
7	.0108	.0130	.0154	.0181	.0212	.0247	.0285	.0327	.0374	.0425	3
8	.0018	.0023	.0028	.0035	.0043	.0052	.0063	.0075	.0090	.0106	2
9	.0002	.0002	.0003	.0004	.0005	.0006	.0008	.0010	.0013	.0016	1
10	.0000	.0000	.0000	.0000	.0000	.0000	.0000	.0001	.0001	.0001	0

| | Q=.69 | Q=.68 | Q=.67 | Q=.66 | Q=.65 | Q=.64 | Q=.63 | Q=.62 | Q=.61 | Q=.60 | |

X1	P=.41	P=.42	P=.43	P=.44	P=.45	P=.46	P=.47	P=.48	P=.49	P=.50	N-X1
0	.0051	.0043	.0036	.0030	.0025	.0021	.0017	.0014	.0012	.0010	10
1	.0355	.0312	.0273	.0238	.0207	.0180	.0155	.0133	.0114	.0098	9
2	.1111	.1017	.0927	.0843	.0763	.0688	.0619	.0554	.0495	.0439	8
3	.2058	.1963	.1865	.1765	.1665	.1564	.1464	.1364	.1267	.1172	7
4	.2503	.2488	.2462	.2427	.2384	.2331	.2271	.2204	.2130	.2051	6
5	.2087	.2162	.2229	.2289	.2340	.2383	.2417	.2441	.2456	.2461	5
6	.1209	.1304	.1401	.1499	.1596	.1692	.1786	.1878	.1966	.2051	4
7	.0480	.0540	.0604	.0673	.0746	.0824	.0905	.0991	.1080	.1172	3
8	.0125	.0147	.0171	.0198	.0229	.0263	.0301	.0343	.0389	.0439	2
9	.0019	.0024	.0029	.0035	.0042	.0050	.0059	.0070	.0083	.0098	1
10	.0001	.0002	.0002	.0003	.0003	.0004	.0005	.0006	.0008	.0010	0

| | Q=.59 | Q=.58 | Q=.57 | Q=.56 | Q=.55 | Q=.54 | Q=.53 | Q=.52 | Q=.51 | Q=.50 | |

N = 11

X1	P=.01	P=.02	P=.03	P=.04	P=.05	P=.06	P=.07	P=.08	P=.09	P=.10	N-X1
0	.8953	.8007	.7153	.6382	.5688	.5063	.4501	.3996	.3544	.3138	11
1	.0995	.1798	.2433	.2925	.3293	.3555	.3727	.3823	.3855	.3835	10
2	.0050	.0183	.0376	.0609	.0867	.1135	.1403	.1662	.1906	.2131	9
3	.0002	.0011	.0035	.0076	.0137	.0217	.0317	.0434	.0566	.0710	8
4	.0000	.0000	.0002	.0006	.0014	.0028	.0048	.0075	.0112	.0158	7
5	.0000	.0000	.0000	.0000	.0001	.0002	.0005	.0009	.0015	.0025	6
6	.0000	.0000	.0000	.0000	.0000	.0000	.0000	.0001	.0002	.0003	5
7	.0	.0000	.0000	.0000	.0000	.0000	.0000	.0000	.0000	.0000	4

| | Q=.99 | Q=.98 | Q=.97 | Q=.96 | Q=.95 | Q=.94 | Q=.93 | Q=.92 | Q=.91 | Q=.90 | |

X1	P=.11	P=.12	P=.13	P=.14	P=.15	P=.16	P=.17	P=.18	P=.19	P=.20	N-X1
0	.2775	.2451	.2161	.1903	.1673	.1469	.1288	.1127	.0985	.0859	11
1	.3773	.3676	.3552	.3408	.3248	.3078	.2901	.2721	.2541	.2362	10
2	.2332	.2507	.2654	.2774	.2866	.2932	.2971	.2987	.2980	.2953	9
3	.0865	.1025	.1190	.1355	.1517	.1675	.1826	.1967	.2097	.2215	8
4	.0214	.0280	.0356	.0441	.0536	.0638	.0748	.0864	.0984	.1107	7
5	.0037	.0053	.0074	.0101	.0132	.0170	.0214	.0265	.0323	.0388	6
6	.0005	.0007	.0011	.0016	.0023	.0032	.0044	.0058	.0076	.0097	5
7	.0000	.0001	.0001	.0002	.0003	.0004	.0006	.0009	.0013	.0017	4
8	.0000	.0000	.0000	.0000	.0000	.0000	.0001	.0001	.0001	.0002	3
9	.0000	.0000	.0000	.0000	.0000	.0000	.0000	.0000	.0000	.0000	2

| | Q=.89 | Q=.88 | Q=.87 | Q=.86 | Q=.85 | Q=.84 | Q=.83 | Q=.82 | Q=.81 | Q=.80 | |

X1	P=.21	P=.22	P=.23	P=.24	P=.25	P=.26	P=.27	P=.28	P=.29	P=.30	N-X1
0	.0748	.0650	.0564	.0489	.0422	.0364	.0314	.0270	.0231	.0198	11
1	.2187	.2017	.1854	.1697	.1549	.1408	.1276	.1153	.1038	.0932	10
2	.2907	.2845	.2768	.2680	.2581	.2474	.2360	.2242	.2121	.1998	9
3	.2318	.2407	.2481	.2539	.2581	.2608	.2619	.2616	.2599	.2568	8
4	.1232	.1358	.1482	.1603	.1721	.1832	.1937	.2035	.2123	.2201	7
5	.0459	.0536	.0620	.0709	.0803	.0901	.1003	.1108	.1214	.1321	6
6	.0122	.0151	.0185	.0224	.0268	.0317	.0371	.0431	.0496	.0566	5
7	.0023	.0030	.0039	.0050	.0064	.0079	.0098	.0120	.0145	.0173	4
8	.0003	.0004	.0006	.0008	.0011	.0014	.0018	.0023	.0030	.0037	3
9	.0300	.0000	.0001	.0001	.0001	.0002	.0002	.0003	.0004	.0005	2
10	.0000	.0000	.0000	.0000	.0000	.0000	.0000	.0000	.0000	.0000	1

| | Q=.79 | Q=.78 | Q=.77 | Q=.76 | Q=.75 | Q=.74 | Q=.73 | Q=.72 | Q=.71 | Q=.70 | |

X1	P=.31	P=.32	P=.33	P=.34	P=.35	P=.36	P=.37	P=.38	P=.39	P=.40	N-X1
0	.0169	.0144	.0122	.0104	.0088	.0074	.0062	.0052	.0044	.0036	11
1	.0834	.0744	.0662	.0587	.0518	.0457	.0401	.0351	.0306	.0266	10
2	.1874	.1751	.1630	.1511	.1395	.1284	.1177	.1075	.0978	.0887	9
3	.2526	.2472	.2408	.2335	.2254	.2167	.2074	.1977	.1876	.1774	8
4	.2269	.2326	.2372	.2406	.2428	.2438	.2436	.2423	.2399	.2365	7
5	.1427	.1533	.1636	.1735	.1830	.1920	.2003	.2079	.2148	.2207	6
6	.0641	.0721	.0806	.0894	.0985	.1080	.1176	.1274	.1373	.1471	5
7	.0206	.0242	.0283	.0329	.0379	.0434	.0493	.0558	.0627	.0701	4
8	.0046	.0057	.0070	.0085	.0102	.0122	.0145	.0171	.0200	.0234	3
9	.0007	.0009	.0011	.0015	.0018	.0023	.0028	.0035	.0043	.0052	2
10	.0001	.0001	.0001	.0001	.0002	.0003	.0003	.0004	.0005	.0007	1
11	.0000	.0000	.0000	.0000	.0000	.0000	.0000	.0000	.0000	.0000	0

| | Q=.69 | Q=.68 | Q=.67 | Q=.66 | Q=.65 | Q=.64 | Q=.63 | Q=.62 | Q=.61 | Q=.60 | |

X1	P=.41	P=.42	P=.43	P=.44	P=.45	P=.46	P=.47	P=.48	P=.49	P=.50	N-X1
0	.0030	.0025	.0021	.0017	.0014	.0011	.0009	.0008	.0006	.0005	11
1	.0231	.0199	.0171	.0147	.0125	.0107	.0090	.0076	.0064	.0054	10
2	.0801	.0721	.0646	.0577	.0513	.0454	.0401	.0352	.0308	.0269	9
3	.1670	.1566	.1462	.1359	.1259	.1161	.1067	.0976	.0888	.0806	8
4	.2321	.2267	.2206	.2136	.2060	.1978	.1892	.1801	.1707	.1611	7
5	.2258	.2299	.2329	.2350	.2360	.2359	.2348	.2327	.2296	.2256	6
6	.1569	.1664	.1757	.1846	.1931	.2010	.2083	.2148	.2206	.2256	5
7	.0779	.0861	.0947	.1036	.1128	.1223	.1319	.1416	.1514	.1611	4
8	.0271	.0312	.0357	.0407	.0462	.0521	.0585	.0654	.0727	.0806	3
9	.0063	.0075	.0090	.0107	.0126	.0148	.0173	.0201	.0233	.0269	2
10	.0009	.0011	.0014	.0017	.0021	.0025	.0031	.0037	.0045	.0054	1
11	.0001	.0001	.0001	.0001	.0002	.0002	.0002	.0003	.0004	.0005	0

| | Q=.59 | Q=.58 | Q=.57 | Q=.56 | Q=.55 | Q=.54 | Q=.53 | Q=.52 | Q=.51 | Q=.50 | |

TABLES

N = 12

X1	P=.01	P=.02	P=.03	P=.04	P=.05	P=.06	P=.07	P=.08	P=.09	P=.10	N-X1
0	.8864	.7847	.6938	.6127	.5404	.4759	.4186	.3677	.3225	.2824	12
1	.1074	.1922	.2575	.3064	.3413	.3645	.3781	.3837	.3827	.3766	11
2	.0060	.0216	.0438	.0702	.0988	.1280	.1565	.1835	.2082	.2301	10
3	.0002	.0015	.0045	.0098	.0173	.0272	.0393	.0532	.0686	.0852	9
4	.0000	.0001	.0003	.0009	.0021	.0039	.0067	.0104	.0153	.0213	8
5	.0000	.0000	.0000	.0001	.0002	.0004	.0008	.0014	.0024	.0038	7
6	.0000	.0000	.0000	.0000	.0000	.0000	.0001	.0001	.0003	.0005	6
7	.0	.0000	.0000	.0000	.0000	.0000	.0000	.0000	.0000	.0000	5
	Q=.99	Q=.98	Q=.97	Q=.96	Q=.95	Q=.94	Q=.93	Q=.92	Q=.91	Q=.90	

X1	P=.11	P=.12	P=.13	P=.14	P=.15	P=.16	P=.17	P=.18	P=.19	P=.20	N-X1
0	.2470	.2157	.1880	.1637	.1422	.1234	.1069	.0924	.0798	.0687	12
1	.3663	.3529	.3372	.3197	.3012	.2821	.2627	.2434	.2245	.2062	11
2	.2490	.2647	.2771	.2863	.2924	.2955	.2960	.2939	.2897	.2835	10
3	.1026	.1203	.1380	.1553	.1720	.1876	.2021	.2151	.2265	.2362	9
4	.0285	.0369	.0464	.0569	.0683	.0804	.0931	.1062	.1195	.1329	8
5	.0056	.0081	.0111	.0148	.0193	.0245	.0305	.0373	.0449	.0531	7
6	.0008	.0013	.0019	.0028	.0040	.0054	.0073	.0096	.0123	.0155	6
7	.0001	.0001	.0002	.0004	.0006	.0009	.0013	.0018	.0025	.0033	5
8	.0000	.0000	.0000	.0000	.0001	.0001	.0002	.0002	.0004	.0005	4
9	.0000	.0000	.0000	.0000	.0000	.0000	.0000	.0000	.0000	.0001	3
10	.0000	.0000	.0000	.0000	.0000	.0000	.0000	.0000	.0000	.0000	2
	Q=.89	Q=.88	Q=.87	Q=.86	Q=.85	Q=.84	Q=.83	Q=.82	Q=.81	Q=.80	

X1	P=.21	P=.22	P=.23	P=.24	P=.25	P=.26	P=.27	P=.28	P=.29	P=.30	N-X1
0	.0591	.0507	.0434	.0371	.0317	.0270	.0229	.0194	.0164	.0138	12
1	.1885	.1717	.1557	.1407	.1267	.1137	.1016	.0905	.0804	.0712	11
2	.2756	.2663	.2558	.2444	.2323	.2197	.2068	.1937	.1807	.1678	10
3	.2442	.2503	.2547	.2573	.2581	.2573	.2549	.2511	.2460	.2397	9
4	.1460	.1589	.1712	.1828	.1936	.2034	.2121	.2197	.2261	.2311	8
5	.0621	.0717	.0818	.0924	.1032	.1143	.1255	.1367	.1477	.1585	7
6	.0193	.0236	.0285	.0340	.0401	.0469	.0542	.0620	.0704	.0792	6
7	.0044	.0057	.0073	.0092	.0115	.0141	.0172	.0207	.0246	.0291	5
8	.0007	.0010	.0014	.0018	.0024	.0031	.0040	.0050	.0063	.0078	4
9	.0001	.0001	.0002	.0003	.0004	.0005	.0007	.0009	.0011	.0015	3
10	.0000	.0000	.0000	.0000	.0000	.0001	.0001	.0001	.0001	.0002	2
11	.0000	.0000	.0000	.0000	.0000	.0000	.0000	.0000	.0000	.0000	1
	Q=.79	Q=.78	Q=.77	Q=.76	Q=.75	Q=.74	Q=.73	Q=.72	Q=.71	Q=.70	

X1	P=.31	P=.32	P=.33	P=.34	P=.35	P=.36	P=.37	P=.38	P=.39	P=.40	N-X1
0	.0116	.0098	.0082	.0068	.0057	.0047	.0039	.0032	.0027	.0022	12
1	.0628	.0552	.0484	.0422	.0368	.0319	.0276	.0237	.0204	.0174	11
2	.1552	.1429	.1310	.1197	.1088	.0986	.0890	.0800	.0716	.0639	10
3	.2324	.2241	.2151	.2055	.1954	.1849	.1742	.1634	.1526	.1419	9
4	.2349	.2373	.2384	.2382	.2367	.2340	.2302	.2254	.2195	.2128	8
5	.1688	.1787	.1879	.1963	.2039	.2106	.2163	.2210	.2246	.2270	7
6	.0885	.0981	.1079	.1180	.1281	.1382	.1482	.1580	.1675	.1766	6
7	.0341	.0396	.0456	.0521	.0591	.0666	.0746	.0830	.0918	.1009	5
8	.0096	.0116	.0140	.0168	.0199	.0234	.0274	.0318	.0367	.0420	4
9	.0019	.0024	.0031	.0038	.0048	.0059	.0071	.0087	.0104	.0125	3
10	.0003	.0003	.0005	.0006	.0008	.0010	.0013	.0016	.0020	.0025	2
11	.0000	.0000	.0000	.0001	.0001	.0001	.0001	.0002	.0002	.0003	1
12	.0000	.0000	.0000	.0000	.0000	.0000	.0000	.0000	.0000	.0000	0
	Q=.69	Q=.68	Q=.67	Q=.66	Q=.65	Q=.64	Q=.63	Q=.62	Q=.61	Q=.60	

X1	P=.41	P=.42	P=.43	P=.44	P=.45	P=.46	P=.47	P=.48	P=.49	P=.50	N-X1
0	.0018	.0014	.0012	.0010	.0008	.0006	.0005	.0004	.0003	.0002	12
1	.0148	.0126	.0106	.0090	.0075	.0063	.0052	.0043	.0036	.0029	11
2	.0567	.0502	.0442	.0388	.0339	.0294	.0255	.0220	.0189	.0161	10
3	.1314	.1211	.1111	.1015	.0923	.0836	.0754	.0676	.0604	.0537	9
4	.2054	.1973	.1886	.1794	.1700	.1602	.1504	.1405	.1306	.1208	8
5	.2284	.2285	.2276	.2256	.2225	.2184	.2134	.2075	.2008	.1934	7
6	.1851	.1931	.2003	.2068	.2124	.2171	.2208	.2234	.2250	.2256	6
7	.1103	.1198	.1295	.1393	.1489	.1585	.1678	.1768	.1853	.1934	5
8	.0479	.0542	.0611	.0684	.0762	.0844	.0930	.1020	.1113	.1208	4
9	.0148	.0175	.0205	.0239	.0277	.0319	.0367	.0418	.0475	.0537	3
10	.0031	.0038	.0046	.0056	.0068	.0082	.0098	.0116	.0137	.0161	2
11	.0004	.0005	.0006	.0008	.0010	.0013	.0016	.0019	.0024	.0029	1
12	.0000	.0000	.0000	.0001	.0001	.0001	.0001	.0001	.0002	.0002	0
	Q=.59	Q=.58	Q=.57	Q=.56	Q=.55	Q=.54	Q=.53	Q=.52	Q=.51	Q=.50	

N = 13

X1	P=.01	P=.02	P=.03	P=.04	P=.05	P=.06	P=.07	P=.08	P=.09	P=.10	N-X1
0	.8775	.7690	.6730	.5882	.5133	.4474	.3893	.3383	.2935	.2542	13
1	.1152	.2040	.2706	.3186	.3512	.3712	.3809	.3824	.3773	.3672	12
2	.0070	.0250	.0502	.0797	.1109	.1422	.1720	.1995	.2239	.2448	11
3	.0003	.0019	.0057	.0122	.0214	.0333	.0475	.0636	.0812	.0997	10
4	.0000	.0001	.0004	.0013	.0028	.0053	.0089	.0138	.0201	.0277	9
5	.0000	.0000	.0000	.0001	.0003	.0006	.0012	.0022	.0036	.0055	8
6	.0000	.0000	.0000	.0000	.0000	.0001	.0001	.0003	.0005	.0008	7
7	.0	.0000	.0000	.0000	.0000	.0000	.0000	.0000	.0000	.0001	6
8	.0	.0	.0000	.0000	.0000	.0000	.0000	.0000	.0000	.0000	5

| | Q=.99 | Q=.98 | Q=.97 | Q=.96 | Q=.95 | Q=.94 | Q=.93 | Q=.92 | Q=.91 | Q=.90 | |

X1	P=.11	P=.12	P=.13	P=.14	P=.15	P=.16	P=.17	P=.18	P=.19	P=.20	N-X1
0	.2198	.1898	.1636	.1408	.1209	.1037	.0887	.0758	.0646	.0550	13
1	.3532	.3364	.3178	.2979	.2774	.2567	.2362	.2163	.1970	.1787	12
2	.2619	.2753	.2849	.2910	.2937	.2934	.2903	.2848	.2773	.2680	11
3	.1187	.1376	.1561	.1737	.1900	.2049	.2180	.2293	.2385	.2457	10
4	.0367	.0469	.0583	.0707	.0838	.0976	.1116	.1258	.1399	.1535	9
5	.0082	.0115	.0157	.0207	.0266	.0335	.0412	.0497	.0591	.0691	8
6	.0013	.0021	.0031	.0045	.0063	.0085	.0112	.0145	.0185	.0230	7
7	.0002	.0003	.0005	.0007	.0011	.0016	.0023	.0032	.0043	.0058	6
8	.0000	.0000	.0001	.0001	.0001	.0002	.0004	.0005	.0008	.0011	5
9	.0000	.0000	.0000	.0000	.0000	.0000	.0000	.0001	.0001	.0001	4
10	.0000	.0000	.0000	.0000	.0000	.0000	.0000	.0000	.0000	.0000	3

| | Q=.89 | Q=.88 | Q=.87 | Q=.86 | Q=.85 | Q=.84 | Q=.83 | Q=.82 | Q=.81 | Q=.80 | |

X1	P=.21	P=.22	P=.23	P=.24	P=.25	P=.26	P=.27	P=.28	P=.29	P=.30	N-X1
0	.0467	.0396	.0334	.0282	.0238	.0200	.0167	.0140	.0117	.0097	13
1	.1613	.1450	.1299	.1159	.1029	.0911	.0804	.0706	.0619	.0540	12
2	.2573	.2455	.2328	.2195	.2059	.1921	.1784	.1648	.1516	.1388	11
3	.2508	.2539	.2550	.2542	.2517	.2475	.2419	.2351	.2271	.2181	10
4	.1667	.1790	.1904	.2007	.2097	.2174	.2237	.2285	.2319	.2337	9
5	.0797	.0909	.1024	.1141	.1258	.1375	.1489	.1600	.1705	.1803	8
6	.0283	.0342	.0408	.0480	.0559	.0644	.0734	.0829	.0928	.1030	7
7	.0075	.0096	.0122	.0152	.0186	.0226	.0272	.0323	.0379	.0442	6
8	.0015	.0020	.0027	.0036	.0047	.0060	.0075	.0094	.0116	.0142	5
9	.0002	.0003	.0005	.0006	.0009	.0012	.0015	.0020	.0026	.0034	4
10	.0000	.0000	.0001	.0001	.0001	.0002	.0002	.0003	.0004	.0006	3
11	.0000	.0000	.0000	.0000	.0000	.0000	.0000	.0000	.0000	.0001	2
12	.0000	.0000	.0000	.0000	.0000	.0000	.0000	.0000	.0000	.0000	1

| | Q=.79 | Q=.78 | Q=.77 | Q=.76 | Q=.75 | Q=.74 | Q=.73 | Q=.72 | Q=.71 | Q=.70 | |

X1	P=.31	P=.32	P=.33	P=.34	P=.35	P=.36	P=.37	P=.38	P=.39	P=.40	N-X1
0	.0080	.0066	.0055	.0045	.0037	.0030	.0025	.0020	.0016	.0013	13
1	.0469	.0407	.0351	.0302	.0259	.0221	.0188	.0159	.0135	.0113	12
2	.1265	.1148	.1037	.0933	.0836	.0746	.0663	.0586	.0516	.0453	11
3	.2084	.1981	.1874	.1763	.1651	.1538	.1427	.1317	.1210	.1107	10
4	.2341	.2331	.2307	.2270	.2222	.2163	.2095	.2018	.1934	.1845	9
5	.1893	.1974	.2045	.2105	.2154	.2190	.2215	.2227	.2226	.2214	8
6	.1134	.1239	.1343	.1446	.1546	.1643	.1734	.1820	.1898	.1968	7
7	.0509	.0583	.0662	.0745	.0833	.0924	.1019	.1115	.1213	.1312	6
8	.0172	.0206	.0244	.0288	.0336	.0390	.0449	.0513	.0582	.0656	5
9	.0043	.0054	.0067	.0082	.0101	.0122	.0146	.0175	.0207	.0243	4
10	.0008	.0010	.0013	.0017	.0022	.0027	.0034	.0043	.0053	.0065	3
11	.0001	.0001	.0002	.0002	.0003	.0004	.0006	.0007	.0009	.0012	2
12	.0000	.0000	.0000	.0000	.0000	.0000	.0001	.0001	.0001	.0001	1
13	.0000	.0000	.0000	.0000	.0000	.0000	.0000	.0000	.0000	.0000	0

| | Q=.69 | Q=.68 | Q=.67 | Q=.66 | Q=.65 | Q=.64 | Q=.63 | Q=.62 | Q=.61 | Q=.60 | |

X1	P=.41	P=.42	P=.43	P=.44	P=.45	P=.46	P=.47	P=.48	P=.49	P=.50	N-X1
0	.0010	.0008	.0007	.0005	.0004	.0003	.0003	.0002	.0002	.0001	13
1	.0095	.0079	.0066	.0054	.0045	.0037	.0030	.0024	.0020	.0016	12
2	.0395	.0344	.0298	.0256	.0220	.0188	.0160	.0135	.0114	.0095	11
3	.1007	.0913	.0823	.0739	.0660	.0587	.0519	.0457	.0401	.0349	10
4	.1750	.1653	.1553	.1451	.1350	.1250	.1151	.1055	.0962	.0873	9
5	.2189	.2154	.2108	.2053	.1989	.1917	.1838	.1753	.1664	.1571	8
6	.2029	.2080	.2121	.2151	.2169	.2177	.2173	.2158	.2131	.2095	7
7	.1410	.1506	.1600	.1690	.1775	.1854	.1927	.1992	.2048	.2095	6
8	.0735	.0818	.0905	.0996	.1089	.1185	.1282	.1379	.1476	.1571	5

										N−X1
9	.0284	.0329	.0379	.0435	.0495	.0561	.0631	.0707	.0788	.0873	4
10	.0079	.0095	.0114	.0137	.0162	.0191	.0224	.0261	.0303	.0349	3
11	.0015	.0019	.0024	.0029	.0036	.0044	.0054	.0066	.0079	.0095	2
12	.0002	.0002	.0003	.0004	.0005	.0006	.0008	.0010	.0013	.0016	1
13	.0000	.0000	.0000	.0000	.0000	.0000	.0001	.0001	.0001	.0001	0

Q=.59 Q=.58 Q=.57 Q=.56 Q=.55 Q=.54 Q=.53 Q=.52 Q=.51 Q=.50

N = 14

X1	P=.01	P=.02	P=.03	P=.04	P=.05	P=.06	P=.07	P=.08	P=.09	P=.10	N−X1
0	.8687	.7536	.6528	.5647	.4877	.4205	.3620	.3112	.2670	.2288	14
1	.1229	.2153	.2827	.3294	.3593	.3758	.3815	.3788	.3697	.3559	13
2	.0081	.0286	.0568	.0892	.1229	.1559	.1867	.2141	.2377	.2570	12
3	.0003	.0023	.0070	.0149	.0259	.0398	.0562	.0745	.0940	.1142	11
4	.0000	.0001	.0006	.0017	.0037	.0070	.0116	.0178	.0256	.0349	10
5	.0000	.0000	.0000	.0001	.0004	.0009	.0018	.0031	.0051	.0078	9
6	.0000	.0000	.0000	.0000	.0000	.0001	.0002	.0004	.0008	.0013	8
7	.0	.0000	.0000	.0000	.0000	.0000	.0000	.0000	.0001	.0002	7
8	.0	.0	.0000	.0000	.0000	.0000	.0000	.0000	.0000	.0000	6

Q=.99 Q=.98 Q=.97 Q=.96 Q=.95 Q=.94 Q=.93 Q=.92 Q=.91 Q=.90

X1	P=.11	P=.12	P=.13	P=.14	P=.15	P=.16	P=.17	P=.18	P=.19	P=.20	N−X1
0	.1956	.1670	.1423	.1211	.1028	.0871	.0736	.0621	.0523	.0440	14
1	.3385	.3188	.2977	.2759	.2539	.2322	.2112	.1910	.1719	.1539	13
2	.2720	.2826	.2892	.2919	.2912	.2875	.2811	.2725	.2620	.2501	12
3	.1345	.1542	.1728	.1901	.2056	.2190	.2303	.2393	.2459	.2501	11
4	.0457	.0578	.0710	.0851	.0998	.1147	.1297	.1444	.1586	.1720	10
5	.0113	.0158	.0212	.0277	.0352	.0437	.0531	.0634	.0744	.0860	9
6	.0021	.0032	.0048	.0068	.0093	.0125	.0163	.0209	.0262	.0322	8
7	.0003	.0005	.0008	.0013	.0019	.0027	.0038	.0052	.0070	.0092	7
8	.0000	.0001	.0001	.0002	.0003	.0005	.0007	.0010	.0014	.0020	6
9	.0000	.0000	.0000	.0000	.0000	.0001	.0001	.0001	.0002	.0003	5
10	.0000	.0000	.0000	.0000	.0000	.0000	.0000	.0000	.0000	.0000	4

Q=.89 Q=.88 Q=.87 Q=.86 Q=.85 Q=.84 Q=.83 Q=.82 Q=.81 Q=.80

X1	P=.21	P=.22	P=.23	P=.24	P=.25	P=.26	P=.27	P=.28	P=.29	P=.30	N−X1
0	.0369	.0309	.0258	.0214	.0178	.0148	.0122	.0101	.0083	.0068	14
1	.1372	.1218	.1077	.0948	.0832	.0726	.0632	.0548	.0473	.0407	13
2	.2371	.2234	.2091	.1946	.1802	.1659	.1519	.1385	.1256	.1134	12
3	.2521	.2520	.2499	.2459	.2402	.2331	.2248	.2154	.2052	.1943	11
4	.1843	.1955	.2052	.2135	.2202	.2252	.2286	.2304	.2305	.2290	10
5	.0980	.1103	.1226	.1348	.1468	.1583	.1691	.1792	.1883	.1963	9
6	.0391	.0466	.0549	.0639	.0734	.0834	.0938	.1045	.1153	.1262	8
7	.0119	.0150	.0188	.0231	.0280	.0335	.0397	.0464	.0538	.0618	7
8	.0028	.0037	.0049	.0064	.0082	.0103	.0128	.0158	.0192	.0232	6
9	.0005	.0007	.0010	.0013	.0018	.0024	.0032	.0041	.0052	.0066	5
10	.0001	.0001	.0001	.0002	.0003	.0004	.0006	.0008	.0011	.0014	4
11	.0000	.0000	.0000	.0000	.0000	.0001	.0001	.0001	.0002	.0002	3
12	.0000	.0000	.0000	.0000	.0000	.0000	.0000	.0000	.0000	.0000	2

Q=.79 Q=.78 Q=.77 Q=.76 Q=.75 Q=.74 Q=.73 Q=.72 Q=.71 Q=.70

X1	P=.31	P=.32	P=.33	P=.34	P=.35	P=.36	P=.37	P=.38	P=.39	P=.40	N−X1
0	.0055	.0045	.0037	.0030	.0024	.0019	.0016	.0012	.0010	.0008	14
1	.0349	.0298	.0253	.0215	.0181	.0152	.0128	.0106	.0088	.0073	13
2	.1018	.0911	.0811	.0719	.0634	.0557	.0487	.0424	.0367	.0317	12
3	.1830	.1715	.1598	.1481	.1366	.1253	.1144	.1039	.0940	.0845	11
4	.2261	.2219	.2164	.2098	.2022	.1938	.1848	.1752	.1652	.1549	10
5	.2032	.2088	.2132	.2161	.2178	.2181	.2170	.2147	.2112	.2066	9
6	.1369	.1474	.1575	.1670	.1759	.1840	.1912	.1974	.2026	.2066	8
7	.0703	.0793	.0886	.0983	.1082	.1183	.1283	.1383	.1480	.1574	7
8	.0276	.0326	.0392	.0443	.0510	.0582	.0659	.0742	.0828	.0918	6
9	.0083	.0102	.0125	.0152	.0183	.0218	.0258	.0303	.0353	.0408	5
10	.0019	.0024	.0031	.0039	.0049	.0061	.0076	.0093	.0113	.0136	4
11	.0003	.0004	.0006	.0007	.0010	.0013	.0016	.0021	.0026	.0033	3
12	.0000	.0000	.0001	.0001	.0001	.0002	.0002	.0003	.0004	.0005	2
13	.0000	.0000	.0000	.0000	.0000	.0000	.0000	.0000	.0000	.0001	1
14	.0000	.0000	.0000	.0000	.0000	.0000	.0000	.0000	.0000	.0000	0

Q=.69 Q=.68 Q=.67 Q=.66 Q=.65 Q=.64 Q=.63 Q=.62 Q=.61 Q=.60

X1	P=.41	P=.42	P=.43	P=.44	P=.45	P=.46	P=.47	P=.48	P=.49	P=.50	N-X1
0	.0006	.0005	.0004	.0003	.0002	.0002	.0001	.0001	.0001	.0001	14
1	.0060	.0049	.0040	.0033	.0027	.0021	.0017	.0014	.0011	.0009	13
2	.0272	.0233	.0198	.0168	.0141	.0118	.0099	.0082	.0068	.0056	12
3	.0757	.0674	.0597	.0527	.0462	.0403	.0350	.0303	.0260	.0222	11
4	.1446	.1342	.1239	.1138	.1040	.0545	.0854	.0768	.0687	.0611	10
5	.2009	.1943	.1869	.1788	.1701	.1610	.1515	.1418	.1320	.1222	9
6	.2094	.2111	.2115	.21C8	.2088	.2057	.2015	.1963	.1902	.1833	8
7	.1663	.1747	.1824	.1892	.1952	.2003	.2042	.2071	.2089	.2095	7
8	.1011	.1107	.1204	.1301	.1398	.1493	.1585	.1673	.1756	.1833	6
9	.0469	.0534	.0605	.0682	.0762	.C848	.0937	.1030	.1125	.1222	5
10	.0163	.0193	.0228	.C063	.0312	.0361	.0415	.0475	.0540	.0611	4
11	.0041	.0051	.0063	.C063	.0076	.0093	.0024	.0112	.0134	.0159	.0189	.0222	3
12	.0007	.0009	.0012	.0015	.0019	.0002	.0003	.0004	.0005	.0007	.0009	1
13	.0001	.0001	.0001	.0002	.0002	.C000	.0000	.0000	.0000	.0000	0
14	.0000	.0000	.0000	.0000	.0000	.C000	.0000	.0000	.0000	.0001	

| | Q=.59 | Q=.58 | Q=.57 | Q=.56 | Q=.55 | Q=.54 | Q=.53 | Q=.52 | Q=.51 | Q=.50 | |

N = 15

X1	P=.01	P=.02	P=.03	P=.04	P=.05	P=.06	P=.07	P=.08	P=.09	P=.10	N-X1
0	.8601	.7386	.6333	.5421	.4633	.3953	.3367	.2863	.2430	.2059	15
1	.1303	.2261	.2938	.3388	.3658	.3785	.3801	.3734	.3605	.3432	14
2	.0092	.0323	.C636	.0988	.1348	.1691	.2003	.2273	.2496	.2669	13
3	.0004	.0029	.0085	.0178	.0307	.0468	.0653	.0857	.1070	.1285	12
4	.0000	.0002	.0008	.0022	.0049	.0090	.0148	.0223	.0317	.0428	11
5	.0000	.0000	.0001	.0002	.0006	.0013	.0024	.0043	.0069	.0105	10
6	.0000	.0000	.0000	.0000	.0000	.0001	.0003	.0006	.0011	.0019	9
7	.0	.0000	.0000	.0000	.0000	.0000	.0000	.0001	.0001	.0003	8
8	.0	.0000	.0000	.0000	.0000	.0000	.0000	.0000	.0000	.0000	7

| | Q=.99 | Q=.98 | Q=.97 | Q=.96 | Q=.95 | Q=.94 | Q=.93 | Q=.92 | Q=.91 | Q=.90 | |

X1	P=.11	P=.12	P=.13	P=.14	P=.15	P=.16	P=.17	P=.18	P=.19	P=.20	N-X1
0	.1741	.1470	.1238	.1041	.0874	.0731	.0611	.0510	.0424	.0352	15
1	.3228	.3006	.2775	.2542	.2312	.2090	.1878	.1678	.1492	.1319	14
2	.2793	.2870	.2903	.2897	.2856	.2787	.2692	.2578	.2449	.2309	13
3	.1496	.1696	.1880	.1880	.2044	.2184	.2300	.2389	.2452	.2489	.2501	12
4	.0555	.0694	.0843	.0998	.1156	.1314	.1468	.1615	.1752	.1876	11
5	.0151	.0208	.0277	.0357	.0449	.0551	.0662	.0780	.0904	.1032	10
6	.0031	.0047	.0069	.0097	.0132	.0175	.0043	.0226	.0059	.0285	.0081	.0353	.0107	.0430	.0138	9
7	.0005	.0008	.0013	.0020	.0030	.0043	.0012	.0018	.0025	.0035	8
8	.0001	.0001	.0002	.0003	.0005	.0008	.0002	.0003	.0005	.0007	7
9	.0000	.0000	.0000	.0000	.0001	.0001	.0002	.0000	.0001	.0001	6
10	.0000	.0000	.0000	.0000	.0000	.0000	.0000	.0000	.0000	.0000	5
11	.0000	.0000	.0000	.0000	.0000	.0000	.0000	.0000	.0000	.0000	4

| | Q=.89 | Q=.88 | Q=.87 | Q=.86 | Q=.85 | Q=.84 | Q=.83 | Q=.82 | Q=.81 | Q=.80 | |

X1	P=.21	P=.22	P=.23	P=.24	P=.25	P=.26	P=.27	P=.28	P=.29	P=.30	N-X1
0	.0291	.0241	.0198	.0163	.0134	.0109	.0089	.0072	.0059	.0047	15
1	.1162	.1018	.C889	.0772	.0668	.0576	.0494	.0423	.0360	.0305	14
2	.2162	.2010	.1858	.17C7	.1559	.1416	.1280	.1150	.1029	.0916	13
3	.2490	.2457	.2405	.2336	.2252	.2156	.2051	.1939	.1821	.1700	12
4	.1986	.2079	.2155	.2213	.2252	.2273	.2276	.2262	.2231	.2186	11
5	.1161	.1290	.1416	.1537	.1651	.1757	.1852	.1935	.2005	.2061	10
6	.0514	.0606	.0705	.0809	.0917	.1029	.1142	.1254	.1365	.1472	9
7	.0176	.0220	.0271	.0328	.0393	.0465	.0543	.0627	.0717	.0811	8
8	.0047	.0062	.0081	.0104	.0131	.0163	.0201	.0244	.0293	.0348	7
9	.0010	.0014	.0019	.0025	.0034	.0045	.0058	.0074	.0093	.0116	6
10	.0002	.0002	.0003	.0005	.0007	.0009	.0013	.0017	.0023	.0030	5
11	.0000	.0000	.0000	.0001	.0001	.0002	.0002	.0003	.0004	.0006	4
12	.0000	.0000	.0000	.0000	.0000	.0000	.0000	.0000	.0001	.0001	3
13	.0000	.0000	.0000	.0000	.0000	.0000	.0000	.0000	.0000	.0000	2

| | Q=.79 | Q=.78 | Q=.77 | Q=.76 | Q=.75 | Q=.74 | Q=.73 | Q=.72 | Q=.71 | Q=.70 | |

X1	P=.31	P=.32	P=.33	P=.34	P=.35	P=.36	P=.37	P=.38	P=.39	P=.40	N-X1
0	.0038	.0031	.0025	.0020	.0016	.0012	.0010	.0008	.0006	.0005	15
1	.0258	.0217	.0182	.0152	.0126	.0104	.0086	.0071	.0058	.0047	14
2	.0811	.0715	.0627	.0547	.0476	.0411	.0354	.0303	.0259	.0219	13
3	.1579	.1457	.1338	.1222	.1110	.1CC2	.0901	.0805	.0716	.0634	12
4	.2128	.2057	.1977	.1888	.1792	.1692	.1587	.1481	.1374	.1268	11
5	.2103	.2130	.2142	.2140	.2123	.2093	.2051	.1997	.1933	.1859	10

X1											N−X1
6	.1575	.1671	.1759	.1837	.1906	.1963	.2008	.2040	.2059	.2066	9
7	.0910	.1011	.1114	.1217	.1319	.1419	.1516	.1608	.1693	.1771	8
8	.0409	.0476	.0549	.0627	.0710	.0798	.0890	.0985	.1082	.1181	7
9	.0143	.0174	.0210	.0251	.0258	.0349	.0407	.0470	.0538	.0612	6
10	.0038	.0049	.0062	.0078	.0096	.0118	.0143	.0173	.0206	.0245	5
11	.0008	.0011	.0014	.0018	.0024	.0030	.0038	.0048	.0060	.0074	4
12	.0001	.0002	.0002	.0003	.0004	.0006	.0007	.0010	.0013	.0016	3
13	.0000	.0000	.0000	.0000	.0001	.0001	.0001	.0001	.0002	.0003	2
14	.0000	.0000	.0000	.0000	.0000	.0000	.0000	.0000	.0000	.0000	1
	Q=.69	Q=.68	C=.67	Q=.66	Q=.65	Q=.64	Q=.63	Q=.62	Q=.61	Q=.60	

X1	P=.41	P=.42	P=.43	P=.44	P=.45	P=.46	P=.47	P=.48	P=.49	P=.50	N−X1
0	.0004	.0003	.0002	.0002	.0001	.0001	.0001	.0001	.0000	.0000	15
1	.0038	.0031	.0025	.0020	.0016	.0012	.0010	.0008	.0006	.0005	14
2	.0185	.0156	.0130	.0108	.0090	.0074	.0060	.0049	.0040	.0032	13
3	.0558	.0489	.0426	.0369	.0318	.0272	.0232	.0197	.0166	.0139	12
4	.1163	.1061	.0963	.0869	.0780	.0696	.0617	.0545	.0478	.0417	11
5	.1778	.1691	.1598	.1502	.1404	.1304	.1204	.1106	.1010	.0916	10
6	.2060	.2041	.2009	.1967	.1914	.1851	.1780	.1702	.1617	.1527	9
7	.1840	.1900	.1949	.1987	.2013	.2028	.2030	.2020	.1997	.1964	8
8	.1279	.1376	.1470	.1561	.1647	.1727	.1800	.1864	.1919	.1964	7
9	.0691	.0775	.0863	.0954	.1048	.1144	.1241	.1338	.1434	.1527	6
10	.0288	.0337	.0390	.0450	.0515	.0585	.0661	.0741	.0827	.0916	5
11	.0091	.0111	.0134	.0161	.0191	.0226	.0266	.0311	.0361	.0417	4
12	.0021	.0027	.0034	.0042	.0052	.0064	.0079	.0096	.0116	.0139	3
13	.0003	.0004	.0006	.0008	.0010	.0013	.0016	.0020	.0026	.0032	2
14	.0000	.0000	.0001	.0001	.0001	.0002	.0002	.0003	.0004	.0005	1
15	.0000	.0000	.0000	.0000	.0000	.0000	.0000	.0000	.0000	.0000	0
	Q=.59	Q=.58	Q=.57	Q=.56	Q=.55	Q=.54	Q=.53	Q=.52	Q=.51	Q=.50	

N = 16

X1	P=.01	P=.02	P=.03	P=.04	P=.05	P=.06	P=.07	P=.08	P=.09	P=.10	N−X1
0	.8515	.7238	.6143	.5204	.4401	.3716	.3131	.2634	.2211	.1853	16
1	.1376	.2363	.3040	.3469	.3706	.3795	.3771	.3665	.3499	.3294	15
2	.0104	.0362	.0705	.1084	.1463	.1817	.2129	.2390	.2596	.2745	14
3	.0005	.0034	.0102	.0211	.0359	.0541	.0748	.0970	.1198	.1423	13
4	.0000	.0002	.0010	.0029	.0061	.0112	.0183	.0274	.0385	.0514	12
5	.0000	.0000	.0001	.0003	.0008	.0017	.0033	.0057	.0091	.0137	11
6	.0000	.0000	.0000	.0000	.0001	.0002	.0005	.0009	.0017	.0028	10
7	.0000	.0000	.0000	.0000	.0000	.0000	.0000	.0001	.0002	.0004	9
8	.0	.0000	.0000	.0000	.0000	.0000	.0000	.0000	.0000	.0001	8
9	.0	.0	.0000	.0000	.0000	.0000	.0000	.0000	.0000	.0000	7
	Q=.99	Q=.98	Q=.97	Q=.96	Q=.95	Q=.94	Q=.93	Q=.92	Q=.91	Q=.90	

X1	P=.11	P=.12	P=.13	P=.14	P=.15	P=.16	P=.17	P=.18	P=.19	P=.20	N−X1
0	.1550	.1293	.1077	.0895	.0743	.0614	.0507	.0418	.0343	.0281	16
1	.3065	.2822	.2575	.2332	.2096	.1873	.1662	.1468	.1289	.1126	15
2	.2841	.2886	.2886	.2847	.2775	.2675	.2554	.2416	.2267	.2111	14
3	.1638	.1837	.2013	.2163	.2285	.2378	.2441	.2475	.2482	.2463	13
4	.0658	.0814	.0977	.1144	.1311	.1472	.1625	.1766	.1892	.2001	12
5	.0195	.0266	.0351	.0447	.0555	.0673	.0799	.0930	.1065	.1201	11
6	.0044	.0067	.0096	.0133	.0180	.0235	.0300	.0374	.0458	.0550	10
7	.0008	.0013	.0020	.0031	.0045	.0064	.0088	.0117	.0153	.0197	9
8	.0001	.0002	.0003	.0006	.0009	.0014	.0020	.0029	.0041	.0055	8
9	.0000	.0000	.0000	.0001	.0001	.0002	.0004	.0005	.0008	.0012	7
10	.0000	.0000	.0000	.0000	.0000	.0000	.0001	.0001	.0001	.0002	6
11	.0000	.0000	.0000	.0000	.0000	.0000	.0000	.0000	.0001	.0002	5
	Q=.89	Q=.88	Q=.87	Q=.86	Q=.85	Q=.84	Q=.83	Q=.82	Q=.81	Q=.80	

X1	P=.21	P=.22	P=.23	P=.24	P=.25	P=.26	P=.27	P=.28	P=.29	P=.30	N−X1
0	.0230	.0188	.0153	.0124	.0100	.0081	.0065	.0052	.0042	.0033	16
1	.0979	.0847	.0730	.0626	.0535	.0455	.0385	.0325	.0273	.0228	15
2	.1952	.1792	.1635	.1482	.1336	.1198	.1068	.0947	.0835	.0732	14
3	.2421	.2359	.2279	.2185	.2079	.1964	.1843	.1718	.1591	.1465	13
4	.2092	.2162	.2212	.2242	.2252	.2243	.2215	.2171	.2112	.2040	12
5	.1334	.1464	.1586	.1699	.1802	.1891	.1966	.2026	.2071	.2099	11
6	.0650	.0757	.0869	.0984	.1101	.1218	.1333	.1445	.1551	.1649	10
7	.0247	.0305	.0371	.0444	.0524	.0611	.0704	.0803	.0905	.1010	9
8	.0074	.0097	.0125	.0158	.0197	.0242	.0293	.0351	.0416	.0487	8
9	.0017	.0024	.0033	.0044	.0058	.0075	.0096	.0121	.0151	.0185	7
10	.0003	.0005	.0007	.0010	.0014	.0019	.0025	.0033	.0043	.0056	6
11	.0000	.0001	.0001	.0002	.0002	.0004	.0005	.0007	.0010	.0013	5
12	.0000	.0000	.0000	.0000	.0000	.0001	.0001	.0001	.0002	.0002	4
13	.0000	.0000	.0000	.0000	.0000	.0000	.0000	.0000	.0000	.0000	3
	Q=.79	Q=.78	C=.77	Q=.76	Q=.75	Q=.74	Q=.73	Q=.72	Q=.71	Q=.70	

X1	P=.31	P=.32	P=.33	P=.34	P=.35	P=.36	P=.37	P=.38	P=.39	P=.40	N-X1
0	.0026	.0021	.0016	.0013	.0010	.0008	.0006	.0005	.0004	.0003	16
1	.0190	.0157	.0130	.0107	.0087	.0071	.0058	.0047	.0038	.0030	15
2	.0639	.0555	.0480	.0413	.0353	.0301	.0255	.0215	.0180	.0150	14
3	.1341	.1220	.1103	.0992	.0888	.0790	.0699	.0615	.0538	.0468	13
4	.1958	.1865	.1766	.1662	.1553	.1444	.1333	.1224	.1118	.1014	12
5	.2111	.2107	.2088	.2054	.2008	.1949	.1879	.1801	.1715	.1623	11
6	.1739	.1818	.1885	.1940	.1982	.2010	.2024	.2024	.2010	.1983	10
7	.1116	.1222	.1326	.1428	.1524	.1615	.1698	.1772	.1836	.1889	9
8	.0564	.0647	.0735	.0827	.0923	.1022	.1122	.1222	.1320	.1417	8
9	.0225	.0271	.0322	.0379	.0442	.0511	.0586	.0666	.0750	.0840	7
10	.0071	.0089	.0111	.0137	.0167	.0201	.0241	.0286	.0336	.0392	6
11	.0017	.0023	.0030	.0038	.0049	.0062	.0077	.0095	.0117	.0142	5
12	.0003	.0004	.0006	.0008	.0011	.0014	.0019	.0024	.0031	.0040	4
13	.0000	.0001	.0001	.0001	.0002	.0003	.0003	.0005	.0006	.0008	3
14	.0000	.0000	.0000	.0000	.0000	.0000	.0000	.0001	.0001	.0001	2
15	.0000	.0000	.0000	.0000	.0000	.0000	.0000	.0000	.0000	.0000	1
	Q=.69	Q=.68	Q=.67	Q=.66	Q=.65	Q=.64	Q=.63	Q=.62	Q=.61	Q=.60	

X1	P=.41	P=.42	P=.43	P=.44	P=.45	P=.46	P=.47	P=.48	P=.49	P=.50	N-X1
0	.0002	.0002	.0001	.0001	.0001	.0001	.0000	.0000	.0000	.0000	16
1	.0024	.0019	.0015	.0012	.0009	.0007	.0005	.0004	.0003	.0002	15
2	.0125	.0103	.0085	.0069	.0056	.0046	.0037	.0029	.0023	.0018	14
3	.0405	.0349	.0299	.0254	.0215	.0181	.0151	.0126	.0104	.0085	13
4	.0915	.0821	.0732	.0649	.0572	.0501	.0436	.0378	.0325	.0278	12
5	.1526	.1426	.1325	.1224	.1123	.1024	.0929	.0837	.0749	.0667	11
6	.1944	.1894	.1833	.1762	.1684	.1600	.1510	.1416	.1319	.1222	10
7	.1930	.1959	.1975	.1978	.1969	.1947	.1912	.1867	.1811	.1746	9
8	.1509	.1596	.1676	.1749	.1812	.1865	.1908	.1939	.1958	.1964	8
9	.0932	.1027	.1124	.1221	.1318	.1413	.1504	.1591	.1672	.1746	7
10	.0453	.0521	.0554	.0672	.0755	.0842	.0934	.1028	.1124	.1222	6
11	.0172	.0206	.0244	.0288	.0337	.0391	.0452	.0518	.0589	.0666	5
12	.0050	.0062	.0077	.0094	.0115	.0139	.0167	.0199	.0236	.0278	4
13	.0011	.0014	.0018	.0023	.0029	.0036	.0046	.0057	.0070	.0085	3
14	.0002	.0002	.0003	.0004	.0005	.0007	.0009	.0011	.0014	.0018	2
15	.0000	.0000	.0000	.0000	.0001	.0001	.0001	.0001	.0002	.0002	1
16	.0000	.0000	.0000	.0000	.0000	.0000	.0000	.0000	.0000	.0000	0
	Q=.59	Q=.58	Q=.57	Q=.56	Q=.55	Q=.54	Q=.53	Q=.52	Q=.51	Q=.50	

N = 17

X1	P=.01	P=.02	P=.03	P=.04	P=.05	P=.06	P=.07	P=.08	P=.09	P=.10	N-X1
0	.8429	.7093	.5958	.4996	.4181	.3493	.2912	.2423	.2012	.1668	17
1	.1447	.2461	.3133	.3539	.3741	.3790	.3726	.3582	.3383	.3150	16
2	.0117	.0402	.0775	.1180	.1575	.1935	.2244	.2492	.2677	.2800	15
3	.0006	.0041	.0120	.0246	.0415	.0618	.0844	.1083	.1324	.1556	14
4	.0000	.0003	.0013	.0036	.0076	.0138	.0222	.0330	.0458	.0605	13
5	.0000	.0000	.0001	.0004	.0010	.0023	.0044	.0075	.0118	.0175	12
6	.0000	.0000	.0000	.0000	.0001	.0003	.0007	.0013	.0023	.0039	11
7	.0000	.0000	.0000	.0000	.0000	.0000	.0001	.0002	.0004	.0007	10
8	.0	.0000	.0000	.0000	.0000	.0000	.0000	.0000	.0000	.0001	9
9	.0	.0	.0000	.0000	.0000	.0000	.0000	.0000	.0000	.0000	8
	Q=.99	Q=.98	Q=.97	Q=.96	Q=.95	Q=.94	Q=.93	Q=.92	Q=.91	Q=.90	

X1	P=.11	P=.12	P=.13	P=.14	P=.15	P=.16	P=.17	P=.18	P=.19	P=.20	N-X1
0	.1379	.1138	.0937	.0770	.0631	.0516	.0421	.0343	.0278	.0225	17
1	.2808	.2638	.2381	.2131	.1893	.1671	.1466	.1279	.1109	.0957	16
2	.2865	.2878	.2846	.2775	.2673	.2547	.2402	.2245	.2081	.1914	15
3	.1771	.1962	.2126	.2259	.2359	.2425	.2460	.2464	.2441	.2393	14
4	.0766	.0937	.1112	.1287	.1457	.1617	.1764	.1893	.2004	.2093	13
5	.0246	.0332	.0432	.0545	.0668	.0801	.0939	.1081	.1222	.1361	12
6	.0061	.0091	.0129	.0177	.0236	.0305	.0385	.0474	.0573	.0680	11
7	.0012	.0019	.0030	.0045	.0065	.0091	.0124	.0164	.0211	.0267	10
8	.0002	.0003	.0006	.0009	.0014	.0022	.0032	.0045	.0062	.0084	9
9	.0000	.0000	.0001	.0002	.0003	.0004	.0006	.0010	.0015	.0021	8
10	.0000	.0000	.0000	.0000	.0000	.0001	.0001	.0002	.0003	.0004	7
11	.0000	.0000	.0000	.0000	.0000	.0000	.0000	.0000	.0000	.0001	6
12	.0000	.0000	.0000	.0000	.0000	.0000	.0000	.0000	.0000	.0000	5
	Q=.89	Q=.88	Q=.87	Q=.86	Q=.85	Q=.84	Q=.83	Q=.82	Q=.81	Q=.80	

X1	P=.21	P=.22	P=.23	P=.24	P=.25	P=.26	P=.27	P=.28	P=.29	P=.30	N-X1
0	.0182	.0146	.0118	.0094	.0075	.0060	.0047	.0038	.0030	.0023	17
1	.0822	.0702	.0597	.0505	.0426	.0357	.0299	.0248	.0206	.0169	16
2	.1747	.1584	.1427	.1277	.1136	.1005	.0883	.0772	.0672	.0581	15
3	.2322	.2234	.2131	.2016	.1893	.1765	.1634	.1502	.1372	.1245	14
4	.2161	.2205	.2228	.2228	.2209	.2170	.2115	.2044	.1961	.1868	13
5	.1493	.1617	.1730	.1830	.1914	.1982	.2033	.2067	.2083	.2081	12

X1										N-X1
6	.0794	.0912	.1034	.1156	.1276	.1393	.1504	.1608	.1701	.1784	11
7	.0332	.0404	.0485	.0573	.0668	.0769	.0874	.0982	.1092	.1201	10
8	.0110	.0143	.0181	.0226	.0279	.0338	.0404	.0478	.0558	.0644	9
9	.0029	.0040	.0054	.0071	.0093	.0119	.0149	.0186	.0228	.0276	8
10	.0006	.0009	.0013	.0018	.0025	.0033	.0044	.0058	.0074	.0095	7
11	.0001	.0002	.0002	.0004	.0005	.0007	.0010	.0014	.0019	.0026	6
12	.0000	.0000	.0000	.0001	.0001	.0001	.0002	.0003	.0004	.0006	5
13	.0000	.0000	.0000	.0000	.0000	.0000	.0002	.0000	.0001	.0001	4
14	.0000	.0000	.0000	.0000	.0000	.0000	.0000	.0000	.0000	.0000	3
15	.0000	.0000	.0000	.0000	.0000	.0000	.0000	.0000	.0000	.0000	2
16	.0000	.0000	.0000	.0000	.0000	.0000	.0000	.0000	.0000	.0000	1
17	.0	.0	.0	.0	.0	.0000	.0000	.0000	.0000	.0000	0

Q=.79 Q=.78 Q=.77 Q=.76 Q=.75 Q=.74 Q=.73 Q=.72 Q=.71 Q=.70

P=	.31	.32	.33	.34	.35	.36	.37	.38	.39	.40	N-X1
X1											
0	.0018	.0014	.0011	.0009	.0007	.0005	.0004	.0003	.0002	.0002	17
1	.0139	.0114	.0093	.0075	.0060	.0048	.0039	.0031	.0024	.0019	16
2	.0500	.0428	.0364	.0309	.0260	.0218	.0182	.0151	.0125	.0102	15
3	.1123	.1007	.0898	.0795	.0701	.0614	.0534	.0463	.0398	.0341	14
4	.1766	.1659	.1547	.1434	.1320	.1208	.1099	.0993	.0892	.0796	13
5	.2063	.2030	.1982	.1921	.1849	.1767	.1677	.1582	.1482	.1379	12
6	.1854	.1910	.1952	.1979	.1991	.1988	.1970	.1939	.1895	.1839	11
7	.1309	.1413	.1511	.1602	.1685	.1757	.1818	.1868	.1904	.1927	10
8	.0735	.0831	.0930	.1032	.1134	.1235	.1335	.1431	.1521	.1606	9
9	.0330	.0391	.0458	.0531	.0611	.0695	.0784	.0877	.0973	.1070	8
10	.0119	.0147	.0181	.0219	.0263	.0313	.0368	.0430	.0498	.0571	7
11	.0034	.0044	.0057	.0072	.0090	.0112	.0138	.0168	.0202	.0242	6
12	.0008	.0010	.0014	.0018	.0024	.0031	.0040	.0051	.0065	.0081	5
13	.0001	.0002	.0003	.0004	.0005	.0007	.0009	.0012	.0016	.0021	4
14	.0000	.0000	.0000	.0001	.0001	.0001	.0002	.0002	.0003	.0004	3
15	.0000	.0000	.0000	.0000	.0000	.0000	.0000	.0000	.0000	.0001	2
16	.0000	.0000	.0000	.0000	.0000	.0000	.0000	.0000	.0000	.0000	1

Q=.69 Q=.68 Q=.67 Q=.66 Q=.65 Q=.64 Q=.63 Q=.62 Q=.61 Q=.60

P=	.41	.42	.43	.44	.45	.46	.47	.48	.49	.50	N-X1
X1											
0	.0001	.0001	.0001	.0001	.0000	.0000	.0000	.0000	.0000	.0000	17
1	.0015	.0012	.0009	.0007	.0005	.0004	.0003	.0002	.0002	.0000	16
2	.0084	.0068	.0055	.0044	.0035	.0028	.0022	.0017	.0013	.0001	15
3	.0290	.0246	.0207	.0173	.0144	.0119	.0097	.0079	.0064	.0052	14
4	.0706	.0623	.0546	.0475	.0411	.0354	.0302	.0257	.0217	.0182	13
5	.1276	.1172	.1070	.0971	.0875	.0784	.0697	.0616	.0541	.0472	12
6	.1773	.1697	.1614	.1525	.1432	.1335	.1237	.1138	.1040	.0944	11
7	.1936	.1932	.1914	.1883	.1841	.1787	.1723	.1650	.1570	.1484	10
8	.1682	.1748	.1805	.1850	.1883	.1903	.1910	.1904	.1886	.1855	9
9	.1169	.1266	.1361	.1453	.1540	.1621	.1694	.1758	.1812	.1855	8
10	.0650	.0733	.0822	.0914	.1008	.1105	.1202	.1298	.1393	.1484	7
11	.0287	.0338	.0394	.0457	.0525	.0599	.0678	.0763	.0851	.0944	6
12	.0100	.0122	.0149	.0179	.0215	.0255	.0301	.0352	.0409	.0472	5
13	.0027	.0034	.0043	.0054	.0068	.0084	.0103	.0125	.0151	.0182	4
14	.0005	.0007	.0009	.0012	.0016	.0020	.0026	.0033	.0041	.0052	3
15	.0001	.0001	.0001	.0002	.0003	.0003	.0005	.0006	.0008	.0010	2
16	.0000	.0000	.0000	.0000	.0000	.0000	.0001	.0001	.0001	.0001	1
17	.0000	.0000	.0000	.0000	.0000	.0000	.0000	.0001	.0000	.0000	0

Q=.59 Q=.58 Q=.57 Q=.56 Q=.55 Q=.54 Q=.53 Q=.52 Q=.51 Q=.50

N = 18

P=	.01	.02	.03	.04	.05	.06	.07	.08	.09	.10	N-X1
X1											
0	.8345	.6951	.5780	.4796	.3972	.3283	.2708	.2229	.1831	.1501	18
1	.1517	.2554	.3217	.3597	.3763	.3772	.3669	.3489	.3260	.3002	17
2	.0130	.0443	.0846	.1274	.1683	.2047	.2348	.2579	.2741	.2835	16
3	.0007	.0048	.0140	.0283	.0473	.0697	.0942	.1196	.1446	.1680	15
4	.0000	.0004	.0016	.0044	.0093	.0167	.0266	.0390	.0536	.0700	14
5	.0000	.0000	.0001	.0005	.0014	.0030	.0056	.0095	.0148	.0218	13
6	.0000	.0000	.0000	.0000	.0002	.0004	.0009	.0018	.0032	.0052	12
7	.0000	.0000	.0000	.0000	.0000	.0000	.0001	.0003	.0005	.0010	11
8	.0	.0000	.0000	.0000	.0000	.0000	.0000	.0000	.0001	.0002	10
9	.0	.0	.0000	.0000	.0000	.0000	.0000	.0000	.0000	.0000	9

Q=.99 Q=.98 Q=.97 Q=.96 Q=.95 Q=.94 Q=.93 Q=.92 Q=.91 Q=.90

P=	.11	.12	.13	.14	.15	.16	.17	.18	.19	.20	N-X1
X1											
0	.1227	.1002	.0815	.0662	.0536	.0434	.0349	.0281	.0225	.0180	18
1	.2731	.2458	.2193	.1940	.1704	.1486	.1288	.1110	.0951	.0811	17
2	.2869	.2850	.2785	.2685	.2556	.2407	.2243	.2071	.1897	.1723	16
3	.1891	.2072	.2220	.2331	.2406	.2445	.2450	.2425	.2373	.2297	15
4	.0876	.1060	.1244	.1423	.1592	.1746	.1882	.1996	.2087	.2153	14
5	.0303	.0405	.0520	.0649	.0787	.0931	.1079	.1227	.1371	.1507	13

6	.0081	.0120	.0168	.0229	.0301	.0384	.0479	.0584	.0697	.0816	12
7	.0017	.0028	.0043	.0064	.0091	.0126	.0168	.0220	.0280	.0350	11
8	.0003	.0005	.0009	.0014	.0022	.0033	.0047	.0066	.0090	.0120	10
9	.0000	.0001	.0001	.0003	.0004	.0007	.0011	.0016	.0024	.0033	9
10	.0000	.0000	.0000	.0000	.0001	.0001	.0002	.0003	.0005	.0008	8
11	.0000	.0000	.0000	.0000	.0000	.0000	.0000	.0001	.0001	.0001	7
12	.0000	.0000	.0000	.0000	.0000	.0000	.0000	.0000	.0000	.0000	6

Q=.89 Q=.88 Q=.87 Q=.86 Q=.85 Q=.84 Q=.83 Q=.82 Q=.81 Q=.80

	P=.21	P=.22	P=.23	P=.24	P=.25	P=.26	P=.27	P=.28	P=.29	P=.30	N−X1
X1											
0	.0144	.0114	.0091	.0072	.0056	.0044	.0035	.0027	.0021	.0016	18
1	.0687	.0580	.0487	.0407	.0338	.0280	.0231	.0189	.0155	.0126	17
2	.1553	.1390	.1236	.1092	.0958	.0836	.0725	.0626	.0537	.0458	16
3	.2202	.2091	.1969	.1839	.1704	.1567	.1431	.1298	.1169	.1046	15
4	.2195	.2212	.2205	.2177	.2130	.2065	.1985	.1892	.1790	.1681	14
5	.1634	.1747	.1845	.1925	.1988	.2031	.2055	.2061	.2048	.2017	13
6	.0941	.1067	.1194	.1317	.1436	.1546	.1647	.1736	.1812	.1873	12
7	.0429	.0516	.0611	.0713	.0820	.0931	.1044	.1157	.1269	.1376	11
8	.0157	.0200	.0251	.0310	.0376	.0450	.0531	.0619	.0713	.0811	10
9	.0046	.0063	.0083	.0109	.0139	.0176	.0218	.0267	.0323	.0386	9
10	.0011	.0016	.0022	.0031	.0042	.0056	.0073	.0094	.0119	.0149	8
11	.0002	.0003	.0005	.0007	.0010	.0014	.0020	.0026	.0035	.0046	7
12	.0000	.0001	.0001	.0001	.0002	.0003	.0004	.0006	.0008	.0012	6
13	.0000	.0000	.0000	.0000	.0000	.0000	.0001	.0001	.0002	.0002	5
14	.0000	.0000	.0000	.0000	.0000	.0000	.0000	.0000	.0000	.0000	4

Q=.79 Q=.78 Q=.77 Q=.76 Q=.75 Q=.74 Q=.73 Q=.72 Q=.71 Q=.70

	P=.31	P=.32	P=.33	P=.34	P=.35	P=.36	P=.37	P=.38	P=.39	P=.40	N−X1
X1											
0	.0013	.0010	.0007	.0006	.0004	.0003	.0002	.0002	.0001	.0001	18
1	.0102	.0082	.0066	.0052	.0042	.0033	.0026	.0020	.0016	.0012	17
2	.0388	.0327	.0275	.0229	.0190	.0157	.0129	.0105	.0086	.0069	16
3	.0930	.0822	.0722	.0630	.0547	.0471	.0404	.0344	.0292	.0246	15
4	.1567	.1450	.1333	.1217	.1104	.0994	.0890	.0791	.0699	.0614	14
5	.1971	.1911	.1838	.1755	.1664	.1566	.1463	.1358	.1252	.1146	13
6	.1919	.1948	.1962	.1959	.1941	.1908	.1862	.1803	.1734	.1655	12
7	.1478	.1572	.1656	.1730	.1792	.1840	.1875	.1895	.1900	.1892	11
8	.0913	.1017	.1122	.1226	.1327	.1423	.1514	.1597	.1671	.1734	10
9	.0456	.0532	.0614	.0701	.0794	.0890	.0988	.1087	.1187	.1284	9
10	.0184	.0225	.0272	.0325	.0385	.0450	.0522	.0600	.0683	.0771	8
11	.0060	.0077	.0097	.0122	.0151	.0184	.0223	.0267	.0318	.0374	7
12	.0016	.0021	.0028	.0037	.0047	.0060	.0076	.0096	.0118	.0145	6
13	.0003	.0005	.0006	.0009	.0012	.0016	.0021	.0027	.0035	.0045	5
14	.0001	.0001	.0001	.0002	.0002	.0003	.0004	.0006	.0008	.0011	4
15	.0000	.0000	.0000	.0000	.0000	.0000	.0001	.0001	.0001	.0002	3
16	.0000	.0000	.0000	.0000	.0000	.0000	.0000	.0000	.0000	.0000	2

Q=.69 Q=.68 Q=.67 Q=.66 Q=.65 Q=.64 Q=.63 Q=.62 Q=.61 Q=.60

	P=.41	P=.42	P=.43	P=.44	P=.45	P=.46	P=.47	P=.48	P=.49	P=.50	N−X1
X1											
0	.0001	.0001	.0000	.0000	.0000	.0000	.0000	.0000	.0000	.0000	18
1	.0009	.0007	.0005	.0004	.0003	.0002	.0002	.0001	.0001	.0001	17
2	.0055	.0044	.0035	.0028	.0022	.0017	.0013	.0010	.0008	.0006	16
3	.0206	.0171	.0141	.0116	.0095	.0077	.0062	.0050	.0039	.0031	15
4	.0536	.0464	.0400	.0342	.0291	.0246	.0206	.0172	.0142	.0117	14
5	.1042	.0941	.0844	.0753	.0666	.0586	.0512	.0444	.0382	.0327	13
6	.1569	.1477	.1380	.1281	.1181	.1081	.0983	.0887	.0796	.0708	12
7	.1869	.1833	.1785	.1726	.1657	.1579	.1494	.1404	.1310	.1214	11
8	.1786	.1825	.1852	.1864	.1864	.1850	.1822	.1782	.1731	.1669	10
9	.1379	.1469	.1552	.1628	.1694	.1751	.1795	.1828	.1848	.1855	9
10	.0862	.0957	.1054	.1151	.1248	.1342	.1433	.1519	.1598	.1669	8
11	.0436	.0504	.0578	.0658	.0742	.0831	.0924	.1020	.1117	.1214	7
12	.0177	.0213	.0254	.0301	.0354	.0413	.0478	.0549	.0626	.0708	6
13	.0057	.0071	.0089	.0109	.0134	.0162	.0196	.0234	.0278	.0327	5
14	.0014	.0018	.0024	.0031	.0039	.0049	.0062	.0077	.0095	.0117	4
15	.0003	.0004	.0005	.0006	.0009	.0011	.0015	.0019	.0024	.0031	3
16	.0000	.0000	.0001	.0001	.0001	.0002	.0002	.0003	.0004	.0006	2
17	.0000	.0000	.0000	.0000	.0000	.0000	.0000	.0000	.0000	.0001	1
18	.0000	.0000	.0000	.0000	.0000	.0000	.0000	.0000	.0000	.0000	0

Q=.59 Q=.58 Q=.57 Q=.56 Q=.55 Q=.54 Q=.53 Q=.52 Q=.51 Q=.50

N = 19

	P=.01	P=.02	P=.03	P=.04	P=.05	P=.06	P=.07	P=.08	P=.09	P=.10	N−X1
X1											
0	.8262	.6812	.5606	.4604	.3774	.3086	.2519	.2051	.1666	.1351	19
1	.1586	.2642	.3294	.3645	.3774	.3743	.3602	.3389	.3131	.2852	18
2	.0144	.0485	.0917	.1367	.1787	.2150	.2440	.2652	.2787	.2852	17
3	.0008	.0056	.0161	.0323	.0533	.0778	.1041	.1307	.1562	.1796	16
4	.0000	.0005	.0020	.0054	.0112	.0199	.0313	.0455	.0618	.0798	15

X1											N-X1
5	.0000	.0000	.0002	.0007	.0018	.0038	.0071	.0119	.0183	.0266	14
6	.0000	.0000	.0000	.0001	.0002	.0006	.0012	.0024	.0042	.0069	13
7	.0000	.0000	.0000	.0000	.0000	.0001	.0002	.0004	.0008	.0014	12
8	.0	.0000	.0000	.0000	.0000	.0000	.0000	.0001	.0001	.0002	11
9	.0	.0	.0000	.0000	.0000	.0000	.0000	.0000	.0000	.0000	10

Q=.99	Q=.98	Q=.97	Q=.96	Q=.95	Q=.94	Q=.93	Q=.92	Q=.91	Q=.90

P=.11	P=.12	P=.13	P=.14	P=.15	P=.16	P=.17	P=.18	P=.19	P=.20

X1											N-X1
0	.1092	.0881	.0709	.0569	.0456	.0364	.0290	.0230	.0182	.0144	19
1	.2555	.2284	.2014	.1761	.1529	.1318	.1129	.0961	.0813	.0685	18
2	.2834	.2803	.2708	.2581	.2428	.2259	.2081	.1898	.1717	.1540	17
3	.1999	.2166	.2293	.2381	.2428	.2439	.2415	.2361	.2282	.2182	16
4	.0988	.1181	.1371	.1550	.1714	.1858	.1979	.2073	.2141	.2182	15
5	.0366	.0483	.0614	.0757	.0907	.1062	.1216	.1365	.1507	.1636	14
6	.0106	.0154	.0214	.0288	.0374	.0472	.0581	.0699	.0825	.0955	13
7	.0024	.0039	.0059	.0087	.0122	.0167	.0221	.0285	.0359	.0443	12
8	.0004	.0008	.0013	.0021	.0032	.0048	.0068	.0094	.0126	.0166	11
9	.0001	.0001	.0002	.0004	.0007	.0011	.0017	.0025	.0036	.0051	10
10	.0000	.0000	.0000	.0001	.0001	.0002	.0003	.0006	.0009	.0013	9
11	.0000	.0000	.0000	.0000	.0000	.0000	.0001	.0001	.0002	.0003	8
12	.0000	.0000	.0000	.0000	.0000	.0000	.0000	.0000	.0000	.0000	7

Q=.89	Q=.88	Q=.87	Q=.86	Q=.85	Q=.84	Q=.83	Q=.82	Q=.81	Q=.80

P=.21	P=.22	P=.23	P=.24	P=.25	P=.26	P=.27	P=.28	P=.29	P=.30

X1											N-X1
0	.0113	.0089	.0070	.0054	.0042	.0033	.0025	.0019	.0015	.0011	19
1	.0573	.0477	.0396	.0326	.0268	.0219	.0178	.0144	.0116	.0093	18
2	.1371	.1212	.1064	.0927	.0803	.0692	.0592	.0503	.0426	.0358	17
3	.2065	.1937	.1800	.1659	.1517	.1377	.1240	.1109	.0985	.0869	16
4	.2196	.2185	.2151	.2096	.2023	.1935	.1835	.1726	.1610	.1491	15
5	.1751	.1849	.1928	.1986	.2023	.2040	.2036	.2013	.1973	.1916	14
6	.1086	.1217	.1343	.1463	.1574	.1672	.1757	.1827	.1880	.1916	13
7	.0536	.0637	.0745	.0858	.0974	.1091	.1207	.1320	.1426	.1525	12
8	.0214	.0270	.0334	.0406	.0487	.0575	.0670	.0770	.0874	.0981	11
9	.0069	.0093	.0122	.0157	.0198	.0247	.0303	.0366	.0436	.0514	10
10	.0018	.0026	.0036	.0050	.0066	.0087	.0112	.0142	.0178	.0220	9
11	.0004	.0006	.0009	.0013	.0018	.0025	.0034	.0045	.0060	.0077	8
12	.0001	.0001	.0002	.0003	.0004	.0006	.0008	.0012	.0016	.0022	7
13	.0000	.0000	.0000	.0000	.0001	.0001	.0002	.0002	.0004	.0005	6
14	.0000	.0000	.0000	.0000	.0000	.0000	.0000	.0000	.0001	.0001	5
15	.0000	.0000	.0000	.0000	.0000	.0000	.0000	.0000	.0000	.0000	4

Q=.79	Q=.78	Q=.77	Q=.76	Q=.75	Q=.74	Q=.73	Q=.72	Q=.71	Q=.70

P=.31	P=.32	P=.33	P=.34	P=.35	P=.36	P=.37	P=.38	P=.39	P=.40

X1											N-X1
0	.0009	.0007	.0005	.0004	.0003	.0002	.0002	.0001	.0001	.0001	19
1	.0074	.0059	.0046	.0036	.0029	.0022	.0017	.0013	.0010	.0008	18
2	.0299	.0249	.0206	.0169	.0138	.0112	.0091	.0073	.0058	.0046	17
3	.0762	.0664	.0574	.0494	.0422	.0358	.0302	.0253	.0211	.0175	16
4	.1370	.1249	.1131	.1017	.0909	.0806	.0710	.0621	.0540	.0467	15
5	.1846	.1764	.1672	.1572	.1468	.1360	.1251	.1143	.1036	.0933	14
6	.1935	.1936	.1921	.1890	.1844	.1785	.1714	.1634	.1546	.1451	13
7	.1615	.1692	.1757	.1808	.1844	.1865	.1870	.1860	.1835	.1797	12
8	.1088	.1195	.1298	.1397	.1489	.1573	.1647	.1710	.1760	.1797	11
9	.0597	.0687	.0782	.0880	.0980	.1082	.1182	.1281	.1375	.1464	10
10	.0268	.0323	.0385	.0453	.0528	.0608	.0694	.0785	.0879	.0976	9
11	.0099	.0124	.0155	.0191	.0233	.0280	.0334	.0394	.0460	.0532	8
12	.0030	.0039	.0051	.0066	.0083	.0105	.0131	.0161	.0196	.0237	7
13	.0007	.0010	.0014	.0018	.0024	.0032	.0041	.0053	.0067	.0085	6
14	.0001	.0002	.0003	.0004	.0006	.0008	.0010	.0014	.0018	.0024	5
15	.0000	.0000	.0000	.0001	.0001	.0001	.0002	.0003	.0004	.0005	4
16	.0000	.0000	.0000	.0000	.0000	.0000	.0000	.0000	.0001	.0001	3
17	.0000	.0000	.0000	.0000	.0000	.0000	.0000	.0000	.0000	.0000	2

Q=.69	Q=.68	Q=.67	Q=.66	Q=.65	Q=.64	Q=.63	Q=.62	Q=.61	Q=.60

P=.41	P=.42	P=.43	P=.44	P=.45	P=.46	P=.47	P=.48	P=.49	P=.50

X1											N-X1
0	.0000	.0000	.0000	.0000	.0000	.0000	.0000	.0000	.0000	.0000	19
1	.0006	.0004	.0003	.0002	.0002	.0001	.0001	.0001	.0001	.0000	18
2	.0037	.0029	.0022	.0017	.0013	.0010	.0008	.0006	.0004	.0003	17
3	.0144	.0118	.0096	.0077	.0062	.0049	.0039	.0031	.0024	.0018	16
4	.0400	.0341	.0289	.0243	.0203	.0168	.0138	.0113	.0092	.0074	15
5	.0834	.0741	.0653	.0572	.0497	.0429	.0368	.0313	.0265	.0222	14
6	.1353	.1252	.1150	.1049	.0949	.0853	.0761	.0674	.0593	.0518	13
7	.1746	.1683	.1611	.1530	.1443	.1350	.1254	.1156	.1058	.0961	12
8	.1820	.1829	.1823	.1803	.1771	.1725	.1668	.1601	.1525	.1442	11
9	.1546	.1618	.1681	.1732	.1771	.1796	.1808	.1806	.1791	.1762	10

10	.1074	.1172	.1268	.1361	.1449	.1530	.1603	.1667	.1721	.1762	9
11	.0611	.0694	.0783	.0875	.0970	.1066	.1163	.1259	.1352	.1442	8
12	.0293	.0335	.0394	.0458	.0529	.0606	.0698	.0775	.0866	.0961	7
13	.0106	.0131	.0160	.0154	.0233	.0278	.0328	.0385	.0448	.0517	6
14	.0032	.0041	.0052	.0065	.0082	.0101	.0125	.0152	.0185	.0222	5
15	.0007	.0010	.0013	.0017	.0022	.0029	.0037	.0047	.0059	.0074	4
16	.0001	.0002	.0002	.0003	.0005	.0006	.0008	.0011	.0014	.0018	3
17	.0000	.0000	.0000	.0000	.0001	.0001	.0001	.0002	.0002	.0003	2
18	.0000	.0000	.0000	.0000	.0000	.0000	.0000	.0000	.0000	.0000	1

Q=.59 Q=.58 Q=.57 Q=.56 Q=.55 Q=.54 Q=.53 Q=.52 Q=.51 Q=.50

N = 20

X1	P=.01	P=.02	P=.03	P=.04	P=.05	P=.06	P=.07	P=.08	P=.09	P=.10	N-X1
0	.8179	.6676	.5438	.4420	.3585	.2901	.2342	.1887	.1516	.1216	20
1	.1652	.2725	.3364	.3683	.3774	.3703	.3526	.3282	.3000	.2702	19
2	.0159	.0528	.0988	.1458	.1887	.2246	.2521	.2711	.2818	.2852	18
3	.0010	.0065	.0183	.0364	.0596	.0860	.1139	.1414	.1672	.1901	17
4	.0000	.0006	.0024	.0065	.0133	.0233	.0364	.0523	.0703	.0898	16
5	.0000	.0000	.0002	.0009	.0022	.0048	.0088	.0145	.0222	.0319	15
6	.0000	.0000	.0000	.0001	.0003	.0008	.0017	.0032	.0055	.0089	14
7	.0000	.0000	.0000	.0000	.0000	.0001	.0002	.0005	.0011	.0020	13
8	.0	.0000	.0000	.0000	.0000	.0000	.0000	.0001	.0002	.0004	12
9	.0	.0	.0000	.0000	.0000	.0000	.0000	.0000	.0000	.0001	11
10	.0	.0	.0	.0000	.0000	.0000	.0000	.0000	.0000	.0000	10

Q=.99 Q=.98 Q=.97 Q=.96 Q=.95 Q=.94 Q=.93 Q=.92 Q=.91 Q=.90

X1	P=.11	P=.12	P=.13	P=.14	P=.15	P=.16	P=.17	P=.18	P=.19	P=.20	N-X1
0	.0972	.0776	.0617	.0490	.0388	.0306	.0241	.0189	.0148	.0115	20
1	.2403	.2115	.1844	.1595	.1368	.1165	.0986	.0829	.0693	.0576	19
2	.2822	.2740	.2618	.2466	.2293	.2109	.1919	.1730	.1545	.1369	18
3	.2093	.2242	.2347	.2409	.2428	.2410	.2358	.2278	.2175	.2054	17
4	.1099	.1299	.1491	.1666	.1821	.1951	.2053	.2125	.2168	.2182	16
5	.0435	.0567	.0713	.0868	.1028	.1189	.1345	.1493	.1627	.1746	15
6	.0134	.0193	.0266	.0353	.0454	.0566	.0689	.0819	.0954	.1091	14
7	.0033	.0053	.0080	.0115	.0160	.0216	.0282	.0360	.0448	.0545	13
8	.0007	.0012	.0019	.0030	.0046	.0067	.0094	.0128	.0171	.0222	12
9	.0001	.0002	.0004	.0007	.0011	.0017	.0026	.0038	.0053	.0074	11
10	.0000	.0000	.0001	.0001	.0002	.0004	.0006	.0009	.0014	.0020	10
11	.0000	.0000	.0000	.0000	.0000	.0001	.0001	.0002	.0003	.0005	9
12	.0000	.0000	.0000	.0000	.0000	.0000	.0000	.0000	.0001	.0001	8
13	.0000	.0000	.0000	.0000	.0000	.0000	.0000	.0000	.0000	.0000	7

Q=.89 Q=.88 Q=.87 Q=.86 Q=.85 Q=.84 Q=.83 Q=.82 Q=.81 Q=.80

X1	P=.21	P=.22	P=.23	P=.24	P=.25	P=.26	P=.27	P=.28	P=.29	P=.30	N-X1
0	.0090	.0069	.0054	.0041	.0032	.0024	.0018	.0014	.0011	.0008	20
1	.0477	.0392	.0321	.0261	.0211	.0170	.0137	.0109	.0087	.0068	19
2	.1204	.1050	.0910	.0783	.0669	.0569	.0480	.0403	.0336	.0278	18
3	.1920	.1777	.1631	.1484	.1339	.1199	.1065	.0940	.0823	.0716	17
4	.2169	.2131	.2070	.1991	.1897	.1790	.1675	.1553	.1429	.1304	16
5	.1845	.1923	.1979	.2012	.2023	.2013	.1982	.1933	.1868	.1789	15
6	.1226	.1356	.1478	.1589	.1686	.1768	.1833	.1879	.1907	.1916	14
7	.0652	.0765	.0883	.1003	.1124	.1242	.1356	.1462	.1558	.1643	13
8	.0282	.0351	.0429	.0515	.0609	.0709	.0815	.0924	.1034	.1144	12
9	.0100	.0132	.0171	.0217	.0271	.0332	.0402	.0479	.0563	.0654	11
10	.0029	.0041	.0056	.0075	.0099	.0128	.0163	.0205	.0253	.0308	10
11	.0007	.0010	.0015	.0022	.0030	.0041	.0055	.0072	.0094	.0120	9
12	.0001	.0002	.0003	.0005	.0008	.0011	.0015	.0021	.0029	.0039	8
13	.0000	.0000	.0001	.0001	.0002	.0002	.0003	.0005	.0007	.0010	7
14	.0000	.0000	.0000	.0000	.0000	.0000	.0001	.0001	.0001	.0002	6
15	.0000	.0000	.0000	.0000	.0000	.0000	.0000	.0000	.0000	.0000	5

Q=.79 Q=.78 Q=.77 Q=.76 Q=.75 Q=.74 Q=.73 Q=.72 Q=.71 Q=.70

X1	P=.31	P=.32	P=.33	P=.34	P=.35	P=.36	P=.37	P=.38	P=.39	P=.40	N-X1
0	.0006	.0004	.0003	.0002	.0002	.0001	.0001	.0001	.0001	.0000	20
1	.0054	.0042	.0033	.0025	.0020	.0015	.0011	.0009	.0007	.0005	19
2	.0229	.0138	.0153	.0124	.0100	.0080	.0064	.0050	.0040	.0031	18
3	.0619	.0531	.0453	.0383	.0323	.0270	.0224	.0185	.0152	.0123	17
4	.1181	.1062	.0947	.0839	.0738	.0645	.0559	.0482	.0412	.0350	16
5	.1698	.1599	.1493	.1384	.1272	.1161	.1051	.0945	.0843	.0746	15
6	.1907	.1881	.1839	.1782	.1712	.1632	.1543	.1447	.1347	.1244	14
7	.1714	.1770	.1811	.1836	.1844	.1836	.1812	.1774	.1722	.1659	13
8	.1251	.1354	.1450	.1537	.1614	.1678	.1730	.1767	.1790	.1797	12
9	.0750	.0849	.0952	.1056	.1158	.1259	.1354	.1444	.1526	.1597	11
10	.0370	.0440	.0516	.0598	.0686	.0779	.0875	.0974	.1073	.1171	10

X1											N-X1
11	.0151	.0188	.0231	.0280	.0336	.0398	.0467	.0542	.0624	.0710	9
12	.0051	.0066	.0085	.0108	.0136	.0168	.0206	.0249	.0299	.0355	8
13	.0014	.0019	.0026	.0034	.0045	.0058	.0074	.0094	.0118	.0146	7
14	.0003	.0005	.0006	.0009	.0012	.0016	.0022	.0029	.0038	.0049	6
15	.0001	.0001	.0001	.0002	.0003	.0004	.0005	.0007	.0010	.0013	5
16	.0000	.0000	.0000	.0000	.0000	.0001	.0001	.0001	.0002	.0003	4
17	.0000	.0000	.0000	.0000	.0000	.0000	.0000	.0000	.0000	.0000	3

Q=.69 Q=.68 Q=.67 Q=.66 Q=.65 Q=.64 Q=.63 Q=.62 Q=.61 Q=.60

X1	P=.41	P=.42	P=.43	P=.44	P=.45	P=.46	P=.47	P=.48	P=.49	P=.50	N-X1
0	.0000	.0000	.0000	.0000	.0000	.0000	.0000	.0000	.0000	.0000	20
1	.0004	.0003	.0002	.0001	.0001	.0001	.0001	.0000	.0000	.0000	19
2	.0024	.0018	.0014	.0011	.0008	.0006	.0005	.0003	.0002	.0002	18
3	.0100	.0080	.0064	.0051	.0040	.0031	.0024	.0019	.0014	.0011	17
4	.0295	.0247	.0206	.0170	.0139	.0113	.0092	.0074	.0059	.0046	16
5	.0656	.0573	.0496	.0427	.0365	.0309	.0260	.0217	.0180	.0148	15
6	.1140	.1037	.0936	.0839	.0746	.0658	.0577	.0501	.0432	.0370	14
7	.1585	.1502	.1413	.1318	.1221	.1122	.1023	.0925	.0830	.0739	13
8	.1790	.1768	.1732	.1683	.1623	.1553	.1474	.1388	.1296	.1201	12
9	.1658	.1707	.1742	.1763	.1771	.1763	.1742	.1708	.1651	.1602	11
10	.1268	.1359	.1446	.1524	.1593	.1652	.1700	.1734	.1755	.1762	10
11	.0801	.0895	.0991	.1089	.1185	.1280	.1370	.1455	.1533	.1602	9
12	.0417	.0486	.0561	.0642	.0727	.0818	.0911	.1007	.1105	.1201	8
13	.0178	.0217	.0260	.0310	.0366	.0429	.0497	.0572	.0653	.0739	7
14	.0062	.0078	.0098	.0122	.0150	.0183	.0220	.0264	.0314	.0370	6
15	.0017	.0023	.0030	.0038	.0049	.0062	.0078	.0098	.0121	.0148	5
16	.0004	.0005	.0007	.0009	.0013	.0017	.0022	.0028	.0036	.0046	4
17	.0001	.0001	.0001	.0002	.0002	.0003	.0005	.0006	.0008	.0011	3
18	.0000	.0000	.0000	.0000	.0000	.0000	.0001	.0001	.0001	.0002	2
19	.0000	.0000	.0000	.0000	.0000	.0000	.0000	.0000	.0000	.0000	1

Q=.59 Q=.58 Q=.57 Q=.56 Q=.55 Q=.54 Q=.53 Q=.52 Q=.51 Q=.50

N = 25

X1	P=.01	P=.02	P=.03	P=.04	P=.05	P=.06	P=.07	P=.08	P=.09	P=.10	N-X1
0	.7778	.6035	.4670	.3604	.2774	.2129	.1630	.1244	.0946	.0718	25
1	.1964	.3079	.3611	.3754	.3650	.3397	.3066	.2704	.2340	.1994	24
2	.0238	.0754	.1340	.1877	.2305	.2602	.2770	.2821	.2777	.2659	23
3	.0018	.0118	.0318	.0600	.0930	.1273	.1598	.1881	.2106	.2265	22
4	.0001	.0013	.0054	.0137	.0269	.0447	.0662	.0899	.1145	.1384	21
5	.0000	.0001	.0007	.0024	.0060	.0120	.0209	.0329	.0476	.0646	20
6	.0000	.0000	.0001	.0003	.0010	.0026	.0052	.0095	.0157	.0239	19
7	.0000	.0000	.0000	.0000	.0001	.0004	.0011	.0022	.0042	.0072	18
8	.0	.0000	.0000	.0000	.0000	.0001	.0002	.0004	.0009	.0018	17
9	.0	.0000	.0000	.0000	.0000	.0000	.0000	.0001	.0002	.0004	16
10	.0	.0	.0000	.0000	.0000	.0000	.0000	.0000	.0000	.0001	15
11	.0	.0	.0	.0000	.0000	.0000	.0000	.0000	.0000	.0000	14

Q=.99 Q=.98 Q=.97 Q=.96 Q=.95 Q=.94 Q=.93 Q=.92 Q=.91 Q=.90

X1	P=.11	P=.12	P=.13	P=.14	P=.15	P=.16	P=.17	P=.18	P=.19	P=.20	N-X1
0	.0543	.0409	.0308	.0230	.0172	.0128	.0095	.0070	.0052	.0038	25
1	.1678	.1395	.1149	.0938	.0759	.0609	.0486	.0384	.0302	.0236	24
2	.2488	.2283	.2060	.1832	.1607	.1392	.1193	.1012	.0851	.0708	23
3	.2358	.2387	.2360	.2286	.2174	.2033	.1874	.1704	.1530	.1358	22
4	.1603	.1790	.1940	.2047	.2110	.2130	.2111	.2057	.1974	.1867	21
5	.0832	.1025	.1217	.1399	.1564	.1704	.1816	.1897	.1974	.1960	20
6	.0343	.0466	.0606	.0759	.0920	.1082	.1240	.1388	.1520	.1633	19
7	.0115	.0173	.0246	.0336	.0441	.0559	.0689	.0827	.0968	.1108	18
8	.0032	.0053	.0083	.0123	.0175	.0240	.0318	.0408	.0511	.0623	17
9	.0007	.0014	.0023	.0038	.0058	.0086	.0123	.0169	.0226	.0294	16
10	.0001	.0003	.0006	.0010	.0016	.0026	.0040	.0059	.0085	.0118	15
11	.0000	.0001	.0001	.0002	.0004	.0007	.0011	.0018	.0027	.0040	14
12	.0000	.0000	.0000	.0000	.0001	.0002	.0003	.0005	.0007	.0012	13
13	.0000	.0000	.0000	.0000	.0000	.0000	.0001	.0001	.0002	.0003	12
14	.0000	.0000	.0000	.0000	.0000	.0000	.0000	.0000	.0000	.0001	11
15	.0000	.0000	.0000	.0000	.0000	.0000	.0000	.0000	.0000	.0000	10

Q=.89 Q=.88 Q=.87 Q=.86 Q=.85 Q=.84 Q=.83 Q=.82 Q=.81 Q=.80

X1	P=.21	P=.22	P=.23	P=.24	P=.25	P=.26	P=.27	P=.28	P=.29	P=.30	N-X1
0	.0028	.0020	.0015	.0010	.0008	.0005	.0004	.0003	.0002	.0001	25
1	.0183	.0141	.0109	.0083	.0063	.0047	.0035	.0026	.0020	.0014	24
2	.0585	.0479	.0389	.0314	.0251	.0199	.0157	.0123	.0096	.0074	23
3	.1192	.1035	.0891	.0759	.0641	.0537	.0446	.0367	.0300	.0243	22
4	.1742	.1606	.1463	.1318	.1175	.1037	.0906	.0785	.0673	.0572	21
5	.1945	.1903	.1836	.1749	.1645	.1531	.1408	.1282	.1155	.1030	20
6	.1724	.1789	.1828	.1841	.1828	.1793	.1736	.1661	.1572	.1472	19

X1										N−X1
7	.1244	.1369	.1482	.1578	.1654	.1709	.1743	.1754	.1743	.1712	18
8	.0744	.0869	.0996	.1121	.1241	.1351	.1450	.1535	.1602	.1651	17
9	.0373	.0463	.0562	.0669	.0781	.0897	.1013	.1127	.1236	.1336	16
10	.0159	.0209	.0269	.0338	.0417	.0504	.0600	.0701	.0808	.0916	15
11	.0058	.0080	.0109	.0145	.0189	.0242	.0302	.0372	.0450	.0536	14
12	.0018	.0026	.0038	.0054	.0074	.0099	.0130	.0169	.0214	.0268	13
13	.0005	.0007	.0011	.0017	.0025	.0035	.0048	.0066	.0088	.0115	12
14	.0001	.0002	.0003	.0005	.0007	.0010	.0015	.0022	.0031	.0042	11
15	.0000	.0000	.0001	.0001	.0002	.0003	.0004	.0006	.0009	.0013	10
16	.0000	.0000	.0000	.0000	.0000	.0001	.0001	.0002	.0002	.0004	9
17	.0000	.0000	.0000	.0000	.0000	.0000	.0000	.0000	.0001	.0001	8
18	.0000	.0000	.0000	.0000	.0000	.0000	.0000	.0000	.0000	.0000	7
	Q=.79	Q=.78	Q=.77	Q=.76	Q=.75	Q=.74	Q=.73	Q=.72	Q=.71	Q=.70	

X1	P=.31	P=.32	P=.33	P=.34	P=.35	P=.36	P=.37	P=.38	P=.39	P=.40	N−X1
0	.0001	.0001	.0000	.0000	.0000	.0000	.0000	.0000	.0000	.0000	25
1	.0011	.0008	.0006	.0004	.0003	.0002	.0001	.0001	.0001	.0000	24
2	.0057	.0043	.0033	.0025	.0018	.0014	.0010	.0007	.0005	.0004	23
3	.0195	.0156	.0123	.0097	.0076	.0058	.0045	.0034	.0026	.0019	22
4	.0482	.0403	.0334	.0274	.0224	.0181	.0145	.0115	.0091	.0071	21
5	.0910	.0797	.0691	.0594	.0506	.0427	.0357	.0297	.0244	.0199	20
6	.1363	.1250	.1134	.1020	.0908	.0801	.0700	.0606	.0520	.0442	19
7	.1662	.1596	.1516	.1426	.1327	.1222	.1115	.1008	.0902	.0800	18
8	.1680	.1690	.1680	.1652	.1607	.1547	.1474	.1390	.1298	.1200	17
9	.1426	.1502	.1563	.1608	.1635	.1644	.1635	.1609	.1567	.1511	16
10	.1025	.1131	.1232	.1325	.1409	.1479	.1536	.1578	.1603	.1612	15
11	.0628	.0726	.0828	.0931	.1034	.1135	.1230	.1319	.1398	.1465	14
12	.0329	.0399	.0476	.0560	.0650	.0745	.0843	.0943	.1043	.1139	13
13	.0148	.0188	.0234	.0288	.0350	.0419	.0495	.0578	.0667	.0760	12
14	.0057	.0076	.0099	.0127	.0161	.0202	.0249	.0304	.0365	.0434	11
15	.0019	.0026	.0036	.0048	.0064	.0083	.0107	.0136	.0171	.0212	10
16	.0005	.0008	.0011	.0015	.0021	.0029	.0039	.0052	.0068	.0088	9
17	.0001	.0002	.0003	.0004	.0006	.0009	.0012	.0017	.0023	.0031	8
18	.0000	.0000	.0001	.0001	.0001	.0002	.0003	.0005	.0007	.0009	7
19	.0000	.0000	.0000	.0000	.0000	.0000	.0001	.0001	.0002	.0002	6
20	.0000	.0000	.0000	.0000	.0000	.0000	.0000	.0000	.0000	.0000	5
	Q=.69	Q=.68	Q=.67	Q=.66	Q=.65	Q=.64	Q=.63	Q=.62	Q=.61	Q=.60	

X1	P=.41	P=.42	P=.43	P=.44	P=.45	P=.46	P=.47	P=.48	P=.49	P=.50	N−X1
1	.0000	.0000	.0000	.0000	.0000	.0000	.0000	.0000	.0000	.0000	24
2	.0003	.0002	.0001	.0001	.0001	.0000	.0000	.0000	.0000	.0000	23
3	.0014	.0011	.0008	.0006	.0004	.0003	.0002	.0001	.0001	.0001	22
4	.0055	.0042	.0032	.0024	.0018	.0014	.0010	.0007	.0005	.0004	21
5	.0161	.0129	.0102	.0081	.0063	.0049	.0037	.0028	.0021	.0016	20
6	.0372	.0311	.0257	.0211	.0172	.0138	.0110	.0087	.0068	.0053	19
7	.0703	.0611	.0527	.0450	.0381	.0319	.0265	.0218	.0178	.0143	18
8	.1099	.0996	.0895	.0796	.0701	.0612	.0529	.0453	.0384	.0322	17
9	.1442	.1363	.1275	.1181	.1084	.0985	.0886	.0790	.0697	.0609	16
10	.1603	.1579	.1539	.1485	.1419	.1342	.1257	.1166	.1071	.0974	15
11	.1519	.1559	.1583	.1591	.1583	.1559	.1521	.1468	.1404	.1328	14
12	.1232	.1317	.1393	.1458	.1511	.1550	.1573	.1581	.1573	.1550	13
13	.0856	.0954	.1051	.1146	.1236	.1320	.1395	.1459	.1512	.1550	12
14	.0510	.0592	.0680	.0772	.0867	.0964	.1060	.1155	.1245	.1328	11
15	.0260	.0314	.0376	.0445	.0520	.0602	.0690	.0782	.0877	.0974	10
16	.0113	.0142	.0177	.0218	.0266	.0321	.0382	.0451	.0527	.0609	9
17	.0042	.0055	.0071	.0091	.0115	.0145	.0179	.0220	.0268	.0322	8
18	.0013	.0018	.0024	.0032	.0042	.0055	.0071	.0090	.0114	.0143	7
19	.0003	.0005	.0007	.0009	.0013	.0017	.0023	.0031	.0040	.0053	6
20	.0001	.0001	.0001	.0002	.0003	.0004	.0006	.0009	.0012	.0016	5
21	.0000	.0000	.0000	.0000	.0001	.0001	.0001	.0002	.0003	.0004	4
22	.0000	.0000	.0000	.0000	.0000	.0000	.0000	.0000	.0000	.0001	3
23	.0000	.0000	.0000	.0000	.0000	.0000	.0000	.0000	.0000	.0000	2
	Q=.59	Q=.58	Q=.57	Q=.56	Q=.55	Q=.54	Q=.53	Q=.52	Q=.51	Q=.50	

N = 50

X1	P=.01	P=.02	P=.03	P=.04	P=.05	P=.06	P=.07	P=.08	P=.09	P=.10	N−X1
0	.6050	.3642	.2181	.1299	.0769	.0453	.0266	.0155	.0090	.0052	50
1	.3056	.3716	.3372	.2706	.2025	.1447	.0999	.0672	.0443	.0286	49
2	.0756	.1858	.2555	.2762	.2611	.2262	.1843	.1433	.1073	.0779	48
3	.0122	.0607	.1264	.1842	.2199	.2311	.2219	.1993	.1698	.1386	47
4	.0015	.0145	.0459	.0902	.1360	.1733	.1963	.2037	.1973	.1809	46
5	.0001	.0027	.0131	.0346	.0658	.1018	.1359	.1629	.1795	.1849	45
6	.0000	.0004	.0030	.0108	.0260	.0487	.0767	.1063	.1332	.1541	44
7	.0000	.0001	.0006	.0028	.0086	.0195	.0363	.0581	.0828	.1076	43
8	.0000	.0000	.0001	.0006	.0024	.0067	.0147	.0271	.0440	.0643	42
9	.0000	.0000	.0000	.0001	.0006	.0020	.0052	.0110	.0203	.0333	41
10	.0	.0000	.0000	.0000	.0001	.0005	.0016	.0039	.0082	.0152	40
11	.0	.0000	.0000	.0000	.0000	.0001	.0004	.0012	.0030	.0061	39
12	.0	.0000	.0000	.0000	.0000	.0000	.0001	.0004	.0010	.0022	38
13	.0	.0	.0000	.0000	.0000	.0000	.0000	.0001	.0003	.0007	37
14	.0	.0	.0000	.0000	.0000	.0000	.0000	.0000	.0001	.0002	36
15	.0	.0	.0	.0000	.0000	.0000	.0000	.0000	.0000	.0001	35
16	.0	.0	.0	.0	.0000	.0000	.0000	.0000	.0000	.0000	34
	Q=.99	Q=.98	Q=.97	Q=.96	Q=.95	Q=.94	Q=.93	Q=.92	Q=.91	Q=.90	

X1	P=.11	P=.12	P=.13	P=.14	P=.15	P=.16	P=.17	P=.18	P=.19	P=.20	N-X1
0	.0029	.0017	.0009	.0005	.0003	.0002	.0001	.0000	.0000	.0000	50
1	.0182	.0114	.0071	.0043	.0026	.0016	.0009	.0005	.0003	.0002	49
2	.0552	.0382	.0259	.0172	.0113	.0073	.0046	.0029	.0018	.0011	48
3	.1091	.0833	.0619	.0449	.0319	.0222	.0151	.0102	.0067	.0044	47
4	.1584	.1334	.1086	.0858	.0661	.0496	.0364	.0262	.0185	.0128	46
5	.1801	.1674	.1493	.1286	.1072	.0869	.0687	.0530	.0400	.0295	45
6	.1670	.1712	.1674	.1570	.1419	.1242	.1055	.0872	.0703	.0554	44
7	.1297	.1467	.1572	.1606	.1575	.1487	.1358	.1203	.1037	.0870	43
8	.0362	.1075	.1262	.1406	.1493	.1523	.1495	.1420	.1307	.1169	42
9	.0497	.0684	.0880	.1068	.1230	.1353	.1429	.1454	.1431	.1364	41
10	.0252	.0383	.0539	.0713	.0890	.1057	.1200	.1309	.1376	.1398	40
11	.0113	.0190	.0293	.0422	.0571	.0732	.0894	.1045	.1174	.1271	39
12	.0045	.0084	.0142	.0223	.0328	.0453	.0595	.0745	.0895	.1033	38
13	.0016	.0034	.0062	.0106	.0169	.0252	.0356	.0478	.0613	.0755	37
14	.0005	.0012	.0025	.0046	.0079	.0127	.0193	.0277	.0380	.0499	36
15	.0002	.0004	.0009	.0018	.0033	.0058	.0095	.0146	.0214	.0299	35
16	.0000	.0001	.0003	.0006	.0013	.0024	.0042	.0070	.0110	.0164	34
17	.0000	.0000	.0001	.0002	.0005	.0009	.0017	.0031	.0052	.0082	33
18	.0000	.0000	.0000	.0001	.0001	.0003	.0007	.0012	.0022	.0037	32
19	.0000	.0000	.0000	.0000	.0000	.0001	.0002	.0005	.0009	.0016	31
20	.0000	.0000	.0000	.0000	.0000	.0000	.0001	.0002	.0003	.0006	30
21	.0000	.0000	.0000	.0000	.0000	.0000	.0000	.0000	.0001	.0002	29
22	.0000	.0000	.0000	.0000	.0000	.0000	.0000	.0000	.0000	.0001	28
23	.0000	.0000	.0000	.0000	.0000	.0000	.0000	.0000	.0000	.0000	27
	Q=.89	Q=.88	Q=.87	Q=.86	Q=.85	Q=.84	Q=.83	Q=.82	Q=.81	Q=.80	

X1	P=.21	P=.22	P=.23	P=.24	P=.25	P=.26	P=.27	P=.28	P=.29	P=.30	N-X1
0	.0000	.0000	.0000	.0000	.0000	.0000	.0000	.0000	.0000	.0000	50
1	.0001	.0001	.0000	.0000	.0000	.0000	.0000	.0000	.0000	.0000	49
2	.0007	.0004	.0002	.0001	.0001	.0000	.0000	.0000	.0000	.0000	48
3	.0028	.0018	.0011	.0007	.0004	.0002	.0001	.0001	.0000	.0000	47
4	.0088	.0059	.0039	.0025	.0016	.0010	.0006	.0004	.0002	.0001	46
5	.0214	.0152	.0106	.0073	.0049	.0033	.0021	.0014	.0009	.0006	45
6	.0427	.0322	.0238	.0173	.0123	.0087	.0060	.0040	.0027	.0018	44
7	.0713	.0571	.0447	.0344	.0259	.0191	.0139	.0099	.0069	.0048	43
8	.1019	.0865	.0718	.0583	.0463	.0361	.0276	.0207	.0152	.0110	42
9	.1263	.1139	.1001	.0859	.0721	.0592	.0476	.0375	.0290	.0220	41
10	.1377	.1317	.1226	.1113	.0985	.0852	.0721	.0598	.0485	.0386	40
11	.1331	.1351	.1332	.1278	.1194	.1089	.0970	.0845	.0721	.0602	39
12	.1150	.1238	.1293	.1311	.1294	.1244	.1166	.1068	.0957	.0838	38
13	.0894	.1021	.1129	.1210	.1260	.1277	.1261	.1215	.1142	.1050	37
14	.0628	.0761	.0891	.1010	.1110	.1186	.1233	.1248	.1233	.1189	36
15	.0400	.0515	.0639	.0766	.0888	.1000	.1094	.1165	.1209	.1223	35
16	.0233	.0318	.0417	.0529	.0648	.0769	.0885	.0991	.1080	.1147	34
17	.0124	.0179	.0249	.0334	.0432	.0540	.0655	.0771	.0882	.0983	33
18	.0060	.0093	.0137	.0193	.0264	.0348	.0444	.0550	.0661	.0772	32
19	.0027	.0044	.0069	.0103	.0148	.0206	.0277	.0360	.0454	.0558	31
20	.0011	.0019	.0032	.0050	.0077	.0112	.0159	.0217	.0288	.0370	30
21	.0004	.0008	.0014	.0023	.0036	.0056	.0084	.0121	.0168	.0227	29
22	.0001	.0003	.0005	.0009	.0016	.0026	.0041	.0062	.0090	.0128	28
23	.0000	.0001	.0002	.0004	.0006	.0011	.0018	.0029	.0045	.0067	27
24	.0000	.0000	.0001	.0001	.0002	.0004	.0008	.0013	.0021	.0032	26
25	.0000	.0000	.0000	.0000	.0001	.0002	.0003	.0005	.0009	.0014	25
26	.0000	.0000	.0000	.0000	.0000	.0001	.0001	.0002	.0003	.0006	24
27	.0000	.0000	.0000	.0000	.0000	.0000	.0000	.0001	.0001	.0002	23
28	.0000	.0000	.0000	.0000	.0000	.0000	.0000	.0000	.0001	.0001	22
29	.0000	.0000	.0000	.0000	.0000	.0000	.0000	.0000	.0000	.0000	21
	Q=.79	Q=.78	Q=.77	Q=.76	Q=.75	Q=.74	Q=.73	Q=.72	Q=.71	Q=.70	

X1	P=.31	P=.32	P=.33	P=.34	P=.35	P=.36	P=.37	P=.38	P=.39	P=.40	N-X1
3	.0000	.0000	.0000	.0000	.0000	.0000	.0000	.0000	.0000	.0000	47
4	.0001	.0000	.0000	.0000	.0000	.0000	.0000	.0000	.0000	.0000	46
5	.0003	.0002	.0001	.0001	.0000	.0000	.0000	.0000	.0000	.0000	45
6	.0011	.0007	.0005	.0003	.0002	.0001	.0001	.0000	.0000	.0000	44
7	.0032	.0022	.0014	.0009	.0006	.0004	.0002	.0001	.0001	.0000	43
8	.0078	.0055	.0037	.0025	.0017	.0011	.0007	.0004	.0003	.0002	42
9	.0164	.0120	.0086	.0061	.0042	.0029	.0019	.0013	.0008	.0005	41
10	.0301	.0231	.0174	.0128	.0093	.0066	.0046	.0032	.0022	.0014	40
11	.0493	.0395	.0311	.0240	.0182	.0136	.0099	.0071	.0050	.0035	39
12	.0719	.0604	.0498	.0402	.0319	.0248	.0189	.0142	.0105	.0076	38
13	.0944	.0831	.0717	.0606	.0502	.0408	.0325	.0255	.0195	.0147	37
14	.1121	.1034	.0933	.0825	.0714	.0607	.0505	.0412	.0330	.0260	36
15	.1209	.1168	.1103	.1020	.0923	.0819	.0712	.0606	.0507	.0415	35
16	.1188	.1202	.1189	.1149	.1087	.1008	.0914	.0813	.0709	.0606	34
17	.1068	.1132	.1171	.1184	.1171	.1133	.1074	.0997	.0906	.0808	33
18	.0880	.0976	.1057	.1118	.1156	.1169	.1156	.1120	.1062	.0987	32
19	.0665	.0774	.0877	.0970	.1048	.1107	.1144	.1156	.1144	.1109	31
20	.0463	.0564	.0670	.0775	.0875	.0965	.1041	.1098	.1134	.1146	30
21	.0297	.0379	.0471	.0570	.0673	.0776	.0874	.0962	.1035	.1091	29
22	.0176	.0235	.0306	.0387	.0478	.0575	.0676	.0777	.0873	.0959	28
23	.0096	.0135	.0183	.0243	.0313	.0394	.0484	.0580	.0679	.0778	27
24	.0049	.0071	.0102	.0141	.0190	.0249	.0319	.0400	.0489	.0584	26
25	.0023	.0035	.0052	.0075	.0106	.0146	.0195	.0255	.0325	.0405	25

X1											N−X1
6	.0369	.0215	.0119	.0063	.0031	.0015	.0007	.0003	.0001	.0001	94
7	.0613	.0394	.0238	.0137	.0075	.0039	.0020	.0009	.0004	.0002	93
8	.0881	.0625	.0414	.0259	.0153	.0086	.0047	.0024	.0012	.0006	92
9	.1112	.0871	.0632	.0430	.0276	.0168	.0098	.0054	.0029	.0015	91
10	.1251	.1080	.0860	.0637	.0444	.0292	.0182	.0108	.0062	.0034	90
11	.1265	.1205	.1051	.0849	.0640	.0454	.0305	.0194	.0113	.0069	89
12	.1160	.1219	.1165	.1025	.0838	.0642	.0463	.0315	.0206	.0128	88
13	.0970	.1125	.1179	.1130	.1001	.0827	.0642	.0470	.0327	.0216	87
14	.0745	.0954	.1094	.1143	.1098	.0979	.0817	.0641	.0476	.0335	86
15	.0528	.0745	.0938	.1067	.1111	.1070	.0960	.0807	.0640	.0431	85
16	.0347	.0540	.0744	.0922	.1041	.1082	.1044	.0941	.0798	.0638	84
17	.0212	.0364	.0549	.0742	.0903	.1019	.1057	.1021	.0924	.0789	83
18	.0121	.0229	.0379	.0557	.0739	.0895	.0998	.1033	.1000	.0909	82
19	.0064	.0135	.0244	.0391	.0563	.0736	.0882	.0979	.1012	.0981	81
20	.0032	.0074	.0148	.0258	.0402	.0567	.0732	.0870	.0962	.0993	80
21	.0015	.0039	.0084	.0160	.0270	.0412	.0571	.0728	.0859	.0945	79
22	.0007	.0019	.0045	.0094	.0171	.0282	.0420	.0574	.0724	.0849	78
23	.0003	.0009	.0023	.0052	.0103	.0182	.0292	.0427	.0575	.0720	77
24	.0001	.0004	.0011	.0027	.0058	.0111	.0192	.0301	.0433	.0577	76
25	.0000	.0002	.0005	.0013	.0031	.0064	.0119	.0201	.0309	.0439	75
26	.0000	.0001	.0002	.0006	.0016	.0035	.0071	.0127	.0209	.0316	74
27	.0000	.0000	.0001	.0003	.0008	.0018	.0040	.0076	.0134	.0217	73
28	.0000	.0000	.0000	.0001	.0004	.0009	.0021	.0044	.0082	.0141	72
29	.0000	.0000	.0000	.0000	.0002	.0004	.0011	.0024	.0048	.0088	71
30	.0000	.0000	.0000	.0000	.0001	.0002	.0005	.0012	.0027	.0052	70
31	.0000	.0000	.0000	.0000	.0000	.0001	.0002	.0006	.0014	.0029	69
32	.0000	.0000	.0000	.0000	.0000	.0000	.0001	.0003	.0007	.0016	68
33	.0000	.0000	.0000	.0000	.0000	.0000	.0000	.0001	.0003	.0008	67
34	.0000	.0000	.0000	.0000	.0000	.0000	.0000	.0001	.0002	.0004	66
35	.0000	.0000	.0000	.0000	.0000	.0000	.0000	.0000	.0001	.0002	65
36	.0	.0000	.0000	.0000	.0000	.0000	.0000	.0000	.0000	.0001	64
37	.0	.0	.0000	.0000	.0000	.0000	.0000	.0000	.0000	.0000	63
Q=	.89	.88	.87	.86	.85	.84	.83	.82	.81	.80	

P=	.21	.22	.23	.24	.25	.26	.27	.28	.29	.30	
X1											**N−X1**
6	.0000	.0000	.0000	.0000	.0000	.0000	.0000	.0000	.0000	.0000	94
7	.0001	.0000	.0000	.0000	.0000	.0000	.0000	.0000	.0000	.0000	93
8	.0003	.0001	.0001	.0001	.0000	.0000	.0000	.0000	.0000	.0000	92
9	.0007	.0003	.0002	.0001	.0000	.0000	.0000	.0000	.0000	.0000	91
10	.0018	.0009	.0004	.0002	.0001	.0000	.0000	.0000	.0000	.0000	90
11	.0038	.0021	.0011	.0005	.0003	.0001	.0001	.0000	.0000	.0000	89
12	.0076	.0043	.0024	.0012	.0006	.0003	.0001	.0001	.0000	.0000	88
13	.0136	.0082	.0048	.0027	.0014	.0007	.0004	.0002	.0001	.0000	87
14	.0225	.0144	.0089	.0052	.0030	.0016	.0009	.0004	.0002	.0001	86
15	.0343	.0233	.0152	.0095	.0057	.0033	.0018	.0010	.0005	.0002	85
16	.0484	.0350	.0241	.0159	.0100	.0061	.0035	.0020	.0011	.0005	84
17	.0636	.0487	.0356	.0248	.0165	.0106	.0065	.0038	.0022	.0012	83
18	.0780	.0634	.0490	.0361	.0254	.0171	.0111	.0069	.0041	.0024	82
19	.0895	.0772	.0631	.0492	.0365	.0259	.0177	.0115	.0072	.0044	81
20	.0963	.0881	.0764	.0629	.0493	.0369	.0264	.0182	.0120	.0076	80
21	.0975	.0947	.0869	.0756	.0625	.0494	.0373	.0269	.0186	.0124	79
22	.0931	.0959	.0932	.0858	.0749	.0623	.0495	.0376	.0273	.0190	78
23	.0839	.0917	.0944	.0919	.0847	.0743	.0621	.0495	.0378	.0277	77
24	.0716	.0830	.0905	.0931	.0906	.0837	.0736	.0613	.0496	.0380	76
25	.0578	.0712	.0822	.0893	.0918	.0894	.0828	.0731	.0615	.0496	75
26	.0444	.0579	.0708	.0814	.0883	.0906	.0883	.0819	.0725	.0613	74
27	.0323	.0448	.0580	.0704	.0806	.0873	.0896	.0873	.0812	.0720	73
28	.0224	.0329	.0451	.0580	.0701	.0799	.0864	.0886	.0864	.0804	72
29	.0148	.0231	.0335	.0455	.0583	.0697	.0793	.0855	.0876	.0856	71
30	.0093	.0154	.0237	.0340	.0458	.0580	.0694	.0787	.0847	.0868	70
31	.0056	.0098	.0160	.0242	.0344	.0460	.0580	.0691	.0781	.0840	69
32	.0032	.0060	.0103	.0165	.0248	.0349	.0462	.0579	.0688	.0775	68
33	.0018	.0035	.0063	.0107	.0170	.0252	.0352	.0464	.0579	.0685	67
34	.0009	.0019	.0037	.0067	.0112	.0175	.0257	.0356	.0466	.0579	66
35	.0005	.0010	.0021	.0040	.0070	.0116	.0179	.0261	.0359	.0468	65
36	.0002	.0005	.0011	.0023	.0042	.0073	.0120	.0183	.0265	.0362	64
37	.0001	.0003	.0006	.0012	.0024	.0045	.0077	.0123	.0187	.0268	63
38	.0000	.0001	.0003	.0006	.0013	.0026	.0047	.0079	.0127	.0191	62
39	.0000	.0001	.0001	.0003	.0007	.0015	.0028	.0049	.0082	.0130	61
40	.0000	.0000	.0000	.0001	.0002	.0005	.0010	.0019	.0033	.0055	60
41	.0000	.0000	.0000	.0000	.0000	.0000	.0000	.0000	.0000	.0001	59
42	.0000	.0000	.0000	.0000	.0000	.0000	.0000	.0000	.0000	.0000	58
Q=	.79	.78	.77	.76	.75	.74	.73	.72	.71	.70	

P=	.31	.32	.33	.34	.35	.36	.37	.38	.39	.40	
X1											**N−X1**
14	.0000	.0000	.0000	.0000	.0000	.0000	.0000	.0000	.0000	.0000	86
15	.0001	.0001	.0000	.0000	.0000	.0000	.0000	.0000	.0000	.0000	85
16	.0003	.0001	.0001	.0000	.0000	.0000	.0000	.0000	.0000	.0000	84
17	.0006	.0003	.0002	.0001	.0000	.0000	.0000	.0000	.0000	.0000	83
18	.0013	.0007	.0004	.0002	.0001	.0000	.0000	.0000	.0000	.0000	82
19	.0025	.0014	.0008	.0004	.0002	.0001	.0000	.0000	.0000	.0000	81
20	.0046	.0027	.0015	.0008	.0004	.0002	.0001	.0001	.0000	.0000	80
21	.0079	.0049	.0029	.0016	.0009	.0005	.0002	.0001	.0001	.0000	79
22	.0127	.0082	.0051	.0030	.0017	.0010	.0005	.0003	.0001	.0000	78
23	.0194	.0131	.0085	.0053	.0032	.0018	.0010	.0006	.0003	.0001	77
24	.0280	.0198	.0134	.0088	.0055	.0033	.0019	.0011	.0006	.0003	76
25	.0382	.0283	.0201	.0137	.0090	.0057	.0035	.0020	.0012	.0006	75

26	.0010	.0016	.0025	.0037	.0055	.0075	.0110	.0150	.0200	.0259	24
27	.0004	.0007	.0011	.0017	.0026	.0039	.0058	.0082	.0113	.0154	23
28	.0001	.0003	.0004	.0007	.0012	.0018	.0028	.0041	.0060	.0084	22
29	.0000	.0001	.0002	.0003	.0005	.0008	.0012	.0019	.0029	.0043	21
30	.0000	.0000	.0001	.0001	.0002	.0003	.0005	.0008	.0013	.0020	20
31	.0000	.0000	.0000	.0000	.0001	.0001	.0002	.0003	.0005	.0009	19
32	.0000	.0000	.0000	.0000	.0000	.0000	.0001	.0001	.0002	.0003	18
33	.0000	.0000	.0000	.0000	.0000	.0000	.0000	.0000	.0001	.0001	17
34	.0000	.0000	.0000	.0000	.0000	.0000	.0000	.0000	.0000	.0000	16

Q=.69 Q=.68 Q=.67 Q=.66 Q=.65 Q=.64 Q=.63 Q=.62 Q=.61 Q=.60

P=.41 P=.42 P=.43 P=.44 P=.45 P=.46 P=.47 P=.48 P=.49 P=.50

X1   N−X1

7	.0000	.0000	.0000	.0000	.0000	.0000	.0000	.0000	.0000	.0000	43
8	.0001	.0001	.0000	.0000	.0000	.0000	.0000	.0000	.0000	.0000	42
9	.0003	.0002	.0001	.0001	.0000	.0000	.0000	.0000	.0000	.0000	41
10	.0009	.0006	.0004	.0002	.0001	.0001	.0001	.0000	.0000	.0000	40
11	.0024	.0016	.0010	.0007	.0004	.0003	.0002	.0001	.0001	.0000	39
12	.0054	.0037	.0026	.0017	.0011	.0007	.0005	.0002	.0002	.0001	38
13	.0109	.0079	.0057	.0040	.0027	.0018	.0012	.0008	.0005	.0003	37
14	.0200	.0152	.0113	.0082	.0059	.0041	.0029	.0019	.0013	.0008	36
15	.0334	.0264	.0204	.0155	.0116	.0085	.0061	.0043	.0030	.0020	35
16	.0508	.0418	.0337	.0267	.0207	.0158	.0118	.0086	.0062	.0044	34
17	.0706	.0605	.0508	.0419	.0339	.0269	.0340	.0270	.0210	.0160	33
18	.0899	.0803	.0703	.0604	.0508	.0420	.0340	.0270	.0210	.0160	32
19	.1053	.0979	.0893	.0799	.0700	.0602	.0507	.0419	.0340	.0270	31
20	.1134	.1099	.1044	.0973	.0888	.0795	.0697	.0600	.0506	.0419	30
21	.1126	.1137	.1126	.1052	.1038	.0967	.0884	.0791	.0695	.0598	29
22	.1031	.1086	.1119	.1131	.1119	.1086	.1033	.0963	.0880	.0788	28
23	.0872	.0957	.1028	.1082	.1115	.1126	.1115	.1082	.1029	.0960	27
24	.0682	.0780	.0872	.0956	.1026	.1079	.1112	.1124	.1112	.1080	26
25	.0493	.0587	.0684	.0781	.0873	.0956	.1026	.1079	.1112	.1123	25
26	.0329	.0409	.0496	.0590	.0687	.0783	.0875	.0957	.1027	.1080	24
27	.0203	.0263	.0333	.0412	.0500	.0593	.0690	.0786	.0877	.0960	23
28	.0116	.0157	.0206	.0266	.0336	.0415	.0502	.0596	.0692	.0788	22
29	.0061	.0086	.0118	.0159	.0208	.0268	.0338	.0417	.0504	.0598	21
30	.0030	.0044	.0062	.0087	.0119	.0160	.0210	.0270	.0339	.0419	20
31	.0013	.0020	.0030	.0044	.0063	.0088	.0120	.0161	.0210	.0270	19
32	.0006	.0009	.0014	.0021	.0031	.0044	.0063	.0088	.0120	.0160	18
33	.0002	.0003	.0006	.0009	.0014	.0021	.0031	.0044	.0063	.0087	17
34	.0001	.0001	.0002	.0003	.0006	.0009	.0014	.0020	.0030	.0044	16
35	.0000	.0000	.0001	.0001	.0002	.0003	.0005	.0009	.0013	.0020	15
36	.0000	.0000	.0000	.0000	.0001	.0001	.0002	.0003	.0005	.0008	14
37	.0000	.0000	.0000	.0000	.0000	.0000	.0001	.0001	.0002	.0003	13
38	.0000	.0000	.0000	.0000	.0000	.0000	.0000	.0000	.0001	.0001	12
39	.0000	.0000	.0000	.0000	.0000	.0000	.0000	.0000	.0000	.0000	11

Q=.59 Q=.58 Q=.57 Q=.56 Q=.55 Q=.54 Q=.53 Q=.52 Q=.51 Q=.50

N = 100

P=.01 P=.02 P=.03 P=.04 P=.05 P=.06 P=.07 P=.08 P=.09 P=.10

X1   N−X1

0	.3660	.1326	.0476	.0169	.0059	.0021	.0007	.0002	.0001	.0000	100
1	.3697	.2707	.1471	.0703	.0312	.0131	.0053	.0021	.0008	.0003	99
2	.1849	.2734	.2252	.1450	.0812	.0414	.0198	.0090	.0039	.0016	98
3	.0610	.1823	.2275	.1973	.1396	.0864	.0486	.0254	.0125	.0059	97
4	.0149	.0902	.1706	.1994	.1781	.1288	.0888	.0536	.0301	.0159	96
5	.0029	.0353	.1013	.1595	.1800	.1639	.1283	.0895	.0571	.0339	95
6	.0005	.0114	.0496	.1052	.1500	.1657	.1529	.1233	.0895	.0596	94
7	.0001	.0031	.0206	.0589	.1060	.1420	.1545	.1440	.1188	.0889	93
8	.0000	.0007	.0074	.0285	.0649	.1054	.1352	.1455	.1366	.1148	92
9	.0000	.0002	.0023	.0121	.0349	.0687	.1040	.1293	.1381	.1304	91
10	.0000	.0000	.0007	.0046	.0167	.0399	.0712	.1024	.1243	.1319	90
11	.0000	.0000	.0002	.0016	.0072	.0209	.0439	.0728	.1006	.1199	89
12	.0000	.0000	.0000	.0005	.0028	.0099	.0245	.0470	.0738	.0988	88
13	.0	.0000	.0000	.0001	.0010	.0043	.0125	.0276	.0494	.0743	87
14	.0	.0000	.0000	.0000	.0003	.0017	.0058	.0149	.0304	.0513	86
15	.0	.0000	.0000	.0000	.0001	.0006	.0025	.0074	.0172	.0327	85
16	.0	.0000	.0000	.0000	.0000	.0002	.0010	.0034	.0090	.0193	84
17	.0	.0	.0000	.0000	.0000	.0001	.0004	.0015	.0044	.0106	83
18	.0	.0	.0000	.0000	.0000	.0000	.0001	.0006	.0020	.0054	82
19	.0	.0	.0	.0000	.0000	.0000	.0000	.0002	.0009	.0026	81
20	.0	.0	.0	.0000	.0000	.0000	.0000	.0001	.0003	.0012	80
21	.0	.0	.0	.0000	.0000	.0000	.0000	.0000	.0001	.0005	79
22	.0	.0	.0	.0	.0000	.0000	.0000	.0000	.0000	.0002	78
23	.0	.0	.0	.0	.0000	.0000	.0000	.0000	.0000	.0001	77
24	.0	.0	.0	.0	.0	.0000	.0000	.0000	.0000	.0000	76

Q=.99 Q=.98 Q=.97 Q=.96 Q=.95 Q=.94 Q=.93 Q=.92 Q=.91 Q=.90

P=.11 P=.12 P=.13 P=.14 P=.15 P=.16 P=.17 P=.18 P=.19 P=.20

X1   N−X1

0	.0000	.0000	.0000	.0000	.0000	.0000	.0000	.0000	.0000	.0000	100
1	.0001	.0000	.0000	.0000	.0000	.0000	.0000	.0000	.0000	.0000	99
2	.0007	.0003	.0001	.0000	.0000	.0000	.0000	.0000	.0000	.0000	98
3	.0027	.0012	.0005	.0002	.0001	.0000	.0000	.0000	.0000	.0000	97
4	.0080	.0038	.0018	.0008	.0003	.0001	.0001	.0000	.0000	.0000	96
5	.0189	.0100	.0050	.0024	.0011	.0005	.0002	.0001	.0000	.0000	95

X1										N-X1
26	.0496	.0384	.0286	.0204	.0140	.0092	.0059	.0036	.0021	.0012	74
27	.0610	.0495	.0386	.0288	.0207	.0143	.0095	.0060	.0037	.0022	73
28	.0715	.0608	.0495	.0387	.0290	.0209	.0145	.0097	.0062	.0038	72
29	.0797	.0710	.0605	.0495	.0388	.0292	.0211	.0147	.0098	.0063	71
30	.0848	.0791	.0706	.0603	.0494	.0389	.0294	.0213	.0149	.0100	70
31	.0860	.0840	.0785	.0702	.0601	.0494	.0389	.0295	.0215	.0151	69
32	.0833	.0853	.0834	.0779	.0693	.0599	.0493	.0390	.0296	.0217	68
33	.0771	.0827	.0846	.0827	.0774	.0694	.0597	.0493	.0390	.0293	67
34	.0683	.0767	.0821	.0840	.0821	.0769	.0691	.0595	.0492	.0391	66
35	.0578	.0680	.0763	.0816	.0834	.0816	.0765	.0688	.0593	.0491	65
36	.0469	.0578	.0678	.0759	.0811	.0829	.0811	.0751	.0685	.0591	64
37	.0365	.0471	.0578	.0676	.0755	.0806	.0824	.0807	.0757	.0632	63
38	.0272	.0367	.0472	.0577	.0674	.0752	.0802	.0820	.0803	.0754	62
39	.0194	.0275	.0369	.0473	.0577	.0672	.0749	.0799	.0816	.0799	61
40	.0086	.0127	.0179	.0240	.0306	.0372	.0433	.0482	.0513	.0524	60
41	.0001	.0001	.0002	.0003	.0004	.0005	.0006	.0007	.0008	.0009	59
42	.0000	.0000	.0000	.0000	.0000	.0000	.0000	.0000	.0000	.0000	58

Q=.69	Q=.68	C=.67	Q=.66	Q=.65	Q=.64	Q=.63	Q=.62	Q=.61	Q=.60

P=.41	P=.42	P=.43	P=.44	P=.45	P=.46	P=.47	P=.48	P=.49	P=.50

X1										N-X1
22	.0000	.0000	.0000	.0000	.0000	.0000	.0000	.0000	.0000	.0000	78
23	.0001	.0000	.0000	.0000	.0000	.0000	.0000	.0000	.0000	.0000	77
24	.0002	.0001	.0000	.0000	.0000	.0000	.0000	.0000	.0000	.0000	76
25	.0003	.0002	.0001	.0000	.0000	.0000	.0000	.0000	.0000	.0000	75
26	.0007	.0003	.0002	.0001	.0000	.0000	.0000	.0000	.0000	.0000	74
27	.0013	.0007	.0004	.0002	.0001	.0000	.0000	.0000	.0000	.0000	73
28	.0023	.0013	.0007	.0004	.0002	.0001	.0000	.0000	.0000	.0000	72
29	.0039	.0024	.0014	.0008	.0004	.0002	.0001	.0000	.0000	.0000	71
30	.0065	.0040	.0024	.0014	.0008	.0004	.0002	.0001	.0001	.0000	70
31	.0102	.0066	.0041	.0025	.0014	.0008	.0004	.0002	.0001	.0001	69
32	.0152	.0103	.0067	.0042	.0025	.0015	.0008	.0004	.0002	.0001	68
33	.0218	.0154	.0104	.0068	.0043	.0026	.0015	.0008	.0004	.0002	67
34	.0298	.0219	.0155	.0105	.0069	.0043	.0026	.0015	.0009	.0005	66
35	.0391	.0299	.0220	.0156	.0106	.0069	.0044	.0026	.0015	.0009	65
36	.0491	.0391	.0300	.0221	.0157	.0107	.0070	.0044	.0027	.0015	64
37	.0590	.0490	.0391	.0300	.0222	.0157	.0107	.0070	.0044	.0027	63
38	.0680	.0588	.0489	.0391	.0301	.0222	.0158	.0108	.0071	.0045	62
39	.0751	.0677	.0587	.0489	.0391	.0301	.0223	.0158	.0108	.0071	61
40	.0513	.0483	.0436	.0378	.0315	.0252	.0194	.0144	.0102	.0070	60
41	.0009	.0009	.0008	.0007	.0006	.0005	.0004	.0003	.0003	.0000	59
42	.0000	.0000	.0000	.0000	.0000	.0000	.0000	.0000	.0000	.0000	58

Q=.59	Q=.58	C=.57	Q=.56	Q=.55	Q=.54	Q=.53	Q=.52	Q=.51	Q=.50

TABLES

Poisson Probability Distribution—
Values of $P(X) = \lambda t^x e^{-\lambda t} / X!$

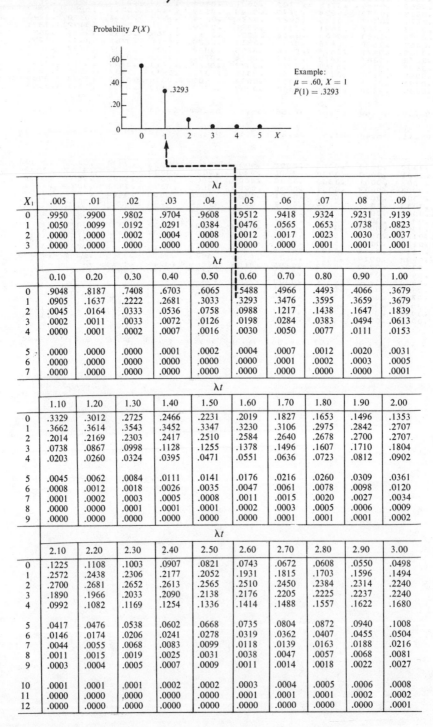

Probability $P(X)$

Example:
$\mu = .60$, $X = 1$
$P(1) = .3293$

X_1	.005	.01	.02	.03	.04	.05	.06	.07	.08	.09
						λt				
0	.9950	.9900	.9802	.9704	.9608	.9512	.9418	.9324	.9231	.9139
1	.0050	.0099	.0192	.0291	.0384	.0476	.0565	.0653	.0738	.0823
2	.0000	.0000	.0002	.0004	.0008	.0012	.0017	.0023	.0030	.0037
3	.0000	.0000	.0000	.0000	.0000	.0000	.0000	.0001	.0001	.0001

	0.10	0.20	0.30	0.40	0.50	0.60	0.70	0.80	0.90	1.00
						λt				
0	.9048	.8187	.7408	.6703	.6065	.5488	.4966	.4493	.4066	.3679
1	.0905	.1637	.2222	.2681	.3033	.3293	.3476	.3595	.3659	.3679
2	.0045	.0164	.0333	.0536	.0758	.0988	.1217	.1438	.1647	.1839
3	.0002	.0011	.0033	.0072	.0126	.0198	.0284	.0383	.0494	.0613
4	.0000	.0001	.0002	.0007	.0016	.0030	.0050	.0077	.0111	.0153
5	.0000	.0000	.0000	.0001	.0002	.0004	.0007	.0012	.0020	.0031
6	.0000	.0000	.0000	.0000	.0000	.0000	.0001	.0002	.0003	.0005
7	.0000	.0000	.0000	.0000	.0000	.0000	.0000	.0000	.0000	.0001

	1.10	1.20	1.30	1.40	1.50	1.60	1.70	1.80	1.90	2.00
						λt				
0	.3329	.3012	.2725	.2466	.2231	.2019	.1827	.1653	.1496	.1353
1	.3662	.3614	.3543	.3452	.3347	.3230	.3106	.2975	.2842	.2707
2	.2014	.2169	.2303	.2417	.2510	.2584	.2640	.2678	.2700	.2707
3	.0738	.0867	.0998	.1128	.1255	.1378	.1496	.1607	.1710	.1804
4	.0203	.0260	.0324	.0395	.0471	.0551	.0636	.0723	.0812	.0902
5	.0045	.0062	.0084	.0111	.0141	.0176	.0216	.0260	.0309	.0361
6	.0008	.0012	.0018	.0026	.0035	.0047	.0061	.0078	.0098	.0120
7	.0001	.0002	.0003	.0005	.0008	.0011	.0015	.0020	.0027	.0034
8	.0000	.0000	.0001	.0001	.0001	.0002	.0003	.0005	.0006	.0009
9	.0000	.0000	.0000	.0000	.0000	.0000	.0001	.0001	.0001	.0002

	2.10	2.20	2.30	2.40	2.50	2.60	2.70	2.80	2.90	3.00
						λt				
0	.1225	.1108	.1003	.0907	.0821	.0743	.0672	.0608	.0550	.0498
1	.2572	.2438	.2306	.2177	.2052	.1931	.1815	.1703	.1596	.1494
2	.2700	.2681	.2652	.2613	.2565	.2510	.2450	.2384	.2314	.2240
3	.1890	.1966	.2033	.2090	.2138	.2176	.2205	.2225	.2237	.2240
4	.0992	.1082	.1169	.1254	.1336	.1414	.1488	.1557	.1622	.1680
5	.0417	.0476	.0538	.0602	.0668	.0735	.0804	.0872	.0940	.1008
6	.0146	.0174	.0206	.0241	.0278	.0319	.0362	.0407	.0455	.0504
7	.0044	.0055	.0068	.0083	.0099	.0118	.0139	.0163	.0188	.0216
8	.0011	.0015	.0019	.0025	.0031	.0038	.0047	.0057	.0068	.0081
9	.0003	.0004	.0005	.0007	.0009	.0011	.0014	.0018	.0022	.0027
10	.0001	.0001	.0001	.0002	.0002	.0003	.0004	.0005	.0006	.0008
11	.0000	.0000	.0000	.0000	.0000	.0001	.0001	.0001	.0002	.0002
12	.0000	.0000	.0000	.0000	.0000	.0000	.0000	.0000	.0000	.0001

X_1					λt					
	3.10	3.20	3.30	3.40	3.50	3.60	3.70	3.80	3.90	4.00
0	.0450	.0408	.0369	.0334	.0302	.0273	.0247	.0224	.0202	.0183
1	.1397	.1304	.1217	.1135	.1057	.0984	.0915	.0850	.0789	.0733
2	.2165	.2087	.2008	.1929	.1850	.1771	.1692	.1615	.1539	.1465
3	.2237	.2226	.2209	.2186	.2158	.2125	.2087	.2046	.2001	.1954
4	.1734	.1781	.1823	.1858	.1888	.1912	.1931	.1944	.1951	.1954
5	.1075	.1140	.1203	.1264	.1322	.1377	.1429	.1477	.1522	.1563
6	.0555	.0608	.0662	.0716	.0771	.0826	.0881	.0936	.0989	.1042
7	.0246	.0278	.0312	.0348	.0385	.0425	.0466	.0508	.0551	.0595
8	.0095	.0111	.0129	.0148	.0169	.0191	.0215	.0241	.0269	.0298
9	.0033	.0040	.0047	.0056	.0066	.0076	.0089	.0102	.0116	.0132
10	.0010	.0013	.0016	.0019	.0023	.0028	.0033	.0039	.0045	.0053
11	.0003	.0004	.0005	.0006	.0007	.0009	.0011	.0013	.0016	.0019
12	.0001	.0001	.0001	.0002	.0002	.0003	.0003	.0004	.0005	.0006
13	.0000	.0000	.0000	.0000	.0001	.0001	.0001	.0001	.0002	.0002
14	.0000	.0000	.0000	.0000	.0000	.0000	.0000	.0000	.0000	.0001

					λt					
	4.10	4.20	4.30	4.40	4.50	4.60	4.70	4.80	4.90	5.00
0	.0166	.0150	.0136	.0123	.0111	.0101	.0091	.0082	.0074	.0067
1	.0679	.0630	.0583	.0540	.0500	.0462	.0427	.0395	.0365	.0337
2	.1393	.1323	.1254	.1188	.1125	.1063	.1005	.0948	.0894	.0842
3	.1904	.1852	.1852	.1798	.1743	.1687	.1631	.1574	.1517	.1460	.1404
4	.1951	.1944	.1933	.1917	.1898	.1875	.1849	.1820	.1789	.1755
5	.1600	.1633	.1662	.1687	.1708	.1725	.1738	.1747	.1753	.1755
6	.1093	.1143	.1191	.1237	.1281	.1323	.1362	.1398	.1432	.1462
7	.0640	.0686	.0732	.0778	.0824	.0869	.0914	.0959	.1002	.1044
8	.0328	.0360	.0393	.0428	.0463	.0500	.0537	.0575	.0614	.0653
9	.0150	.0168	.0188	.0209	.0232	.0255	.0280	.0307	.0334	.0363
10	.0061	.0071	.0081	.0092	.0104	.0118	.0132	.0147	.0164	.0181
11	.0023	.0027	.0032	.0037	.0043	.0049	.0056	.0064	.0073	.0082
12	.0008	.0009	.0011	.0014	.0016	.0019	.0022	.0026	.0030	.0034
13	.0002	.0003	.0004	.0005	.0006	.0007	.0008	.0009	.0011	.0013
14	.0001	.0001	.0001	.0001	.0002	.0002	.0003	.0003	.0004	.0005
15	.0000	.0000	.0000	.0000	.0001	.0001	.0001	.0001	.0001	.0002

					λt					
	5.10	5.20	5.30	5.40	5.50	5.60	5.70	5.80	5.90	6.00
0	.0061	.0055	.0050	.0045	.0041	.0037	.0033	.0030	.0027	.0025
1	.0311	.0287	.0265	.0244	.0225	.0207	.0191	.0176	.0162	.0149
2	.0793	.0746	.0701	.0659	.0618	.0580	.0544	.0509	.0477	.0446
3	.1348	.1293	.1239	.1185	.1133	.1082	.1033	.0985	.0938	.0892
4	.1719	.1681	.1641	.1600	.1558	.1515	.1472	.1428	.1383	.1339
5	.1753	.1748	.1740	.1728	.1714	.1697	.1678	.1656	.1632	.1606
6	.1490	.1515	.1537	.1555	.1571	.1584	.1594	.1601	.1605	.1606
7	.1086	.1125	.1163	.1200	.1234	.1267	.1298	.1326	.1353	.1377
8	.0692	.0731	.0771	.0810	.0849	.0887	.0925	.0962	.0998	.1033
9	.0392	.0423	.0454	.0486	.0519	.0552	.0586	.0620	.0654	.0688
10	.0200	.0220	.0241	.0262	.0285	.0309	.0334	.0359	.0386	.0413
11	.0093	.0104	.0116	.0129	.0143	.0157	.0173	.0190	.0207	.0225
12	.0039	.0045	.0051	.0058	.0065	.0073	.0082	.0092	.0102	.0113
13	.0015	.0018	.0021	.0024	.0028	.0032	.0036	.0041	.0046	.0052
14	.0006	.0007	.0008	.0009	.0011	.0013	.0015	.0017	.0019	.0022
15	.0002	.0002	.0003	.0003	.0004	.0005	.0006	.0007	.0008	.0009
16	.0001	.0001	.0001	.0001	.0001	.0002	.0002	.0002	.0003	.0003
17	.0000	.0000	.0000	.0000	.0000	.0001	.0001	.0001	.0001	.0001

					λt					
X_1	6.10	6.20	6.30	6.40	6.50	6.60	6.70	6.80	6.90	7.00
0	.0022	.0020	.0018	.0017	.0015	.0014	.0012	.0011	.0010	.0009
1	.0137	.0126	.0116	.0106	.0098	.0090	.0082	.0076	.0070	.0064
2	.0417	.0390	.0364	.0340	.0318	.0296	.0276	.0258	.0240	.0223
3	.0848	.0806	.0765	.0726	.0688	.0652	.0617	.0584	.0552	.0521
4	.1294	.1249	.1205	.1162	.1118	.1076	.1034	.0992	.0952	.0912
5	.1579	.1549	.1519	.1487	.1454	.1420	.1385	.1349	.1314	.1277
6	.1605	.1601	.1595	.1586	.1575	.1562	.1546	.1529	.1511	.1490
7	.1399	.1418	.1435	.1450	.1462	.1472	.1480	.1486	.1489	.1490
8	.1066	.1099	.1130	.1160	.1188	.1215	.1240	.1263	.1284	.1304
9	.0723	.0757	.0791	.0825	.0858	.0891	.0923	.0954	.0985	.1014
10	.0441	.0469	.0498	.0528	.0558	.0588	.0618	.0649	.0679	.0710
11	.0245	.0265	.0285	.0307	.0330	.0353	.0377	.0401	.0426	.0452
12	.0124	.0137	.0150	.0164	.0179	.0194	.0210	.0227	.0245	.0264
13	.0058	.0065	.0073	.0081	.0089	.0098	.0108	.0119	.0130	.0142
14	.0025	.0029	.0033	.0037	.0041	.0046	.0052	.0058	.0064	.0071
15	.0010	.0012	.0014	.0016	.0018	.0020	.0023	.0026	.0029	.0033
16	.0004	.0005	.0005	.0006	.0007	.0008	.0010	.0011	.0013	.0014
17	.0001	.0002	.0002	.0002	.0003	.0003	.0004	.0004	.0005	.0006
18	.0000	.0001	.0001	.0001	.0001	.0001	.0001	.0002	.0002	.0002
19	.0000	.0000	.0000	.0000	.0000	.0000	.0000	.0001	.0001	.0001

					λt					
	7.10	7.20	7.30	7.40	7.50	7.60	7.70	7.80	7.90	8.00
0	.0008	.0007	.0007	.0006	.0006	.0005	.0005	.0004	.0004	.0003
1	.0059	.0054	.0049	.0045	.0041	.0038	.0035	.0032	.0029	.0027
2	.0208	.0194	.0180	.0167	.0156	.0145	.0134	.0125	.0116	.0107
3	.0492	.0464	.0438	.0413	.0389	.0366	.0345	.0324	.0305	.0286
4	.0874	.0836	.0799	.0764	.0729	.0696	.0663	.0632	.0602	.0573
5	.1241	.1204	.1167	.1130	.1094	.1057	.1021	.0986	.0951	.0916
6	.1468	.1445	.1420	.1394	.1367	.1339	.1311	.1282	.1252	.1221
7	.1489	.1486	.1481	.1474	.1465	.1454	.1442	.1428	.1413	.1396
8	.1321	.1337	.1351	.1363	.1373	.1382	.1388	.1392	.1395	.1396
9	.1042	.1070	.1096	.1121	.1144	.1167	.1187	.1207	.1224	.1241
10	.0740	.0770	.0800	.0829	.0858	.0887	.0914	.0941	.0967	.0993
11	.0478	.0504	.0531	.0558	.0585	.0613	.0640	.0667	.0695	.0722
12	.0283	.0303	.0323	.0344	.0366	.0388	.0411	.0434	.0457	.0481
13	.0154	.0168	.0181	.0196	.0211	.0227	.0243	.0260	.0278	.0296
14	.0078	.0086	.0095	.0104	.0113	.0123	.0134	.0145	.0157	.0169
15	.0037	.0041	.0046	.0051	.0057	.0062	.0069	.0075	.0083	.0090
16	.0016	.0019	.0021	.0024	.0026	.0030	.0033	.0037	.0041	.0045
17	.0007	.0008	.0009	.0010	.0012	.0013	.0015	.0017	.0019	.0021
18	.0003	.0003	.0004	.0004	.0005	.0006	.0006	.0007	.0008	.0009
19	.0001	.0001	.0001	.0002	.0002	.0002	.0003	.0003	.0003	.0004
20	.0000	.0000	.0001	.0001	.0001	.0001	.0001	.0001	.0001	.0002
21	.0000	.0000	.0000	.0000	.0000	.0000	.0000	.0000	.0001	.0001

					λt					
X_1	8.10	8.20	8.30	8.40	8.50	8.60	8.70	8.80	8.90	9.00
0	.0003	.0003	.0002	.0002	.0002	.0002	.0002	.0002	.0001	.0001
1	.0025	.0023	.0021	.0019	.0017	.0016	.0014	.0013	.0012	.0011
2	.0100	.0092	.0086	.0079	.0074	.0068	.0063	.0058	.0054	.0050
3	.0269	.0252	.0237	.0222	.0208	.0195	.0183	.0171	.0160	.0150
4	.0544	.0517	.0491	.0466	.0443	.0420	.0398	.0377	.0357	.0337
5	.0882	.0849	.0816	.0784	.0752	.0722	.0692	.0663	.0635	.0607
6	.1191	.1160	.1128	.1097	.1066	.1034	.1003	.0972	.0941	.0911
7	.1378	.1358	.1338	.1317	.1294	.1271	.1247	.1222	.1197	.1171
8	.1395	.1392	.1388	.1382	.1375	.1366	.1356	.1344	.1332	.1318
9	.1256	.1269	.1280	.1290	.1299	.1306	.1311	.1315	.1317	.1318
10	.1017	.1040	.1063	.1084	.1104	.1123	.1140	.1157	.1172	.1186
11	.0749	.0776	.0802	.0828	.0853	.0878	.0902	.0925	.0948	.0970
12	.0505	.0530	.0555	.0579	.0604	.0629	.0654	.0679	.0703	.0728
13	.0315	.0334	.0354	.0374	.0395	.0416	.0438	.0459	.0481	.0504
14	.0182	.0196	.0210	.0225	.0240	.0256	.0272	.0289	.0306	.0324
15	.0098	.0107	.0116	.0126	.0136	.0147	.0158	.0169	.0182	.0194
16	.0050	.0055	.0060	.0066	.0072	.0079	.0086	.0093	.0101	.0109
17	.0024	.0026	.0029	.0033	.0036	.0040	.0044	.0048	.0053	.0058
18	.0011	.0012	.0014	.0015	.0017	.0019	.0021	.0024	.0026	.0029
19	.0005	.0005	.0006	.0007	.0008	.0009	.0010	.0011	.0012	.0014
20	.0002	.0002	.0002	.0003	.0003	.0004	.0004	.0005	.0005	.0006
21	.0001	.0001	.0001	.0001	.0001	.0002	.0002	.0002	.0002	.0003
22	.0000	.0000	.0000	.0000	.0001	.0001	.0001	.0001	.0001	.0001

					λt					
	9.10	9.20	9.30	9.40	9.50	9.60	9.70	9.80	9.90	10.00
0	.0001	.0001	.0001	.0001	.0001	.0001	.0001	.0001	.0001	.0000
1	.0010	.0009	.0009	.0008	.0007	.0007	.0006	.0005	.0005	.0005
2	.0046	.0043	.0040	.0037	.0034	.0031	.0029	.0027	.0025	.0023
3	.0140	.0131	.0123	.0115	.0107	.0100	.0093	.0087	.0081	.0076
4	.0319	.0302	.0285	.0269	.0254	.0240	.0226	.0213	.0201	.0189
5	.0581	.0555	.0530	.0506	.0483	.0460	.0439	.0418	.0398	.0378
6	.0881	.0851	.0822	.0793	.0764	.0736	.0709	.0682	.0656	.0631
7	.1145	.1118	.1091	.1064	.1037	.1010	.0982	.0955	.0928	.0901
8	.1302	.1286	.1269	.1251	.1232	.1212	.1191	.1170	.1148	.1126
9	.1317	.1315	.1311	.1306	.1300	.1293	.1284	.1274	.1263	.1251
10	.1198	.1210	.1219	.1228	.1235	.1241	.1245	.1249	.1250	.1251
11	.0991	.1012	.1031	.1049	.1067	.1083	.1098	.1112	.1125	.1137
12	.0752	.0776	.0799	.0822	.0844	.0866	.0888	.0908	.0928	.0948
13	.0526	.0549	.0572	.0594	.0617	.0640	.0662	.0685	.0707	.0729
14	.0342	.0361	.0380	.0399	.0419	.0439	.0459	.0479	.0500	.0521
15	.0208	.0221	.0235	.0250	.0265	.0281	.0297	.0313	.0330	.0347
16	.0118	.0127	.0137	.0147	.0157	.0168	.0180	.0192	.0204	.0217
17	.0063	.0069	.0075	.0081	.0088	.0095	.0103	.0111	.0119	.0128
18	.0032	.0035	.0039	.0042	.0046	.0051	.0055	.0060	.0065	.0071
19	.0015	.0017	.0019	.0021	.0023	.0026	.0028	.0031	.0034	.0037
20	.0007	.0008	.0009	.0010	.0011	.0012	.0014	.0015	.0017	.0019
21	.0003	.0003	.0004	.0004	.0005	.0006	.0006	.0007	.0008	.0009
22	.0001	.0001	.0002	.0002	.0002	.0002	.0003	.0003	.0004	.0004
23	.0000	.0001	.0001	.0001	.0001	.0001	.0001	.0001	.0002	.0002
24	.0000	.0000	.0000	.0000	.0000	.0000	.0000	.0001	.0001	.0001

X_1	\(\lambda t\) 11.	12.	13.	14.	15.	16.	17.	18.	19.	20.
0	.0000	.0000	.0000	.0000	.0000	.0000	.0000	.0000	.0000	.0000
1	.0002	.0001	.0000	.0000	.0000	.0000	.0000	.0000	.0000	.0000
2	.0010	.0004	.0002	.0001	.0000	.0000	.0000	.0000	.0000	.0000
3	.0037	.0018	.0008	.0004	.0002	.0001	.0000	.0000	.0000	.0000
4	.0102	.0053	.0027	.0013	.0006	.0003	.0001	.0001	.0000	.0000
5	.0224	.0127	.0070	.0037	.0019	.0010	.0005	.0002	.0001	.0001
6	.0411	.0255	.0152	.0087	.0048	.0026	.0014	.0007	.0004	.0002
7	.0646	.0437	.0281	.0174	.0104	.0060	.0034	.0019	.0010	.0005
8	.0888	.0655	.0457	.0304	.0194	.0120	.0072	.0042	.0024	.0013
9	.1085	.0874	.0661	.0473	.0324	.0213	.0135	.0083	.0050	.0029
10	.1194	.1048	.0859	.0663	.0486	.0341	.0230	0150	.0095	.0058
11	.1194	.1144	.1015	.0844	.0663	.0496	.0355	.0245	.0164	.0106
12	.1094	.1144	.1099	.0984	.0829	.0661	.0504	.0368	.0259	.0176
13	.0926	.1056	.1099	.1060	.0956	.0814	.0658	.0509	.0378	.0271
14	.0728	.0905	.1021	.1060	.1024	.0930	.0800	.0655	.0514	.0387
15	.0534	.0724	.0885	.0989	.1024	.0992	.0906	.0786	.0650	.0516
16	.0367	.0543	.0719	.0866	.0960	.0992	.0963	.0884	.0772	.0646
17	.0237	.0383	.0550	.0713	.0847	.0934	.0963	.0936	.0863	.0760
18	.0145	.0256	.0397	.0554	.0706	.0830	.0909	.0936	.0911	.0844
19	.0084	.0161	.0272	.0409	.0557	.0699	.0814	.0887	.0911	.0888
20	.0046	.0097	.0177	.0286	.0418	.0559	.0692	.0798	.0866	.0888
21	.0024	.0055	.0109	.0191	.0299	.0426	.0560	.0684	.0783	.0846
22	.0012	.0030	.0065	.0121	.0204	.0310	.0433	.0560	.0676	.0769
23	.0006	.0016	.0037	.0074	.0133	.0216	.0320	.0438	.0559	.0669
24	.0003	.0008	.0020	.0043	.0083	.0144	.0226	.0329	.0442	.0557
25	.0001	.0004	.0010	.0024	.0050	.0092	.0154	.0237	.0336	.0446
26	.0000	.0002	·0005	.0013	.0029	.0057	.0101	.0164	.0246	.0343
27	.0000	.0001	.0002	.0007	.0016	.0034	.0063	.0109	.0173	.0254
28	.0000	.0000	.0001	.0003	.0009	.0019	.0038	.0070	.0117	.0181
29	.0000	.0000	.0001	.0002	.0004	.0011	.0023	.0044	.0077	.0125
30	.0000	.0000	.0000	.0001	.0002	.0006	.0013	.0026	.0049	.0083
31	.0000	.0000	.0000	.0000	.0001	.0003	.0007	.0015	.0030	.0054
32	.0000	.0000	.0000	.0000	.0001	.0001	.0004	.0009	.0018	.0034
33	.0000	.0000	.0000	.0000	.0000	.0001	.0002	.0005	.0010	.0020
34	.0000	.0000	.0000	.0000	.0000	.0000	.0001	.0002	.0006	.0012
35	.0000	.0000	.0000	.0000	.0000	.0000	.0000	.0001	.0003	.0007
36	.0000	.0000	.0000	.0000	.0000	.0000	.0000	.0001	.0002	.0004
37	.0000	.0000	.0000	.0000	.0000	.0000	.0000	.0000	.0001	.0002
38	.0000	.0000	.0000	.0000	.0000	.0000	.0000	.0000	.0000	.0001
39	.0000	.0000	.0000	.0000	.0000	.0000	.0000	.0000	.0000	.0001

SOURCE: From Stephen P. Shao, *Statistics for Business and Economics*, 3rd ed. (Columbus, Ohio: Charles E. Merrill, 1976), pp. 782–86. Used with permission.

Standard Normal Distribution Table

To illustrate: 43.45 percent of the area under a normal curve lies between the mean, μ_x, and a point 1.51 standard deviation units away.

Example:
$z = 0.52$ (or -0.52),
$A(z) = 0.19847$ or 19.847%

Z	.00	.01	.02	.03	.04	.05	.06	.07	.08	.09
0.0	.00000	.00399	.00798	.01197	.01595	.01994	.02392	.02790	.03188	.03586
0.1	.03983	.04380	.04776	.05172	.05567	.05962	.06356	.06749	.07142	.07535
0.2	.07926	.08317	.08706	.09095	.09483	.09871	.10257	.10642	.11026	.11409
0.3	.11791	.12172	.12552	.12930	.13307	.13683	.14058	.14431	.14803	.15173
0.4	.15542	.15910	.16276	.16640	.17003	.17364	.17724	.18082	.18439	.18793
0.5	.19146	.19497	.19847	.20194	.20540	.20884	.21226	.21566	.21904	.22240
0.6	.22575	.22907	.23237	.23565	.23891	.24215	.24537	.24857	.25175	.25490
0.7	.25804	.26115	.26424	.26730	.27035	.27337	.27637	.27935	.28230	.28524
0.8	.28814	.29103	.29389	.29673	.29955	.30234	.30511	.30785	.31057	.31327
0.9	.31594	.31859	.32121	.32381	.32639	.32894	.33147	.33398	.33646	.33891
1.0	.34134	.34375	.34614	.34850	.35083	.35314	.35543	.35769	.35993	.36214
1.1	.36433	.36650	.36864	.37076	.37286	.37493	.37698	.37900	.38100	.38298
1.2	.38493	.38686	.38877	.39065	.39251	.39435	.39617	.39796	.39973	.40147
1.3	.40320	.40490	.40658	.40824	.40988	.41149	.41309	.41466	.41621	.41774
1.4	.41924	.42073	.42220	.42364	.42507	.42647	.42786	.42922	.43056	.43189
1.5	.43319	.43448	.43574	.43699	.43822	.43943	.44062	.44179	.44295	.44408
1.6	.44520	.44630	.44738	.44845	.44950	.45053	.45154	.45254	.45352	.45449
1.7	.45543	.45637	.45728	.45818	.45907	.45994	.46080	.46164	.46246	.46327
1.8	.46407	.46485	.46562	.46638	.46712	.46784	.46856	.46926	.46995	.47062
1.9	.47128	.47193	.47257	.47320	.47381	.47441	.47500	.47558	.47615	.47670
2.0	.47725	.47778	.47831	.47882	.47932	.47982	.48030	.48077	.48124	.48169
2.1	.48214	.48257	.48300	.48341	.48382	.48422	.48461	.48500	.48537	.48574
2.2	.48610	.48645	.48679	.48713	.48745	.48778	.48809	.48840	.48870	.48899
2.3	.48928	.48956	.48983	.49010	.49036	.49061	.49086	.49111	.49134	.49158
2.4	.49180	.49202	.49224	.49245	.49266	.49286	.49305	.49324	.49343	.49361
2.5	.49379	.49396	.49413	.49430	.49446	.49461	.49477	.49492	.49506	.49520
2.6	.49534	.49547	.49560	.49573	.49585	.49598	.49609	.49621	.49632	.49643
2.7	.49653	.49664	.49674	.49683	.49693	.49702	.49711	.49720	.49728	.49736
2.8	.49744	.49752	.49760	.49767	.49774	.49781	.49788	.49795	.49801	.49807
2.9	.49813	.49819	.49825	.49831	.49386	.49841	.49846	.49851	.49856	.49861
3.0	.49865	.49869	.49874	.49878	.49882	.49886	.49889	.49893	.49897	.49900
3.1	.49903	.49906	.49910	.49913	.49916	.49918	.49921	.49924	.49926	.49929
3.2	.49931	.49934	.49936	.49938	.49940	.49942	.49944	.49946	.49948	.49950
3.3	.49952	.49953	.49955	.49957	.49958	.49960	.49961	.49962	.49964	.49965
3.4	.49966	.49968	.49969	.49970	.49971	.49972	.49973	.49974	.49975	.49976
3.5	.49977	.49978	.49978	.49979	.49980	.49981	.49981	.49982	.49983	.49983
3.6	.49984	.49985	.49985	.49986	.49986	.49987	.49987	.49988	.49988	.49989
3.7	.49989	.49990	.49990	.49990	.49991	.49991	.49992	.49992	.49992	.49992
3.8	.49993	.49993	.49993	.49994	.49994	.49994	.49994	.49995	.49995	.49995
3.9	.49995	.49995	.49996	.49996	.49996	.49996	.49996	.49996	.49997	.49997
4.0	.49997									

SOURCE: From Stephen P. Shao, *Statistics for Business and Economics*, 3rd ed. (Columbus, Ohio: Charles E. Merrill, 1976), p. 788. Used with permission.

Values of *t* for Selected Probabilities

Example.
d.f. (Number of degrees of freedom) = 6:
One tail above $t = 1.134$ *or* below $t = -1.134$ represents 0.15 or 15% of the area under the curve.
Two tails above $t = 1.134$ *and* below $t = -1.134$ represent 0.30 or 30%.

	Probabilities (or Areas Under *t*-Distribution Curve)								
One tail	.45	.35	.25	.15	.10	.05	.025	.01	.005
Two tails	.90	.70	.50	.30	.20	.10	.05	.02	.01
Conf. Level	.10	.30	.50	.70	.80	.90	.95	.98	.99
d.f.				Values of *t*					
1	.158	.510	1.000	1.963	3.078	6.314	12.706	31.821	63.657
2	.142	.445	.816	1.386	1.886	2.920	4.303	6.965	9.925
3	.137	.424	.765	1.250	1.638	2.353	3.182	4.541	5.841
4	.134	.414	.741	1.190	1.533	2.132	2.776	3.747	4.604
5	.132	.408	.727	1.156	1.476	2.015	2.571	3.365	4.032
6	.131	.404	.718	**1.134**	1.440	1.943	2.447	3.143	3.707
7	.130	.402	.711	1.119	1.415	1.895	2.365	2.998	3.499
8	.130	.399	.706	1.108	1.397	1.860	2.306	2.896	3.355
9	.129	.398	.703	1.100	1.383	1.833	2.262	2.821	3.250
10	.129	.397	.700	1.093	1.372	1.812	2.228	2.764	3.169
11	.129	.396	.697	1.088	1.363	1.796	2.201	2.718	3.106
12	.128	.395	.695	1.083	1.356	1.782	2.179	2.681	3.055
13	.128	.394	.694	1.079	1.350	1.771	2.160	2.650	3.012
14	.128	.393	.692	1.076	1.345	1.761	2.145	2.624	2.977
15	.128	.393	.691	1.074	1.341	1.753	2.131	2.602	2.947
16	.128	.392	.690	1.071	1.337	1.746	2.120	2.583	2.921
17	.128	.392	.689	1.069	1.333	1.740	2.110	2.567	2.898
18	.127	.392	.688	1.067	1.330	1.734	2.101	2.552	2.878
19	.127	.391	.688	1.066	1.328	1.729	2.093	2.539	2.861
20	.127	.391	.687	1.064	1.325	1.725	2.086	2.528	2.845
21	.127	.391	.686	1.063	1.323	1.721	2.080	2.518	2.831
22	.127	.390	.686	1.061	1.321	1.717	2.074	2.508	2.819
23	.127	.390	.685	1.060	1.319	1.714	2.069	2.500	2.807
24	.127	.390	.685	1.059	1.318	1.711	2.064	2.492	2.797
25	.127	.390	.684	1.058	1.316	1.708	2.060	2.485	2.787
26	.127	.390	.684	1.058	1.315	1.706	2.056	2.479	2.779
27	.127	.389	.684	1.057	1.314	1.703	2.052	2.473	2.771
28	.127	.389	.683	1.056	1.313	1.701	2.048	2.467	2.763
29	.127	.389	.683	1.055	1.311	1.699	2.045	2.462	2.756
30	.127	.389	.683	1.055	1.310	1.697	2.042	2.457	2.750
40	.126	.388	.681	1.050	1.303	1.684	2.021	2.423	2.704
60	.126	.387	.679	1.046	1.296	1.671	2.000	2.390	2.660
120	.126	.386	.677	1.041	1.289	1.658	1.980	2.358	2.617
∞	.126	.385	.674	1.036	1.282	1.645	1.960	2.326	2.576

SOURCE: From Stephen P. Shao, *Statistics for Business and Economics*, 3rd ed. (Columbus, Ohio: Charles E. Merrill, 1976), p. 789. Used with permission.

Values of χ^2 for Selected Probabilities

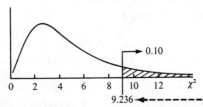

Example.
d.f. (Number of degrees of freedom) = 5,
the tail above $\chi^2 = 9.236$ represents 0.10 or 10% of the area under the curve.

	Probabilities (or Areas Under χ^2 Distribution Curve Above Given χ^2 Values)								
	.90	.70	.50	.30	.20	.10	.05	.02	.01
d.f	Values of χ^2								
1	.016	.148	.455	1.074	1.642	2.706	3.841	5.412	6.635
2	.211	.713	1.386	2.408	3.219	4.605	5.991	7.824	9.210
3	.584	1.424	2.366	3.665	4.642	6.251	7.815	9.837	11.345
4	1.064	2.195	3.357	4.878	5.989	7.779	9.488	11.668	13.277
5	1.610	3.000	4.351	6.064	7.289	9.236	11.070	13.388	15.086
6	2.204	3.828	5.348	7.231	8.558	10.645	12.592	15.033	16.812
7	2.833	4.671	6.346	8.383	9.803	12.017	14.067	16.622	18.475
8	3.490	5.527	7.344	9.524	11.030	13.362	15.507	18.168	20.090
9	4.168	6.393	8.343	10.656	12.242	14.684	16.919	19.679	21.666
10	4.865	7.267	9.342	11.781	13.442	15.987	18.307	21.161	23.209
11	5.578	8.148	10.341	12.899	14.631	17.275	19.675	22.618	24.725
12	6.304	9.034	11.340	14.011	15.812	18.549	21.026	24.054	26.217
13	7.042	9.926	12.340	15.119	16.985	19.812	22.362	25.472	27.688
14	7.790	10.821	13.339	16.222	18.151	21.064	23.685	26.873	29.141
15	8.547	11.721	14.339	17.322	19.311	22.307	24.996	28.259	30.578
16	9.312	12.624	15.338	18.418	20.465	23.542	26.296	29.633	32.000
17	10.085	13.531	16.338	19.511	21.615	24.769	27.587	30.995	33.409
18	10.865	14.440	17.338	20.601	22.760	25.989	28.869	33.346	34.805
19	11.651	15.352	18.338	21.689	23.900	27.204	30.144	33.687	36.191
20	12.443	16.266	19.337	22.775	25.038	28.412	31.410	35.020	37.566
21	13.240	17.182	20.337	23.858	26.171	29.615	32.671	36.343	38.932
22	14.041	18.101	21.337	24.939	27.301	30.813	33.924	37.659	40.289
23	14.848	19.021	22.337	26.018	28.429	32.007	35.172	38.968	41.638
24	15.659	19.943	23.337	27.096	29.553	33.196	36.415	40.270	42.980
25	16.473	20.867	24.337	28.172	30.675	34.382	37.652	41.566	44.314
26	17.292	21.792	25.336	29.246	31.795	35.563	38.885	42.856	45.642
27	18.114	22.719	26.336	30.319	32.912	36.741	40.113	44.140	46.963
28	18.939	23.647	27.336	31.391	34.027	37.916	41.337	45.419	48.278
29	19.768	24.577	28.336	32.461	35.139	39.087	42.557	46.693	49.588
30	20.599	25.508	29.336	33.530	36.250	40.256	43.773	47.962	50.892

SOURCE: From Stephen P. Shao, *Statistics for Business and Economics*, 3rd ed. (Columbus, Ohio: Charles E. Merrill, 1976), p. 790. Used with permission.

Values of F

Upper 5% Probability
(or 5% Area under F- Distribution Curve)

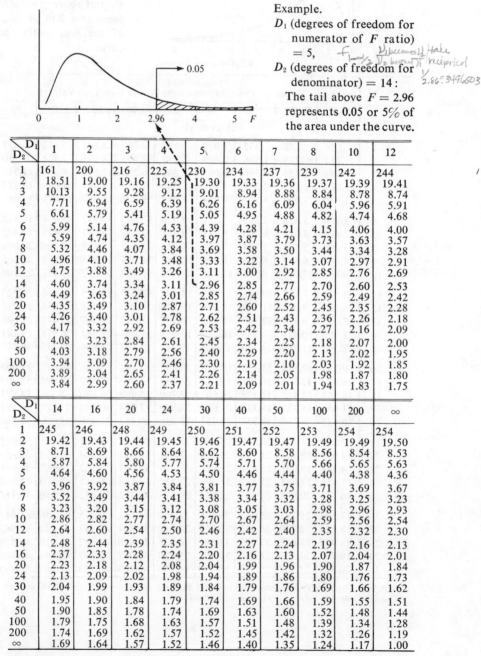

Example.
D_1 (degrees of freedom for numerator of F ratio) = 5,
D_2 (degrees of freedom for denominator) = 14:
The tail above $F = 2.96$ represents 0.05 or 5% of the area under the curve.

D_1 / D_2	1	2	3	4	5	6	7	8	10	12
1	161	200	216	225	230	234	237	239	242	244
2	18.51	19.00	19.16	19.25	19.30	19.33	19.36	19.37	19.39	19.41
3	10.13	9.55	9.28	9.12	9.01	8.94	8.88	8.84	8.78	8.74
4	7.71	6.94	6.59	6.39	6.26	6.16	6.09	6.04	5.96	5.91
5	6.61	5.79	5.41	5.19	5.05	4.95	4.88	4.82	4.74	4.68
6	5.99	5.14	4.76	4.53	4.39	4.28	4.21	4.15	4.06	4.00
7	5.59	4.74	4.35	4.12	3.97	3.87	3.79	3.73	3.63	3.57
8	5.32	4.46	4.07	3.84	3.69	3.58	3.50	3.44	3.34	3.28
10	4.96	4.10	3.71	3.48	3.33	3.22	3.14	3.07	2.97	2.91
12	4.75	3.88	3.49	3.26	3.11	3.00	2.92	2.85	2.76	2.69
14	4.60	3.74	3.34	3.11	2.96	2.85	2.77	2.70	2.60	2.53
16	4.49	3.63	3.24	3.01	2.85	2.74	2.66	2.59	2.49	2.42
20	4.35	3.49	3.10	2.87	2.71	2.60	2.52	2.45	2.35	2.28
24	4.26	3.40	3.01	2.78	2.62	2.51	2.43	2.36	2.26	2.18
30	4.17	3.32	2.92	2.69	2.53	2.42	2.34	2.27	2.16	2.09
40	4.08	3.23	2.84	2.61	2.45	2.34	2.25	2.18	2.07	2.00
50	4.03	3.18	2.79	2.56	2.40	2.29	2.20	2.13	2.02	1.95
100	3.94	3.09	2.70	2.46	2.30	2.19	2.10	2.03	1.92	1.85
200	3.89	3.04	2.65	2.41	2.26	2.14	2.05	1.98	1.87	1.80
∞	3.84	2.99	2.60	2.37	2.21	2.09	2.01	1.94	1.83	1.75

D_1 / D_2	14	16	20	24	30	40	50	100	200	∞
1	245	246	248	249	250	251	252	253	254	254
2	19.42	19.43	19.44	19.45	19.46	19.47	19.47	19.49	19.49	19.50
3	8.71	8.69	8.66	8.64	8.62	8.60	8.58	8.56	8.54	8.53
4	5.87	5.84	5.80	5.77	5.74	5.71	5.70	5.66	5.65	5.63
5	4.64	4.60	4.56	4.53	4.50	4.46	4.44	4.40	4.38	4.36
6	3.96	3.92	3.87	3.84	3.81	3.77	3.75	3.71	3.69	3.67
7	3.52	3.49	3.44	3.41	3.38	3.34	3.32	3.28	3.25	3.23
8	3.23	3.20	3.15	3.12	3.08	3.05	3.03	2.98	2.96	2.93
10	2.86	2.82	2.77	2.74	2.70	2.67	2.64	2.59	2.56	2.54
12	2.64	2.60	2.54	2.50	2.46	2.42	2.40	2.35	2.32	2.30
14	2.48	2.44	2.39	2.35	2.31	2.27	2.24	2.19	2.16	2.13
16	2.37	2.33	2.28	2.24	2.20	2.16	2.13	2.07	2.04	2.01
20	2.23	2.18	2.12	2.08	2.04	1.99	1.96	1.90	1.87	1.84
24	2.13	2.09	2.02	1.98	1.94	1.89	1.86	1.80	1.76	1.73
30	2.04	1.99	1.93	1.89	1.84	1.79	1.76	1.69	1.66	1.62
40	1.95	1.90	1.84	1.79	1.74	1.69	1.66	1.59	1.55	1.51
50	1.90	1.85	1.78	1.74	1.69	1.63	1.60	1.52	1.48	1.44
100	1.79	1.75	1.68	1.63	1.57	1.51	1.48	1.39	1.34	1.28
200	1.74	1.69	1.62	1.57	1.52	1.45	1.42	1.32	1.26	1.19
∞	1.69	1.64	1.57	1.52	1.46	1.40	1.35	1.24	1.17	1.00

SOURCE: Reproduced by permission from *Statistical Methods*, 5th ed., by George W. Snedecor, © 1956 by the Iowa State University Press.

Upper 1% Probability
(or 1% Area under *F*-Distribution Curve)

Example.
D_1 (degrees of freedom for numerator of F ratio) = 5,
D_2 (degrees of freedom for denominator) = 14:
The tail above $F = 4.69$ represents 0.01 or 1% of the area under the curve.

D_1 D_2	1	2	3	4	5	6	7	8	10	12
1	4,052	4,999	5,403	5,625	5,764	5,859	5,928	5,981	6,056	6,106
2	98.49	99.00	99.17	99.25	99.30	99.33	99.34	99.36	99.40	99.42
3	34.12	30.82	29.46	28.71	28.24	27.91	27.67	27.49	27.23	27.05
4	21.20	18.00	16.69	15.98	15.52	15.21	14.98	14.80	14.54	14.37
5	16.26	13.27	12.06	11.39	10.97	10.67	10.45	10.27	10.05	9.89
6	13.74	10.92	9.78	9.15	8.75	8.47	8.26	8.10	7.87	7.72
7	12.25	9.55	8.45	7.85	7.46	7.19	7.00	6.84	6.62	6.47
8	11.26	8.65	7.59	7.01	6.63	6.37	6.19	6.03	5.82	5.67
10	10.04	7.56	6.55	5.99	5.64	5.39	5.21	5.06	4.85	4.71
12	9.33	6.93	5.95	5.41	5.06	4.82	4.65	4.50	4.30	4.16
14	8.86	6.51	5.56	5.03	4.69	4.46	4.28	4.14	3.94	3.80
16	8.53	6.23	5.29	4.77	4.44	4.20	4.03	3.89	3.69	3.55
20	8.10	5.85	4.94	4.43	4.10	3.87	3.71	3.56	3.37	3.23
24	7.82	5.61	4.72	4.22	3.90	3.67	3.50	3.36	3.17	3.03
30	7.56	5.39	4.51	4.02	3.70	3.47	3.30	3.17	2.98	2.84
40	7.31	5.18	4.31	3.83	3.51	3.29	3.12	2.99	2.80	2.66
50	7.17	5.06	4.20	3.72	3.41	3.18	3.02	2.88	2.70	2.56
100	6.90	4.82	3.98	3.51	3.20	2.99	2.82	2.69	2.51	2.36
200	6.76	4.71	3.88	3.41	3.11	2.90	2.73	2.60	2.41	2.28
∞	6.64	4.60	3.78	3.32	3.02	2.80	2.64	2.51	2.32	2.18

D_1 D_2	14	16	20	24	30	40	50	100	200	∞
1	6,142	6,169	6,208	6,234	6,258	6,286	6,302	6,334	6,352	6,366
2	99.43	99.44	99.45	99.46	99.47	99.48	99.48	99.49	99.49	99.50
3	26.92	26.83	26.69	26.60	26.50	26.41	26.35	26.23	26.18	26.12
4	14.24	14.15	14.02	13.93	13.83	13.74	13.69	13.57	13.52	13.46
5	9.77	9.68	9.55	9.47	9.38	9.29	9.24	9.13	9.07	9.02
6	7.60	7.52	7.39	7.31	7.23	7.14	7.09	6.99	6.94	6.88
7	6.35	6.27	6.15	6.07	5.98	5.90	5.85	5.75	5.70	5.65
8	5.56	5.48	5.36	5.28	5.20	5.11	5.06	4.96	4.91	4.86
10	4.60	4.52	4.41	4.33	4.25	4.17	4.12	4.01	3.96	3.91
12	4.05	3.98	3.86	3.78	3.70	3.61	3.56	3.46	3.41	3.36
14	3.70	3.62	3.51	3.43	3.34	3.26	3.21	3.11	3.06	3.00
16	3.45	3.37	3.25	3.18	3.10	3.01	2.96	2.86	2.80	2.75
20	3.13	3.05	2.94	2.86	2.77	2.69	2.63	2.53	2.47	2.42
24	2.93	2.85	2.74	2.66	2.58	2.49	2.44	2.33	2.27	2.21
30	2.74	2.66	2.55	2.47	2.38	2.29	2.24	2.13	2.07	2.01
40	2.56	2.49	2.37	2.29	2.20	2.11	2.05	1.94	1.88	1.81
50	2.46	2.39	2.26	2.18	2.10	2.00	1.94	1.82	1.76	1.68
100	2.26	2.19	2.06	1.98	1.89	1.79	1.73	1.59	1.51	1.43
200	2.17	2.09	1.97	1.88	1.79	1.69	1.62	1.48	1.39	1.28
∞	2.07	1.99	1.87	1.79	1.69	1.59	1.52	1.36	1.25	1.00

SOURCE: Reproduced by permission from *Statistical Methods*, 5th ed., by George W. Snedecor, © 1956 by the Iowa State University Press.

TABLES

Distribution of the Studentized Range

Percentage points of the studentized range, $q = (x_n - x_1)/s_\nu$.

$p = 0.95$

ν \ n	2	3	4	5	6	7	8	9	10
1	17·97	26·98	32·82	37·08	40·41	43·12	45·40	47·36	49·07
2	6·08	8·33	9·80	10·88	11·74	12·44	13·03	13·54	13·99
3	4·50	5·91	6·82	7·50	8·04	8·48	8·85	9·18	9·46
4	3·93	5·04	5·76	6·29	6·71	7·05	7·35	7·60	7·83
5	3·64	4·60	5·22	5·67	6·03	6·33	6·58	6·80	6·99
6	3·46	4·34	4·90	5·30	5·63	5·90	6·12	6·32	6·49
7	3·34	4·16	4·68	5·06	5·36	5·61	5·82	6·00	6·16
8	3·26	4·04	4·53	4·89	5·17	5·40	5·60	5·77	5·92
9	3·20	3·95	4·41	4·76	5·02	5·24	5·43	5·59	5·74
10	3·15	3·88	4·33	4·65	4·91	5·12	5·30	5·46	5·60
11	3·11	3·82	4·26	4·57	4·82	5·03	5·20	5·35	5·49
12	3·08	3·77	4·20	4·51	4·75	4·95	5·12	5·27	5·39
13	3·06	3·73	4·15	4·45	4·69	4·88	5·05	5·19	5·32
14	3·03	3·70	4·11	4·41	4·64	4·83	4·99	5·13	5·25
15	3·01	3·67	4·08	4·37	4·59	4·78	4·94	5·08	5·20
16	3·00	3·65	4·05	4·33	4·56	4·74	4·90	5·03	5·15
17	2·98	3·63	4·02	4·30	4·52	4·70	4·86	4·99	5·11
18	2·97	3·61	4·00	4·28	4·49	4·67	4·82	4·96	5·07
19	2·96	3·59	3·98	4·25	4·47	4·65	4·79	4·92	5·04
20	2·95	3·58	3·96	4·23	4·45	4·62	4·77	4·90	5·01
24	2·92	3·53	3·90	4·17	4·37	4·54	4·68	4·81	4·92
30	2·89	3·49	3·85	4·10	4·30	4·46	4·60	4·72	4·82
40	2·86	3·44	3·79	4·04	4·23	4·39	4·52	4·63	4·73
60	2·83	3·40	3·74	3·98	4·16	4·31	4·44	4·55	4·65
120	2·80	3·36	3·68	3·92	4·10	4·24	4·36	4·47	4·56
∞	2·77	3·31	3·63	3·86	4·03	4·17	4·29	4·39	4·47

ν \ n	11	12	13	14	15	16	17	18	19	20
1	50·59	51·96	53·20	54·33	55·36	56·32	57·22	58·04	58·83	59·56
2	14·39	14·75	15·08	15·38	15·65	15·91	16·14	16·37	16·57	16·77
3	9·72	9·95	10·15	10·35	10·52	10·69	10·84	10·98	11·11	11·24
4	8·03	8·21	8·37	8·52	8·66	8·79	8·91	9·03	9·13	9·23
5	7·17	7·32	7·47	7·60	7·72	7·83	7·93	8·03	8·12	8·21
6	6·65	6·79	6·92	7·03	7·14	7·24	7·34	7·43	7·51	7·59
7	6·30	6·43	6·55	6·66	6·76	6·85	6·94	7·02	7·10	7·17
8	6·05	6·18	6·29	6·39	6·48	6·57	6·65	6·73	6·80	6·87
9	5·87	5·98	6·09	6·19	6·28	6·36	6·44	6·51	6·58	6·64
10	5·72	5·83	5·93	6·03	6·11	6·19	6·27	6·34	6·40	6·47
11	5·61	5·71	5·81	5·90	5·98	6·06	6·13	6·20	6·27	6·33
12	5·51	5·61	5·71	5·80	5·88	5·95	6·02	6·09	6·15	6·21
13	5·43	5·53	5·63	5·71	5·79	5·86	5·93	5·99	6·05	6·11
14	5·36	5·46	5·55	5·64	5·71	5·79	5·85	5·91	5·97	6·03
15	5·31	5·40	5·49	5·57	5·65	5·72	5·78	5·85	5·90	5·96
16	5·26	5·35	5·44	5·52	5·59	5·66	5·73	5·79	5·84	5·90
17	5·21	5·31	5·39	5·47	5·54	5·61	5·67	5·73	5·79	5·84
18	5·17	5·27	5·35	5·43	5·50	5·57	5·63	5·69	5·74	5·79
19	5·14	5·23	5·31	5·39	5·46	5·53	5·59	5·65	5·70	5·75
20	5·11	5·20	5·28	5·36	5·43	5·49	5·55	5·61	5·66	5·71
24	5·01	5·10	5·18	5·25	5·32	5·38	5·44	5·49	5·55	5·59
30	4·92	5·00	5·08	5·15	5·21	5·27	5·33	5·38	5·43	5·47
40	4·82	4·90	4·98	5·04	5·11	5·16	5·22	5·27	5·31	5·36
60	4·73	4·81	4·88	4·94	5·00	5·06	5·11	5·15	5·20	5·24
120	4·64	4·71	4·78	4·84	4·90	4·95	5·00	5·04	5·09	5·13
∞	4·55	4·62	4·68	4·74	4·80	4·85	4·89	4·93	4·97	5·01

n: size of sample from which range obtained. ν: degrees of freedom of independent s_ν.

$p = 0.99$

ν \ n	2	3	4	5	6	7	8	9	10
1	90·03	135·0	164·3	185·6	202·2	215·8	227·2	237·0	245·6
2	14·04	19·02	22·29	24·72	26·63	28·20	29·53	30·68	31·69
3	8·26	10·62	12·17	13·33	14·24	15·00	15·64	16·20	16·69
4	6·51	8·12	9·17	9·96	10·58	11·10	11·55	11·93	12·27
5	5·70	6·98	7·80	8·42	8·91	9·32	9·67	9·97	10·24
6	5·24	6·33	7·03	7·56	7·97	8·32	8·61	8·87	9·10
7	4·95	5·92	6·54	7·01	7·37	7·68	7·94	8·17	8·37
8	4·75	5·64	6·20	6·62	6·96	7·24	7·47	7·68	7·86
9	4·60	5·43	5·96	6·35	6·66	6·91	7·13	7·33	7·49
10	4·48	5·27	5·77	6·14	6·43	6·67	6·87	7·05	7·21
11	4·39	5·15	5·62	5·97	6·25	6·48	6·67	6·84	6·99
12	4·32	5·05	5·50	5·84	6·10	6·32	6·51	6·67	6·81
13	4·26	4·96	5·40	5·73	5·98	6·19	6·37	6·53	6·67
14	4·21	4·89	5·32	5·63	5·88	6·08	6·26	6·41	6·54
15	4·17	4·84	5·25	5·56	5·80	5·99	6·16	6·31	6·44
16	4·13	4·79	5·19	5·49	5·72	5·92	6·08	6·22	6·35
17	4·10	4·74	5·14	5·43	5·66	5·85	6·01	6·15	6·27
18	4·07	4·70	5·09	5·38	5·60	5·79	5·94	6·08	6·20
19	4·05	4·67	5·05	5·33	5·55	5·73	5·89	6·02	6·14
20	4·02	4·64	5·02	5·29	5·51	5·69	5·84	5·97	6·09
24	3·96	4·55	4·91	5·17	5·37	5·54	5·69	5·81	5·92
30	3·89	4·45	4·80	5·05	5·24	5·40	5·54	5·65	5·76
40	3·82	4·37	4·70	4·93	5·11	5·26	5·39	5·50	5·60
60	3·76	4·28	4·59	4·82	4·99	5·13	5·25	5·36	5·45
120	3·70	4·20	4·50	4·71	4·87	5·01	5·12	5·21	5·30
∞	3·64	4·12	4·40	4·60	4·76	4·88	4·99	5·08	5·16

ν \ n	11	12	13	14	15	16	17	18	19	20
1	253·2	260·0	266·2	271·8	277·0	281·8	286·3	290·4	294·3	298·0
2	32·59	33·40	34·13	34·81	35·43	36·00	36·53	37·03	37·50	37·95
3	17·13	17·53	17·89	18·22	18·52	18·81	19·07	19·32	19·55	19·77
4	12·57	12·84	13·09	13·32	13·53	13·73	13·91	14·08	14·24	14·40
5	10·48	10·70	10·89	11·08	11·24	11·40	11·55	11·68	11·81	11·93
6	9·30	9·48	9·65	9·81	9·95	10·08	10·21	10·32	10·43	10·54
7	8·55	8·71	8·86	9·00	9·12	9·24	9·35	9·46	9·55	9·65
8	8·03	8·18	8·31	8·44	8·55	8·66	8·76	8·85	8·94	9·03
9	7·65	7·78	7·91	8·03	8·13	8·23	8·33	8·41	8·49	8·57
10	7·36	7·49	7·60	7·71	7·81	7·91	7·99	8·08	8·15	8·23
11	7·13	7·25	7·36	7·46	7·56	7·65	7·73	7·81	7·88	7·95
12	6·94	7·06	7·17	7·26	7·36	7·44	7·52	7·59	7·66	7·73
13	6·79	6·90	7·01	7·10	7·19	7·27	7·35	7·42	7·48	7·55
14	6·66	6·77	6·87	6·96	7·05	7·13	7·20	7·27	7·33	7·39
15	6·55	6·66	6·76	6·84	6·93	7·00	7·07	7·14	7·20	7·26
16	6·46	6·56	6·66	6·74	6·82	6·90	6·97	7·03	7·09	7·15
17	6·38	6·48	6·57	6·66	6·73	6·81	6·87	6·94	7·00	7·05
18	6·31	6·41	6·50	6·58	6·65	6·73	6·79	6·85	6·91	6·97
19	6·25	6·34	6·43	6·51	6·58	6·65	6·72	6·78	6·84	6·89
20	6·19	6·28	6·37	6·45	6·52	6·59	6·65	6·71	6·77	6·82
24	6·02	6·11	6·19	6·26	6·33	6·39	6·45	6·51	6·56	6·61
30	5·85	5·93	6·01	6·08	6·14	6·20	6·26	6·31	6·36	6·41
40	5·69	5·76	5·83	5·90	5·96	6·02	6·07	6·12	6·16	6·21
60	5·53	5·60	5·67	5·73	5·78	5·84	5·89	5·93	5·97	6·01
120	5·37	5·44	5·50	5·56	5·61	5·66	5·71	5·75	5·79	5·83
∞	5·23	5·29	5·35	5·40	5·45	5·49	5·54	5·57	5·61	5·65

SOURCE: Reprinted with permission from E. S. Pearson and H. O. Hartley, *Biometrika Tables for Statisticians* (New York: Cambridge University Press, 1954).

TABLES

Critical Values of D in the Kolmogorov-Smirnov Two-Sample, Two-tailed Test

Alpha	Critical Value of D
0.10	$1.22\sqrt{\dfrac{n_1 + n_2}{n_1 n_2}}$
0.05	$1.36\sqrt{\dfrac{n_1 + n_2}{n_1 n_2}}$
0.025	$1.48\sqrt{\dfrac{n_1 + n_2}{n_1 n_2}}$
0.01	$1.63\sqrt{\dfrac{n_1 + n_2}{n_1 n_2}}$
0.005	$1.73\sqrt{\dfrac{n_1 + n_2}{n_1 n_2}}$
0.001	$1.95\sqrt{\dfrac{n_1 + n_2}{n_1 n_2}}$

SOURCE: From N. Smirnov, "Tables for Estimating Goodness of Fit of Empirical Distributions," *Annals of Mathematical Statistics* 19 (1948): 280–81.

Random Number Table

1260	5529	9540	3569	8381	9742	2590	2516	4243	8130
8979	2446	7606	6948	4519	2636	6655	1166	2096	1137
5470	0061	1760	5993	4319	0825	6874	3753	8362	1237
9733	0297	0804	4942	7694	9340	2502	3597	7691	5000
7492	6719	6816	7567	0364	2306	2217	5626	6526	6166
3715	2248	2337	6530	1660	7441	1598	0477	6620	1250
9491	4842	6210	9140	0180	5935	7218	4966	0537	4416
5192	7719	0654	4428	9771	4677	3291	4459	7432	2054
1714	3725	9397	9648	2550	0704	1239	5263	1601	5177
1052	8415	3686	4239	3272	7135	5768	8718	7582	8366
6998	3891	9352	6056	3621	5395	4551	4017	2405	0831
4216	4724	2898	1050	2164	8020	5274	6688	8636	2438
7115	2637	3828	0810	4598	2329	7953	4913	0033	2661
6647	4252	1869	9634	1341	7958	9460	1712	6060	0638
3475	2925	5097	8258	8343	7264	3295	8021	6318	0454
0415	1533	7670	0618	5193	9291	2205	2046	0890	6997
7064	4946	2618	7116	3784	2007	2326	4361	4695	8612
9772	4445	0343	5238	0317	7531	5916	8229	3296	2321
5365	4306	4036	9873	9669	8505	8675	3116	1484	9975
6395	4681	9319	6908	8154	5415	5728	1593	9452	8213
1554	0411	3436	1101	0966	5188	0225	5615	8568	5169
6745	7372	6984	0228	1920	6710	3459	0663	1407	9211
2217	3801	5860	1673	0264	6911	7623	7137	0774	5898
9255	9297	4305	5060	8312	9192	6016	7238	5193	4908
1615	9761	5744	2733	5314	6985	6670	0975	5487	6107
6679	4951	5716	5889	4413	1513	5023	7313	0317	0517
5221	3207	0351	2452	1072	3830	5518	5972	6111	9352
7720	5131	4867	6501	6970	4075	4869	4798	7104	5342
2767	6055	2801	7033	5305	9382	2354	4135	5975	7830
9202	6815	8211	5274	2303	1437	8995	2514	6515	0049
9359	2754	8587	2790	9524	5068	1230	4165	7025	4365
1078	7661	0999	4413	3446	2971	7576	3385	3308	0557
5732	4853	2025	1145	9743	8646	4918	3674	8049	1622
1596	0578	1493	4681	3806	8837	4241	4123	6513	3083
0602	8144	8976	9195	9012	7700	1708	5724	0315	8032
3624	8592	7942	4289	0736	4986	4839	4507	4997	7407
9328	3001	1462	1101	0804	9724	4082	2384	9631	8334
1981	1295	1963	3391	5757	4403	5857	4329	3682	3823
6001	4295	1488	6702	9954	6980	4027	8492	8195	7934
3743	0097	4798	6390	6465	6449	7990	3774	3577	3895

TABLES

3343	6936	1449	2915	6668	8543	3147	1442	6022	0056
9208	3820	5165	3445	2642	2910	8336	1244	6346	0487
4581	3768	1559	7558	6660	0116	7949	5609	2887	4156
3206	5146	7191	8420	2319	4650	3734	0501	0739	2025
5662	8315	4226	8395	2931	1812	3575	9341	5894	3691
4631	6278	9444	4058	0505	4449	5959	7483	8641	1311
1046	6653	8333	5813	6586	9820	0190	9214	1947	9677
5231	2788	7198	5904	8370	8347	6599	7304	6430	3495
5349	6641	3234	8692	4424	9179	2767	9517	1173	4160
8363	9625	3329	5262	5360	8181	9298	6629	2433	9414
5967	1261	2470	8867	1962	1630	8360	6024	6232	8386
6716	3916	8712	6673	1156	0001	6760	6287	4546	5743
0323	3514	8550	3709	6614	5764	0600	7444	2795	4426
1765	0918	8972	9924	5941	0331	6909	4872	5693	8957
4345	6886	2032	4817	2725	9471	2443	9532	4770	1271
9614	8571	2174	0071	7824	0504	9600	6414	5734	1371
1291	8246	5019	6559	5051	5265	1184	9030	6689	2776
3867	3915	8311	2430	1235	7283	6481	2012	8487	0226
1056	1880	3610	7796	7192	6663	5810	3512	1572	2921
1691	2237	6713	4048	8865	5794	3419	4372	6996	8342
7920	8490	2822	2647	1700	5335	0732	9987	7501	1223
1646	4251	7732	2136	4339	0331	9293	1061	2663	4821
7928	2575	2139	2825	3806	2082	9285	7640	6166	5758
3563	9078	1979	1141	7911	6981	0183	7479	1146	8949
6546	3459	3824	2151	3313	0178	9143	7854	3935	7300
8315	8778	1296	3434	9420	3622	3521	0807	5719	7764
8442	4933	8173	6427	4354	3523	3492	2816	9191	6261
7446	1576	2520	6120	8546	3146	6084	1260	3737	1333
2619	1261	2028	7505	0710	4589	9632	2347	1975	6839
0814	8542	1526	1202	8091	9441	0456	0603	8297	1412

SOURCE: Reprinted from Richard J. Hopeman, *Production and Operations Management*, 4th ed. (Columbus, Ohio: Charles E. Merrill, 1980), pp. 569–70. Used with permission.

Complete worked-out solutions to all the following problems are presented in the *Solutions Manual*. This manual contains the solutions to all odd-numbered problems, as well as a summary of each chapter in this text.

Chapter 1

7. $X = 43.06$

9. a. 0.1787
   b. 3.16
   c. 0.0005
   d. 0.20
   e. 0.016783

11. a. $X = \dfrac{12}{10}$

   b. $Y = -\dfrac{4}{8}$

   c. $X = 5Y + 10$

   d. $Y = \dfrac{4}{24X - 2X^4 + X^2}$

   e. $Z = \dfrac{AX^2 + 4YX + 6 - 4A}{4 - A}$

13. a. 164
   b. 560
   c. 24
   d. $r = 0.9230$

15. 0

Chapter 2

5. a. interval
   b. ordinal

Chapter 3

7.

Class		Midpoint
1,000 and under	3,500	2,250
3,500 and under	6,000	4,750
.	.	.
.	.	.
.	.	.
23,500 and under	26,000	24,750

9.

Class		Cumulative Frequency
Less than	3,500	40
Less than	6,000	120
.		.
.		.
.		.
Less than	26,000	2,800

11.

Class		Cumulative Relative Frequency
Less than	3,500	0.01429
Less than	6,000	0.04286
.		.
.		.
.		.
Less than	26,000	1.00000

13.

Class	Frequency
0.00– 19.99	15
20.00– 39.99	11
.	.
.	.
.	.
140.00–159.99	3

Chapter 4

9. Sawyer $CV = 8.33\%$; Horton $CV = 48.29\%$

11. $\bar{X} = \$1,569.50$; $S_x = \$627.48$

13. 2.94 standard deviations

17. height $CV = 3.6\%$; weight $CV = 6.78\%$

Chapter 5

1. **a.** relative frequency
   b. subjective
   c. relative frequency
   d. subjective

3. **a.** not mutually exclusive
   b. yes; independent

5. $\dfrac{5}{126}$

7. $\dfrac{1}{126}$

9. 0.42

11. 0.73728

13. no chance

15. **a.** 0.440
    b. 0.194
    c. 0.927

17. **a.** 0.596
    b. 0.087
    c. $0.087 \neq 0.404(0.375)$; no

19. 0.250

21. 0.333

23. 0.533

25. 0.275

27. reliever since $0.75 > 0.70$

29. **a.** 0.0225
    b. 0.556

31. 120 ways

33. 720 teams

35. 1,000,000; 17,576,000; 6,760,000

37. $P(A|\text{bad}) = 0.444$; $P(B|\text{bad}) = 0.370$;
    $P(C|\text{bad}) = 0.186$

Chapter 6

7. 0.0159

9.

| No. Ruined | |
X	P(X)
0	0.6561
1	0.2916
2	0.0468
3	0.0036
4	0.0001

11. $\sigma_x = 0.60$; $\sigma_x^2 = 0.36$

13. 0.0916

15. $\mu_x = 3$ accidents

17. approximately 0.0001

21. $P(\text{fired}) = 0.6572$;
    $P(\text{bonus}) = 0.0213$;
    $P(\text{neither fired nor bonus}) = 0.3215$

23. profit = \$0.88/bolt

Chapter 7

3. **a.** 0.50
   b. $0.4082 + 0.1293 = 0.5375$
   c. $0.5000 - 0.4082 = 0.0918$

5. $X = 8.96$

7. **a.** $0.5000 - 0.4938 = 0.0062$
   b. $0.4938 + 0.4938 = 0.9876$
   c. 0.25

9. **a.** 0.008
   b. essentially 0
   c. 2,000 sales

11. **a.** assume normal distribution
    b. $\sigma_x = 15.24$; $P(X \geq 160) = 0.2546$

13. $E[\text{cost no schedule}] = \$41.50$;
    $E[\text{cost schedule}] = \$125.10$

15. $(0.0062)(0.0062) = 0.00003844$

17. 0.0668

19. $0.5000 - 0.5000 = 0$

21. $0.5000 - 0.4713 = 0.0287$

23. no; $0.8664 < 0.9900$

25. for $X = 74$, $p = 0.1271$;
    for $X = 90$, $p = 0.0110$

27. $X = 75.92$ h ≈ 76 h

29. $0.5000 - 0.4633 = 0.0367$

31. $X = \$9.06/\text{ft}^2$

33. **a.** $0.5000 - 0.3413 = 0.1587$
    b. $0.2734 + 0.5000 = 0.7734$
    c. $0.3413 + 0.2734 = 0.6147$
    d. $Z = -6.5$; probability $\simeq 0$

Chapter 9

9. $1.0000 - 0.1867 = 0.8133$

11. $0.5000 - 0.1554 = 0.3446$

13. essentially 0; $Z = -10.66$

15. essentially 0; $Z = 15.4$

17. $0.0675 + 0.2486 = 0.3161$;
$0.3413 + 0.5000 = 0.8413$

19. $0.0714 + 0.1736 = 0.2450$

21. $Z = 4.83$; probability $\simeq 0$

23. $0.5000 - 0.2517 = 0.2483$

25. $A = -1.645(500) + 18,100 = \$17,277.50$

27. a. $Z = 4.51$; probability $\simeq 0$
b. $Z = -5.26$; probability $\simeq 0$
c. $Z = 2.25$

Chapter 10

5. $\bar{X} = Z\dfrac{\sigma_x}{\sqrt{n}}$

11. disagree; parameter is not a random variable

13. $14,205 \pm 207.68$

15. $n = 301$

17. sample 3 is largest; sample 2 is smallest

19. sample 3

21. sample 1; $\bar{X} = 44.5$

23. a. 17.56 ± 0.77
b. $Z = 5.92$; probability $\simeq 0$

25. a. 0.28 ± 0.052
b. 0.271 ± 0.036

27. $n = 2,390$

29. $n = 1,111$

31. 0.02 ± 0.0249

33. a. 0.036 ± 0.024
b. -0.028 ± 0.038

35. a. 0.76 ± 0.1398
b. majority prefer Chrysler

37. $-1,258 \pm 1,109.17$

Chapter 11

3. a. 11.3 ± 1.19
b. 11.3 ± 0.758

5. a. 5.93 ± 0.12
b. no; the interval includes 0
c. $t = -1.70$; probability $\simeq 0.07$

9. -0.13 ± 0.33

11. sample 2; precision $= 173.57$

13. $n = 30$

19. b. -2.40 ± 3.31; no difference

21. 0.30 ± 0.42; no difference

23. 88.00 ± 11.70

25. paired samples; 0.70 ± 1.176

27. 95% interval is wider

Chapter 12

1. $H_0: \mu \le 20$; $H_A: \mu > 20$

5. $P(20 \text{ tails}) = 0.00000095$

7. beta smaller than alpha

9. reduce Type I, increase Type II

11. $3.2 < 3.359$; reject H_0

13. $3.34 < 3.35$; reject H_0

15. a. $H_0: \mu \ge 50$; $H_A: \mu < 50$
b. $48 > 46.8$; don't reject H_0

17. $82 < 82.355$; reject H_0

19. beta $= 0.5000 - 0.4906 = 0.0094$

21. beta $= 0.5000 + 0.1700 = 0.6700$

23. beta $= 0.3413 + 0.5000 = 0.8413$

25. $114 > 109.91$; reject H_0

27. beta $= 0.5000 - 0.2324 = 0.2676$

29. beta $= 0.5000 + 0.4394 = 0.9394$

31. power $= 0.5000 - 0.4162 = 0.0838$

Chapter 13

7. $n = 3,459$; $A_L = \$8,053.14$

9. $0.85 < 0.8616$; reject H_0

11. $n = 628$

15. beta $= 0.3315 + 0.4992 = 0.8307$

17. $n = 286$

19. $t = -0.027 > -1.833$; don't reject H_0

21. $t = -1.01 > -2.074$; don't reject H_0

23. $\chi^2 = 56.84 > 42.557$; reject H_0

25. $Z = -0.50 > -1.28$; don't reject H_0

27. $0.367 > 0.237$; don't reject H_0

29. beta $= 0.5000 + 0.4803 = 0.9803$

31. $n = 119$

33. $\chi^2 = 28.5 > 27.204$; reject H_0

Chapter 14

3. $F = 7.528 < 8.02$; don't reject H_0

5. $F = 34.4 > 3.89$; reject H_0

7. use two-factor factorial design

9. difference between 2 and 3; S range = 1.325

11. T range = 137.7; brand 2

13. $F = 108.42 > 3.47$; reject H_0

15. S range = 2.416

17. T range = 1.8265; S range = 2.416

19. $t = 1.16 < 2.12$; don't reject H_0

21. $F = t^2$; $(1.16)^2 = 1.35$

23. T range = 10.45; pick brand A

25. S range Canadian vs. Austrian = 1,025.02;
S range Canadian vs. German = 1,055.93;
S range Austrian vs. German = 982.29

Chapter 15

3. $\chi^2 = 27.176 > 7.81$; reject H_0

5. $\chi^2 = 15.25 > 7.81$; reject H_0

7. $\chi^2 = 4.952 < 7.78$; don't reject H_0

9. $H = 12.252 > 4.61$; reject H_0

11. $H = 5.857 > 4.61$; reject H_0

13. $\chi^2 = 10.58 > 5.99$; reject H_0

15. $|D| = |-0.200| > 0.192$; reject H_0

17. $Z = 0.605 < 1.645$; don't reject H_0

19. $|D| = |0.545| < 0.5799$; don't reject H_0

21. $|D| = |0.114| < 0.2062$; don't reject H_0

23. $\chi^2 = 38.02 > 10.644$; reject H_0

Chapter 16

3. **a.** scatter plot indicates strong linear relationship is present
   b. $\hat{Y} = 287.91 + 143.79X$

5. $1,006.86 \pm 26.78$; $1,006.86 \pm 79.21$

9. $\hat{Y} = 166.46 + 0.0537X_1$; $\hat{Y} = 239.66 + 0.0639X_2$

11. 603.04 ± 18.81

13. 0.0537 ± 0.00665; 0.0639 ± 0.0435

19. -0.0015 ± 0.00064

21. **b.** $\hat{Y} = 8.004 + 0.9002$ (student average)

23. 93.52 ± 17.75

27. $r = 0.95065$

29. $R^2 = 0.90374$

31. 2.154 ± 0.571; $\$21.54 \pm \$5.71$

Chapter 17

3. $\hat{Y} = 21.4805 + 2.364X_1 + 1.531X_2 + 3.807X_3$

5. $F = 16.70 > 3.59$; reject H_0

7. $R^2 = 0.819$

9. $\hat{Y} = 92.78$

11. 1.531 ± 3.902

13. Y, X_1: $t = -2.12 < -2.069$; reject H_0
Y, X_2: $t = 2.47 > 2.069$; reject H_0
Y, X_3: $t = -1.21 > -2.069$; don't reject H_0
X_1X_2: $t = 0.246 < 2.069$; don't reject H_0
X_1X_3: $t = 2.79 > 2.069$; reject H_0
X_2X_3: $t = 1.35 < 2.069$; don't reject H_0

15. $F = 6.14 > 4.26$; reject H_0
$t = 2.478 > 2.069$; reject H_0

17. $\hat{Y} = 82.8875 - 1.31807X_1 + 0.00208X_2$;
R^2 increased; S_e declined; R_A^2 increased

19. $H_0: B_1 = 0$; $t = -1.665 > -2.08$; don't reject H_0
$H_0: B_2 = 0$; $t = 3.132 > 2.08$; reject H_0
$H_0: B_3 = 0$; $t = 1.163 < 2.08$; don't reject H_0
$H_0: B_1 = B_2 = B_3 = 0$; $F = 5.33 > 2.84$; reject H_0

23. $\hat{Y} = -125,307.8 + 175.89X_1 - 1,573.77X_2 + 1.59X_3 + 1,613.74X_4$

25. B_1: 175.89 ± 88.60
B_2: $-1,573.77 \pm 4,446.75$
B_3: 1.59 ± 0.9892
B_4: $1,613.74 \pm 1,392.54$

27. Y, X_5: $r = 0.89579$
Y, X_6: $r = 0.66045$

29. $\hat{Y} = -100,834.50 + 94.77X_1 - 346.62X_2 + 1.004X_3 + 1,042.96X_4 + 0.836X_5 + 596.02X_6$
B_1: $t = 1.101 < 2.306$; don't reject H_0
B_2: $t = 0.130 < 2.306$; don't reject H_0
B_3: $t = 0.1369 < 2.306$; don't reject H_0
B_4: $t = 1.334 < 2.306$; don't reject H_0
B_5: $t = 0.902 < 2.306$; don't reject H_0
B_6: $t = 0.104 < 2.306$; don't reject H_0

Chapter 18

1. **a.** cyclical
   b. seasonal; irregular
   c. seasonal
   d. trend; cyclical
   e. trend; cyclical
   f. trend

3. Ratio to moving average

July 1979	119.75
Aug 1979	129.19
.	.
.	.
.	.
June 1979	111.20

7. **a.** down trend from 1950 on
   b. start of a long-term cycle
   c. not if there is a true cycle
   d. time is only a surrogate measure

9. Models to be explored:

$$\hat{Y} = b_o + b_1 t$$
$$\hat{Y} = b_o + b_1 t + b_2 t^2$$

15.

Quarter	Index
1st 1976	100.00
2nd 1976	101.80
3rd 1976	97.80
.	.
.	.
.	.
4th 1980	131.80

17.

Quarter	Index
Fall 1976	100.0
Winter 1976	85.1
.	.
.	.
.	.
Spring 1980	46.9

21. $\hat{Y} = 80.1299 + 2.103t$; $R^2 = 0.271$;
    $S_e = 22.91$; November forecast = 128.5;
    December forecast = 130.6

Chapter 19

7. payoff functions:

   if demand exceeds supply:

   Payoff = −$220,000 − $48,000(units built) + $80,000(units sold)

   if supply exceeds demand:

   Payoff = −$220,000 − $48,000(units built) + $80,000(units sold) + $21,000(excess supply)

9. build 50; payoff = $1.38 million

11.

	Successful	Failure
Market	$1,000,000	−$625,000
Not Market	$ 0	$ 0

13.

	Market	Not Market
Excellent	$1,200,000	$0
Good	$1,000,000	$0
Fair	$ 100,000	$0
Poor	−$ 625,000	$0

15.

	State of Nature			
	500	1,000	1,500	2,000
500	$1,875	$1,875	$1,875	$1,875
1,000	$1,125	$3,750	$3,750	$3,750
1,500	$ 375	$3,000	$5,625	$5,625
2,000	−$ 375	$2,250	$4,875	$7,500

17. Profit function:

    Profit = (8.00 − 2.25)(demand) − (2.25 − 1.25)(excess supply)

19. Part 1: TV-radio because $20,000 > $15,150

21. don't make new offer because $47,280 > $44,440

Chapter 20

5.

	Demand Level				
	10,000	15,000	20,000	30,000	50,000
Old Machine	$14,700	$23,450	$32,200	$49,700	$ 84,700
New Machine	$11,500	$22,750	$34,000	$56,500	$101,500

7.

Demand Level	Posterior Probability
10,000	0.0057
15,000	0.0343
20,000	0.1371
30,000	0.4571
50,000	0.3657

9.

Proportion Defective	Posterior Probability
0.05	0.06
0.20	0.94

11.

Proportion Defective	Posterior Probability
0.05	0.00
0.20	1.00

13.

Posterior Probability	Quality Level	
0	0.80	good
0.997	0.60	fair
0.003	0.30	poor

17. take project; $E[\text{profit}] = \$285,000 > \$125,000$

19. \$530,000 maximum

21. \$78,500

23. −\$10,280; sample has increased the uncertainty

index

STANDARD NORMAL DISTRIBUTION TABLE

To illustrate: 43.45 percent of the area under a normal curve lies between the mean, μ_x, and a point 1.51 standard deviation units away.

Example :
$z = 0.52 \,(\text{or} -0.52)$,
$A(z) = 0.19847$ or 19.847%

Z	.00	.01	.02	.03	.04	.05	.06	.07	.08	.09
0.0	.00000	.00399	.00798	.01197	.01595	.01994	.02392	.02790	.03188	.03586
0.1	.03983	.04380	.04776	.05172	.05567	.05962	.06356	.06749	.07142	.07535
0.2	.07926	.08317	.08706	.09095	.09483	.09871	.10257	.10642	.11026	.11409
0.3	.11791	.12172	.12552	.12930	.13307	.13683	.14058	.14431	.14803	.15173
0.4	.15542	.15910	.16276	.16640	.17003	.17364	.17724	.18082	.18439	.18793
0.5	.19146	.19497	.19847	.20194	.20540	.20884	.21226	.21566	.21904	.22240
0.6	.22575	.22907	.23237	.23565	.23891	.24215	.24537	.24857	.25175	.25490
0.7	.25804	.26115	.26424	.26730	.27035	.27337	.27637	.27935	.28230	.28524
0.8	.28814	.29103	.29389	.29673	.29955	.30234	.30511	.30785	.31057	.31327
0.9	.31594	.31859	.32121	.32381	.32639	.32894	.33147	.33398	.33646	.33891
1.0	.34134	.34375	.34614	.34850	.35083	.35314	.35543	.35769	.35993	.36214
1.1	.36433	.36650	.36864	.37076	.37286	.37493	.37698	.37900	.38100	.38298
1.2	.38493	.38686	.38877	.39065	.39251	.39435	.39617	.39796	.39973	.40147
1.3	.40320	.40490	.40658	.40824	.40988	.41149	.41309	.41466	.41621	.41774
1.4	.41924	.42073	.42220	.42364	.42507	.42647	.42786	.42922	.43056	.43189
1.5	.43319	.43448	.43574	.43699	.43822	.43943	.44062	.44179	.44295	.44408
1.6	.44520	.44630	.44738	.44845	.44950	.45053	.45154	.45254	.45352	.45449
1.7	.45543	.45637	.45728	.45818	.45907	.45994	.46080	.46164	.46246	.46327
1.8	.46407	.46485	.46562	.46638	.46712	.46784	.46856	.46926	.46995	.47062
1.9	.47128	.47193	.47257	.47320	.47381	.47441	.47500	.47558	.47615	.47670
2.0	.47725	.47778	.47831	.47882	.47932	.47982	.48030	.48077	.48124	.48169
2.1	.48214	.48257	.48300	.48341	.48382	.48422	.48461	.48500	.48537	.48574
2.2	.48610	.48645	.48679	.48713	.48745	.48778	.48809	.48840	.48870	.48899
2.3	.48928	.48956	.48983	.49010	.49036	.49061	.49086	.49111	.49134	.49158
2.4	.49180	.49202	.49224	.49245	.49266	.49286	.49305	.49324	.49343	.49361
2.5	.49379	.49396	.49413	.49430	.49446	.49461	.49477	.49492	.49506	.49520
2.6	.49534	.49547	.49560	.49573	.49585	.49598	.49609	.49621	.49632	.49643
2.7	.49653	.49664	.49674	.49683	.49693	.49702	.49711	.49720	.49728	.49736
2.8	.49744	.49752	.49760	.49767	.49774	.49781	.49788	.49795	.49801	.49807
2.9	.49813	.49819	.49825	.49831	.49386	.49841	.49846	.49851	.49856	.49861
3.0	.49865	.49869	.49874	.49878	.49882	.49886	.49889	.49893	.49897	.49900
3.1	.49903	.49906	.49910	.49913	.49916	.49918	.49921	.49924	.49926	.49929
3.2	.49931	.49934	.49936	.49938	.49940	.49942	.49944	.49946	.49948	.49950
3.3	.49952	.49953	.49955	.49957	.49958	.49960	.49961	.49962	.49964	.49965
3.4	.49966	.49968	.49969	.49970	.49971	.49972	.49973	.49974	.49975	.49976
3.5	.49977	.49978	.49978	.49979	.49980	.49981	.49981	.49982	.49983	.49983
3.6	.49984	.49985	.49985	.49986	.49986	.49987	.49987	.49988	.49988	.49989
3.7	.49989	.49990	.49990	.49990	.49991	.49991	.49992	.49992	.49992	.49992
3.8	.49993	.49993	.49993	.49994	.49994	.49994	.49994	.49995	.49995	.49995
3.9	.49995	.49995	.49996	.49996	.49996	.49996	.49996	.49996	.49997	.49997
4.0	.49997									

SOURCE: From Stephen P. Shao, *Statistics for Business and Economics*, 3rd ed. (Columbus, Ohio: Charles E. Merrill, 1976), p. 788. Used with permission.